Noncommutative Structures in Mathematics and Physics

NATO Science Series

A Series presenting the results of scientific meetings supported under the NATO Science Programme.

The Series is published by IOS Press, Amsterdam, and Kluwer Academic Publishers in conjunction with the NATO Scientific Affairs Division

Sub-Series

I. **Life and Behavioural Sciences**	IOS Press
II. **Mathematics, Physics and Chemistry**	Kluwer Academic Publishers
III. **Computer and Systems Science**	IOS Press
IV. **Earth and Environmental Sciences**	Kluwer Academic Publishers

The NATO Science Series continues the series of books published formerly as the NATO ASI Series.

The NATO Science Programme offers support for collaboration in civil science between scientists of countries of the Euro-Atlantic Partnership Council. The types of scientific meeting generally supported are "Advanced Study Institutes" and "Advanced Research Workshops", and the NATO Science Series collects together the results of these meetings. The meetings are co-organized bij scientists from NATO countries and scientists from NATO's Partner countries – countries of the CIS and Central and Eastern Europe.

Advanced Study Institutes are high-level tutorial courses offering in-depth study of latest advances in a field.
Advanced Research Workshops are expert meetings aimed at critical assessment of a field, and identification of directions for future action.

As a consequence of the restructuring of the NATO Science Programme in 1999, the NATO Science Series was re-organized to the four sub-series noted above. Please consult the following web sites for information on previous volumes published in the Series.

http://www.nato.int/science
http://www.wkap.nl
http://www.iospress.nl
http://www.wtv-books.de/nato-pco.htm

Series II: Mathematics, Physics and Chemistry – Vol. 22

Noncommutative Structures in Mathematics and Physics

KAP ARCHIEF

edited by

Steven Duplij

Theory Group,
Nuclear Physics Laboratory,
Kharkov National University,
Kharkov, Ukraine

and

Julius Wess

Sektion Physik,
Ludwig-Maximillians-Universität,
München, Germany

Kluwer Academic Publishers

Dordrecht / Boston / London

Published in cooperation with NATO Scientific Affairs Division

Proceedings of the NATO Advanced Research Workshop on
Noncommutative Structures in Mathematics and Physics
Kiev, Ukraine
September 24–28, 2000

A C.I.P. Catalogue record for this book is available from the Library of Congress.

ISBN 0-7923-6998-X (HB)
ISBN 0-7923-6999-8 (PB)

Published by Kluwer Academic Publishers,
P.O. Box 17, 3300 AA Dordrecht, The Netherlands.

Sold and distributed in North, Central and South America
by Kluwer Academic Publishers,
101 Philip Drive, Norwell, MA 02061, U.S.A.

In all other countries, sold and distributed
by Kluwer Academic Publishers,
P.O. Box 322, 3300 AH Dordrecht, The Netherlands.

Printed on acid-free paper

Printed in the Netherlands.

CONTENTS

PREFACE   vii

J. Wess Gauge Theories Beyond Gauge Theory   1

D. Leites, V. Serganova Symmetries Wider Than Supersymmetry   13

K. Stelle Tensions in Supergravity Braneworlds   31

P. Grozman, D. Leites An Unconventional Supergravity   41

E. Bergshoeff, R. Kallosh, A. Van Proeyen Supersymmetry Of RS Bulk And Brane   49

D. Galtsov, V. Dyadichev D-branes And Vacuum Periodicity   61

P. Kosiński, J. Lukierski, P. Maślanka Quantum Deformations Of Space-Time SUSY And Noncommutative Superfield Theory   79

D. Leites, I. Shchepochkina The Howe Duality And Lie Superalgebras   93

A. Sergeev Enveloping Algebra Of GL(3) And Orthogonal Polynomials   113

S. Duplij, W. Marcinek Noninvertibility, Semisupermanifolds And Categories Regularization   125

F. Brandt An Overview Of New Supersymmetric Gauge Theories With 2-Form Gauge Potentials   141

K. Peeters, P. Vanhove, A. Westerberg Supersymmetric R^4 Actions And Quantum Corrections To Superspace Torsion Constraints   153

S. Fedoruk, V. G. Zima Massive Superparticle With Spinorial Central Charges   161

A. Burinskii Rotating Super Black Hole as Spinning Particle   181

F. Toppan Classifying N-extended 1-dimensional Super Systems   195

C. Quesne Para, Pseudo, And Orthosupersymmetric Quantum Mechanics And Their Bosonization   203

A. Frydryszak Supersymmetric Odd Mechanical Systems And Hilbert Q-module Quantization   215

S. Vacaru, I. Chiosa, N. Vicol Locally Anisotropic Supergravity And Gauge Gravity On Noncommutative Spaces   229

T. Kobayashi, J. Kubo, M. Mondragón, G. Zoupanos Finiteness In Conventional $N = 1$ GUTs   245

J. Simon World Volume Realization Of Automorphisms 259

G. Fiore, M. Maceda, J. Madore Some Metrics On The Manin Plane 271

V. Lyubashenko Coherence Isomorphisms For A Hopf Category 283

A. Ganchev Fusion Rings And Tensor Categories 295

V. Mazorchuk On Categories Of Gelfand-Zetlin Modules 299

D. Shklyarov, S. Sinel'shchikov, L. Vaksman Hidden Symmetry Of
Some Algebras Of q-differential Operators 309

P. Jorgensen, D. Proskurin, Y. Samoilenko A Family Of *-Algebras
Allowing Wick Ordering: Fock Representations And Universal
Enveloping C^*-Algebras 321

A. U. Klimyk Nonstandard Quantization Of The Enveloping Algebra
U(so(n)) And Its Applications 331

A. Gavrilik Can the Cabibbo mixing originate from noncommutative
extra dimensions? 343

N. Iorgov Nonclassical Type Representations Of Nonstandard
Quantization Of Enveloping Algebras U(so(n)), U(so(n,1)) and
U(iso(n)) 357

K. Landsteiner Quasiparticles In Non-commutative Field Theory 369

A. Sergyeyev Time Dependence And (Non)Commutativity Of
Symmetries Of Evolution Equations 379

B. Dragovich, I. V. Volovich p-Adic Strings And Noncommutativity 391

G. Djordjević, B. Dragovich, L. Nešić Adelic Quantum Mechanics:
Nonarchimedean And Noncommutative Aspects 401

Y. Kozitsky Gibbs States Of A Lattice System Of Quantum Anharmonic
Oscillators 415

D. Vassiliev A Metric-Affine Field Model For The Neutrino 427

M. Visinescu Generalized Taub-NUT Metrics And Killing-Yano
Tensors 441

V. Dzhunushaliev An Effective Model Of The Spacetime Foam 453

A. Higuchi Possible Constraints On String Theory In Closed Space
With Symmetries 465

A. Alscher, H. Grabert Semiclassical Dynamics Of $SU(2)$ Models 475

LIST OF SPEAKERS AND THEIR E-PRINTS 481

PREFACE

The concepts of noncommutative space-time and quantum groups have found growing attention in quantum field theory and string theory. The mathematical concepts of quantum groups have been far developed by mathematicians and physicists of the Eastern European countries. Especially, V. G. Drinfeld from Ukraine, S. Woronowicz from Poland and L. D. Faddeev from Russia have been pioneering the field. It seems to be natural to bring together these scientists with researchers in string theory and quantum field theory of the Western European countries. From another side, supersymmetry, as one of examples of noncommutative structure, was discovered in early 70's in the West by J. Wess (one of the co-Directors) and B. Zumino and in the East by physicists from Ukraine V. P. Akulov and D. V. Volkov. Therefore, Ukraine seems to be a natural place to meet.

Supersymmetry is a very important and intriguing mathematical concept which has become a basic ingredient in many branches of modern theoretical physics. In spite of its still lacking physical evidence, its far-reaching theoretical implications uphold the belief that supersymmetry plays a prominent role in the fundamental laws of nature. At present the most promising hope for a truly supersymmetric unified and finite description of quantum field theory and general relativity is superstring theory and its latest formulation, Witten's M-theory. Superstrings possess by far the largest set of gauge symmetries ever found in physics, perhaps even large enough to eliminate all divergences in quantum gravity. Not only does superstring's symmetry include that of Einstein's theory of general relativity and the Yang-Mills theory, it also includes supergravity and the Grand Unified Theories.

One of the exciting new approaches to nonperturbative string theory involves M-theory and duality, which, in fact, force theoretical physicists to reconsider the central role played by strings in supersymmetry. In this revised new picture all five superstring theories, which on first glance have entirely different properties and spectra, are now seen as different vacua of a same theory, M-theory. This unification cannot, however, occur at the perturbative level, because it is precisely the perturbative analysis which singles out the five different string theories. The hope is that when one goes beyond this perturbative limit, and takes into account all non-perturbative effects, the five string theories turn out to be five different descriptions of the same physics. In this context a duality is a particular relation applying to string theories, which can map for instance the strong coupling region of a theory to the weak coupling region of the same theory or of another one, and vice versa, thus being an intrinsically non-perturbative relation. In the recent years, the structure of M-theory has begun to be uncovered, with the essential tool provided by supersymmetry. Its most striking characteristic is that it indicates that space-time should be eleven dimensional. Because of the intrinsic non-perturbative nature of any approach to M-theory, the study of the p-brane solitons, or more simply 'branes', is a natural step to take. The branes are extended objects present in M-theory or in string theories, generally associated to classical solutions of the respective supergravities.

Quantum groups arise as the abstract structure underlying the symmetries of integrable systems. Then the theory of quantum inverse scattering gives rise to some deformed algebraic structures which were first explained by Drinfeld as deformations of the envelopping algebras of the classical Lie algebras. An analogous structure was obtained by Woronowicz in the context of noncommutative C^*-algebras. There is a third approach, due to Yu. I. Manin, where quantum groups are interpreted as the endomorphisms of certain noncommutative algebraic varieties defined by quadratic algebras, called quantum linear spaces. L. D. Faddeev and his collaborators had also interpreted the quantum groups from the point of view of corepresentations and quantum spaces, furnishing a connection with the quantum deformations of the universal enveloping algebras and the quantum double of Hopf algebras. From the algebraic point of view, quantum groups are Hopf algebras and the relation with the endomorphism algebra of quantum linear spaces comes from their corepresentations on tensor product spaces. The usual construction of the coaction on the tensor product space involves the flip operator interchanging factors of the tensor product of the quantum linear spaces with the bialgebra. This fact implies the commutativity between the matrix elements of a representation of the endomorphism and the coordinates of the quantum linear spaces. Moreover, the flip operator for the tensor product is also involved in many steps of the construction of quantum groups. In the braided approach to q-deformations the flip operator is replaced with a braiding giving rise to the quasi-tensor category of k-modules, where a natural braided coaction appears.

The study of differential geometry and differential calculus on quantum groups that Woronowicz initiated is also very important and worthwile to investigate. Next step in this direction is consideration of noncommutative space-time as a possible realistic picture of how space-time behaves at short distances. Starting from such a noncommutative space as configuration space, one can generalize it to a phase space where noncommutativity is already intrinsic for a quantum mechanical system. The definition of this noncommutative phase space is derived from the noncommutative differential structure on the configuration space. The noncommutative phase space is a q-deformation of the quantum mechanical phase space and one can apply all the machinery learned from quantum mechanics. If one demands that space-time variables are modules or co-modules of the q-deformed Lorentz group, then they satisfy commutation relations that make them elements of a non-commutative space. The action of momenta on this space is non-commutative as well. The full structure is determined by the (co-)module property. It can serve as an explicit example of a non-commutative structure for space-time. This has the advantages that the q-deformed Lorentz group plays the role of a kinematical group and thus determines many of the properties of this space and allows explicit calculations. One can explicitly construct Hilbert space representations of the algebra and find that the vectors in the Hilbert space can be determined by measuring the time, the three-dimensional distance, the q-deformed angular momentum and its third component. The eigenvalues of these observables form a q-lattice with accumulation points on the light-cone. In a way physics on the light-cone is best approximated by this q-deformation. One can consider the simplest version of a q-deformed Heisenberg algebra as an example of a noncommutative structure, first derive a calculus entirely based on the algebra and then formulate laws of physics based on this calculus.

Bringing together scientists from quantum field theory, string theory and quantum gravity with researchers in noncommutative geometry, Hopf algebras and quantum groups as well as experts on representation theory of these algebras had a stimulating effect on each side and will lead to new developments. In each field there is a highly developed knowledge by experts which can only be transformed to another field only by having close personal contact through discussions, talks and reports. We hope that common projects can be found such that working in these projects the detailed techniques can be learned from each other. The Workshop has promoted the development of new directions in the field of modern theoretical and mathematical physics combining the efforts of scientists from NATO, East European countries and NIS.

We are greatly indebted to the NATO Division of Scientific Affairs for funding of our meeting and to the National Academy of Sciences of Ukraine for help in its local organizing. It is also a great pleasure to thank all the people who contributed to the successful organization of the Workshop, especially members of the Local Organizing Committee Profs. N. Chashchyn and P. Smalko. Finally, we would like to thank all the participants for creating an excellent working atmosphere and for outstanding contributions to this volume.

Editors

GAUGE THEORIES BEYOND GAUGE THEORY

JULIUS WESS
Sektion Physik der Ludwig-Maximilians-Universität Theresienstr. 37, D-80333 München, Germany
and
Max-Planck-Institut für Physik (Werner-Heisenberg-Institut) Föhringer Ring 6, D-80805 München, Germany

1. Algebraic preliminaries

In gauge theories we consider differentiable manifolds as base manifolds and fibres that carry a representation of a Lie group. In the following we shall show that it is possible to replace the differentiable manifold by a non-commutative algebra, ref. [1]. For this purpose we first focus our attention on algebraic properties. The coordinates x^i

$$x^1, \ldots, x^n \in \mathbb{R}, \tag{1}$$

are considered as elements of an algebra over $\mathbb{C}$ subject to the relations:

$$\mathcal{R}: \quad x^i x^j - x^j x^i = 0. \tag{2}$$

This characterizes $\mathbb{R}^n$ as a commutative space. The relations generate a 2-sided ideal $I_\mathcal{R}$. From the algebraic point of view, we deal with the algebra freely generated by the elements x^i and divided by the ideal $I_\mathcal{R}$:

$$\mathcal{A}_x = \frac{\mathbb{C}\left[\left[x^1, \ldots, x^n\right]\right]}{I_\mathcal{R}}. \tag{3}$$

Formal power series are accepted, this is indicated by the double bracket. The elements of the algebra are the functions in $\mathbb{R}^n$ that have a formal power series expansion at the origin:

$$f(x^1, \ldots, x^n) \in \mathcal{A}_x, \tag{4}$$

$$f(x^1, \ldots, x^n) = \sum_{r_i=0}^{\infty} f_{r_1 \ldots r_n} (x^1)^{r_1} \cdot \ldots \cdot (x^n)^{r_n}.$$

1

S. Duplij and J. Wess (eds.), Noncommutative Structures in Mathematics and Physics, 1–11.

Multiplication is the pointwise multiplication of these functions.

The monomials of fixed degree form a finite-dimensional subspace of the algebra. This algebraic concept can be easily generalized to non-commutative spaces. We consider algebras freely generated by elements $\hat{x}^1, \ldots \hat{x}^n$, again calling them coordinates. But now we change the relations to arrive at non-commutative spaces:

$$\mathcal{R}_{\hat{x},\hat{x}}: \quad [\hat{x}^i, \hat{x}^j] = i\theta^{ij}(\hat{x}). \tag{5}$$

Following L.Landau, non-commutativity carries a hat. Now we deal with the algebra:

$$\mathcal{A}_{\hat{x}} = \frac{\mathbb{C} << \hat{x^1}, \ldots, \hat{x^n} >>}{I_{\mathcal{R}_{\hat{x},\hat{x}}}}, \tag{6}$$

$$\hat{f} \in \mathcal{A}_{\hat{x}}.$$

In the following we impose one more condition on the algebra: the dimension of the subspace of homogeneous polynomials should be the same as for commuting coordinates. This is the so called Poincare-Birkhof-Witt property (PBW). Only algebras with this property will be considered, among them are the algebras where θ^{ij} is a constant:
<u>Canonical structure</u>, ref. [2]:

$$[\hat{x}^i, \hat{x}^j] = i\theta^{ij}, \tag{7}$$

where θ^{ij} is linear in $\hat{x}$:
<u>Lie structure</u>, ref. [3]:

$$[\hat{x}^i, \hat{x}^j] = i\theta_k^{ij}\hat{x}^k, \tag{8}$$

where θ^{ij} is quadratic in $\hat{x}$:
<u>Quantum space structure</u>, ref.[4]:

$$[\hat{x}^i, \hat{x}^j] = i\theta_{kl}^{ij}\hat{x}^k\hat{x}^l, \tag{9}$$

The constants θ_k^{ij} and θ_{kl}^{ij} are subject to conditions to guarantee PBW. For Lie structures this will be the Jacobi identity, for the quantum space structure the Yang-Baxter equation. There is a natural vector space isomorphism between $\mathcal{A}_x$ and $\mathcal{A}_{\hat{x}}$. It is based on the isomorphism of the vector spaces of homogeneous polynomials that have the same degree due to the PBW property.

In order to establish the isomorphism we choose a particular basis in the vector space of homogeneous polynomials in the non-commuting variables $\hat{x}$ and characterize the elements of $\mathcal{A}_{\hat{x}}$ by the coefficient functions in this basis. The corresponding element in the algebra $\mathcal{A}_x$ of commuting variables is supposed to have the same coefficient function. The particular form of this isomorphism depends on the basis chosen. The vector space isomorphism can be extended to an algebra isomorphism. To establish it we compute the coefficient function of the product of two elements in $\mathcal{A}_{\hat{x}}$ and map it to $\mathcal{A}_x$. This defines a product in $\mathcal{A}_{\S}$ that we denote as diamond product ($\Diamond$ product). The algebra with this $\Diamond$ product we call $^{\Diamond}\mathcal{A}_x$. There is a natural isomorphism:

$$\mathcal{A}_{\hat{x}} \longleftrightarrow \,^{\Diamond}\mathcal{A}_x. \tag{10}$$

The three structures that we have mentioned above have an even stronger property than PBW. It turns out that monomials in any well-defined ordering of the coordinates form a basis. Among them is an ordering as we have used it before or the completely symmetrized ordering of monomials as well. For such structures we shall denote the $\Diamond$ product as * product (star product), ref. [5]. For the <u>canonical structure</u> we obtain the Moyal-Weyl * product, ref. [6], if we start from the basis of completely symmetrized monomials:

$$(f * g)(x) = e^{\frac{i}{2}\frac{\partial}{\partial x^i}\theta^{ij}\frac{\partial}{\partial y^j}} f(x)g(y)\Big|_{y \Rightarrow x} \tag{11}$$

$$= \int d^n y \, \delta^n(x - y) e^{\frac{i}{2}\frac{\partial}{\partial x^i}\theta^{ij}\frac{\partial}{\partial y^j}} f(x)g(y).$$

For the <u>Lie structure</u> we can use the Baker-Campbell-Hausdorf formula:

$$e^{ik\cdot\hat{x}}e^{ip\cdot\hat{x}} = e^{i(k+p+\frac{1}{2}g(k,p))\cdot\hat{x}}. \tag{12}$$

This defines $g(k,p)$.

$$(f * g)(x) = e^{\frac{i}{2}x\cdot g(i\frac{\partial}{\partial y}, i\frac{\partial}{\partial z})} f(y)g(z)\Big|_{\substack{y \to x \\ z \to x}}. \tag{13}$$

For the <u>quantum plane</u> we consider the example of the Manin plane

$$\hat{x}\hat{y} = q\hat{y}\hat{x}, \tag{14}$$

$$(f * g)(x) = q^{-x'\frac{\partial}{\partial x'}y\frac{\partial}{\partial y}} f(x,y)g(x',y')\Big|_{\substack{x' \to x \\ y' \to y}}.$$

It is natural to use the elements of $^{\Diamond}\mathcal{A}_x$ as objects in physics. Fields of a field theory will be such objects.

$$\phi(x) \in \,^{\Diamond}\mathcal{A}_x. \tag{15}$$

The product of fields will always be the * product. To formulate field equations we introduce derivatives. On the algebra $\mathcal{A}_{\hat{x}}$ this can be done on purely algebraic grounds. We have to extend the algebra $\mathcal{A}_{\hat{x}}$ by algebraic elements $\hat{\partial}_i$, ref. [7]. A generalized Leibniz rule will play the role of algebraic relations.

Leibniz rule:

$$(\hat{\partial}_i \hat{f} \hat{g}) = (\hat{\partial}_i \hat{f})\hat{g} + O_i^l(\hat{f})\hat{\partial}_l\hat{g} \quad : \mathcal{R}_{\hat{x},\hat{\partial}}. \tag{16}$$

From the law of associativity in $\mathcal{A}_{\hat{x}}$ follows that the operation O has to be an algebra homomorphism:

$$O_j^i(\hat{f}\hat{g}) = O_l^i(\hat{f})O_j^l(\hat{g}). \tag{17}$$

But we shall restrict the Leibniz rule by an even stronger requirement. The ideal generated by the $\mathcal{R}_{\hat{x},\hat{x}}$ relations has to remain a two-sided ideal in the larger algebra generated by $\hat{x}$ and $\hat{\partial}$. This leads to so called consistency relations.

Finally $\mathcal{R}_{\hat{\partial},\hat{\partial}}$ relations have to be defined. As conditions we consider the $\hat{\partial}$ subalgebra, demand PBW and derive consistency relations from $\mathcal{R}_{\hat{\partial},\hat{\partial}}$ and the Leibniz rule as before. Derivatives defined that way induce a map from $\mathcal{A}_{\hat{x}}$ to $\mathcal{A}_{\hat{x}}$:

$$\hat{f} \in \mathcal{A}_{\hat{x}} \quad , \quad (\hat{\partial}_i \hat{f}) \in \mathcal{A}_{\hat{x}}, \tag{18}$$
$$(\hat{\partial}_i \hat{f}) = \hat{\partial}_i \hat{f} - O_i^l(\hat{f})\hat{\partial}_l.$$

This algebraic concept of derivatives has been explained in ref[] and applied to quantum planes. Following the same strategy derivatives can be defined for the canonical structure as well.

For the rest of this talk we will restrict ourselves to the canonical case only. The Leibniz rule for the canonical case is the usual one:

$$\hat{\partial}_i \hat{x}^j = \delta_i^j + \hat{x}^j \hat{\partial}_i. \tag{19}$$

It satisfies all the consistency relations. As explained above, the derivatives induce a map on the algebra $\mathcal{A}_{\hat{x}}$:

$$\hat{f} \in \mathcal{A}_{\hat{x}} : \hat{f} \to [\hat{\partial}_i, \hat{f}] \in \mathcal{A}_{\hat{x}}. \tag{20}$$

This is the relation that we shall use to define derivatives on fields. For this purpose we map $\hat{\partial}$ to $^\Diamond \mathcal{A}_x$. From (20) follows that it becomes the usual derivative in $^\Diamond \mathcal{A}_x$:

$$f(x) \to \partial_i f(x). \tag{21}$$

From the definition of the * product follows:

$$\partial_i(f * g) = \partial_i f * g + f * \partial_i g. \tag{22}$$

This is the Leibniz rule (20) when mapped to the $^\Diamond \mathcal{A}_x$ algebra. As a consequence of (20) we find that

$$\hat{x}^i - i\theta^{ij}\hat{\partial}_j \tag{23}$$

commutes with all coordinates. For invertible θ^{ij} this can be used to define the action of the derivative entirely in $\mathcal{A}_{\hat{x}}$

$$\hat{\partial}_i = -i\theta_{ij}^{-1}\hat{x}^j. \tag{24}$$

Translated to the $^\diamond\mathcal{A}_x$ algebra this implies:

$$\partial_i f(x) = -i\theta_{ij}^{-1}[x^j \stackrel{*}{,} f]. \tag{25}$$

As a consequence we derive

$$\hat{\partial}_j\hat{\partial}_k - \hat{\partial}_k\hat{\partial}_j = -i\theta_{jk}^{-1} \quad : \mathcal{R}_{\hat{\partial},\hat{\partial}}. \tag{26}$$

This $\mathcal{R}_{\hat{\partial},\hat{\partial}}$ relation satisfies all the requirements of (ref7).

To formulate a Lagrangian field theory we have to learn how to integrate. Whereas it was easier to formulate derivatives on objects of $\mathcal{A}_{\hat{x}}$ it is easier to formulate integration on objects of $^\diamond\mathcal{A}_x$. For the canonical structure we define:

$$\int \hat{f} = \int d^n x \, f(x), \qquad \hat{f} \in \mathcal{A}_{\hat{x}}, f \in \,^\diamond\mathcal{A}_x. \tag{27}$$

This is a linear map of the algebra $\mathcal{A}_{\hat{x}}$ into $\mathbb{C}$

$$S \; : \; \mathcal{A}_{\hat{x}} \to \mathbb{C}, \tag{28}$$

$$S(c_1\hat{f} + c_2\hat{g}) \; = \; c_1 \int \hat{f} + c_2 \int \hat{g},$$

and it has the trace property:

$$\int \hat{f}\hat{g} = \int \hat{g}\hat{f}. \tag{29}$$

This can be verified explicitly using the definition of the $*$ product:

$$\int f * g = \int g * f = \int d^n x \, f(x)g(x). \tag{30}$$

For the quantum space structure the definition (30) for the integral does not have the trace property. There is, however, a measure for the integration that leads to an integral with the trace property.

$$\int \hat{f} \equiv \int d^n x \, \mu(x)f(x) \tag{31}$$

For the Manin plane we can verify explicitly that the measure

$$\mu(x,y) = \frac{1}{xy} \tag{32}$$

has this property.

In general we can construct Hilbert space representations of the algebra and define the integral as the trace. This will lead to infinite sums that can be interpreted as Riemannian sums for an integral and lead to the respective measure for the integration.

2. Gauge theories

Our aim is to formulate gauge theories. They will be based on a Lie algebra:

$$[T^a, T^b] = if_c^{ab}T^c. \tag{33}$$

In a usual gauge theorie on $\mathbb{R}^n$ the fields will span a representation of the Lie algebra and transform under an infinitesimal gauge transformation:

$$\delta_{\alpha^0}\psi(x) = i\alpha^0(x)\psi(x). \tag{34}$$

The transformation parameters are Lie algebra valued:

$$\alpha^0(x) = \alpha_a^0 T^a \tag{35}$$

and consequently:

$$
\begin{aligned}
(\delta_{\alpha^0}\delta_{\beta^0} - \delta_{\beta^0}\delta_{\alpha^0})\psi &= -(\beta^0\alpha^0 - \alpha^0\beta^0)\psi \\
&= i(\alpha^0 \times \beta^0)\psi = \delta_{\alpha^0 \times \beta^0}\psi, \\
\alpha^0 \times \beta^0 &\equiv \alpha_a^0 \beta_b^0 f_c^{ab} T^c.
\end{aligned}
\tag{36}
$$

covariant derivatives are defined with the help of a Lie algebra valued gauge field a:

$$
\begin{aligned}
\mathcal{D}_i \psi &= (\partial_i - ia_i)\psi, \\
a_i &= a_i^a T_a.
\end{aligned}
\tag{37}
$$

To obtain:

$$\delta_{\alpha^0}\mathcal{D}_i\psi = i\alpha^0 \mathcal{D}_i\psi \tag{38}$$

we have to demand:

$$
\begin{aligned}
\delta a_i &= \partial_i \alpha^0 + i[\alpha^0, a_i], \\
\delta a_{i,a} &= \partial_i \alpha_a^0 - \alpha_b^0 f_a^{bc} a_{i,c}.
\end{aligned}
\tag{39}
$$

To formulate a gauge theory on a non-commutative space we start with fields $\psi(x)$ that are elements of $^\diamond\mathcal{A}_{\hat{x}}$ and again span a representation of the Lie algebra (33). We demand the transformation law:

$$\delta_\alpha \psi(x) = i\alpha(x) * \psi(x) \tag{40}$$

in analogy to (34). But now we cannot demand α to be Lie algebra valued, we shall assume it to be enveloping algebra valued:

$$\alpha(x) = \alpha_a^0(x)T^a + \alpha_{ab}^1(x) : T^aT^b : + \cdots + \alpha_{a_1\ldots a_n}^{n-1}(x) : T^{a_1} \cdot \ldots \cdot T^{a_n} : + \cdots \tag{41}$$

This is in analogy to (35). We have adopted the :: notation for the basis elements of the enveloping algebra. We shall use the symmetrized polynomials as a basis:

$$: T^a : = T^a, \tag{42}$$

$$: T^aT^b : = \frac{1}{2}(T^aT^b + T^bT^a) \text{ etc.}$$

In analogy to (36) we find

$$(\delta_\alpha\delta_\beta - \delta_\beta\delta_\alpha)\psi = [\alpha \overset{*}{,} \beta] * \psi. \tag{43}$$

Naturally, $[\alpha \overset{*}{,} \beta]$ will be an enveloping algebra valued element of $^\Diamond \mathcal{A}_x$.

The unpleasant fact of the definition (41) of an enveloping algebra valued transformation parameter is that it depends on an infinite set of parameter fields $\alpha^n(x)$. In physics we would have to deal with an infinite set of fields when defining a covariant derivative, something we try to avoid. However, a gauge transformation can be realized by transformation parameters that depend on x via the parameter field $\alpha^0(x)$, the gauge field $a_{i,a}(x)$ and their derivatives only. In the notation of eqn (41) we have

$$\alpha_{a_1\ldots a_{n+1}}^n(x) = \alpha_{a_1\ldots a_{n+1}}^n(\alpha_a^0(x), a_{i,a}^0(x), \partial_i\alpha_a^0(x), \ldots). \tag{44}$$

Transformation parameters that are restricted that way we shall denote $\Lambda_{\alpha^0}(x)$. These parameters can be constructed in such a way that eqn (36) holds:

$$\delta_{\alpha^0}\psi(x) = i\Lambda_{\alpha^0(x)}(x) * \psi(x),$$

$$(\delta_{\alpha^0}\delta_{\beta^0} - \delta_{\beta^0}\delta_{\alpha^0})\psi = \delta_{\alpha^0 \times \beta^0}\psi, \tag{45}$$

$$(\alpha^0 \times \beta^0)_a = \alpha_b^0\beta_c^0 f_a^{bc}.$$

This together with the * product is the defining equations for the gauge transformations. That such parameters $\Lambda_{\alpha^0}(x)$ can be found is not obvious, it's rather a miracle in our present understanding of such gauge theories. Their existence is a consequence of the Seiberg-Witten map [2].

In the second variation of ψ we also have to account for the variation of Λ_{α^0} as it depends on $a_{i,a}$:

$$(\delta_{\alpha^0}\delta_{\beta^0} - \delta_{\beta^0}\delta_{\alpha^0})\psi = i(\delta_{\alpha^0}\Lambda_{\beta^0} - \delta_{\beta^0}\Lambda_{\alpha^0}) * \psi + [\Lambda_{\alpha^0} \overset{*}{,} \Lambda_{\beta_0}] * \psi, \tag{46}$$

$$= \delta_{\alpha^0 \times \beta^0}\psi = i\Lambda_{\alpha^0 \times \beta^0} * \psi$$

We shall construct Λ_{α^0} in a power series expansion in θ. To illustrate the method we expand Λ_{α^0} to first order in θ

$$\Lambda_{\alpha^0} = \alpha_a^0 T^a + \theta^{ij} \Lambda_{\alpha^0,ij}^1 + \cdots, \tag{47}$$

To be consistent we expand the $*$ product in (46) also to first order in θ and compare powers of θ. the θ-independent term defines $\alpha^0 \times \beta^0$ as we have used it in (45). This had to be expected, this order is exactly the commutative case. To first order we obtain the equation:

$$\theta^{ij} \left((\delta_{\alpha^0} \Lambda_{\beta^0,ij}^1 - \delta_{\beta^0} \Lambda_{\alpha^0,ij}^1) - i([\alpha^0, \Lambda_{\beta^0,ij}^1] - \tag{48}$$
$$- [\beta^0, \Lambda_{\alpha^0,ij}^1]) \right) + \frac{1}{2} \partial_i \alpha_a^0 \partial_j \beta_b^0 : T^a T^b := \theta^{ij} \Lambda_{\alpha^0 \times \beta^0,ij}^1.$$

This equation has the solution:

$$\theta^{ij} \Lambda_{\alpha^0,ij}^1 = \frac{1}{2} \theta^{ij} (\partial_i \alpha_a^0) a_{j,b} : T^a T^b : . \tag{49}$$

We see that Λ^1 is of second order in the generators T of the Lie algebra. The structure of eqn (46) allows a solution where Λ^n, the term in (47) of order $n - 1$ in θ, is a polynomial of order n in T.

$$\Lambda_{\alpha^0} = \alpha_a^0 T^a + \frac{1}{2} \theta^{ij} (\partial_i \alpha_a^0) a_{j,b} : T^a T^b : + \dots \tag{50}$$

In a next step in the formulation of a gauge theory we introduce covariant derivatives. Eqn (24) shows that we can relate this problem to the construction of covariant coordinates. We try to define such coordinates with the help of a gauge field, in the same way as we did it for derivatives in eqn (37):

$$X^i = x^i + A^i(x), \tag{51}$$
$$\delta_{\alpha^0} X^i * \psi = i\Lambda_{\alpha^0} * X^i * \psi. \tag{52}$$

This leads to a transformation law for the gauge field $A^i(x)$:

$$\delta A^i = -i[x^i \overset{*}{,} \Lambda_{\alpha^0}] + i[\Lambda_{\alpha^0} \overset{*}{,} A^i]. \tag{53}$$

We have to assume that A^i is enveloping algebra valued but we try to make an ansatz where all the coefficient functions only depend on $a_{i,a}$ and its derivatives:

$$A^i(x) = A_a^{i,0}(x) T^a + A_{ab}^{i,1}(x) : T^a T^b : + \dots \tag{54}$$
$$+ A_{a_1 \dots a_n}^{i,n-1}(x) : T^{a_1} \cdots \cdots T^{a_n} : + \dots,$$
$$A^{i,n} = A^{i,n}(a_{i,a}, \partial a_{i,a}, \dots).$$

Now we expand (53) in θ, demand $A^{i,n}$ to be a polynomial of order n in θ and solve eqn (53),

$$A^i(x) = \theta^{ij} V_j,$$

$$V_j(x) = a_{j,a} T^a - \frac{1}{2}\theta^{ln} a_{l,a}(\partial_n a_{j,b} + F_{nj,b} : T^a T^b : + \dots, \quad (55)$$

$$F_{nj,b} = \partial_n a_{j,b} - \partial_j a_{n,b} + f_b^{cd} a_{n,c} a_{j,d}.$$

This together with (41) is known as Seiberg-Witten map for an abelian gauge group. We have constructed it for an arbitrary non-abelian gauge group as well. Covariant derivatives follow from (37)

$$\mathcal{D}_i * \psi = (\partial_i - iV_i) * \psi, \quad (56)$$

$$\delta_{\alpha^0} \mathcal{D}_i * \psi = i\Lambda_{\alpha^0} * \mathcal{D}_i * \psi.$$

We now proceed with the definition of tensors as in a usual gauge theory, keeping in mind (27)

$$\tilde{F}_{ij} = \mathcal{D}_i * \mathcal{D}_j - \mathcal{D}_j * \mathcal{D}_i - i\theta_{ij}^{-1}. \quad (57)$$

The transformation law of the tensor is

$$\delta_{\alpha^0} \tilde{F}_{ij} = i[\Lambda_{\alpha^0} \overset{*}{,} \tilde{F}_{ij}]. \quad (58)$$

This can be verified from (53) and the definition of $\tilde{F}$.

To first order in θ we find:

$$\tilde{F}_{ij} = F_{ij,a} T^a + \theta^{ln}(F_{il,a} F_{jn,l} - \quad (59)$$

$$\frac{1}{2} a_{l,a}(2\partial_n F_{ij,b} + a_{n,c} F_{ij,d} f_e^{cd})) : T^a T^b : + \dots. \quad (60)$$

We see that new "contact" terms appear in the field strength $\tilde{F}$.

A good candidate for a Lagrangian is

$$L = \frac{1}{4} \mathrm{Tr} F_{ij} * F^{ij}. \quad (61)$$

The trace is taken in the representation space of the generators T. The Lagrangian (61) is not invariant because the $*$ product is not commutative:

$$\delta L = \frac{1}{4} \mathrm{Tr}\, i[\Lambda_{\alpha^0} \overset{*}{,} L]. \quad (62)$$

We know, however, that the integral has the trace property (31). This allows us to define the invariant action:

$$W = \frac{1}{4} \int \mathrm{Tr} F_{ij} * F^{ij} \quad (63)$$

$$= \frac{1}{4} \int \mathrm{Tr} F_{ij} F^{ij}.$$

This action depends on the gauge field $a_{i,a}$ and its derivatives only. It can be considered as a gauge-invariant object if $a_{i,a}$ transforms according to (39). this implies that W satisfies the Ward identities.

$$\delta_{\alpha^0(x)} W = 0, \tag{64}$$

$$\delta_{\alpha^0(x)} = -\alpha_a^0(x) \left(\frac{\partial}{\partial x^i} \delta_d^a + a_{i,b}(x) f_d^{ab} \right) \frac{\delta}{\delta a_{i,d}(x)}.$$

The Lagrangian expanded to all orders in θ, is a non-local object. It remains to be seen if it is acceptable for a quantum field theory or if it has to be viewed as an effective Lagrangian, ref. [8].

References

1. B. Jurčo, S. Schraml, P. Schupp and J. Wess, *Enveloping algebra-valued gauge transformations for non-abelian gauge groups on non-commutative spaces*, Eur. Phys. J. **C 17**, (2000) 521, hep-th/0006246.
    J. Madore, S. Schraml, P. Schupp and J. Wess, *Gauge theory on noncommutative spaces*, Eur. Phys. J. **C 16**, (2000), 161, hep-th/0001203.
    B. Jurčo, P. Schupp, *Noncommutative Yang-Mills from equivalence of star products*, Eur. Phys. J. **C 14**, 367 (2000), hep-th/0001032.
    B. Jurčo, P. Schupp and J. Wess, *Noncommutative gauge theory for Poisson manifolds*, Nucl. Phys. **B 584**, (2000), 784, hep-th/0005005.
    B. Jurčo, P. Schupp and J. Wess, *Nonabelian noncommutative gauge theory and Seiberg-Witten map*, in preparation.
2. N. Seiberg and E. Witten, *String theory and noncommutative geometry*, JHEP **9909** (1999) 032, hep-th/9908142.
3. A. Dimakis, J. Madore, *Differential Calculi and Linear Connections*, J.Math.Phys. **37** (1996) 4647.
    M. Dubois-Violette, R. Kerner, J. Madore, *Gauge bosons in a noncommutative geometry*, Phys.Lett. **B217** (1989) 485.
    J. Hoppe, *Diffeomorphism groups, Quantization and $SU(\infty)$*, Int.J.Mod.Phys. A **4** (1989) 5235.
    J. Madore, *An Introduction to Noncommutative Differential Geometry and it Physical Applications*, 2nd Edition, Cambridge University Press, 1999.
    B. deWit, J. Hoppe, H. Nicolai, Nucl.Phys. **B305** [FS 23] (1988) 545.
    D. Kabat, W. Taylor IV, *Spherical membranes in Matrix theory*, Adv.Theor.Phys. 2 (1998) 181-206, (hep-th 9711078).
4. J. Wess, *q-deformed Heisenberg Algebras*, in H. Gausterer, H. Grosse and L. Pittner, eds., Proceedings of the 38. Internationale Universitätswochen für Kern- und Teilchenphysik, no. 543 in Lect. Notes in Phys., Springer-Verlag, 2000, Schladming, January 1999, math-ph/9910013.
5. F. Bayen, M. Flato, C. Fronsdal, A. Lichnerowicz, D. Sternheimer, *Deformation theory and quantization. I. Deformations of symplectic structures*, Ann. Physics **111**, 61 (1978).
    M. Kontsevitch, *Deformation quantization of Poisson manifolds, I,* q-alg/9709040.
    D. Sternheimer, *Deformation Quantization: Twenty Years After*, math/9809056.
6. H. Weyl, *Quantenmechanik und Gruppentheorie*, Z. Physik **46**, 1 (1927); *The theory of groups and quantum mechanics*, Dover, New-York (1931), translated from *Gruppentheorie und Quantenmechanik*, Hirzel Verlag, Leipzig (1928).

J. E. Moyal, *Quantum mechanics as a statistical theory*, Proc. Cambridge Phil. Soc. **45**, 99 (1949).

7. J. Wess and B. Zumino, *Covariant differential calculus on the quantum hyperplane*, Nucl. Phys. Proc. Suppl. **18B** (1991) 302.

8. L. Bonora, M. Schnabl, M. M. Sheikh-Jabbari and A. Tomasiello, *Noncommutative SO(n) and Sp(n) gauge theories*, hep-th/0006091.

I. Chepelev, R. Roiban, *Convergence Theorem for Non-commutative Feynman Graphs and Renormalization*, hep-th/0008090.

A. Bichl, J.M. Grimstrup, V. Putz, M. Schweda, *Perturbative Chern-Simons Theory on non-commutative* $\mathbb{R}^3$, hep-th/0004071.

A. Bichl, J.M. Grimstrup, H. Grosse, L. Popp, M. Schweda, R. Wulkenhaar, *The Superfield Formalism Applied to the Non-commutative Wess-Zumino Model*, hep-th/0007050.

SYMMETRIES WIDER THAN SUPERSYMMETRY *

DIMITRY LEITES[††‡]
Department of Mathematics, University of Stockholm, Roslags-vägen. 101, Kräftriket hus 6, S-106 91, Stockholm, Sweden

VERA SERGANOVA[§]
Department of Mathematics, University of California at Berkeley, Berkeley, CA 94720, USA

Abstract. We observe that supersymmetries do not exhaust all the symmetries of the super-manifolds. On a generalization of supermanifolds (called *metamanifolds*), the "functions" form a metaabelean algebra, i.e., the one for which $[[x,y],z] = 0$ with respect to the usual commutator. The superspaces considered as metaspaces admit symmetries wider than supersymmetries. Conjecturally, infinitesimal transformations of these metaspaces constitute *Volichenko algebras* which we introduce as inhomogeneous subalgebras of Lie superalgebras. The Volichenko algebras are natural generalizations of Lie superalgebras being 2-step filtered algebras. They are non-conventional deformations of Lie algebras bridging them with Lie superalgebras.

1. Introduction: Towards noncommutative geometry

This is an elucidation of our paper [31]. In 1990 we were unaware of [42] to which we now would like to add later papers [14], and [2], and papers cited therein pertaining to this topic. Observe also an obvious connection of Volichenko algebras with structures that become more and more fashionable lately, see [22]; Volichenko algebras are one of the ingredients in the construction of simple Lie algebras over fields of characteristic 2, cf. [23]

1.1. The gist of idea. To describe physical models, the least one needs is a triple $(X, F(X), L)$, consisting of the "phase space" X, the sheaf of functions on it, locally represented by the algebra $F(X)$ of sections of this sheaf, and a Lie subalgebra L of the Lie algebra of of differentiations of $F(X)$ considered

---

* Instead of J. Naudts contribution by the editor S. Duplij's request

† D.L. is thankful to an NFR grant for partial financial support, to V. Molotkov, A. Premet and S. Majid for help.

‡ mleites@matematik.su.se

§ serganov@math.berkeley.edu

S. Duplij and J. Wess (eds.), Noncommutative Structures in Mathematics and Physics, 13–30.
© 2001 *Kluwer Academic Publishers. Printed in the Netherlands.*

as vector fields on X. Here X can be recovered from $F(X)$ as the collection $\mathrm{Spec}(F(X))$, called the *spectrum* and consisting of maximal or prime ideals of $F(X)$. Usually, X is endowed with a suitable topology.

After the discovery of quantum mechanics the attempts to replace $F(X)$ with the noncommutative ("quantum") algebra A became more and more popular. The first successful attempt was superization [25], [5] the road to which was prepared in the works of A. Weil, Leray, Grothendiek and Berezin, see [11]. It turns out that having suitably generalized the notion of the tensor product and differentiation (by inserting certain signs in the conventional formulas) we can reproduce on supermanifolds all the characters of differential geometry and actually obtain a much reacher and interesting plot than on manifolds. This picture proved to be a great success in theoretical physics since the language of supermanifolds and supergroups is a "natural" for a uniform description of bose and fermi particles. Today there is no doubt that this is the language of the Grand Unified Theories of all known fundamental forces.

Observe that physicists who, being unaware of [25], rediscovered super-groups and superspaces (Golfand–Likhtman, Volkov–Akulov, Neveu–Schwartz, Stavraki) were studying possibilities to enlarge the group of symmetries (or rather the Lie algebra of infinitesimal symmetries) of the known objects (in particular, objects described by Maxwell and Dirac equations). Their efforts did not draw much attention (like our [25] and [31]) until Wess and Zumino [43] understood and showed to others some of the whole series of wonders one can obtain by means of supersymmetries.

Here we show that the supergroups are not the largest possible symmetries of superspaces; there are transformations that preserve more noncommutativity than just a "mere" supercommutativity. To be able to observe that there are symmetries that unify bose and fermi particles we had to admit a broader point of view on our Universe and postulate that we live on a supermanifold. Here (and in [31]) we suggest to consider our supermanifolds as paticular case of *metamanifolds*, introduced in what follows.

How noncommutative should $F(X)$ be? To define the space correspond-ing to an arbitrary algebra is very hard, see Manin's gloomy remarks in [33], where he studies quadratic algebras as functions on "perhaps, nonexisting" noncommutative projective spaces.

Manin's idea that there hardly exists one uniform definition suitable for any noncommutative algebra (because there are several quite distinct types of them) was supported by A. Rosenberg's studies; he managed to define several types of spectra in order to interpret ANY algebra as the algebra of functions on a suitable spectrum, see preprints of his two books [27], no. 25, and nos. 26, 31 (the latter be-ing expanded as [35]). In particular, there IS a space corresponding to a quadratic (or "quadraticizable") algebra such as the so-called "quantum" deformation $U_q(\mathfrak{g})$ of $U(\mathfrak{g})$, see [12].

Observe that in [33] Manin also introduced and studied symmetries of super-commutative superalgebras wider than supersymmetries, but he only considered them in the context of quadratic algebras (not all relations of a supercommutative suepralgebra are quadratic or quadraticizable). Regrettably, nobody, as far as we know, investigated consequences of Manin's approach to enlarging supersymmetries.

Unlike numerous previous attempts, Rosenberg's theory is more natural; still, it is algebraic, without any real geometry (no differential equations, integration, etc.). For some noncommutative algebras certain notions of differential geometry can be generalized: such is, now well-known, A. Connes geometry, see [10], and [34]. Arbitrary algebras seem to be too noncommutative to allow to do any physics.

In contrast, the experience with the simplest non-commutative spaces, the superspaces, tells us that all constructions expressible in the language of differential geometry (these are particularly often used in physics) can be carried over to the super case. Still, supersymmetry has, as we will show, certain shortcomings, which disappear in the theory we propose.

Specifically, we continue the study started under Berezin's influence in [25] (later suppressed under the same influence in [5], [26]), of algebras just slightly more general than supercommutative superalgebras, namely their arbitrary, not necessarily homogeneous, subalgebras and quotients. Thanks to Volichenko's theorem F (F is for "functions", see [27], no. 17 and Appendix below) such algebras are precisely *metaabelean* ones, i.e., those that satisfy the identity

$$[x, [y, z]] = 0 \text{ (here } [\cdot, \cdot] \text{ is the usual commutator).} \qquad (1.1)$$

As in noncommutative geometries, we think of metaabelean algebras as "functions" on a what we will call *metaspace*.

Observe that the conventional superspaces considered as metaspaces and Lagrangians on them have *additional* symmetries as compared with supersymmetry.

1.2. The notion of Volichenko algebras. Volichenko's Theorem F gives us a natural generalization of the supercommutativity. It remains to define the analogs of the tensor product and study differentiation (e.g., Volichenko's approach, see §3). We conjecture that the analogs of Lie algebras in the new setting are *Volichenko algebras* defined here as nonhomogeneous subalgebras of Lie superalgebras.

Supersymmetry had been already justified for physicists when mathematicians' attention was drawn to it by the list of simple finite dimensional Lie superalgebras: bar one exception it was discrete and looked miraculously like the list of simple Lie algebras. Our list of simple Volichenko algebras is similar. Our main mathematical result is the classification (under a technical hypothesis) of simple finite dimensional (and vectorial) Volichenko algebras, see [40], [31].

Remarkably, Volichenko algebras are just deformations of Lie algebras though in an entirely new sense: in a category broader than that of Lie algebras or Lie superalgebras. This feature of Volichenko algebras could be significant for parastatistics because once we abandon bose-fermi statistics, there seem to be too many *ad hoc* ways to generalize. Our classification asserts that within the natural context of simple Volichenko algebras the set of possibilities is discrete or has at most 1-parameter (hence, anyway, describable!). It is important because it suggests the possibility of associating distinct types of particles to representations of these structures.

Our generalization of supersymmetry and its implications for parastatistics appear to be complementary to works on braid statistics in two dimensions [15] in the context of [13], see also [19]. We expect them to tie up at some stage.

Examples of what looks like nonsimple Volichenko algebras recently appeared in another context in [2], [36], [42] and [14].

1.3. An intriguing example: the general Volichenko algebra $\mathfrak{vgl}_\mu(p|q)$. Let the space $\mathfrak{h}$ of $\mathfrak{vgl}_\mu(p|q)$ be the space of $(p+q) \times (p+q)$-matrices divided into the two subspaces as follows:

$$\mathfrak{h}_{\hat{0}} = \left\{ \begin{matrix} A & 0 \\ 0 & D \end{matrix} \right\}; \quad \mathfrak{h}_{\hat{1}} = \left\{ \begin{matrix} 0 & B \\ C & 0 \end{matrix} \right\}. \tag{1.3.1}$$

Here $\mathfrak{h}_{\hat{1}}$ is a natural $\mathfrak{h}_{\hat{0}}$-module with respect to the bracket of matrices; fix $a, b \in \mathbb{C}$ such that $a : b = \mu \in \mathbb{C}P^1$ and define the multiplication $\mathfrak{h}_{\hat{1}} \times \mathfrak{h}_{\hat{1}} \longrightarrow \mathfrak{h}_{\hat{0}}$ by the formula

$$[X, Y] = a[X, Y]_- + b[X, Y]_+ \text{ for any } X, Y \in \mathfrak{h}_{\hat{1}}. \tag{1.3.2}$$

(The subscript $-$ or $+$ indicates the commutator and the anticommutator, respectively.) As we sill see, $\mathfrak{h}$ is a simple Volichenko algebra for any a, b except for $ab = 0$ when it becomes isomorphic to either the Lie algebra $\mathfrak{gl}(p+q)$ or the Lie superalgebra $\mathfrak{gl}(p|q)$. To show that $\mathfrak{vgl}_\mu(p|q)$ is indeed a Volichenko algebra, we have to realize it as a subalgebra of a Lie superalgebra. This is done in heading 2 of Theorem 2.7.

2. Metaabelean algebra as the algebra of "functions". Volichenko algebra as an analog of Lie algebra

2.1. Symmetries broader than supersymmetries. It was the desire to broaden the notion of a group that lead physicists to supersymmetry. However, in viewing supergroups as transformations of superspaces we consider only even, "statistics-preserving", maps: nonhomogeneous "statistics-mixing" maps between super-algebras are explicitly excluded and this is why and how odd parameters of supergroups appear, cf. [3], [11].

On the one hand, this is justified: since we consider graded objects why should we consider transformations that preserve these objects as abstract ones but destroy the grading? It would be inconsistent on our part, unless we decide to consider the grading or "parity" as one considers the electric charge of a nucleon: in certain problems we ignore it.

On the other hand, if such parity violating transformations exist, they deserve to be studied, to disregard them is physically and mathematically an artificial restriction.

We would like to broaden the notion of supergroups and superalgebras to allow for the possibility of statistics-changing maps. Soon after Berezin published his description of automorphisms of the Grassmann algebra [4] it became clear that Berezin missed nonhomogeneous automorphisms, but the complete description of automorphisms was unknown for a while. In 1977, L. Makar-Limanov gave us a correct description of such automorphisms (private communication). A. Kirillov rediscovered it while editing [3], Ch.1; for automorphisms in presence of even variables see [28].

Recall the answer: the generic finite transformation of a supercommutative superalgebra $\mathcal{F}$ of functions in n even generators $x_1, ..., x_n$ and m odd ones $\theta_1, ..., \theta_m$ is of the form (here p_m is the parity of m, i.e., either 0 or 1)

$$x_i \mapsto [(f_i + \sum_k f_i^{i_1...i_{2k}}\theta_{i_1}...\theta_{i_{2k}}) + \sum_k f_i^{i_1...i_{2k+1}}\theta_{i_1}...\theta_{i_{2k+1}}]\underline{\underline{(1 + F_i\theta_1...\theta_m p_m)}}$$

$$\theta_j \mapsto [(\sum_k g_j^{i_1...i_{2k+1}}\theta_{i_1}...\theta_{i_{2k+1}}) + g_j + \sum_k g_j^{i_1...i_{2k}}\theta_{i_1}...\theta_{i_{2k}}]\underline{\underline{(1 + g)}}$$

$$(2.1)$$

where f_i, F_i and $f_i^{i_1...i_{2k}}$, and also $g_j^{i_1...i_{2k+1}}$ are even superfields, whereas $f_i^{i_1...i_{2k+1}}, g_j$ and $g_j^{i_1...i_{2k}}$ and also g, F_i are odd superfields. (A mathematician, see [11], would say that the odd superfields (<u>underlined</u> once) represent the parameters corresponding to Λ-points with nonzero odd part of the background supercommutative superalgebra Λ.) Notice that one g serves all the θ_j. The <u>twice</u> underlined factors account for the extra symmetry of $\mathcal{F}$ as compared with supersymmetry.

Comment. When the number of odd variables is even, as is usually the case in modern models of Minkowski superspace, there is only one extra functional parameter, g. Therefore, on such supermanifolds, the *notion of a boson is coordinate-free, whereas that of a fermion depends on coordinates.*

Summing up, (this is our main message to the reader)

supersymmetry is not the most broad symmetry of supercommutative superalgebras

2.2. Two complexifications. Another quite unexpected flaw of supersymmetry is that the category of supercommutative superalgebras is *not* closed with respect to complexification. It certainly is if $\mathbb{C}$ is understood naively, as a purely even space. Declaring $\sqrt{-1}$ to be odd, we make $\mathbb{C}$ into a nonsupercommutative superalgebra. This associative superalgebra over $\mathbb{R}$ is denoted by $Q(1; \mathbb{R})$, see [26], [6].

The complex structure given by an odd operator gives rise to a "queer" superanalogue of the matrix algebra, $Q(n; \mathbb{K})$ over any field $\mathbb{K}$. Its Lie version, the projectivization of its queertraceless subalgebra (first discovered by Gell-Mann, Mitchel and Radicatti, cf. [9]) is one of main examples of simple Lie superalgebras, whereas $Q(1)$ corresponds to one of the two cases of Schur's Lemma for superalgebras. An infinite dimensional representation of $Q(1)$ is crucial in A. Connes' noncommutative differential geometry. In short, the odd complex structure on superspaces is an important one.

How to modify definition of supermanifold to incorporate the above structures?

Conjecturally, the answer is to consider arbitrary, not necessarily homogeneous subalgebras and quotients of supercommutative superalgebras. These algebras are, clearly, metaabelean algebras. But how to describe arbitrary metaabelean algebras? In 1975 D.L. discussed this with V. Kac and Kac conjectured (see [26]) that considering metaabelean algebras we do not digress far from supercommutative superalgebras, namely, every metaabelean algebra is a subalgebra of a supercommutative superalgebra. Therefore, the most broad notion of morphisms of supercommutative superalgebras should only preserve their metaabeleanness but not parity. (Since $\mathbb{C}$, however understood, is metaabelean, we get a category of algebras closed with respect to all algebra morphisms and complexifications.)

Volichenko proved more than Kac' conjecture (Appendix). Namely, he proved that any finitely generated metaabelean algebra admits an embedding into a universal supercommutative superalgebra and developed an analogue of Taylor series expansion.

Until Volichenko's results, it was unclear how to work with metaabelean algebras: are there any analogues of differential equations, or integral, in other words, is there any "real life" on metaspaces [26]? Thanks to Volichenko, we can now consider pairs

(a metaabelean algebra, its ambient supercommutative superalgebra)

and corresponding projections "superspace $\longrightarrow$ metaspace" when we consider these algebras as algebras of functions.

It is interesting to characterize metaabelean algebras which are *quotients* of supercommutative superalgebras: in this case the corresponding metaspace can be embedded into the superspace and we can consider the induced structures (Lagrangeans, various differential equations, etc.).

But even if we would have been totally unable to work with metaspaces which are not superspaces, it is manifestly useful to consider superspaces as metaspaces. In so doing, we retain all the paraphernalia of the differential geometry for sure, and in addition get more transformations of the same entities.

For example, it is desirable to make use of the formula (first applied by Arnowitt, Coleman and Nath)

$$\text{Ber } X = \exp \text{ str } \log X$$

which extends the domain of the berezinian (superdeterminant) to nonhomogeneous matrices X. Then we can consider the additional nonhomogeneous transformations, like the ones described in (2.1). All supersymmetric Lagrangeans admit metasymmetry wider than supersymmetry.

Remark. In mathematics and physics, spaces are needed almost exclusively to integrate over them or consider limits in analytic questions. In problems where integration is not involved we need sheaves of sections of various bundles over the spaces rather than the spaces themselves. Gauge fields, Lagrangeans, etc. are all sections of coherent sheaves, corresponding to sections of vector bundles. Now, almost 30 years after the definition of the scheme of a metaabelean algebra (metavariety or metaspace) had been delivered at A. Kirillov's seminar ([25]), there is still no accepted definition of nice ("morally coherent" as Manin says) sheaves over such a scheme even for superspaces (for a discussion see [8]). As to candidates for such sheaves see Rosenberg's books on noncommutative geometry [27], nos. 25, 26, 31 and [35]) and §9 in [8]. This §9 is, besides all, a possible step towards "compactification in odd directions".

2.3. A description of Volichenko algebras. It seemed natural [26] to get for Lie superalgebras a result similar to Volichenko's theorem F, i.e., to describe arbitrary subalgebras of Lie superalgebras. Shortly before his untimely death I. Volichenko (1955-88) announced such a description (Theorem A, here A is for (Lie) "algebra"). In his memory then, a *Volichenko algebra* is a nonhomogeneous subalgebra $\mathfrak{h}$ of a Lie superalgebra $\mathfrak{g}$. The adjective "Lie" before a (super)algebra indicates that the algebra is not associative, likewise the adjective "Volichenko" reminds that the algebra is neither associative nor should it satisfy Jacobi or super-Jacobi identities. Thus, a Volichenko algebra $\mathfrak{h}$ is a non-homogeneous subspace of a Lie superalgebra $\mathfrak{g}$ closed with respect to the superbracket of $\mathfrak{g}$. How to describe $\mathfrak{h}$ by identities, i.e., in inner terms, without appealing to any ambient?

Theorem. A (I. Volichenko, 1987) *Let A be an algebra with multiplication denoted by juxtaposition. Define the Jordan elements $a \circ b := ab + ba$ and Jacobi elements $J(a, b, c) := a(bc) + c(ab) + b(ca)$. Suppose that*

(a) *A is Lie admissible, i.e., A is a Lie algebra with respect to the new product defined by the bracket (not superbracket) $[a, b] = ab - ba$;*

(b) *the subalgebra $A^{(JJ)}$ generated by all Jordan and Jacobi elements belongs to the anticenter of A, in other words*

$$ax + xa = 0 \text{ for any } a \in A^{(JJ)}, \ x \in A;$$

(c) *$a(xy) = (ax)y + x(ay)$ for any $a \in A^{(JJ)}$, $x, y \in A$.*
Then

(1) *Any (not necessarily homogeneous) subalgebra $\mathfrak{h}$ of a Lie superalgebra $\mathfrak{g}$ satisfies the above conditions (a) — (c).*

(2) *If A satisfies (a) — (c), then there exists a Lie superalgebra $\text{SLie}(A)$ such that A is a subsuperalgebra (closed with respect to the superbracket) of $\text{SLie}(A)$.*

Heading (1) is subject to a direct verification.

Clearly, the parts of conditions (b) and (c) which involve Jordan (resp. Jacobi) elements replace the superskew-commutativity (resp. Jacobi identity). Condition (a) ensures that A is closed in $\mathrm{SLie}(A)$ with respect to the bracket in the ambient.

Discussion. If true, Volichenko's theorem A would have disproved a pessimistic conjecture of V. Markov cited in [26]: *the minimal set of polynomial identities that single out nonhomogeneous subalgebras of Lie superalgebras is infinite.* I. Volichenko did not investigate under which conditions a finite dimensional Volichenko algebra A can be embedded into a finite dimensional Lie superalgebra $\mathfrak{g}$; which is, perhaps, the quotient of $\mathrm{SLie}(A)$ modulo an ideal.

Volichenko's scrap papers were destroyed after his death and no hint of his ideas remains. Several researchers tried to refute it and A. Baranov succeeded. He showed [1] that Volichenko's theorem V is wrong as stated: one should add at least one more relation of degree 5. First, following Volichenko, Baranov introduced instead of $J(a, b, c)$ more convenient linear combinations of the Jacobi elements

$$j(a, b, c) = [a, b \circ c] + [b, c \circ a] + [c, a \circ b] \text{ for } a, b, c \in A.$$

Then Baranov rewrote identities (a)–(c) in the following equivalent but more transparent form (i)–(v):

(i) $[a, [b, c]] + [b, [c, a]] + [c, [a, b]] = 0$;

(ii) $a \circ b \circ c = 0$;

(iii) $j(a, b, c) \circ d = 0$;

(iv) $[a \circ b, c \circ d] = [a \circ b, c] \circ d + [a \circ b, d] \circ c$;

(v) $[j(a, b, c), c \circ d] = [j(a, b, c), c] \circ d + [j(a, b, c), d] \circ c$.

Baranov's new identity independent of (i) – (v) is of degree 5 and is somewhat implicit; it involves 49 monomials and no lucid expression for it is found yet.

True or false, Volichenko's theorem A does not affect our results, since we do not appeal to an intrinsic definition of Volichenko algebras.

2.4. On simplicity of Volichenko algebras. As we will see, the notion of Volichenko algebra is a totally new type of deformation of the usual Lie algebra. It also generalizes the notion of a Lie superalgebra in a sence that the Lie superalgebras are $\mathbb{Z}/2$-graded algebras (i.e., they are of the form $\mathfrak{g} = \bigoplus_{i=\bar{0},\bar{1}} \mathfrak{g}_i$ such that $[\mathfrak{g}_i, \mathfrak{g}_j] \subset \mathfrak{g}_{i+j}$) whereas Volichenko algebras are only 2-step filtered ones (i.e., they are of the form $\mathfrak{h} = \bigoplus_{i=\bar{0},\bar{1}} \mathfrak{h}_i$ as *spaces* and $\mathfrak{h}_{\bar{0}}$ is a subalgebra. There are, however, several series of examples when Volichenko algebras are $\mathbb{Z}/2$-graded (e.g., $\mathfrak{vgl}_\mu(p|q)$).

Hereafter $\mathfrak{g}$ is a Lie superalgebra over $\mathbb{C}$ and $\mathfrak{h} \subset \mathfrak{g}$ a subspace which is not a subsuperspace closed with respect to the superbracket in $\mathfrak{g}$. For notations of simple complex finite dimensional Lie superalgebras, the list of known simple $\mathbb{Z}$-graded infinite dimensional Lie superalgebras of polynomial growth over $\mathbb{C}$ and $\mathbb{R}$, and their gradings see [20], [38], [27], [37], [30]. A Volichenko algebra is said to be *simple* if it has no two-sided ideals and its dimension is $\neq 1$.

Remark. P. Deligne argued that for an algebra such as a Volichenko one, modules over which have no natural two-sided structure, the above definition seems to be too restrictive: one should define simplicity by requiring the absence of *one-sided* ideals. As it turns out, none of the simple Volichenko algebras we list in what follows has one-sided ideals, so we will stick to the above (at first glance, preliminary) definition: it is easier to work with.

Lemma. *For any* simple *Volichenko algebra* $\mathfrak{h}$, $\mathfrak{h} \subset \mathfrak{g}'$, *there exists a simple Lie subsuperalgebra* $\mathfrak{g} \subset \mathfrak{g}'$ *that contains* $\mathfrak{h}$.

So, we can (and will) assume that the ambient $\mathfrak{g}$ of a simple Volichenko algebra is simple. In what follows we will see that under a certain condition for a simple Volichenko algebra $\mathfrak{h}$ its simple ambient Lie superalgebra $\mathfrak{g}$ is unique. here is this condition:

2.5. The "epimorphy" condition. Denote by $p_i : \mathfrak{g} \longrightarrow \mathfrak{g}_i$, where $i = \bar{0}, \bar{1}$, the projections to homogeneous components. A Volichenko algebra $\mathfrak{h} \subset \mathfrak{g}$ will be called *epimorphic* if $p_0(\mathfrak{h}) = \mathfrak{g}_{\bar{0}}$. Not every Volichenko subalgebra is epimorphic: for example, the two extremes, Volichenko algebras with the zero bracket and free Volichenko algebras, are not epimorphic, generally. All simple finite dimensional Volichenko algebras known to us are, however, epimorphic.

Hypothesis. *Every simple Volichenko algebra is epimorphic.*

A case study of various simple Lie superalgebras of low dimensions reveals that they do not contain non-epimorphic simple Volichenko algebra. Still, we can not prove this hypothesis but will adopt it for it looks very natural at the moment.

Lemma. *Let* $\mathfrak{h} \subset \mathfrak{g}$ *be an epimorphic Volichenko algebra and* $f : \mathfrak{g}_{\bar{0}} \longrightarrow \mathfrak{g}_{\bar{1}}$ *a linear map that determines* $\mathfrak{h}$, *i.e.,*

$$\mathfrak{h} = \mathfrak{h}_f := \{a + f(a) \mid a \text{ runs over } \mathfrak{g}_{\bar{0}}\}.$$

Then
1) *f is a 1-cocycle from* $C^1(\mathfrak{g}_{\bar{0}}; \mathfrak{g}_{\bar{1}})$;
2) *f can be uniquely extended to a derivation of* $\mathfrak{g}$ *(also denoted by* f*) such that* $f(f(\mathfrak{g}_{\bar{0}})) = 0$.

Example. Recall, that the odd element x of any Lie superalgebra is called a *homologic* one if $[x, x] = 0$, cf. [41]. Let $x \in \mathfrak{g}_{\bar{1}}$ be such that

$$[x, x] \in C(\mathfrak{g}), \tag{2.5.1}$$

where $C(\mathfrak{g})$ is the center of $\mathfrak{g}$. Clearly, the map $f = \text{ad}\,(x)$ satisfies Lemma 2.4 if x satisfies (2.5.1), i.e., is homologic modulo center.

A homologic modulo center element x will be said to *ensure nontriviality* (of the algebra

$$\mathfrak{h}_x = \{a + [a, x] \mid a \text{ runs over } \mathfrak{g}_{\bar{0}}\}) \tag{2.5.2}$$

if

$$[[\mathfrak{g}_{\bar{0}}, x], [\mathfrak{g}_{\bar{0}}, x]] \neq 0,$$

i.e., if there exist elements $a, b \in \mathfrak{g}_{\bar{0}}$ such that

$$[[a, x], [b, x]] \neq 0. \tag{2.5.3}$$

The meaning of this notion is as follows. Let $a, b \in \mathfrak{h}$, $a = a_0 + a_1$, $b = b_0 + b_1$, where $a_1 = [a_0, x]$, $b_1 = [b_0, x]$ for some $x \in \mathfrak{g}_{\bar{1}}$. Notice that for x satisfying (2.5.1) we have

$$[[a_1, b_1], x] = 0. \tag{2.5.4}$$

If (2.5.3) holds, we have

$$[a, b] = [a_0, b_0] + [a_1, b_1] + [a_0, b_1] + [a_1, b_0] = ([a_0, b_0] + [a_1, b_1]) + [[a_0, b_0], x]. \tag{2.5.5}$$

It follows from (2.5.4) and (2.5.5) that if x is homologic modulo center, then $\mathfrak{h}_x$ is closed under the bracket of $\mathfrak{g}$; if this x does not ensure nontriviality, then $\mathfrak{h}_x$ is just isomorphic to $\mathfrak{g}_{\bar{0}}$.

In other words, an epimorphic Volichenko algebra is a deformation of the Lie algebra $\mathfrak{g}_{\bar{0}}$ in a totally new sence: not in the class of Lie algebras, nor in that of Lie superalgebras but in the class of Volichenko algebras whose intrinsic description is to be given. (To see that an epimorphic Volichenko algebra $\mathfrak{h}_x$ is a result of a deformation of sorts, multiply x by an even parameter, t. If t were odd, we would have obtained a deformation of $\mathfrak{g}_{\bar{0}}$ in the class of Lie superalgebras.)

Remark. It is easy to show making use of formula (2.5.5) why it is impossible to consider any other (inconsistent with parity) $\mathbb{Z}/2$-grading (call it deg) of $\mathfrak{g}$ and deform in a similar way the Lie subsuperalgebra of elements of degree 0 with respect to deg.

Any epimorphic Volichenko algebra $\mathfrak{h}_x \subset \mathfrak{g}$ is naturally filtered: it contains as as subalgebra the Lie algebra $\mathfrak{ann}\,(x) = \{a \in \mathfrak{g}_{\bar{0}} \mid [x, a] = 0\}$.

Problems. 1) We have a sandwich: between Hopf (super)algebras, $U(\mathfrak{h}_x)$ and $U(\mathfrak{g})$, a non-Hopf algebra, $U(\mathfrak{h})$ (the subalgebra of $U(\mathfrak{g})$ generated by $\mathfrak{h}$), is squeezed. How to measure its "non-Hopfness"? This invariant seems to be of interest.

2) It is primarily real algebras and their representations that arise in applications. So what are these notions for Volichenko algebras?

We do not know at the moment the definition of a *representation of a Volichenko algebra* even for epimorphic ones. To say "a *representation of a Volichenko algebra* is a through map: the composition of an embedding $\mathfrak{h} \subset \mathfrak{g}$ into a minimal ambient and a representation $\mathfrak{g} \longrightarrow \mathfrak{gl}(V)$" is too restrictive: the adjoint representation and homomorphisms of Volichenko algebras are ruled out.

3) If we abandon the technical hypothesis on epimorphy, do we obtain any simple Volichenko algebras? (Conjecture: we do not.)

4) Describe Volichenko algebras intrinsically, via polynomial identities. This seems to be a difficult problem.

5) Classify simple Volichenko subalgebras of the other known simple Lie superalgebras of interest, e.g., of polynomial growth, cf. [16], [17].

2.6. Vectorial Volichenko superalgebras. For a vector field $D = \sum f_r \partial_r$ from $\mathfrak{vect}(m|n) = \mathfrak{der}\mathbb{C}[x, \theta]$, define its *inverse order* with respect to the nonstandard (if $m \neq 0$) grading induced by the grading of $\mathbb{C}[x, \theta]$ (for which deg $x_i = 0$ and deg $\theta_j = 1$ for all i and j) and inv.ord(f_r) is the least of the degrees of monomials in the power series expansion of f_r.

There are two major types of Lie (super)algebras and their subalgebras: the ones realized by matrices and the ones realized by vector fields. The former ones will be refered to as matrix ones, the latter ones as vectorial algebras.

2.6.1. Lemma. *Let $\mathfrak{h} \subset \mathfrak{g}$ be a simple epimorphic vectorial Volichenko algebra, i.e., a subalgebra of a simple vectorial Lie superalgebra. Then in the representation $\mathfrak{h} = \mathfrak{h}_f$ we have $f(\cdot) = [\cdot, x]$, where x is homologic and* inv.ord$(x) = -1$.

2.6.2. Lemma. *Let G be the Lie group with the Lie algebra $\mathfrak{g}_{\bar{0}}$, let G_0 be the Lie group with the Lie algebra $\mathfrak{g}_0$ of linear vector fields with respect to the standard (see [37]) grading; let Aut G_0 be the group of automorphisms of G_0. Table 2.7.2 contains all, up to (Aut G_0)-action, homologic elements of the minimal inverse order in the vectorial Lie superalgebras. In particular, for $\mathfrak{svect}'(2n)$ there are none.*

2.7. Theorem. *A simple epimorphic finite dimensional Volichenko algebra $\mathfrak{h} \subset \mathfrak{g}$ can be only one of the following $\mathfrak{h} = \mathfrak{h}_x$, where:*

1) *x is an element from Table 2.7.2 or an element from Table 2.7.1 satisfying the condition ensuring non-triviality if $\mathfrak{g} \neq \mathfrak{psq}(n)$;*

2) *if $\mathfrak{g} = \mathfrak{psq}(n)$, then either x is an element from Table 2.7.1 satisfying the condition ensuring non-triviality or $x = $ antidiag (X, X), where*

$$X = \text{diag }(a1_p, b1_{n-p}) \text{ with } ap + b(n - p) = 0.$$

Now, the final touch:

Proposition. *Simple epimorphic Volichenko algebras from Tables 1, 2 have no one-sided ideals.*

Table 2.7.1. Homologic elements x and the condition when x ensures nontriviality of $\mathfrak{h}$ for matrix Lie superalgebras $\mathfrak{g}$

$\mathfrak{g}$	x (**when** does x ensure nontriviality)
$\mathfrak{sl}(m\|n), m \leq n$	$x_q^p = \operatorname{antidiag}(B,C)$, where $B = \operatorname{diag}(1_p, 0)$ $C = \operatorname{diag}(0, 1_q)$ $(p, q > 0, p + q \leq m)$
$\mathfrak{psl}(n\|n)$	same as for $\mathfrak{sl}(n\|n)$ and also $\operatorname{antidiag}(1_n, 1_n)$ (**as above**)
$\mathfrak{osp}(2m\|2n)$	the image of the above $x_p^p \in \mathfrak{sl}(m\|n) \subset \mathfrak{osp}(2m\|2n)$ $2p \leq \min(m,n)$ $(p > 0)$
$\mathfrak{osp}(2m+1\|2n)$	the image of the above x under the embedding $\mathfrak{osp}(2m\|2n) \subset \mathfrak{osp}(2m+1\|2n)$
$\mathfrak{spe}(n)$	$\operatorname{antidiag}(B,C)$, where $B = \operatorname{diag}(1_p, 0_{n-p})$, $C = \operatorname{diag}(0_{n-2q}, J_{2q})$, $p + 2q \leq n$ $(p, q > 0)$
$\mathfrak{psq}(n)$	$\operatorname{antidiag}(X, X)$, where $X = \operatorname{diag}(J_2(0), ..., J_2(0), 0, ..., 0)$ with k-many $J_2(0)$'s, where $J_2(0) = \operatorname{antidiag}(1, 0), 2k \leq n$ $(k > 0)$
$\mathfrak{ag}_2, \mathfrak{ab}_3,$ $\mathfrak{osp}(4\|2; \alpha)$	the root vector corresponding to an isotropic (odd) simple root (**never**)

In Table 2.7.2 we have listed not only homologic elements — that is to say Volichenko subalgebras — of finite dimensional simple Lie superalgebras of vector fields but also simple Volichenko subalgebras of all nonexceptional simple Lie superalgebras of vector fields, for their list see [30].

Table 2.7.2. Homologic elements x of minimal inverse order in simple Lie superalgebras $\mathfrak{g}$ of vector fields

$\mathfrak{vect}(m\|n)$, where $mn \neq 0, n > 1$ or $m = 0, n > 2$;	$\frac{\partial}{\partial \theta_1}$
$\mathfrak{svect}(m\|n), \mathfrak{le}(n), \mathfrak{sle}^\circ(n)$ for $n > 1$	
$\mathfrak{k}(2m+1\|n)$, where $n > 1$	K_{θ_1}
$\mathfrak{h}(2m\|n)$, where $mn \neq 0, n > 1$ and $\mathfrak{sh}(n), n > 3$	$\frac{\partial}{\partial \theta_1}$ and $\frac{\partial}{\partial \theta_1} + \sqrt{-1}\frac{\partial}{\partial \theta_2}$
$\mathfrak{m}(n), n > 1$, and $\mathfrak{sm}_\lambda(n), \lambda \neq 0, n > 1$	M_1 and $M_{1+\theta_1...\theta_{2k}}$ for $\mathfrak{sm}_\lambda(2k)$
$\mathfrak{svect}(0\|2n+1), n > 1$	$\frac{\partial}{\partial \theta_1}$ and $(1 + t\theta_2...\theta_{2n+1})\frac{\partial}{\partial \theta_1}, t \in \mathbb{C}$

3. Appendix. Volichenko's theorem F and elements of Calculus on matamenifolds

3.1. In what follows all the algebras are associative with unit over a field K, char $K \neq 2$. We will deal with two important PI-varieties of algebras (the varieties singled out by polynomial identities):

– the variety C of supercommutative superalgebras;

– the variety G generated (by tensoring and passing to quotients) by the Grassmann algebra $\Lambda(\infty)$ of countably many indeterminates (its natural $\mathbb{Z}/2$-grading ignored).

The variety G plays a significant role in the theory of varieties of associative algebras ([21]). It is known that if char $K = 0$ it is distinguished by the identity (1.1). If char $K \neq 0$, the identity $X^p = 0$ should be added.

I. Volochenko wrote: "As pointed out by D. Leites [26], in the conventional supermanifold theory *it seems too restrictive that not all subalgebras or quotients of superalgebras are considered as algebras of functions on supermanifolds but only the graded (homogenous) ones. It is tempting to construct a variant of Calculus which enables one to operate with arbitrary subalgebras, ideals and quotients. ... Definition of the category of topological spaces ringed by such general algebras is obvious, cf. [25], where the algebraic case is considered.*

It remained unclear, however, how to uniformly describe such algebras. For instance, do they constitute a variety? Leites recalls a conjecture of Kac (1975) that such algebras are *metàabelean*, i.e., satisfy the identity (1.1). The conjecture is a well-known fact of the theory of varieties of associative algebras, cf. [24]. From the context of [25], however, it is clear that the actual problem is, first of all, how to describe a variety of not necessarily homogeneous subalgebras which a priori can be less than G.

Actually, I will not only prove that any algebra $G \in G$ can be embedded into a commutative superalgebra but will also prove the existence of a universal (in a natural sence) enveloping algebra $U_C(G)$ from the class C of all the supercommutative superalgebras and give an explicit realization of $U_C(G)$. Therefore, we can, in principle, reduce the study of homomorphisms of algebras from G to that of their enveloping superalgebras from C.

I hope that this is (at least partly) an answer to Leites' question *how to work with algebras from G and the corresponding 'supermanifolds'*".

3.2. Let $K_C[X, Y]$ be the algebra determined by the system of indeterminates $X \cup Y = (X_i)_{i \in I} \cup (Y_j)_{j \in J}$ and relations

$$X_{i_1} X_{i_2} - X_{i_2} X_{i_1} = 0, \ X_i Y_j - Y_j X_i = 0, \ Y_{j_1} Y_{j_2} + Y_{j_2} Y_{j_1} = 0$$

for $i, i_1, i_2 \in I$, and $j, j_1, j_2 \in J$. This algebra possesses a natural parity: $p(X_i) = \bar{0}, p(Y_j) = \bar{1}$ for $i \in I, j \in J$.

Let $I = J$; let $K_\mathcal{G}[Z]$ be a non-graded subalgebra in $K_C[X, Y]$ generated by all the elements $Z_i = X_i + Y_i$ ($i \in I$).

Statement. $K_\mathcal{G}[Z]$ is a free algebra in the variety $\mathcal{G}$ and the elements Z_i ($i \in I$) are its free generators. In other words, let $K_A[T]$ be a free associative algebra with free generators $T_1, T_2, \ldots$. If $f(Z_1, \ldots, Z_n) = 0$ in $K_\mathcal{G}[Z]$ for some $f(T_1, \ldots, T_n) \in K_A[T]$, then $f(a_1, \ldots, a_n) = 0$ for any $a_1, \ldots, a_n \in K_C[X, Y]$.

3.3. Set $d = \sum_{i \in I} Y_i \frac{\partial}{\partial X_i}$.

Statement. The polynomial $f(X, Y) \in K_C[X, Y]$ belongs to $K_\mathcal{G}[Z]$ if and only if $df = f_{\bar{1}}$, or, equivalently, $df_{\bar{0}} = f_{\bar{1}}$.

3.4. A relation between ideals of $K_\mathcal{G}[Z]$ and $K_C[X, Y]$.

Statement. Let A be an ideal of $K_\mathcal{G}[Z]$ and $\bar{A}$ the ideal of $K_C[X, Y]$ generated by $A_{\bar{0}} \cup A_{\bar{1}} = \{f_{\bar{0}}, f_{\bar{1}} : f \in A\}$. Then $\bar{A} \cap K_\mathcal{G}[Z] = A$.

Now, let $\tilde{G} = G_{\bar{0}} \oplus G_{\bar{1}}$ be a linear superspace, where each G_i is a copy of our algebra G from $\mathcal{G}$. Consider the subalgebra $K_\mathcal{G}[G] \subset K_C[\tilde{G}]$ generated by all the elements $g_{\bar{0}} + g_{\bar{1}}$, where $g \in G$. Clearly, $K_C[\tilde{G}] \simeq K_C[X, Y]$, where X and Y are bases in $G_{\bar{0}}$ and $G_{\bar{1}}$, respectively, and $K_\mathcal{G}[G] \simeq K_\mathcal{G}[Z]$. Then G is isomorphic to the quotient of $K_\mathcal{G}[Z]$ modulo the ideal A generated by all the elements of the form

$$(g_{\bar{0}} + g_{\bar{1}})(h_{\bar{0}} + h_{\bar{1}}) - ((gh)_{\bar{0}} + (gh)_{\bar{1}}).$$

The *universal C-enveloping of G* is the quotient $U_C[G]$ of $K_C[\tilde{G}]$ modulo the ideal $\bar{A}$ generated by the elements of the form

$$g_{\bar{0}}h_{\bar{0}} + g_{\bar{1}}h_{\bar{1}} - (gh)_{\bar{0}}, \quad g_{\bar{0}}h_{\bar{1}} + g_{\bar{1}}h_{\bar{0}} - (gh)_{\bar{1}}.$$

Any element $g \in G$ is identified with the image of $g_{\bar{0}} + g_{\bar{1}}$ under the canonical epimorphism $K_C[\tilde{G}] \to U_C[G]$.

In $K_C[\tilde{G}]$, same as in $K_C[X, Y]$, there is defined the derivation: $d(g_{\bar{0}}) = g_{\bar{1}}$, $d(g_{\bar{1}}) = 0$ for any $g \in G$. Since $\bar{A}$ is d-invariant, it follows that d induces a canonical derivation of $U_C[G]$ which we will also denote by d.

Proposition. The element f of $U_C[G]$ belongs to G if and only if $df_{\bar{0}} = f_{\bar{1}}$.

3.5. An explicit description of the supercommutative envelope: Theorem F. The universal C-enveloping $U_C(G)$ of the algebra G of $\mathcal{G}$ is isomorphic to the supercommutative superalgebra $S = G^{(+)} \oplus \Omega^1_{G^{(+)}/C}$ whose even component $G^{(+)}$ is G considered with the Jordan product $x \circ y = \frac{1}{2}(xy + yx)$ and the odd component $\Omega^1_{G^{(+)}/C}$ considered as a $G^{(+)}$-module is the module of differentials, i.e., the quotient of the free $G^{(+)}$-module with basis $(dx)_{x \in G}$ modulo the submodule generated by

$$d(x + y) - dx - dy, \; d(x \circ y) - x \, dy - y \, dx \text{ for } x, y \in G^{(+)}, \text{ and } dc \text{ for } c \in C,$$

where C be the subalgebra (with unit) in G and in $G^{(+)}$ generated by the elements of the form $[x, y]$ for $x, y \in G$. The product of odd elements is determined by the *formula*

$$dx \cdot dy = \frac{1}{2}[x, y] \quad (x, y \in G).$$

3.6. The Taylor formula. Hereafter char $K = 0$, the set of indices I is either $\mathbb{N}$ or $\{1, 2, \ldots, n\}$. For arbitrary $c_1, \ldots, c_p \in K_{\mathcal{G}}[Z]$ $(p \in \mathbb{N})$ set

$$\text{symm}(c_1, \ldots, c_p) = \frac{1}{p!} \sum_{\sigma \in S_p} c_{\sigma(1)} \cdots c_{\sigma(p)}.$$

The expressions of this form will be called an *s-monomial* (in $c_1, \ldots, c_p$). Determine also an *a-monomial* in $c_1, \ldots, c_{2q}$ by setting

$$\text{alt}(c_1, \ldots, c_{2q}) = \frac{1}{(2q)!} \sum_{\tau \in S_{2q}} (-1)^{\text{sign}\tau} c_{\tau(1)} \cdots c_{\tau(2q)} = 2^{-q}[c_1, c_2] \cdots [c_{2q-1}, c_{2q}].$$

(The last equality is a nontrivial statement.) Let M be the set of all the pairs of the form $m = (\alpha, \beta)$, where

$$\alpha = (\alpha_1, \ldots, \alpha_p), \ \alpha_1 \leq \ldots \leq \alpha_p, \ \alpha_\nu \in I \ \text{for} \ 1 \leq \nu \leq p$$

$$\beta = \{\beta_1, \ldots, \beta_{2q}\}, \ \beta_1 < \ldots < \beta_{2q}, \ \beta_\mu \in I \ \text{for} \ 1 \leq \mu \leq 2q.$$

In these notations for an arbitrary family $c = (c_i)_{i \in I}$ of elements from $K_{\mathcal{G}}[Z]$ set

$$c^m = \text{symm}(c_{\alpha_1}, \ldots, c_{\alpha_p})\text{alt}(c_{\beta_1}, \ldots, c_{\beta_{2q}}).$$

The elements of the form c^m ($m \in M$) will be called *sa-monomials* in c_i ($i \in I$).

Proposition *The sa-monomials Z^m ($m \in M$) constitute a basis of $K_{\mathcal{G}}[Z]$.*

Set

$$\frac{\partial}{\partial Z_i} = \frac{\partial}{\partial X_i} + \frac{\partial}{\partial Y_i} \quad (i \in I)$$

and for an arbitrary $m \in M$ set

$$\frac{\partial^m}{\partial Z^m} = \text{symm}\left(\frac{\partial}{\partial Z_{\alpha_1}}, \ldots, \frac{\partial}{\partial Z_{\alpha_p}}\right)\text{alt}\left(\frac{\partial}{\partial Z_{\beta_1}}, \ldots, \frac{\partial}{\partial Z_{\beta_{2q}}}\right).$$

Hereafter we assume that $I = \{1, 2, \ldots, n\}$. For $m = (\alpha, \beta)$ set $\delta(m) = q$ and let $m! = (-1)^{\delta(m)} d_1! \ldots d_n!$, where d_i is the degree of $\text{symm}(Z_{\alpha_1}, \ldots, Z_{\alpha_p})$ in Z_i ($i \in I$).

Theorem (The Taylor series expansion.) *For an arbitrary $f(Z) \in K_{\mathcal{G}}[Z]$ and an arbitrary $a = (a_1, \ldots, a_n) \in K^n$ we have*

$$f(Z) = \sum_{m \in M} \frac{1}{m!} \frac{\partial^m f_0(a)}{\partial Z^m} (Z - a)^m.$$

References

1. Baranov, A. A., *Volichenko algebras and nonhomogeneous subalgebras of Lie superalgebras*. (Russian) Sibirsk. Mat. Zh. **36** (1995), no. 5, 998–1009; translation in Siberian Math. J. **36** (1995), no. 5, 859–868

2. Quesne C., Vansteenkiste N. C_λ-*extended oscillator algebra and parasupersymmetric quantum mechanics*. Czechoslovak J. Phys. 48 (1998), no. 11, 1477–1482; Beckers J., Debergh N., Quesne C., *Parasupersymmetric quantum mechanics with generalized deformed parafermions*, hep-th/9604132; Helv. Phys. Acta **69**, 1996, 60–68 e-Print Archive: hep-th/9604132; Beckers J., Debergh N., Nikitin A.G., *On parasupersymmetries and relativistic description for spin one particles. 1, 2*. Fortsch. Phys. **43**, 1995, 67–96; id., *More on parasupersymmetry of the Schroedinger equation*, Mod. Phys. Lett. **A8**, 1993, 435–444; Beckers J., Debergh N., *From relativistic vector mesons in constant magnetic fields to nonrelativistic (pseudo)supersymmetries*, Int. J. Mod. Phys. **A10**, 1995, 2783–2797

3. Berezin F., *Analysis with anticommuting variables*, Kluwer, 1987

4. Berezin F., *Automorphisms of the Grassmann algebra*. Mat. Zametki **1**, 1967, 269–276

5. Berezin F. A., Leites D. A., *Supermanifolds*, Sov. Math. Doklady **16** (1975), 1976, 1218–1222

6. Bernstein J., Leites D., *The superalgebra $Q(n)$, the odd trace and the odd determinant*, C. R. Acad. Bulgare Sci., **35** (1982), no. 3, 285–286

7. Bokut' L. A., Makar-Limanov L. G., *A base of the free metaabelean associative algebra*, Sib. Math. J., **32**, 1991, 6, 910–915

8. Buchweitz R.-O., *Cohen–Macauley modules and Tate-cohomology over Gorenstein rings*. Univ. Toronto, Dept. Math. preprint, 1988

9. Corwin L., Ne'eman Y., Sternberg S., *Graded Lie algebras in mathematics and physics*, Rev. Mod. Phys. **47**, 1975, 573–609

10. Connes A., *Noncommutative geometry*. Academic Press, Inc., San Diego, CA, 1994. xiv+661 pp.

11. Deligne P. et al (eds.) *Quantum fields and strings: a course for mathematicians*. Vol. 1, 2. Material from the Special Year on Quantum Field Theory held at the Institute for Advanced Study, Princeton, NJ, 1996–1997. AMS, Providence, RI; Institute for Advanced Study (IAS), Princeton, NJ, 1999. Vol. 1: xxii+723 pp.; Vol. 2: pp. i–xxiv and 727–1501

12. Drinfeld V.G., *Quantum groups*, Proc. Int. Congress Math., Berkeley, **1**, 1986, 798–820

13. Doplicher S., Roberts J.E., *Why is there a field algebra with compact gauge group describing the superselection structure in particle physics*, Commun. Math. Phys. **131**, 1990, 51–107

14. Filippov A. T., Isaev A. P., Kurdikov A. B., *Para-Grassmann analysis and quantum groups*, Modern Phys. Lett. **A7**, no. 23, 1992, 2129–2214; id., *Para-Grassmann differential calculus*, Teoret. Mat. Fiz. **94**, no. 2, 1993, 213–231; translation in Theoret. and Math. Phys. 94, no. 2, 1993,150–165; id., Para-Grassmann extensions of the Virasoro algebra. Internat. J. Modern Phys. A 8,1993, no. 28, 4973–5003

15. Fredenhagen K., Rehren K.H., Schroer B., *Superselection sectors with braid statistics and exchange algebras*, Commun. Math. Phys. 125, 1989, 201–226

16. Feigin B., Leites D., Serganova V., *Kac–Moody superalgebras*. In: Markov M. et al (eds) *Group–theoretical methods in physics* (Zvenigorod, 1982), v. 1, Nauka, Moscow, 1983, 274–278 (Harwood Academic Publ., Chur, 1985, Vol. 1–3 , 631–637)

17. Grozman P., Leites D., *From supergravity to ballbearings*. In: Wess J., Ivanov E. (eds.) Procedings of the Internatnl seminar in the memory of V. Ogievetsky, Dubna 1997, Springer Lect. Notes in Physics, **524**,1999, 58–67

18. Grozman P., Leites D., *Lie superalgebras of supermatrices of complex size. Their generalizations and related integrable systems*. In: Vasilevsky N. et. al. (eds.) Proc. Internatnl. Symp. Complex Analysis and related topics, Mexico, 1996, Birkhauser Verlag, 1999, 73–105

19. Jing N., Ge Mo-Lin, Wu Yong-Shi, New quantum groups associated with a "nonstandard" braid group representation, Lett. Math. Phys., 21, 1991, 193–204

20. Kac V., *Lie superalgebras*, Adv. Math., **26**, 1977, 8–96

21. Kemer A. R. *Varieties of $\mathbb{Z}$-graded algebras.* Math. USSR Izvestiya, 1984, **v. 48**, No. 5, 1042–1059 (Russian)

22. Kinyon M., Weinstein A., *Leibniz Algebras, Courant Algebroids, and Multiplications on Reductive Homogeneous Spaces*, 24 pages. math.DG/0006022

23. Kochetkov Yu., Leites D., *Simple Lie algebras in characteristic 2 recovered from superalgebras and on the notion of a simple finite group.* Proceedings of the International Conference on Algebra, Part 2 (Novosibirsk, 1989), 59–67, Contemp. Math., 131, Part 2, Amer. Math. Soc., Providence, RI, 1992

24. Krakowski D., Reyev A. *The polynomial identities of the Grassmann algebra.* Trans. Amer. Math. Soc., 1973, **v. 181**, 429–438.

25. Leites D., *Spectra of graded-commutative rings.* Uspehi Matem. Nauk, **29**, no. 3, 1974, 157–158 (in Russian)

26. Leites D., *Introduction to supermanifold theory*, Russian Math. Surveys, **35**, 1980, 1, 1-57. An expanded version is: *Supermanifold theory*, Karelia Branch of the USSR Acad. of Sci., Petrozavodsk, 1983, 200p. (in Russian, an expanded version in English is [27])

27. Leites D. (ed.), *Seminar on supermanifolds*, no. 1–34, 2100 pp. Reports of Dept. of Math. of Stockholm Univ., 1986–1990; Leites D., *Lie superalgebras*, Modern Problems of Mathematics. Recent developments, **25**, VINITI, Moscow, 1984, 3–49 (Russian; English transl. in: JOSMAR **30 (6)**, 1985, 2056).

28. Leites D., *Selected problems of supermanifold theory.* Duke Math J., **54**, no. 2, 1987, 649–656

29. Leites D., Poletaeva E., *Supergravities and contact type structures on supermanifolds.* Second International Conference on Algebra (Barnaul, 1991), 267–274, Contemp. Math., **184**, Amer. Math. Soc., Providence, RI, 1995

30. Leites D., Shchepochkina I., *Classification of simple Lie superalgebras of vector fields*, to appear; id., *How to quantize antibracket*, Theor. and Math. Physics, to appear

31. Leites D., Serganova I., *Metasymmetry and Volichenko algebras*, Phys. Lett. B, **252**, 1990, no. 1, 91–96

32. Majid S., *Quasitriangular Hopf algebras and Yang-Baxter equations*, Int. J. Mod. Phys. **A5**, 1990, 1–91

33. Manin Yu. *Quantum groups and non-commutative geometry*, CRM, Montreal, 1988

34. Manin Yu. *Topics in non-commutative geometry*, Rice Univ., 1989

35. Rosenberg A., *Noncommutative algebraic geometry and representations of quantized algebras.* Mathematics and its Applications, 330. Kluwer, Dordrecht, 1995. xii+315 pp.

36. Rubakov V., Spiridonov V., *Parasupersymmetric quantum mechanics*, Mod. Phys. Lett. A, **3**, no. 14, 1988, 1337–1347

37. Shchepochkina I., *The five exceptional simple Lie superalgebras of vector fields* (Russian) Funktsional. Anal. i Prilozhen. **33** (1999), no. 3, 59–72, 96 translation in Funct. Anal. Appl. **33** (1999), no. 3, 208–219 (hep-th 9702121); id., *The five exceptional simple Lie superalgebras of vector fields and their fourteen regradings.* Represent. Theory, (electronic) **3** (1999), 373–415

38. Serganova V., *Classification of real forms of simple finite-dimensional Lie superalgebras and symmetric superspaces.* Funct. Anal Appl. 1983, **17**, no. 3, 46–54

39. Serganova V., Automorphisns of Lie superalgebras of string theories. Funct. Anal Appl. 1985, v. 19, no.3, 75–76

40. Serganova V., Simple Volichenko algebras. Proceedings of the International Conference on Algebra, Part 2 (Novosibirsk, 1989), 155–160, Contemp. Math., 131, Part 2, Amer. Math. Soc., Providence, RI, 1992

41. Shander V., *Vector fields and differential equations on supermanifolds*. (Russian) Funktsional. Anal. i Prilozhen. **14** (1980), no. 2, 91–92

42. Spiridonov V., *Dynamical parasypersymmetries in Quantum systems*. In: Tavkhelidze et. al. Proc. Int. Seminar "Quarcs 1990" , Telavi, May 1990, World Sci., 1991

43. Wess, J., Zumino, B., *Supergauge transformations in four dimensions*. Nuclear Phys. **B70** (1974), 39–50; Wess J., *Supersymmetry-supergravity. Topics in quantum field theory and gauge theories* (Proc. VIII Internat. GIFT Sem. Theoret. Phys., Salamanca, 1977), pp. 81–125, Lecture Notes in Phys., 77, Springer, Berlin-New York, 1978; Wess J., Zumino B., *Superspace formulation of supergravity*. Phys. Lett. **B 66** (1977), no. 4, 361–364; Wess J., *Supersymmetry/supergravity. Concepts and trends in particle physics* (Schladming, 1986), 29–58, Springer, Berlin, 1987; Wess J., Bagger J., *Supersymmetry and supergravity*. Princeton Series in Physics. Princeton University Press, Princeton, N.J., 1983. i+180 pp

TENSIONS IN SUPERGRAVITY BRANEWORLDS

KELLOGG STELLE *

Theoretical Physics Group, Imperial College. London SW7 2BW, UK

Abstract. We show how the Randall-Sundrum geometry, which has been proposed as a scenario for the universe realized as a 3-brane embedded in a 5-dimensional spacetime, arises naturally as an S^5 dimensional reduction of a supersymmetric 3-brane of type IIB supergravity. However, a closer inspection of the $D = 10$ delta-function sources for this solution reveals a more complex situation: in addition to the anticipated positive and negative shells of 3-brane source, there is also a non-brane stress-tensor delta-function. The latter singularity may be interpreted as arising from a patching of two discs of $D = 10$ spacetime coincident with the inner and outer brane locations.

The idea that our universe might be realized as a 3-brane embedded in a higher-dimensional spacetime has been considered at various times in recent years [1–5]. In the context of string duality, it was specifically the construction of Hořava and Witten [6, 7] realizing heterotic to M-theory duality via an orbifold compactification that set a pattern for this scenario. In particular, one may obtain a 3-brane solution to M-theory reduced on a Calabi-Yau manifold down to five spacetime dimensions [8–10]. This solution has parallel 3-brane universes facing each other across a transverse fifth dimension, located at the fixed planes of the Hořava-Witten $S^1/\mathbb{Z}_2$ orbifold. The 3-branes are magnetically charged and saturate a BPS bound, so are supersymmetric. The solution is supported by a $D = 5$ scalar field which has a higher-dimensional interpretation as the volume modulus, or "breathing mode" of the compactifying space. This scalar field acquires a potential as a result of 4-form fluxes being turned on in the compactified dimensions. The dimensional reduction is thus an example of a generalized (aka Scherk-Schwarz) reduction with non-trivial field strengths turned on in the compactifying space.

Interest in such pictures became very much heightened when Randall and Sundrum showed [11, 12] that in such a brane-world universe, gravity could behave as if it were effectively 4-dimensional even though the distance between the two 3-branes might be taken to infinity, provided the bulk geometry near the brane we live on is a $D = 5$ anti de Sitter space. Specifically, a model was considered that

---

* k.stelle@ic.ac.uk

S. Duplij and J. Wess (eds.), Noncommutative Structures in Mathematics and Physics, 31–40.
© 2001 *Kluwer Academic Publishers. Printed in the Netherlands.*

involved two segments of a pure AdS$_5$ spacetime patched together,

$$ds_5^2 = e^{\frac{-2|z|}{L}} dx^\mu dx_\mu + dz^2 \qquad (1)$$

with a "kink" at $z = 0$ corresponding to a positive-tension δ-function stress-tensor source. In Ref. [12] it was shown that this gives rise to a "binding" of gravity to the $D = 5$ spacetime region near the (3+1) dimensional braneworld, with an effective Newtonian gravitational potential plus eventual measurable corrections,

$$V_{\text{grav}} \sim \frac{1}{r} + \frac{L^2}{r^3} . \qquad (2)$$

It was the potential measurability of these corrections to Newtonian gravity that attracted such strong attention within the scientific community.

No specific supergravity realization of such a construction was given, although clearly it seems natural to try to embed the RS braneworld into a $D = 5$ dimensional reduction of type IIB supergravity. Realizing the RS brane in a supergravity context ran into certain difficulties, however, primarily concerning the behavior of the scalar field that would need to be used to support the 3-brane solution. No known scalar field in any of the dimensionally-reduced versions of $D = 5$ supergravity has the properties needed to flow correctly to a fixed point at locations far from the RS brane, and this was encoded in a "no-go theorem" [13, 14]. As is frequently the case with no-go theorems, however, the main result may be to direct attention towards the underlying assumptions that need to be relaxed. The key one in this case concerns the nature of the supporting scalar.

Even before the Randall-Sundrum work on our universe as a braneworld embedded in a $D = 5$ spacetime, a general study had been made [15] of the spherical dimensional reductions of various supergravity theories and of the branes and domain walls that exist in these reduced theories. For the specific case of the S^5 reduction of type IIB supergravity down to $D = 5$, it was shown that the familiar D3-brane geometry of $D = 10$ type IIB theory dimensionally reduces to a 3-brane in $D = 5$, supported by the "breathing mode" scalar modulus that determines the volume of the compactifying S^5. This works in a very similar way to the breathing-mode supported 3-brane in the Calabi-Yau reduction of M-theory [8–10]. It should be noted, however, that the breathing mode for an S^5 reduction does not itself belong to the massless supergravity multiplet. Instead, the breathing mode belongs to a massive spin-two multiplet, as is appropriate, since the dimensional reduction turns on a flux in the internal S^5 directions, and this gives the breathing mode φ a scalar potential that allows this mode to support a 3-brane solution. Without this potential, the breathing mode could not support a 3-brane solution. But the massive character of this mode places it outside the class of modes normally considered in $D = 5$ compactifications of supergravity theories. The importance of this mode for realizing the Randall-Sundrum braneworld as a supergravity construction was recognized in Refs [16, 17], although a main focus was still on the difficulty of realizing RS geometries as a fully "smoothed-out" solitonic solution.

To see how a construction analogous to the M-theory 3-brane solution can be made in S^5 reduced type IIB theory, consider a simplified theory just retaining the $D = 10$ metric and the self-dual 5-form field strength $H_{[5]}$,

$$R_{\mu\nu} = \frac{1}{96}(H_{[5]})^2_{\mu\nu}$$

$$H^{[5]} = {}^*H_{[5]} \qquad dH_{[5]} = 0 , \tag{3}$$

where the equations of motion for the five-form are implied by the Bianchi identity $dH \equiv 0$ taken together with the $H_{[5]} = {}^*H_{[5]}$ duality relation. Dimensionally reducing on S^5, one makes the Kaluza-Klein ansatz

$$ds^2_{10} = e^{2\alpha\varphi}ds^2_5 + e^{2\beta\varphi}ds^2(S^5) \tag{4}$$

$$H_{[5]} = 4me^{8\alpha\varphi}\varepsilon_{[5]} + 4m\varepsilon_{[5]}(S^5) \tag{5}$$

$$\alpha = \frac{1}{4}\sqrt{\frac{5}{3}} \qquad \beta = \frac{-3\alpha}{5} .$$

This reduction yields the $D = 5$ bosonic theory

$$\mathcal{L}_5 = eR - \tfrac{1}{2}e\partial_\mu\varphi\partial^\mu\varphi - 8m^2ee^{8\alpha\varphi} + R_5ee^{\frac{16\alpha\varphi}{5}} + \text{more} \tag{6}$$

where the terms represented by "more" include bosonic fields that are not relevant for the 3-brane solution, plus all the fermions. Note that there are two potential terms in (6): the one with coefficient $-8m^2$ comes from the $H_{[5]}$ fluxes turned on in the reduction ansatz (5), while the one with the coefficient R_5 comes from the Einstein-Hilbert action in the five compactified directions, since S^5 is not Ricci-flat. The coefficient R_5 is equal to the constant Ricci scalar of the internal S^5.

The presence of two potential terms in (6) with opposite signs enables a particularly simple and maximally symmetric solution to the $D = 5$ reduced theory. In this case, one can find a solution with a constant breathing-mode scalar $\varphi = \varphi_*$, with

$$e^{\frac{24\alpha\varphi_*}{5}} = \frac{R_5}{20m^2} \qquad R_{\mu\nu} = -4m^2e^{8\alpha\varphi_*}g_{\mu\nu} . \tag{7}$$

Solving this $D = 5$ Einstein equation with a cosmological term, one finds the AdS$_5 \times S^5$ "vacuum" of the S^5 compactified theory. The existence of this vacuum makes this a simpler situation than the one obtained in M-theory reduced on a Calabi-Yau manifold, where only a single potential term is obtained, and where no maximally-symmetric solution in $D = 5$ is found.

In addition to the AdS$_5 \times S^5$ solution (7), one can also search for brane solutions with less symmetry, but which tend asymptotically in appropriate regions

to the above solution. Before pursuing this search, let us make a small change
to the reduction ansatz (5) which is frequently made when considering domain-
wall solutions (*i.e.* for codimension-one branes). The original type IIB theory in
$D = 10$ has a $\mathbb{Z}_2$ symmetry (actually, it is just a discrete $D = 10$ proper Lorentz
transformation) that couples an orientation-reversing transformation on the S^5
coordinates together with a sign flip on one of the lower $D = 5$ coordinates, say
$y \to -y$. This symmetry is broken by the original ansatz (5), but will be restored
if one generalizes the ansatz by inclusion of θ functions ($\theta(y) = 1$ for $y > 0$ and
$\theta(y) = -1$ for $y < 0$):

$$H_{[5]} = 4m\theta(y)e^{8\alpha\varphi}\varepsilon_{[5]} + 4m\theta(y)\varepsilon_{[5]}(S^5) . \tag{8}$$

Note that both terms in (8) need to have θ functions in order to satisfy the $H_{[5]}$ self-
duality condition in (3). With this modified ansatz, one has traded in translation
invariance in the y coordinate for this preserved $\mathbb{Z}_2$ symmetry. Although the field
strength $H_{[5]}$ is discontinuous in (8), the underlying four-form gauge potential $A_{[4]}$
can still be continuous. We shall adopt a basic boundary condition requirement of
continuity for the metric and the gauge potentials at such "kink" locations.

Adopting the ansatz (8) and searching for a domain-wall solution, one finds
the following [15]:

$$ds_5^2 = e^{2A}dx^\mu dx^\nu \eta_{\mu\nu} + e^{2B}dy^2$$
$$e^{4A} = e^{-B} = \tilde{b}_1 H^{\frac{2}{7}} + \tilde{b}_2 H^{\frac{5}{7}}$$
$$e^{\frac{-7\varphi}{\sqrt{15}}} = H = e^{\frac{-7\varphi_0}{\sqrt{15}}} + k|y|$$
$$\tilde{b}_1 = \pm\frac{28m}{3k} \qquad \tilde{b}_2 = \pm\frac{14}{15k}\sqrt{5R_5} . \tag{9}$$

Of the sign choices allowed in (9), we shall pick $\tilde{b}_2 > 0$, $\tilde{b}_1 < 0$ in order to
ensure reality of the metric and to permit a $k \to 0$ limit so as to recover the pure
AdS Randall-Sundrum bulk spacetime [18]. We shall also choose the integration
constant φ_0 so that $H(0) > H_* = e^{\frac{-7\varphi_*}{\sqrt{15}}}$ and we shall take the slope parameter
k to be negative. Then the "kink" at $y = 0$ faces downward, so that the function
$H(y)$ reaches H_* at some finite value y_*. The solution (9) may then be considered
to be a "semi-interpolating soliton" in the sense that, although the point $y = 0$ at
which the domain-wall kink is located is not null, *i.e.* not a horizon, the solution
evolves as one moves away from $y = 0$ through either positive or negative y
values towards the AdS$_5 \times S^5$ vacuum solution (7) at $y = y_*$. With the "kink-
down" structure selected here, the surface at $y = 0$ corresponds to an extended
object of positive tension.

The solution (9) is a fully supersymmetric solution, despite its kink singularity.
The bulk geometry admits a 16-component Killing spinor, since it is none other
than the regular D3-brane geometry of type IIB theory. Moreover, the $\mathbb{Z}_2$ invariant

structure of (9) is precisely what is needed for the Killing spinor equation to be valid at all points, including at the kink location $y = 0$, with a continuous Killing spinor. The flip of sign in the 5-form flux value m as given in the modified Kaluza-Klein ansatz (8) is essential for the Killing spinor equations to be solved in this way.

The positive-tension nature of the $y = 0$ surface and the approach to the AdS$_5 \times S^5$ vacuum solution at y_* suggests that one should be able to take a limit of the solution (9) and obtain the Randall-Sundrum spacetime [18]. This limit needs to be taken conjointly in both the integration constants φ_0 and k. We let $H_0 = e^{\frac{-7\varphi_0}{\sqrt{15}}} = H_* + \beta|k|$ and then take the limit $k \to 0_-$. In this conjoint limit, factors of k^{-1} cancel against factors of k, and the limiting metric becomes

$$ds^2 = \frac{2}{\sqrt{L}}(\beta - |y|)^{\frac{1}{2}} dx^\mu dx^\nu \eta_{\mu\nu} + \frac{L^2}{16} \frac{dy^2}{(\beta - |y|)^2}, \qquad L = m^{-1} \left(\frac{20m^2}{R_5}\right)^{\frac{5}{6}} \tag{10}$$

This solution is a patched $D = 5$ anti de Sitter space with the horizon at $y = y_* = \pm\beta$. To recognize it in a more standard form, make a final coordinate change: $\beta - |y| = \beta e^{\frac{-4|\tilde{y}|}{L}}$, thus obtaining AdS$_5$ spacetime in Poincaré coordinates:

$$ds^2 = e^{\frac{-2|\tilde{y}|}{L}} dx^\mu dx^\nu \eta_{\mu\nu} + d\tilde{y}^2. \tag{11}$$

Let us now consider how this Randall-Sundrum metric has been successfully obtained as a solution of type IIB supergravity theory, despite the apparent implications of the various "no-go" theorems for the necessary scalar flows [13, 14]. Consider a theory consisting of gravity coupled to a scalar field ϕ with a potential $V(\phi)$:

$$\mathcal{L} = e[R - \tfrac{1}{2}\nabla_\mu \phi \nabla^\mu \phi - V(\phi)], \tag{12}$$

where the minimum of the potential is taken to occur at $\phi = \phi_0$. Then expand $V(\phi)$ near ϕ_0: $V(\phi) = -12g^2 + \tfrac{1}{2}\mu^2(\phi - \phi_0)^2 + \ldots$ (the constant g is chosen to make the AdS curvature equal to $-g^2(g_{MP}g_{NQ} - g_{MQ}g_{NP})$. Writing the metric as $ds^2 = e^{2A(y)}dx^\mu dx^\nu \eta_{\mu\nu} + e^{2B(y)}dy^2$ and solving the Einstein equations up to linear order in ϕ, one finds $A(y) = \pm gy$. Then one finds for static $\phi(y)$ near ϕ_0 the approximate field equation

$$\phi'' \pm 4g\phi' - \mu^2\phi \approx 0. \tag{13}$$

This equation has two solutions:

$$\phi \approx \phi_0 + ce^{-E_0 A(y)} \tag{14}$$

$$\phi \approx \phi_0 + ce^{-(4-E_0)A(y)} \tag{15}$$

$$E_0 = 2 + \sqrt{\left(\frac{\mu}{g}\right)^2 + 4} \geq 2 \tag{16}$$

where E_0 is the AdS energy. Requiring a stable infrared flow to the vacuum value ϕ_0 as $A \to -\infty$ (*i.e.* $e^{2A} \to 0$, so one moves in to the horizon), one must take the second solution (15) and also impose a restriction that the scalar field's AdS energy be bounded below by 4: $E_0 > 4$. Now, the AdS energy is a fixed constant for a given field, determined by the Lagrangian. General fields in $D = 5$ AdS spacetime carry AdS representations $D(E_0, j_1, j_2)$, where j_1 and j_2 are spins. For "standard" supergravities in $D = 5$ (*i.e.* supergravities containing the massless graviton and vector multiplets, plus hypermultiplets and tensor multiplets), one finds scalars $D(E_0, 0, 0)$ with $E_0 = 2, 3, 4$ only, so an infrared stable flow of the above type is not possible. However, the solution (9) is supported by the breathing-mode scalar ϕ, obtained from the S^5 dimensional reduction down from $D = 10$. This mode belongs to a short massive multiplet of $D = 5$, $N = 4$ supergravity, which contains a massive spin-two mode, so it does not belong to one of the supermultiplets customarily considered in $D = 5$ massless supergravity models. Comparison of the breathing-mode potential $V(\phi) = 8m^2 e^{8\alpha\varphi} - e^{\frac{16\alpha\varphi}{5}} R_5$, $\alpha = \frac{1}{4}\sqrt{\frac{5}{3}}$, with the formula (16) for E_0 gives $E_0 = 8$, clearly satisfying the required bound for a stable flow to $\varphi_0 = \varphi_*$.

Since the Kaluza-Klein ansatz (4,8) constitutes a consistent truncation of the $D = 10$ theory down to $D = 5$, one may automatically oxidize the solution (9) back up to $D = 10$ and consider its structure there. In this case, it becomes a patched set of domains of a standard type IIB D3-brane geometry. Each patch runs from a horizon at isotropic-coordinate radius $r = 0 \leftrightarrow y = y_*$ out to an outer radius $r = r_{RS} \leftrightarrow y = 0$, where $r_{RS} = \sqrt{\frac{20}{R_5}}[e^{-\sqrt{\frac{3}{5}}\varphi_0} - e^{-\sqrt{\frac{3}{5}}\varphi_*}]$. At this outer radius $r = r_{RS}$, the solution is patched onto a $\mathbb{Z}_2$ mirror solution on another sheet of spacetime, corresponding to the $D = 5$ region with $y < 0$. The Randall-Sundrum limit $k \to 0_-$, $\varphi_0 \to \varphi_*$ corresponds to shrinking down to zero the radius r_{RS} at which the patch to the second sheet is made. Alternately, one could take a limit $m \to \infty$ for the flux parameter in the reduction ansatze (4,8). In either case, one obtains a spacetime that has a uniform AdS structure: in the first case, because one is restricting the spacetime ever more narrowly down to a solid annulus around the horizon, which is asymptotically $AdS_5 \times S^5$; in the second case because this asymptotic region spreads out to fill the whole spacetime. Regardless of the perspective one takes on this limit, the proper length running from a given radius $0 < r < r_{RS}$ down to the horizon at $r = 0$ diverges. So, in this sense, the horizon is an infinite proper distance away along a radial (*i.e.* spacelike) geodesic. However, as is generally the case with extremal geometry horizons, one may also reach the horizon along a timelike or lightlike geodesic within a finite affine parameter interval. So the question of whether this Randall-Sundrum spacetime is really infinite or not requires careful interpretation.

At the horizon itself, one has a choice of interpretations for the structure of the solution (9) when oxidized back up to $D = 10$. The D3 brane geometry is actually non-singular and $\mathbb{Z}_2$ symmetric at the $r = 0$ horizon [19]. If one takes the horizons in the two sheets patched together at $r = r_{RS}$ to be distinct, then one considers a patched-brane realization of RSII geometry [12], which in $D = 5$ consists of a single kinked warp-factor AdS metric as in (11), extending out then to infinite proper distances in the $y > 0$ and $y < 0$ regions. On the other hand, if one decides to exploit the $\mathbb{Z}_2$ symmetry of the D3 brane solution at the horizon, one may alternatively make a second patch of the horizon at $y = y_*$ onto the second sheet horizon at $y = -y_*$. This produces a second, upwards-facing kink in the $D = 5$ geometry, corresponding to an extended object of negative tension, reproducing the RSI geometry [11] with two branes of opposite tension, facing each other across a compact dimension. This situation is clearly a type IIB analogue of the M-theory 3-brane solution obtained in a Calabi-Yau compactification [8–10]. The second patching surface can equally well be moved off from the horizon by moving the inner patching radius away from $r = 0$, corresponding to moving the second $D = 5$ brane in to a finite proper distance from the $y = 0$ surface.

Whatever the interpretation given to the horizon region, the kink surface at $y = 0 \leftrightarrow r = r_{RS}$ possesses the essential properties of the Randall-Sundrum solution. This surface has a positive tension $\sigma_{RS} > 0$, as can be verified using the Israel matching conditions

$$\Delta K_{\mu\nu} = K^+_{\mu\nu} - K^-_{\mu\nu} = -\frac{8\pi G}{3}\sigma_{RS}g_{\mu\nu} , \tag{17}$$

for the discontinuity in the extrinsic curvature $K_{\mu\nu} = \frac{1}{2}n^\lambda\partial_\lambda g_{\mu\nu}$, where n^λ is the outward-pointing surface normal. Consequently, in accordance with the results of Ref. [12] this surface has the property of "binding" gravity to it: matter on this 3+1 dimensional surface gravitationally interacts as if the theory were in $D = 4$.

The above picture of the Randall-Sundrum spacetime as a patching of type IIB 3-brane geometries leaves some important questions unaddressed. The principal one of these is the nature of the singular sources that must be present as a result of the curvature delta-functions arising from the patching process. An immediate appreciation of this may be had by considering the signs of the source brane delta functions. The bulk geometry between the inner and outer patching radii in the $D = 10$ perspective is a normal D3-brane geometry with a positive energy. At the same time, if the outermost source is of positive tension, as it must be in order to agree with the Randall-Sundrum tension as obtained from (17) in $D = 5$, then the inner source would have to be of opposite, i.e. negative, tension. This is clearly inconsistent with the positive-energy D3-brane geometry in the solid annulus between the inner and outer sources. A related problem is that not only the sign, but also the magnitude of the tensions do not agree with D3-brane tensions: the D3-brane tension is only $\frac{2}{3}$ of the Randall-Sundrum value as determined by (17) [20].

Both of the above problems are resolved by a recognition that the sources at the inner and outer radii in $D = 10$ cannot simply be D3-brane sources alone

[21]. A brane stress tensor in $D = 10$ would have nonzero components *only* in the brane worldvolume directions, $\hat{T}_{\mu\nu} = -\sigma g_{\mu\nu}\delta(z)$. However, the $D = 5$ stress tensor for the limiting solution (10) oxidizes up to $D = 10$ in the form

$$\hat{T}_{\mu\nu} = -56m^2\beta\left(\frac{20m^2}{R_5}\right)^{-\frac{25}{12}}\delta(y)g_{\mu\nu} + \text{Reg.}$$

$$\hat{T}_{55} = 0 + \text{Reg.}$$

$$\hat{T}_{ab} = -\frac{224}{3}m^2\beta\left(\frac{20m^2}{R_5}\right)^{-\frac{25}{12}}\delta(y)g_{ab} + \text{Reg.}, \qquad (18)$$

where the a, b indices lie in the compact S^5 directions. It is immediately apparent that this is not of the form of a brane stress tensor, notwithstanding the fact that the surrounding spacetime is a limit of a normal type IIB 3-brane solution. One may understand what is going on by taking the difference between the stress tensor (18) and that expected from the 3-brane bulk geometry. Alternatively (and this is much simpler in practice), one may find the structure of the difference stress tensor by keeping the general domain-wall form of the $D = 5$ solution (9, 10) with the $|y|$ modulus, but turning off the magnetic flux parameter m. The result of this analysis is a stress tensor of the form

$$\hat{T}^{\text{Diff.}}_{\mu\nu} = 3\kappa\delta(y)g_{\mu\nu}$$

$$\hat{T}^{\text{Diff.}}_{55} = 0$$

$$\hat{T}^{\text{Diff.}}_{ab} = \frac{12}{5}\kappa\delta(y)g_{ab}, \qquad (19)$$

where κ is a constant. This singular stress tensor occurs even in the absence of the 3-brane, *i.e.* it is a singularity occurring between patches of flat space. The $D = 5$ interval $-\beta < y < \beta \sim -\infty < \tilde{y} < \infty$ lifts to two copies of a disc in the flat $D = 10$ spacetime, with an outermost patch corresponding to $y = 0$, and another patch at the horizon, $y = y_* = \pm\beta$.

Although the stress-tensor (19) is not of the form of a brane stress tensor, one can still compare its $\hat{T}_{00}$ component to that of the 3-brane. Comparing the value of κ obtained with that of the D3 brane source for the bulk geometry shows that the stress-tensor (19) has an effective "tension" related to that of a 3-brane by

$$\sigma_{\text{flatpatch}} = -\frac{5}{2}\sigma_{\text{D3}}. \qquad (20)$$

This explains what is happening in the relationship between the type IIB 3-brane solution and the Randall-Sundrum solution. The $D = 5$ Randall-Sundrum solution (prior to taking the pure AdS limit) lifts to a $D = 10$ solution that is composed of two copies of the 3-brane geometry, patched together at a radius r_{RS} and at the horizon. The extra stress-tensor component (19), related to that of the 3-brane by (20), combines with the 3-brane stress tensor to make a composite singular stress-tensor which when viewed from a $D = 5$ viewpoint appears to be a brane stress tensor of sign *opposite* to that of the 3-brane in $D = 10$, and with a magnitude $\frac{3}{2}$ that of the 3-brane, explaining the discrepancy noted in Ref. [20].

The overall solution lifted to $D = 10$ is still $\mathbb{Z}_2$ symmetric, and if one demands that this discrete symmetry be respected, together with the S^5 spherical symmetry required for a spherical dimensional reduction down to $D = 5$, then the location of the "patch" stress-tensor singularity (19) is fixed by the symmetry. This is not the case, however, with the 3-brane itself. There is no symmetry principle that restricts this to be superposed on the patch singularity (19) – it may freely move inwards from the patch. For static solutions, this has the effect of joing the D3 brane spacetime continuously onto an outermost solid annulus of flat space. In generalized solutions, however, this boundary may also become dynamical. Owing to the sign flip inherent in (20), it is clear that what looks like a positive tension brane from the $D = 5$ perspective actually contains a *negative* tension 3-brane from the $D = 10$ perspective. Establishing the stability or otherwise of this configuration clearly remains an essential task for future analysis of braneworld scenarios like that of Randall and Sundrum.

References

1. K. Akama, *Pregeometry*, Lect. Notes Phys. **176**, 267 (1982) [hep-th/0001113].
2. V. A. Rubakov and M. E. Shaposhnikov, *Do We Live Inside A Domain Wall?*, Phys. Lett. **B125**, 136 (1983).
3. V. A. Rubakov and M. E. Shaposhnikov, *Extra Space-Time Dimensions: Towards A Solution To The Cosmological Constant Problem*, Phys. Lett. **B125**, 139 (1983).
4. G. W. Gibbons and K. Maeda, *Black Holes And Membranes In Higher Dimensional Theories With Dilaton Fields*, Nucl. Phys. **B298**, 741 (1988).
5. G. T. Horowitz and A. Strominger, *Black strings and P-branes*, Nucl. Phys. **B360**, 197 (1991).
6. P. Hořava and E. Witten, *Heterotic and type I string dynamics from eleven dimensions*, Nucl. Phys. **B460**, 506 (1996) [hep-th/9510209].
7. P. Horava and E. Witten, *Eleven-Dimensional Supergravity on a Manifold with Boundary*, Nucl. Phys. **B475**, 94 (1996) [hep-th/9603142].
8. A. Lukas, B. Ovrut, K. S. Stelle and D. Waldram, *The Universe as a Domain Wall*, Phys. Rev. **D59**, 086001 (1999) [hep-th/9803235].
9. A. Lukas, B. Ovrut, K. S. Stelle and D. Waldram, *Heterotic M–theory in Five Dimensions*, Nucl. Phys. **B552**, 246 (1999) [hep-th/9806051].
10. A. Lukas and K. S. Stelle, *Heterotic anomaly cancellation in five dimensions*, JHEP **0001**, 010 (2000) [hep-th/9911156].
11. L. Randall and R. Sundrum, *A large mass hierarchy from a small extra dimension*, Phys. Rev. Lett. **83**, 3370 (1999) [hep-ph/9905221].
12. L. Randall and R. Sundrum, *An alternative to compactification*, Phys. Rev. Lett. **83**, 4690 (1999) [hep-th/9906064].
13. R. Kallosh and A. Linde, *Supersymmetry and the brane world*, JHEP **0002**, 005 (2000) [hep-th/0001071].

14. K. Behrndt and M. Cvetič, *Anti-de Sitter vacua of gauged supergravities with 8 supercharges*, Phys. Rev. **D61**, 101901 (2000) [hep-th/0001159].

15. M. S. Bremer, M. J. Duff, H. Lü, C. N. Pope and K. S. Stelle, *Instanton cosmology and domain walls from M-theory and string theory*, Nucl. Phys. **B543**, 321 (1999) [hep-th/9807051].

16. M. Cvetič, H. Lü and C. N. Pope, *Domain walls and massive gauged supergravity potentials*, hep-th/0001002.

17. M. Cvetič, H. Lü and C. N. Pope, *Localised gravity in the singular domain wall background?* hep-th/0002054.

18. M.J. Duff, J.T. Liu and K.S. Stelle, *A supersymmetric type IIB Randall-Sundrum realization*, hep-th/0007120.

19. G.W. Gibbons, G.T. Horowitz and P.K. Townsend, Class. Quantum Grav. **12**, 297 (1995) [hep-th/9410073].

20. P. Kraus, *Dynamics of anti-de Sitter domain walls*, JHEP **9912**, 011 (1999) [hep-th/9910149].

21. M. Cvetič, M.J. Duff, J.T. Liu, C.N. Pope and K.S. Stelle, *Randall-Sundrum Brane Tensions*, hep-th/0011167.

AN UNCONVENTIONAL SUPERGRAVITY

PAVEL GROZMAN and DIMITRY LEITES [††]
Department of Mathematics, University of Stockholm, Roslagsv. 101, Kräftriket hus 6, S-106 91, Stockholm, Sweden

Abstract. We introduce and completely describe the analogues of the Riemann curvature tensor for the curved supergrassmannian of the passing through the origin $(0|2)$-dimensional subsupermanifolds in the $(0|4)$-dimensional supermanifold with the preserved volume form. The underlying manifold of this supergrassmannian is the conventional Penrose's complexified and compactified version of the Minkowski space, i.e. the Grassmannian of 2-dimensional subspaces in the 4-dimensional space.

The result provides with yet another counterexample to Coleman–Mandula theorem.

1. New supertwistors. Penrose suggested an unusual description of our space-time, namely to compactify the Minkowski space-time model of the Universe (nontrivially: with a light cone at the infinity) and complexify this compactification. The final result is Gr_2^4, the Grassmanian of 2-dimensional subspaces in the 4-dimensional (complex) space (of so-called twistors). There are many papers and several monographs on advantages of this interpretation of the space-time in various problems of mathematical physics; we refer the reader to Manin's book [5], where an original Witten's idea to incorporate supervarieties and consider infinitesimal neighborhoods for interpretation of the "usual", i.e., non-super, Yang-Mills equations is thouroughly investigated together with several ways to superize Minkowski space. Ours is one more, distinct, way.

Observe that the supermanifold of $(0|2)$-dimensional subsuperspaces in the $(0|4)$-dimensional superspace is identical with Gr_2^4, only the tautological bundle is different: the fiber is purely odd. In this work we consider not subsuper*spaces* but subsuper*manifolds*.

We considered the structure functions — analogs of the Riemann tensor — for the *curved supergrassmannian* $CGr_{0|2}^{0|4}$ of $(0|2)$-dimensional subsuper*manifolds* in the $(0|4)$-dimensional supermanifold. Recall that the "usual" grassmannian consists of linear subspaces of the linear space passing through the origin whereas

---

[‡] We gratefully acknowledge partial financial support of The Swedish Institute and an NFR grant, respectively.

[†] mleites@matematik.su.se

S. Duplij and J. Wess (eds.), Noncommutative Structures in Mathematics and Physics, 41–47.
© 2001 *Kluwer Academic Publishers. Printed in the Netherlands.*

the curved one consists of submanifolds, in other words, nonlinear embeddings are allowed and the submanifolds do not *have* to pass through a fixed point. Obviously, the curved Grassmannian is infinite dimensional, but the curved *supergrassmannian* $CGr_{0,k}^{0,n}$ is of finite superdimension: it is a quotient of the supergroup of superdiffeomorphisms of the linear supermanifold $C^{0,n}$ (the Lie superalgebra of this Lie supergroup is $\mathfrak{vect}(0|n) = \mathfrak{der}\,\mathbb{C}[\theta_1, \ldots, \theta_n]$). For the list of classical superspaces including curved supergrassmannians see [4].) The underlying manifold of $CGr_{0|2}^{0|4}$ is the conventional Gr_2^4 but $CGr_{0,2}^{0,4}$ has also odd coordinates.

On $CGr_{0|2}^{0|4}$, we have expanded the curvature supertensor in components with respect to the (complexification of the) Lorentz group and saw that it does not contain the components used for the ordinary Einstein equations (EE), namely, there is no Ricci curvature Ric and no scalar curvature Scalar (in what follows $R(22)$ and $R(00)$, respectively).

So we decided to amend the initial model and consider the supergrassmanian $CGr_{0|2}^{0|4}(0)$ of subsupermanifolds through the origin. It turns out that this does not help: no Ric and Scalar, either.

We decided not to give up, and took for the model of Minkowski superspace the supergrassmannian $SCGr_{0|2}^{0|4}(0)$ of subsupermanifolds through the origin with the volume element of the ambient and the subsupermanifolds preserved. On $SCGr_{0|2}^{0|4}(0)$, the expansion of the curvature supertensor does contain $R(22)$ and $R(00)$! There are no analogs of conformal (off shell) structure functions.

Our model and its supergroup of motion — an analogue of the Poincaré group — do not contradict the restrictions of the famous no-go theorems by Haag–Łopuszanski–Sohnius and Coleman–Mandula (for further discussions see [1]) and provides us with a new, missed so far, version of the Poincaré supergroup. The analogues of Einstein equations we suggest are a totally new version of SUGRA. Equating to zero other conformally non-invariant components we get extra conditions; we do not know how to interpret them.

We do not see any reason for discarding this and similar models. In particular, we suggest to analyze the structurre functions (definition below) on $CGr_{0|2}^{0|4}$ and $CGr_{0|2}^{0|4}(0)$ which we have abandoned above.

The conventional reading of Coleman–Mandula's theorem (cf. [6]) assumes that the complexified Lorentz Lie algebra $\mathfrak{L} = \mathfrak{sl}(2)_L \oplus \mathfrak{sl}(2)_R$ *commutes* with the Lie algebra of internal symmetries i (for us i is equal to $\mathfrak{sl}(2)_L \otimes \mathbb{C}\xi_1\xi_2$, see sec. 4).

In our case $\mathfrak{L}$ acts on i and forms a semidirect sum with it; the bracket on i is identically zero. This possibility does not contradict assumptions of Coleman–Mandula's theorem but was not considered.

The odd parameters have a correct statistics with respect to the Lorentz Lie algebra.

We represent Einstein's equations as conditions on conformally noninvariant components of the analog of the Riemann tensor, and represent the Riemann tensor as a section of the bundle on the (locally) Minkowski space whose fiber is certain **Lie algebra cohomology**. This is a more user-friendly description of the Riemannian tensor than the classical treatment of obstructions to nonflatness in differential geometry. We have in mind *Spencer homology*, cf. [7], where the case of any G-structure, not only $G = O(n)$ is considered. Superization of the definitions from [7] is the routine straightforward application of the Sign Rule.

Remark. It is interesting to test the whole list of curved supergrassmannians with the simple Lie supergroup of motion (see Tables in [4]) and similarly to the above sacrify the simplicity of the supergroup of motion in order to get EE. Grozman's package SuperLie (see [2]) is a useful tool in this research problem: without a computer (and a good code) this task is hardly feasible.

2. Structure functions: recapitulation ([7]). Let $F(M)$ be the frame bundle over a manifold M, i.e., the principal $GL(n)$-bundle. Let $G \subset GL(n)$ be a Lie group. A G-*structure on* M is a reduction of the principal $GL(n)$-bundle to the principal G-bundle.

The simplest G-structure is the *flat* G-structure defined as follows. Let V be $\mathbb{R}^n$ (or $\mathbb{C}^n$) with a fixed frame. The flat structure is the bundle over V whose fiber over $v \in V$ consists of all frames obtained from the fixed one under the G-action, V being identified with $T_v V$ by means of the parallel translation by v.

Examples of flat structures. The classical spaces, e.g., compact Hermitian symmetric spaces, provide us with examples of manifolds with nontrivial topology but flat G-structure.

In [7] the obstructions to identification of the kth infinitesimal neighbourhood of a point m on a manifold M with G-structure with the kth infinitesimal neighbourhood of a point of the flat manifold V with the above described flat G-structure are called *structure functions of order* k. In [7] it is shown further that the tensors that constitute these obstructions are well-defined *provided* the structure functions of all orders $< k$ vanish. (In supergravity the conditions that structure functions of lesser orders vanish are called *Wess-Zumino constraints*.)

The classical description of the structure functions uses the notion of the *Spencer cochain complex*. Let us recall it. Let S^i denote the operator of the i-th symmetric power. Set $\mathfrak{g}_{-1} = T_m M$, let $\mathfrak{g}_0$ be the Lie algebra of G; for $i > 0$ set:

$$\mathfrak{g}_i = \{X \in \text{Hom}\,(\mathfrak{g}_{-1}, \mathfrak{g}_{i-1}) \mid X(v_0)(v_1, \dots, v_i) = X(v_1)(v_0, \dots, v_i) \tag{2.1}$$
$$\text{for any } v_0, v_1, \dots, v_i \in \mathfrak{g}_{-1}\}.$$

Finally, set $(\mathfrak{g}_{-1}, \mathfrak{g}_0)_* = \underset{i \geq -1}{\oplus}\, \mathfrak{g}_i$. This is the Lie algebra of *all* transformations that preserve on $\mathfrak{g}_{-1}$ the same structure which is preserved by the linear transformations from $\mathfrak{g}_0$.

Suppose that the $\mathfrak{g}_0$-module $\mathfrak{g}_{-1}$ is faithful, i.e., each nonzero element from $\mathfrak{g}_0$ acts nontrivially. Then, clearly,

$$(\mathfrak{g}_{-1}, \mathfrak{g}_0)_* \subset \mathfrak{vect}(n) = \mathfrak{der}\, \mathbb{R}\, [x_1, ..., x_n],$$

where $n = \dim \mathfrak{g}_{-1}$, with

$$\mathfrak{g}_i = \{X \in \mathfrak{vect}(n)_i : [X, D] \in \mathfrak{g}_{i-1} \text{for any} D \in \mathfrak{g}_{-1}\}$$

for $i \geq 1$. It is easy to check that $(\mathfrak{g}_{-1}, \mathfrak{g}_0)_*$ is a Lie subalgebra of $\mathfrak{vect}(n)$.

The Lie algebra $(\mathfrak{g}_{-1}, \mathfrak{g}_0)_*$ will be called the *Cartan's prolong* (the result of *Cartan's prolongation*) of the pair $(\mathfrak{g}_{-1}, \mathfrak{g}_0)$.

Let E^i be the operator of the i-th exterior power; set (prime denotes dualization)

$$C^{k,s}_{(\mathfrak{g}_{-1}, \mathfrak{g}_0)} = \mathfrak{g}_{k-s} \otimes E^s(\mathfrak{g}'_{-1}).$$

Define the differential $\partial_s : C^{k,s}_{(\mathfrak{g}_{-1}, \mathfrak{g}_0)} \longrightarrow C^{k,s+1}_{(\mathfrak{g}_{-1}, \mathfrak{g}_0)}$ by setting for any $v_1, \ldots, v_{s+1} \in V$ (as usual, the slot with the hatted variable is to be ignored):

$$(\partial_s f)(v_1, \ldots, v_{s+1}) = \sum (-1)^i [f(v_1, \ldots, \hat{v}_{s+1-i}, \ldots, v_{s+1}), v_{s+1-i}]. \quad (2.2)$$

As expected, $\partial_s \partial_{s+1} = 0$, and the homology $H^{k,s}_{(\mathfrak{g}_{-1}, \mathfrak{g}_0)}$ of the bicomplex $\bigoplus_{k,s} C^{k,s}_{(\mathfrak{g}_{-1}, \mathfrak{g}_0)}$ is called the (k, s)-th *Spencer cohomology* of $(\mathfrak{g}_{-1}, \mathfrak{g}_0)_*$. (Observe that we use a grading of the Spencer complex different form that in [7]. Ours is a more natural one.)

Proposition ([7]) *The structure functions of order k constitute the space of the $(k, 2)$-th Spencer cohomology of the $(\mathfrak{g}_{-1}, \mathfrak{g}_0)_*$.*

3. Spencer cohomology in terms of Lie algebra cohomology. We observe that

$$\bigoplus_k H^{k,2}_{(\mathfrak{g}_{-1}, \mathfrak{g}_0)} = H^2(\mathfrak{g}_{-1}; (\mathfrak{g}_{-1}, \mathfrak{g}_0)_*). \quad (3)$$

The advantage of this reformulation: the Lie algebra cohomology (the right hand side of (3)) is easier to compute (e.g., by means of the package SupeLie when the general theory fails, or with the help of various theorem). At the same time the fine grading of Spencer homology is not lost: the $\mathbb{Z}$-grading of $(\mathfrak{g}_{-1}, \mathfrak{g}_0)_*$ which induces the grading (3) of $H^2(\mathfrak{g}_{-1}; (\mathfrak{g}_{-1}, \mathfrak{g}_0)_*)$ coincides (up to a shift) with the oder of the structure functions.

Analogs of Weyl and Riemann tensors. Suppose $\mathfrak{g}_0$ contains a center (like in the case when a metric is preserved up to a conformal factor). Then the elements of $H^2(\mathfrak{g}_{-1}; (\mathfrak{g}_{-1}, \mathfrak{g}_0)_*)$ are analogs of the Weyl tensor.

Let $\hat{\mathfrak{g}}_0$ be the semisimple part of $\mathfrak{g}_0$ and let $\hat{\mathfrak{g}}_*$ be a shorthand for $(\mathfrak{g}_{-1}, \hat{\mathfrak{g}}_0)_*$. The elements of $H^2(\mathfrak{g}_{-1}; \hat{\mathfrak{g}}_*)$ are analogs of the Riemann tensor.

The relation between $\hat{H} = H^2(\mathfrak{g}_{-1}; \hat{\mathfrak{g}}_*)$ and $H = H^2(\mathfrak{g}_{-1}; (\mathfrak{g}_{-1}, \mathfrak{g}_0)_*)$ is more intricate for the general $\hat{\mathfrak{g}}_0$ than in the Riemannian case ($\hat{\mathfrak{g}}_0 = \mathfrak{o}(n)$) when $\hat{H}$ strictly contains H. In general, these spaces have common components (conformally invariant, "on shell" ones) and have other components, analogs of "off shell" components, cf. [3].

In the Riemann case, there are two "off shell" components: with the highest weights $(2, 2)$ (the traceless Ricci tensor) and $(0, 0)$ (the scalar curvature). Here the highest weights are given with respect to the complexification $\mathfrak{L} = \mathfrak{sl}(2)_L \oplus \mathfrak{sl}(2)_R$ of the $\mathfrak{o}(1, 3)$. The Einstein equaton (in vacum) is a vanishing condition of these components. Remarkably, there are no structure functions of lesser order. If they had existed, we would have to impose analogs of Wess-Zumino constraints to be able to define the usual Riemann curvature tensor.

4. The description of $(\mathfrak{g}_{-1}, \mathfrak{g}_0)_*$ for the curved supergrassmannians. For the general curved supergrassmannian of $(0, k)$-dimensional subsupermanifolds S in the $(0, n)$-dimensional supermanifold $\mathcal{T}$ let $\xi_1, ..., \xi_k$ be the coordinates of S and $\theta_1, ..., \theta_{n-k}$ the remaining coordinates of $\mathcal{T}$. Then setting $\deg xi_i = 0$ for all i and $\deg \theta_j = 1$ for all j we get a $\mathbb{Z}$-grading of $\mathfrak{vect}(0|n)$ of the form

$$\mathfrak{g}_0 = (\mathfrak{gl}(V) \otimes \mathbb{C}[\xi]) \ni \mathfrak{vect}(\xi); \quad \mathfrak{g}_{-1} = V \otimes \mathbb{C}[\xi]; \tag{4}$$

where $V = \mathrm{Span}(\frac{\partial}{\partial \theta_1}, ..., \frac{\partial}{\partial \theta_{n-k}})$ is the identity $\mathfrak{gl}(V)$-module, and $\ni$ is the sign of a semidirect sum of algebras: $\mathfrak{a} \ni \mathfrak{b}$ with the ideal $\mathfrak{a}$.

For $n = 4$ we computed $H^2(\mathfrak{g}_{-1}; (\mathfrak{g}_{-1}, \mathfrak{g}_0)_*)$ in the following cases:

(a) the general curved supergrassmannians;

(b) the supergrassmannians of subspaces through 0, i.e., we removed from $\mathfrak{vect}$ all partial derivatives (since this is not an invariant formulation, it is better to say: we only considered the vector fields that vanish at the origin);

(c) in case (b) we only considered volume-preserving transformations, i.e., we diminished $\mathfrak{g}_0$ as well:

$$\mathfrak{g}_0 = (\mathfrak{sl}(V) \otimes \mathbb{C}[\xi]) \ni \mathfrak{sl}(\mathrm{Span}(\xi)); \quad \mathfrak{g}_{-1} = V \otimes \mathbb{C}\xi.$$

In particular, since $\mathfrak{g}_{-1}$ is isomorphic to the tangent space at a point of the curved supergrassmannian, we see that its even part in cases (a) – (c) is the same Gr_2^4 while the tangent space to the whole supermanifold at the "origin" is $\mathrm{Span}(\xi_i \frac{\partial}{\partial \theta_j} : 1 \leq i, j \leq 2)$. So the number of odd coordinates of our model varies from 4 in case (a) to 2 in cases (b) and (c).

Table. In the first line there are indicated the degrees, i.e., orders, of all nonzero structure functions and the rest of the table lists their the weights (with respect to $\mathfrak{L}$) (superscript denotes the multiplicity of the weight the subscript the degree of the corresponding structure function). The $\mathfrak{g}_0$-action is nontrivial and glues distinct irreducible $(\mathfrak{g}_0)_{\bar{0}}$-modules. (We did not show the action though we have computed it.)

Odd structure functions	Even structure functions

-2	-1	0
(11)	$(01)^2$	(11)
(13)	(23)	
	(03)	
	(21)	

0	1	2
$(00)^2$	(10)	(00)
(02)	(12)	(02)
$(04)^2$	(14)	(04)
(22)	(32)	(22)
(24)		
(40)		

The $(\mathfrak{g}_0)_{\bar{0}}$-modules whose highest weights are given in the table are glued into $\mathfrak{g}_0$-modules as follows (an arrow indicates a submodule). The even tensors:

$$(00)_0^2 \longrightarrow (02)_2; \quad (04)_0 \longrightarrow (04)_2;$$
$$\searrow (12)_1 \nearrow; \quad \searrow (14)_1 \nearrow;$$

$$(22)_0 \longrightarrow (14)_1 \longrightarrow (22)_2; \quad (22)_0 \longrightarrow (32)_1 \longrightarrow (22)_2;$$

$$(24)_0 \longrightarrow (32)_1; \quad (12)_1 \longrightarrow (04)_2; \quad (40)_0 \longrightarrow (32)_1; \quad (12)_1 \longrightarrow (22)_2.$$

The odd tensors:

$$(11)_{-2} \longrightarrow (23)_{-1}; \quad (01)^2_{-1} \longrightarrow (11)_0;$$
$$(13)_{-2} \longrightarrow (23)_{-1}.$$

5. The Einstein equations. The conventional EE in vacuum are the conditions on the two tensors of degree 2 and weight (00) and (22), namely,

$$R(22) = 0 \quad \text{and} \quad R(00) = \lambda g, \tag{5}$$

where $\lambda \in \mathbb{C}$ is interepreted in terms of the cosmological constant and g is the metric preserved.

For an analog of the Einstein equations on the curved supergrassmannian we may take the same vanishing conditions of the 2-nd order structure functions of weights (00) and (22) with respect to $\mathfrak{L}$. However, unlike the Einstein's case, we have to vanish the constraints, the structure functions of lesser orders, both even and odd. The meaning of these analogs of Wess-Zumino constraints is unclear to us.

References

1. Deligne P. et al (eds.) *Quantum fields and strings: a course for mathematicians.* Vol. 1, 2. Material from the Special Year on Quantum Field Theory held at the Institute for Advanced Study, Princeton, NJ, 1996–1997. AMS, Providence, RI; Institute for Advanced Study (IAS), Princeton, NJ, 1999. Vol. 1: xxii+723 pp.; Vol. 2: pp. i–xxiv and 727–1501

2. Grozman P., Leites D., Mathematica-*aided study of Lie algebras and their cohomology. From supergravity to ballbearings and magnetic hydrodynamics* In: Keränen V. (ed.) *The second International Mathematica symposium*, Rovaniemi, 1997, 185–192

3. Grozman P., Leites D., *Supergravities and N-extended Minkowski superspaces for any N.* In: Wess J., Ivanov E. (eds.) *Supersymmetries and quantum symmetries.* Proc. International Conference in memory of V. Ogievetsky, June 1997, Lecture Notes in Physics **524**, Springer, 1999, 58–67

4. Leites D., Serganova V., Vinel G., *Classical superspaces and related structures.* In: Bartocci C. et al. (eds) *Differential Geometric Methods in Theoretical Physics.* Proc. DGM-XIX, 1990, Springer, LN Phys. **375**, 1991, 286–297

5. Manin, Yu. *Gauge field theory and complex geometry*, Springer-Verlag, Berlin, 1997.

6. Salam A., Sezgin E., *Supergravities in diverse dimensions*, v.v. 1, 2, World Scientific, 1989

7. Sternberg S., *Lectures on differential geometry*, Chelsey, 2nd edition, 1985

SUPERSYMMETRY OF RS BULK AND BRANE

ERIC BERGSHOEFF
*Institute for Theoretical Physics, Nijenborgh 4, 9747 AG Groningen,
The Netherlands*

RENATA KALLOSH
*Department of Physics, Stanford University, Stanford, CA 94305,
USA*

ANTOINE VAN PROEYEN
*Instituut voor Theoretische Fysica, Katholieke Universiteit Leuven,
Celestijnenlaan 200D B-3001 Leuven, Belgium*

Abstract. We review the construction of actions with supersymmetry on spaces with a domain wall. The latter objects act as sources inducing a jump in the gauge coupling constant. Despite these singularities, supersymmetry can be formulated, maintaining its role as a square root of translations in this singular space. The setup is designed for the application in five dimensions related to the Randall–Sundrum (RS) scenario. The space has two domain walls. We discuss the solutions of the theory with fixed scalars and full preserved supersymmetry, in which case one of the branes can be pushed to infinity, and solutions where half of the supersymmetries are preserved.

1. Introduction

It is not obvious how supersymmetry can be implemented in a space with domain walls. The wall is at a fixed place and its presence seems to lead to a breaking of translations orthogonal to the plane. Supersymmetry, being the square root of translations, seems rather difficult to realize in this context. It is interesting to see how this obstacle has been avoided in [1], which we summarize here.

The work is mostly motivated by the Randall–Sundrum (RS) scenarios [2]. The simplest form of the situation that is under investigation consists of a 3-brane in a 5-dimensional bulk. The solution can be generalized e.g. to 8-branes in $D = 10$, but the full implementation of that situation is still under investigation.

When the RS scenarios appeared, supersymmetrisation was soon investigated. After initial attempts, it was found that no smooth supersymmetric RS single-brane scenario is possible [3]. This scenario with one brane was put forward as an alternative to compactification.

S. Duplij and J. Wess (eds.), Noncommutative Structures in Mathematics and Physics, 49–59.

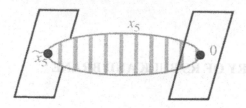

Figure 1.   Two-brane scenario. The fifth dimension is a circle with branes at opposite ends and a $\mathbb{Z}_2$ identification of points symmetric w.r.t. $x_5 = 0$.

This lead us to the original RS setup with two branes. The 2-brane scenario has a compactified fifth dimension, $x_5 \simeq x_5 + 2\tilde{x}_5$, with two branes fixed at $x_5 = 0$ and $x_5 = \tilde{x}_5$. There is moreover an orbifold condition relating points x_5 and $-x_5$. Thus, the five-dimensional manifold has the form $\mathbf{M} = \mathbf{M}_4 \times \frac{S^1}{\mathbb{Z}_2}$. This is similar to the Hořava–Witten [4] scenario. The latter one embeds 10-dimensional manifolds in an 11-dimensional space. They obtain the supersymmetry by a cancellation between anomalies of the bulk theory and a non-invariance of the classical brane action. Lukas, Ovrut, Stelle and Waldram [5] reduced this on a Calabi–Yau manifold to five dimensions, and further developed this setup in five dimensions. Further steps have been taken by [6–9]. In [7, 9] the gauge coupling constant does not change when crossing the branes, while in [6, 8] this coupling constant changes sign. In that respect, our approach is most close to the latter. In these papers, the action in the bulk is modified, such that it is not supersymmetric any more by itself, but the non-invariance is compensated by the brane action to obtain invariance of the total action. We [1] obtain separate invariance of bulk and brane action.

The first part of this report will treat the construction of the action with local supersymmetry on the singular space. In that part, we will show how the bulk and brane action are separately invariant under supersymmetry. The supersymmetry that we are considering is the one with 8 real components, i.e. minimal ($\mathcal{N} = 2$) supersymmetry in 5 dimensions. The algebra is preserved despite the discontinuity. The second part treats background solutions. The Killing spinors are discussed. There are solutions with fixed scalars and 8 Killing spinors, and solutions of 1/2 supersymmetry, i.e. with 4 Killing spinors. Finally a summary is given, discussing open issues.

2. The action for bulk and brane

The construction of the action involves three steps. First, we consider the bulk action. That is the action of supergravity in $D = 5$ with matter couplings. A quite general action has been given in [10] based on the general methods developed in 4 dimensions in [11]. But it may not be excluded that further generalizations are possible [12]. We will restrict ourselves to the couplings of vector multiplets, for which the general couplings were found in [13]. One can separate the ungauged part, and the part dependent on a gauge coupling constant g. We will consider only the gauging of a $U(1)$ R-symmetry group.

In the second step, the gauge coupling constant g is replaced by a field $G(x)$. A Lagrange multiplier field, a $(D-1)$-form (4-form for our application), is introduced, whose field equation imposes the constancy of $G(x)$ such that effectively it is still a constant.

The third step introduces the brane action. That action has extra terms for the Lagrange multiplier $(D-1)$-form, which allows $G(x)$ to vary crossing the brane. We will show how every step preserves the supersymmetry!

Before embarking on that programme, we want to repeat the fundamental algebraic relation between the cosmological constant and the gauge coupling constant of R-symmetry. The super-anti-de Sitter algebra for $\mathcal{N} = 2$ in $D = 5$ is $SU(2,2|1)$. It involves the anti-de Sitter algebra $SO(4,2) \simeq SU(2,2)$ with translations P_a and Lorentz rotations M_{ab}, the supersymmetries Q^i, with $i = 1, 2$, a symplectic Majorana spinor, and a $U(1)$ generator as R-symmetry. The most characteristic (anti)commutator relations are

$$\left\{ Q^i, Q^j \right\} = \tfrac{1}{2} \varepsilon^{ij} \gamma_a P^a + ig Q^{ij} \gamma^{ab} M_{ab} + i\varepsilon^{ij} U \,,$$

$$\left[U, Q^i \right] = g Q^i{}_j Q^j \,,$$

$$[P_a, P_b] = g^2 Q^i{}_j Q^j{}_i M_{ab} \,,$$

$$\left[P_a, Q^i \right] = i\gamma_a g Q^i{}_j Q^j \,. \tag{1}$$

Q_{ij} satisfies

$$Q_{ij} = Q_{ji} \,, \qquad Q^i{}_j \equiv \varepsilon^{ik} Q_{kj} = i \left(q_1 \sigma_1 + q_2 \sigma_2 + q_3 \sigma_3 \right) \,,$$

$$q_1, q_2, q_3 \in \mathbb{R} \,, \qquad (q_1)^2 + (q_2)^2 + (q_3)^2 = 1 \,. \tag{2}$$

This matrix determines the embedding of $U(1)$ in the automorphism group of the supersymmetries $SU(2)$. This choice is not physically relevant in itself. The second of the commutators in (1) implies that g is the coupling constant of R-symmetry. But the third equation says that g^2 determines the curvature of spacetime, i.e. it determines the cosmological constant. This fact is the cornerstone of the situation that we describe. The gauge coupling and the cosmological constant are related. However, one can change the coupling constant from $+g$ to $-g$, not affecting the cosmological constant. That is what will happen going through the branes. This jump in the sign of g will thus occur together with the action of the $\mathbb{Z}_2$. This $\mathbb{Z}_2$ acts on the fields, which therefore live on an orbifold. One can distinguish odd and even fields. The circle condition on the fields and the orbifold condition are then

$$\Phi(x^5) = \Phi(x^5 + 2\tilde{x}^5),$$
$$\Phi_{even}(-x^5) = \Phi_{even}(x^5), \qquad \Phi_{odd}(-x^5) = -\Phi_{odd}(x^5). \qquad (3)$$

These conditions imply that odd fields vanish on the branes: at $x^5 = 0$ and at $x^5 = \tilde{x}^5$.

Also the supersymmetries split. Half of them are even, and half are odd. Therefore, on the brane one has 4 supersymmetries, i.e. $\mathcal{N} = 1$ in 4 dimensions. This splitting of the fermions requires a projection matrix in SU(2) space. Now the relative choice of this projection matrix and Q in (2) matters. If they anticommute, the choice that has been taken in [7, 9], then g does not change when one crosses the brane. If they commute, as in [6, 8], then g jumps over the brane. And the latter is what we will take further.

After these general remarks, we come to **step 1**. We thus consider the action of supergravity coupled to n vector multiplets [13]. The fields are

$$e_\mu^a, \ \psi_\mu^i, \ A_\mu^I, \ \varphi^x, \ \lambda^{ix}, \qquad (4)$$

i.e. the graviton, gravitini, $n + 1$ gauge fields ($I = 0, 1, \dots, n$), including the graviphoton, n scalars ($x = 1, \dots, n$), and n doublets of spinors. The scalars describe a manifold structure that has been called very special geometry [14]. That geometry, and the complete action, is determined by a symmetric tensor C_{IJK}. The scalars are best described as living in an n-dimensional scalar manifold embedded in an $(n + 1)$-dimensional space. h^I are the coordinates of this larger space. The submanifold is defined by an embedding condition such that the h^I as functions of the independent coordinates φ^x should satisfy

$$h^I(\varphi)h^J(\varphi)h^K(\varphi)C_{IJK} = 1. \qquad (5)$$

The metric and all relevant quantities of this bulk theory is thus so far only dependent on C_{IJK}.

Then we add the gauging of a U(1) group. That means that we take a linear combination of the vectors as gauge field for this R-symmetry. The linear combination is defined by real constants V_I:

$$A_\mu^{(R)} \equiv V_I A_\mu^I. \qquad (6)$$

The action and the transformation laws are then modified by terms that all depend on $gQ^i{}_j$.

In **step 2**, the coupling constant g is replaced by a coupling field $G(x)$. In the Günaydin–Sierra–Townsend (GST) action, the coupling constant appears up to terms in g^2. We thus replace

$$S_{GST}(g) = S_0 + gS_1 + g^2 S_2 \Rightarrow S_{GST}(G(x)) = S_0 + G(x)S_1 + G(x)^2 S_2. \qquad (7)$$

Another term is added to the bulk action that forces $G(x)$ to be a constant, using a Lagrange-multiplier 4-form $A_{\mu\nu\rho\sigma}$:

$$S_{\text{bulk}} = S_{GST}(G(x)) + \int d^5x \, e \, \frac{1}{4!} \varepsilon^{\mu\nu\rho\sigma} A_{\mu\nu\rho\sigma} \partial_\tau G(x)$$

$$= S_0 - \int d^5x \, e \, V - \int d^5x \, e \, \hat{F}(x) G(x) + \text{fermionic terms.} \qquad (8)$$

In the second line, the terms have been reordered. The potential V originates from S_2 in (7), and leads to the potential

$$V = -6G^2 \left[W^2 - \frac{3}{4} \left(\frac{\partial W}{\partial \varphi^x} \right)^2 \right], \qquad W \equiv \sqrt{\tfrac{2}{3}} h^I V_I, \qquad (9)$$

where the linear combination W appears, analogous to (6). The third term in (8) appears from integrating by part the term with the Lagrange multiplier, leading to the flux

$$\hat{F} \equiv \tfrac{1}{4!} e^{-1} \varepsilon^{\mu\nu\rho\sigma\tau} \partial_\mu A_{\nu\rho\sigma\tau} + \text{covariantization.} \qquad (10)$$

The covariantization terms come from S_1 in (7). This method of describing a constant using a $(D-1)$-form is in fact an old method that was already used in [15].

It is easy to understand how supersymmetry is preserved. Indeed, the GST action is known to be invariant:

$$\delta(\epsilon) S_{GST}(g) = 0. \qquad (11)$$

Therefore, the only non-invariance for $S_{GST}(G(x))$ appears, if we define $\delta(\epsilon)G = 0$, from the x-dependence of $G(x)$. It is thus proportional to its spacetime derivative

$$\delta(\epsilon) S_{GST}(G(x)) = B^\mu \, \partial_\mu G(x), \qquad (12)$$

where B^μ is some expression of the other fields and parameters, whose exact form is not important for the argument here. One immediately sees then that invariance of (8) is obtained by defining the transformation law of the 4-form as

$$\delta(\epsilon) \tfrac{1}{4!} \varepsilon^{\mu\nu\rho\sigma\tau} A_{\mu\nu\rho\sigma} = B^\tau =$$

$$e \left[-i\tfrac{3}{2} \overline{\psi}^i_\mu \gamma^{\mu\tau} \epsilon^j W - \overline{\psi}^i_\mu \gamma^{\mu\tau\rho} \epsilon^j A^{(R)}_\rho + \tfrac{3}{2} \overline{\lambda}^i_x W^{,x} \gamma^\tau \epsilon^j \right] Q_{ij}, \qquad (13)$$

where we gave also the explicit form for our case. However, it is clear that the method is also valid in other theories.

Step 3 introduces the brane action, such that the total action is

$$S_{\text{new}} = S_{\text{bulk}} + S_{\text{brane}}. \qquad (14)$$

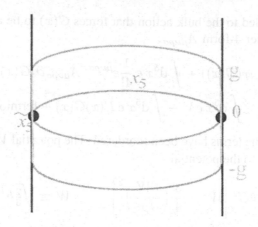

Figure 2.   *The coupling constant g jumps at $x_5 = 0$ and at $x_5 = \tilde{x}_5$.*

The brane action has the form

$$S_{\text{brane}} = -2g \int \mathrm{d}^5 x \left(\delta(x^5) - \delta(x^5 - \tilde{x}^5) \right) \left(e_{(4)} 3W + \tfrac{1}{4!} \varepsilon^{\underline{\mu\nu\rho\sigma}} A_{\underline{\mu\nu\rho\sigma}} \right)$$
$$= S_{brane,1} - S_{brane,2} . \tag{15}$$

Underlined indices refer to the values in the brane directions: $\underline{\mu} = 0, 1, 2, 3$. The action is presented as an integral over 5 dimensions, but the delta functions imply that it is a four-dimensional action for each brane separately. The action of each brane consists of a Dirac–Born–Infeld (DBI) term and a Wess–Zumino (WZ) term. However, both parts depend only on the pullback of the bulk fields to the branes. There are no fields living on the brane. The function W appears in the DBI term, and plays the role of the central charge of the brane. But most importantly, the 4-form Lagrange multiplier appears in the WZ term, and this thus modifies its field equation. The new field equation is

$$\partial_5 G(x^5) = 2g \left(\delta(x^5) - \delta(x^5 - \tilde{x}^5) \right) , \tag{16}$$

and leads to the solution (taking into account the cyclicity condition)

$$G(x) = g\, \varepsilon(x^5) . \tag{17}$$

The function $\varepsilon(x^5)$ jumps as well at $x^5 = 0$ as at $x^5 = \tilde{x}^5$, see figure 2. It is clear from this picture that we need the second brane. Indeed, one has to come back to the original value of g, in order that total derivatives in x^5 do not contribute to the action. The flux, which is determined by the field equation of $G(x)$, is

$$\hat{F} = 12G \left[W^2 - \frac{3}{4} \left(\frac{\partial W}{\partial \varphi^x} \right)^2 \right] + \text{fermionic terms.} \tag{18}$$

The overall factor changes when crossing each brane due to (17). These jumps imply that the *wall acts as a sink for the fluxes*.

That supersymmetry is still preserved by the addition of the brane is less obvious and is the non-trivial part of the construction. It turns out that the supersymmetry is preserved thanks to the projections. One finds (indices m are tangent space indices in brane directions)

$$\delta S_{\text{brane}} = -3g \int d^5x \left(\delta(x^5) - \delta(x^5 - \tilde{x}^5)\right) e_{(4)} \left[W\bar{\epsilon}^i \gamma^m e_m^\mu \left(\psi_{\mu i} - i\gamma_5 Q_{ij}\psi_\mu^j\right) + \right.$$
$$\left. + W_{,x}\bar{\epsilon}^i \left(i\lambda_i^x - \gamma_5 Q_{ij}\lambda^{xj}\right)\right] \cdot \quad (19)$$

The combinations of the gravitino and the gauginos that are in brackets are the components that are odd under the $\mathbb{Z}_2$ projection, and thus vanish on the brane. This leads to the invariance. Remark that in each case one of the two terms comes from the DBI (mass) term and the other from the WZ (charge) term. This therefore determines the relative weight of the two terms, and is the mass = charge relation, that says that the brane is BPS. We thus see, indeed, that the brane action is separately invariant. Note, that if we would not use (or eliminate) the Lagrange multiplier, then this would relate bulk and brane, and only the sum would be invariant.

3. The background: BPS solutions

We consider solutions with a warped metric, i.e.

$$ds^2 = a^2(x^5)\, dx^{\underline{\mu}} dx^{\underline{\nu}} \eta_{\underline{\mu\nu}} + (dx^5)^2 \,. \quad (20)$$

The energy density for solutions that depend only on x^5 is

$$E(x^5) = -6a^2 a'^2 + \tfrac{1}{2}a^4(\varphi^{x\prime})^2 + a^4 V - \tfrac{1}{4!}\varepsilon^{\mu\nu\rho\sigma 5} A_{\mu\nu\rho\sigma} G' +$$
$$+ 2g\left(\delta(x^5) - \delta(x^5 - \tilde{x}^5)\right)\left(3a^4 W + \tfrac{1}{4!}\varepsilon^{\underline{\mu\nu\rho\sigma}} A_{\underline{\mu\nu\rho\sigma}}\right), \quad (21)$$

where the prime denotes a derivative w.r.t. x^5. The first three terms come from the GST action, the last one on the first line from the term that we added with the Lagrange multiplier. The second line comes from the brane action. For this type of brane actions, one can rewrite it using squares and total derivatives:

$$E = \frac{1}{2}a^4\left\{[\varphi^{x\prime} - 3GW^{,x}]^2 - 12[\frac{a'}{a} + GW]^2\right\} + 3[a^4 GW]' +$$
$$+ \left[2g\left(\delta(x^5) - \delta(x^5 - \tilde{x}^5)\right) - G'\right]\left(3a^4 W + \tfrac{1}{4!}\varepsilon^{\underline{\mu\nu\rho\sigma}} A_{\underline{\mu\nu\rho\sigma}}\right). \quad (22)$$

The expression in square brackets in the second line is the field equation of the Lagrange multiplier, and this line can thus be omitted. The last term of the first line is a total derivative in x^5 and thus also does not contribute to the energy due to the continuity of the fields. The vanishing of the squared terms gives thus the minimum of the energy, and this minimum is even zero, as the zero energy of a closed universe. The BPS conditions are thus

$$\varphi^{x\prime} = 3G\,W^{,x}\,, \qquad \frac{a'}{a} = -G\,W\,. \tag{23}$$

These equations are also called stabilization equations. These equations are important to investigate the preserved supersymmetries. The transformations of the fermions are

$$\delta(\epsilon)\lambda_i^x = -i\tfrac{1}{2}\gamma_5\varphi^{x\prime}\epsilon_i - \tfrac{3}{2}GQ_{ij}W^{,x}\epsilon^j\,,$$
$$\delta(\epsilon)\psi_{\mu i} = \partial_\mu\epsilon_i + \tfrac{1}{2}\delta_\mu^m\gamma_m\left(a'\gamma_5\epsilon_i + iaGQ_{ij}W\epsilon^j\right)\,,$$
$$\delta(\epsilon)\psi_{5i} = \epsilon_i' + \tfrac{1}{2}iGQ_{ij}W\gamma_5\epsilon^j\,. \tag{24}$$

To solve these, we split the supersymmetries in their even and odd parts:

$$\epsilon_i = \epsilon_i^+ + \epsilon_i^-\,, \qquad \epsilon_i^\pm = \tfrac{1}{2}\left(\epsilon_i \pm i\gamma_5 Q_{ij}\epsilon^j\right) = \pm i\gamma_5 Q_{ij}\epsilon^{\pm j}\,. \tag{25}$$

The vanishing of the last transformation of (24) determines the x^5 dependence of both parts. We have $\epsilon_i^\pm = a^{\pm 1/2}\epsilon_i^\pm(x^\mu)$. The transformations of the other components of the gravitino then determines the dependence on the other four spacetime variables. This gives the general solution,

$$\epsilon_i = a^{1/2}\epsilon_i^{+(0)} + a^{-1/2}\left(1 - \frac{a'}{a}x^\mu\gamma_\mu\gamma_5\right)\epsilon_i^{-(0)}\,, \tag{26}$$

as function of $\epsilon_i^{\pm(0)}$, which are constant spinors with each only 4 real components. There remains the transformations of the gaugino, which lead to

$$\varphi^{x\prime}\epsilon_i^{-(0)} = 0\,. \tag{27}$$

This leaves two possibilities. The first factor can be zero, which implies that we have constant scalars. In that case 8 Killing spinors survive. The other possibility allows non-constant scalars. Then the second factor should be zero, and this thus eliminates 4 supersymmetries. There remain 4 Killing spinors, $\epsilon_i^{+(0)}$, which are the 4 that are non-vanishing also on the brane.

We consider both possibilities. First, let us look at the situation with *fixed scalars*. The BPS equations are then

$$(\varphi^y)' = 0\,, \qquad \left(\frac{\partial W}{\partial\varphi^x}\right)_{\text{crit}} = 0\,, \qquad \frac{a'}{a} = -g\varepsilon(x^5)W\,. \tag{28}$$

The constancy of W is translated by formulae of very special geometry in a 'supersymmetric attractor equation'

$$C_{IJK}\bar{h}^J\bar{h}^K = q_I, \qquad \bar{h}^K \equiv \sqrt{W_{crit}}h^K, \qquad q_I \equiv \sqrt{\tfrac{2}{3}}V_I. \qquad (29)$$

This equation is well-known from black-hole physics [16]. A solution gives rise to a metric of the form

$$ds^2 = e^{-2gW_{crit}|x^5|}dx^{\underline{\mu}}dx^{\underline{\nu}}\eta_{\mu\nu} + (dx^5)^2, \qquad \text{or} \qquad a = e^{-2gW_{crit}|x^5|}. \qquad (30)$$

In this case, the negative-tension brane can be pushed to infinity. Indeed, there is no obstruction as a never vanishes.

To consider supersymmetric domain walls with *non-constant scalars*, we use another coordinate, y, such that $\frac{\partial}{\partial x^5} = a^2\frac{\partial}{\partial y}$. The metric is then

$$ds^2 = a^2(y)dx^{\underline{\mu}}dx^{\underline{\nu}}\eta_{\mu\nu} + a^{-4}(y)dy^2. \qquad (31)$$

The stabilization equations take the form

$$a^2\frac{d}{dy}\varphi^x = 3G(y)W^{,x}, \qquad a\frac{d}{dy}a = -G(y)W. \qquad (32)$$

These $n+1$ equations are combined, using relations of very special geometry, to

$$\frac{d}{dy}(C_{IJK}\tilde{h}^J\tilde{h}^K) = -2G(y)q_I \qquad \text{where} \qquad \tilde{h}^I \equiv a(y)h^I, \qquad (33)$$

whose solutions are given in terms of harmonic functions $H_I(y)$:

$$C_{IJK}\tilde{h}^J\tilde{h}^K = H_I(y) = c_I - 2gq_I|y|, \qquad (34)$$

where c_I are integration constants, while q_I are the constants that were introduced in the gauging (V_I up to a normalization). They are harmonic in the sense that

$$\frac{d}{dy}\frac{d}{dy}H_I = -4gq_I[\delta(y) - \delta(y - \tilde{y})]. \qquad (35)$$

The warp factor is

$$a^2(y) = h^I H_I. \qquad (36)$$

In this case the distance between the branes is restricted. There can be two types of restrictions:

1. There can be fundamental restrictions due to the origin of the functions h^I. E.g. these are in various applications related to integrals over Calabi–Yau cycles. Their vanishing can put a restriction on the distance.

2. The vanishing of the harmonic functions also puts a restriction. Indeed, these harmonic functions enter in the warp factor, which should be non-vanishing.

In each case this restricts the distance to be smaller than a critical distance

$$|\tilde{y}| < |y|_{\text{sing}}. \tag{37}$$

4. Summary and outlook

The RS scenario in 5 dimensions can be made supersymmetric despite the singularities of the space. The action and transformation laws can be obtained using a 4-form, such that bulk and brane are separately supersymmetric. Supersymmetric solutions exist with fixed scalars or 1/2 supersymmetry.

Half of the supersymmetries vanish on the branes. Also the translation generator in the fifth direction vanishes on the brane. That is how the algebra can be realized. These algebraic aspects could still be clarified further. Also the extension to hypermultiplets deserves further study. The same mechanism could be applied to study 8-branes in $D = 10$ and other similar situations. It is furthermore an intriguing question how supersymmetric matter can live on the branes.

Acknowledgments.

This work was supported by the European Commission RTN programme HPRN-CT-2000-00131, in which E.B. is associated with Utrecht University. The work of R.K. was supported by NSF grant PHY-9870115.

References

1. E. Bergshoeff, R. Kallosh and A. Van Proeyen, *Supersymmetry in singular spaces*, JHEP **10** (2000) 033 [hep-th/0007044].
2. L. Randall and R. Sundrum, *A large mass hierarchy from a small extra dimension*, Phys. Rev. Lett. **83**, 3370 (1999) [hep-ph/9905221]; *An alternative to compactification*, Phys. Rev. Lett. **83**, 4690 (1999) [hep-th/9906064].
3. R. Kallosh and A. Linde, *Supersymmetry and the brane world*, JHEP **0002**, 005 (2000) [hep-th/0001071];
   K. Behrndt and M. Cvetič, *Anti-de Sitter vacua of gauged supergravities with 8 supercharges*, Phys. Rev. **D61**, 101901 (2000) [hep-th/0001159].
4. P. Hořava and E. Witten, *Eleven-dimensional supergravity on a manifold with boundary*, Nucl. Phys. **B475** (1996) 94 [hep-th/9603142].
5. A. Lukas, B.A. Ovrut, K.S. Stelle and D. Waldram, *The universe as a domain wall*, Phys. Rev. **D59** (1999) 086001 [hep-th/9803235]; *Heterotic M-theory in five dimensions*, Nucl. Phys. **B552** (1999) 246 [hep-th/9806051].
6. T. Gherghetta and A. Pomarol, *Bulk fields and supersymmetry in a slice of AdS*, Nucl. Phys. **B586** (2000) 141 [hep-ph/0003129].

7. R. Altendorfer, J. Bagger and D. Nemeschansky, *Supersymmetric Randall–Sundrum scenario*, hep-th/0003117.

8. A. Falkowski, Z. Lalak and S. Pokorski, *Supersymmetrizing branes with bulk in five-dimensional supergravity*, Phys. Lett. **B491** (2000) 172 [hep-th/0004093].

9. M. Zucker, *Supersymmetric brane world scenarios from off-shell supergravity*, hep-th/0009083.

10. A. Ceresole and G. Dall'Agata, *General matter coupled $\mathcal{N} = 2$, $D = 5$ gauged supergravity*, Nucl. Phys. **B585** (2000) 143 [hep-th/0004111].

11. L. Andrianopoli, M. Bertolini, A. Ceresole, R. D'Auria, S. Ferrara, P. Frè and T. Magri, $N = 2$ *supergravity and* $N = 2$ *super Yang–Mills theory on general scalar manifolds: Symplectic covariance, gaugings and the momentum map*, J. Geom. Phys. **23** (1997) 111 [hep-th/9605032].

12. K. Behrndt, C. Herrmann, J. Louis and S. Thomas, *Domain walls in five dimensional supergravity with non-trivial hypermultiplets*, hep-th/0008112.

13. M. Günaydin, G. Sierra and P.K. Townsend, *The geometry of $N = 2$ Maxwell–Einstein supergravity and Jordan algebras*, Nucl. Phys. **B242** (1984) 244; *Gauging the $D = 5$ Maxwell–Einstein supergravity theories: more on Jordan algebras*, Nucl. Phys. **B253** (1985) 573.

14. B. de Wit and A. Van Proeyen, *Broken sigma model isometries in very special geometry*, Phys. Lett. **B293** (1992) 94 [hep-th/9207091].

15. A. Aurilia, H. Nicolai and P.K. Townsend, *Hidden constants: the theta parameter of QCD and the cosmological constant of $N = 8$ supergravity*, Nucl. Phys. **B176** (1980) 509.

16. S. Ferrara, R. Kallosh and A. Strominger, $N = 2$ *extremal black holes*, Phys. Rev. **D52**, 5412 (1995) [hep-th/9508072]; S. Ferrara and R. Kallosh, *Supersymmetry and attractors*, Phys. Rev. **D54**, 1514 (1996) [hep-th/9602136]; *Universality of supersymmetric attractors*, Phys. Rev. **D54**, 1525 (1996) [hep-th/9603090].

D-BRANES AND VACUUM PERIODICITY

DMITRI GALTSOV *
Laboratoire de Physique Théorique LAPTH (CNRS), B.P.110, F-74941 Annecy-le-Vieux cedex, France,
and
Department of Theoretical Physics, Moscow State University, 119899, Moscow, Russia

VLADIMIR DYADICHEV †
Department of Theoretical Physics, Moscow State University, 119899, Moscow, Russia

Abstract. The superstring/M-theory suggests the Born-Infeld type modification of the classical gauge field lagrangian. We discuss how this changes topological issues related to vacuum periodicity in the $SU(2)$ theory in four spacetime dimensions. A new feature, which is due to the breaking of scale invariance by the non-Abelian Born-Infeld (NBI) action, is that the potential barrier between the neighboring vacua is lowered to a finite height. At the top of the barrier one finds an infinite family of sphaleron-like solutions mediating transitions between different topological sectors. We review these solutions for two versions of the NBI action: with the ordinary and symmetrized trace. Then we show the existence of sphaleron excitations of monopoles in the NBI theory with the triplet Higgs. Soliton solutions in the constant external Kalb-Ramond field are also discussed which correspond to monopoles in the gauge theory on non-commutative space. A non-perturbative monopole solution for the non-commutative $U(1)$ theory is presented.

1. Introduction

Recent development in the superstring theory [1, 2] suggests that the low-energy dynamics of a Dp-brane moving in a flat D-dimensional spacetime $z^M = z^M(x^\mu)$, $M = 0, ..., D - 1$, $\mu = 0, ... p$ is governed by the Dirac-Born-Infeld (DBI) action

$$S_p = \int \left(1 - \sqrt{-\det(g_{\mu\nu} + F_{\mu\nu})}\right) d^{p+1}x, \qquad (1)$$

* galtsov@grg.phys.msu.su
† rkf@mail.ru

61

S. Duplij and J. Wess (eds.), Noncommutative Structures in Mathematics and Physics, 61–78.
© *2001 Kluwer Academic Publishers. Printed in the Netherlands.*

where

$$g_{\mu\nu} = \partial_\mu z^M \partial_\nu z^N \eta_{MN}, \tag{2}$$

is an induced metric on the brane and $F_{\mu\nu}$ is a $U(1)$ gauge field strength. Using the gauge freedom under diffeomorphisms of the world-volume, one can choose coordinates $z^M = (x^\mu, X^m)$, where X^m are transverse to the brane, and rewrite the action as

$$S_p = \int \left(1 - \sqrt{-\det(\eta_{\mu\nu} + \partial_\mu X^m \partial_\nu X^m + F_{\mu\nu})} \right) d^{p+1}x. \tag{3}$$

A trivial solution to this action is $X^m = 0$, $F_{\mu\nu} = 0$, what means that the p-brane is flat and there is no electromagnetic field. Because of the symmetry $X^m \to -X^m$, the planar solution remains true when $F_{\mu\nu}$ does not vanish, in which case the electromagnetic field is governed by the Born-Infeld (BI) action. Moreover, as was noticed by Gibbons [3], the only regular static source-free solution of the BI electrodynamics which falls off at spatial infinity is a trivial one.

This is no longer true in the case of N coincident Dp-branes whose low-energy dynamics is described by the non-Abelian generalization of the DBI action involving the $SU(N)$ Yang-Mills (YM) field. Namely, for flat $D3$-branes the regular sourceless finite energy configurations of the YM field were found to exist [4, 5]. The topological reason for this lies in the vacuum periodicity of the $SU(2)$ gauge field in four dimensions. Neighboring YM vacua are separated by potential barriers which in the case of the BI action are lowered down to a finite height due to the breaking of the scale invariance in the BI theory. This removes the well-known obstruction for classical glueballs [6–8], which can be summarized as follows. Scale invariance of the usual quadratic Yang-Mills action implies that the YM field stress–energy tensor is traceless: $T^\mu_\mu = 0 = -T_{00} + T_{ii}$, where $\mu = 0, ..., 3$, $i = 1, 2, 3$. Since the energy density is positive, $T_{00} > 0$, the sum of the principal pressures T_{ii} is also everywhere positive, *i.e.* the Yang–Mills matter is repulsive. Consequently, mechanical equilibrium within the localized static YM field configuration is impossible [9]. In the spontaneously broken gauge theories scale invariance is broken by scalar fields, what opens the possibility of particle-like solutions: magnetic monopoles (in the theory with the real triplet Higgs) and sphalerons (in the theory with the complex doublet Higgs).

The role of the Higgs field in these two cases is somewhat different. For monopoles the topological significance of the Higgs field is essential: indeed, monopoles interpolate between the unbroken and broken Higgs phases. In the case of sphalerons, the Higgs field plays mostly a role of an attractive agent which is able to glue the repulsive YM matter. Historically, topological significance of the Dashen-Hasslacher-Neveu (DHN) solution in the $SU(2)$ theory with the doublet Higgs [10] was first explained by Manton [11] as a consequence of non–triviality of the *third* homotopy group of the Higgs broken phase manifold $\pi_3(G/H)$. This is equivalent to existence of non-contractible loops in the space of field configurations passing through the vacuum. Then by the minimax argument one finds that a saddle point exists on the energy surface which is a proper place for the sphaleron. Later it became clear that similar solutions arise in some models without Higgs, such as Einstein-Yang-Mills [12] or Yang-Mills with dilaton [13] (for a review and further references see [14]). The main common feature of these theories is that the conformal invariance of the classical YM equations is broken, what removes the "mechanical" obstruction for existence of particle-like configurations. As far as the topological argument is concerned, it is worth noting that $H = 1$ for the DHN solution, so the same third homotopy group argument applies to the gauge group G itself, that is, it works equally in the theories without Higgs.

Breaking of the scale invariance in the NBI theory also gives rise to sphaleron glueballs which mediate transitions between different topological sectors of the theory. Their mass is related to the BI field-strength parameter which for the D-branes is $2\pi\alpha'$. We will discuss here the difference between glueball solutions in two versions of the NBI theory: with the ordinary and symmetrized trace. We also show that, when the triplet Higgs field is added, the theory admits, apart from the usual magnetic monopoles, the hybrid solutions which can be interpreted as sphaleron excitations of monopoles. At the end we briefly discuss monopole solutions in gauge theories on non-commutative spaces and give an explicit solution for the $U(1)$ monopole with Higgs in the D-brane picture with the Kalb-Ramond field.

2. NBI action with ordinary and symmetrized trace

A precise definition of the NBI action was actively discussed during past few years [15–20], for an earlier discussion see [21]. An ambiguity is encoded in specifying the trace operation over the gauge group generators. Formally a number of possibilities can be envisaged. Starting with the determinant form of the $U(1)$ Dirac-Born-Infeld action

$$S = \frac{1}{4\pi} \int \left\{ 1 - \sqrt{-\det(g_{\mu\nu} + F_{\mu\nu})} \right\} d^4x, \tag{4}$$

one can use the usual trace, the symmetrized or antisymmetrized [15] ones, or evaluate the determinant both with respect to Lorentz and the gauge matrix indices [19]. Alternatively one can start with the 'square root' form, which is most easily derived from (4) using the identities

$$\det(g_{\mu\nu} + F_{\mu\nu}) = \det(g_{\mu\nu} - F_{\mu\nu}) = \det(g_{\mu\nu} + i\tilde{F}_{\mu\nu}) =$$
$$= \det(g_{\mu\nu} - i\tilde{F}_{\mu\nu}) = \left[\det(g_{\mu\nu} - F_{\mu\nu}^2)(g_{\mu\nu} + \tilde{F}_{\mu\nu}^2)\right]^{1/4}, \tag{5}$$

where $F_{\mu\nu}^2 = F_{\mu\alpha}F^\alpha{}_\nu$ (similarly for $\tilde{F}_{\mu\nu}$), and

$$F_{\mu\alpha}F^\alpha{}_\nu - \tilde{F}_{\mu\alpha}\tilde{F}^\alpha{}_\nu = \frac{1}{2}g_{\mu\nu}F_{\alpha\beta}F^{\alpha\beta},$$

$$F_{\mu\alpha}\tilde{F}^\alpha{}_\nu = -\frac{1}{4}g_{\mu\nu}F_{\alpha\beta}\tilde{F}^{\alpha\beta}. \tag{6}$$

This gives the relation

$$\sqrt{-\det(g_{\mu\nu}+F_{\mu\nu})} = \sqrt{-\det(g)}\sqrt{1+\frac{1}{2}F^2-\frac{1}{16}(F\tilde{F})^2}, \tag{7}$$

with $F^2 = F_{\mu\nu}F^{\mu\nu}$, $F\tilde{F} = F_{\mu\nu}\tilde{F}^{\mu\nu}$.

For a non-Abelian gauge group the relations (6) are no longer valid, so there is no direct connection between the 'determinant' and the 'square root' form of the lagrangian. Therefore the latter can be chosen as an independent starting point for a non-Abelian generalization.

There is, however, a particular trace operation – symmetrized trace – under which generators commute, so both forms of the lagrangian remain equivalent. This definition is favored by the no-derivative argument, as was clarified by Tseytlin [15]. Restricting the validity of the non-Abelian effective action by the constant field approximation, one has to drop commutators of the matrix-valued $F_{\mu\nu}$ since these can be reexpressed through the derivatives of $F_{\mu\nu}$. This corresponds to the following definition

$$S = \frac{1}{4\pi}\,\mathrm{Str}\int\left\{1-\sqrt{-\det(g_{\mu\nu}+F_{\mu\nu})}\right\}d^4x, \tag{8}$$

where symmetrization applies to the field strength (not to potentials). This action reproduces an exact string theory result for non-Abelian fields up to α'^2 order. Although there is no reason to believe that this will be true in higher orders in α', the Str action is an interesting model providing minimal generalization of the Abelian action [15].

An explicit form of the SU(2) NBI action with the symmetrized trace for static $SO(3)$-symmetric magnetic type configurations was found only recently [5]. One starts with the definition

$$L_{NBI} = \frac{\beta^2}{4\pi}\,\mathrm{Str}\left(1-\sqrt{-\det\left(g_{\mu\nu}+\frac{1}{\beta}F_{\mu\nu}\right)}\right) = k\frac{\beta^2}{4\pi}\,\mathrm{Str}(1-\mathcal{R}), \tag{9}$$

where

$$\mathcal{R} = \sqrt{1+\frac{1}{2\beta^2}F_{\mu\nu}F^{\mu\nu}-\frac{1}{16\beta^4}(F_{\mu\nu}\tilde{F}_{\mu\nu})^2}, \tag{10}$$

and β of the dimension of length^{-2} is the BI 'critical field'. The normalization of the gauge group generators is unusual and is chosen as follows

$$F_{\mu\nu} = F_{\mu\nu}^a t_a, \quad \operatorname{tr} t_a t_b = \delta_{ab}. \tag{11}$$

The symmetrized trace of the product of p matrices is defined as

$$\operatorname{Str}(t_{a_1} \ldots t_{a_p}) \equiv \frac{1}{p!} \operatorname{tr}\left(t_{a_1} \ldots t_{a_p} + \text{all permutations}\right), \tag{12}$$

and it is understood that the general matrix function like (9) has to be series expanded. It has to be noted that under the Str operation the generators can be treated as commuting objects, and the gauge algebra should not be applied, (e.g. the square of the Pauli matrix $\tau_x^2 \neq 1$) until the symmetrization in the series expansion is completed.

A general $SO(3)$ symmetric $SU(2)$ gauge field is described by the Witten's ansatz

$$\sqrt{2}A = a_0 t_1 \, dt + a_1 t_1 \, dr + \{w_2 \, t_2 - (1 - w) \, t_3\} \, d\theta + \tag{13}$$
$$\{(1 - w) \, t_2 + \tilde{w} \, t_3\} \sin\theta \, d\phi,$$

where the functions a_0, a_1, w, $\tilde{w}$ depend on r, t and $\sqrt{2}$ is introduced to maintain the standard normalization. Here we use a rotating basis t_i, $i = 1, 2, 3$ for the $SU(2)$ generators defined as

$$t_1 = n^a \tau^a / \sqrt{2}, \quad t_2 = \partial_\theta t_1, \quad \sin\theta t_3 = \partial_\varphi t_1, \tag{14}$$

where $n^a = (\sin\theta \cos\varphi, \sin\theta \sin\varphi, \cos\theta)$, with τ^a being the Pauli matrices. These generators obey the commutation relations $[t_i, t_j] = \frac{1}{\sqrt{2}}\epsilon_{ijk}t_k$.

From four functions entering this ansatz one can be gauged away. In the static case we can further reduce the number of independent functions to two, while the static purely magnetic configurations are fully described by a single function $w(r)$:

$$\sqrt{2}A_\theta = -(1 - w)t_3, \quad \sqrt{2}A_\varphi = \sin\theta(1 - w)t_2. \quad A_t = A_r = 0. \tag{15}$$

The field strength tensor has the following non-zero components

$$\sqrt{2}F_{r\theta} = w't_3, \quad \sqrt{2}F_{r\varphi} = -\sin\theta w't_2, \quad \sqrt{2}F_{\theta\varphi} = \sin\theta(w^2 - 1)t_1, \tag{16}$$

where prime denotes derivatives with respect to r.

For purely magnetic configurations the second term under the square root is zero, and the substitution of (16) gives

$$\mathcal{R}^2 = 1 + \frac{(1 - w^2)^2}{\beta^2 r^4}t_1^2 + \frac{w'^2}{\beta^2 r^2}(t_2^2 + t_3^2). \tag{17}$$

To find an explicit expression for the lagrangian one has to expand the square root in a triple series in terms of the even powers of generators t_1, t_2, t_3, then to calculate the symmetrized trace of the powers of generators in all orders, and finally to make a resummation of the series. The result reads

$$L_{NBI} = \frac{\beta^2}{4\pi} \left(1 - \frac{1 + V^2 + K^2 \mathcal{A}}{\sqrt{1 + V^2}} \right), \tag{18}$$

where

$$V^2 = \frac{(1 - w^2(r))^2}{2\beta^2 r^4}, \quad K^2 = \frac{w'^2(r)}{2\beta^2 r^2},$$

$$\mathcal{A} = \sqrt{\frac{1 + V^2}{V^2 - K^2}} \operatorname{arctanh} \sqrt{\frac{V^2 - K^2}{1 + V^2}}. \tag{19}$$

Here we assumed that $V^2 > K^2$, otherwise an arctan form is more appropriate. Note that when the difference $V^2 = K^2$ changes sign, the k function $\mathcal{A}$ remains real valued. It can be checked that when $\beta \to \infty$, the standard Yang-Mills lagrangian (restricted to monopole ansatz) is recovered. In the strong field region our expression differs essentially from the square root/ordinary trace lagrangian.

The corresponding explicit action defined in a square root form with an ordinary trace reads:

$$L_{NBI} = \frac{\beta^2}{4\pi} \left(1 - \sqrt{1 + V^2 + 2K^2} \right) \tag{20}$$

3. Topological vacua and sphalerons

As is well-known, vacuum in the $SU(2)$ YM theory in the four-dimensional spacetime splits into an infinite number of disjoint classes which can not be deformed into each other by 'small' (contractible to a point) gauge transformations. Writing the pure gauge vacuum YM potentials as $A = iUdU^{-1}$, where $U \in SU(2)$ and imposing an asymptotic condition

$$\lim_{r \to \infty} U(x^i) = 1, \tag{21}$$

we can interpret $U(x^i)$ as mappings $S_3 \to SU(2)$. All sets of such U's falls into the sequence of homotopy classes characterized by the winding number

$$k[U] = \frac{1}{24\pi^2} \operatorname{tr} \int_{R_3} UdU^{-1} \wedge UdU^{-1} \wedge UdU^{-1}. \tag{22}$$

A representative of the k-th class can be chosen as

$$U_k = \exp\{i\alpha(r)t_1/\sqrt{2}\}, \quad \text{where} \quad \alpha(0) = 0, \alpha(\infty) = -2\pi k. \quad (23)$$

The corresponding potential will be given by the Witten ansatz with $a = 0, w = exp(i\alpha(r))$. The asymptotic condition (21) leads to the following fall-off requirements.

$$A_a = o(r^{-1}) \quad \text{for} \quad r \to \infty. \quad (24)$$

The representatives of different vacuum classes with different k cannot be continuously deformed into each other within the class of the purely vacuum fields. But there exists an interpolating sequence of nonvacuum field configurations of finite energy (the latter can be defined on shell and then continued off-shell) satisfying the required boundary conditions (24) that connects different vacuum classes. Finite energy solutions for the actions (18) or (20) should satisfy the following boundary conditions near the origin

$$w = 1 + br^2 + O(r^4), \quad (25)$$

and at the infinity

$$w = \pm 1 + \frac{c}{r} + O(\frac{1}{r^2}), \quad (26)$$

where b and c are free parameters. (The value $w(\infty) = 0$ together with finiteness of the energy implies that $w \equiv 0$.) The leading terms are the same as required for the vacuum configurations. These solutions, if exists, can be shown to lie on the path in the solution space connecting two topologically distinct vacua. Consider a one-parameter sequence of field configurations (off shell generally) depending on a continuous parameter $\lambda \in [0, \pi]$ [22]

$$A[\lambda] = i\frac{1-w}{2}U_+dU_+^{-1} + i\frac{1+w}{2}U_-dU_-^{-1}, \quad (27)$$

where

$$U_\pm = \exp\left\{i\lambda(w \pm 1)t_1/\sqrt{2}\right\}. \quad (28)$$

This field vanishes for $\lambda = 0$, whereas for $\lambda = \pi$ it can be represented as

$$A[\pi] = iUdU^{-1}, \quad \text{with} \quad U = \exp\{i\pi(w - 1)t_1/\sqrt{2}\}. \quad (29)$$

In view of the above boundary condition for w, in the case $w(\infty) = -1$ one has the $k = 1$ vacuum. Now, the crucial thing is that for $\lambda = \pi/2$ we come back to the configuration (15). So if the solution to the classical field equations with the required asymptotics exists indeed, this can be interpreted as a manifestation of the finiteness of the potential barrier between distinct vacua.

Note that the same reasoning holds for the ordinary Yang–Mills system. But due to the scale invariance of this theory there is no function w which minimizes the energy functional.

Both the analysis of the equations following from NBI lagrangians (18,20) using the methods of dynamical systems [4] and numerical experiments [5] shows that such solutions exist in both NBI models — with ordinary and symmetrized trace. They form a discrete sequence labeled by the number of nodes of the function $w(r)$, and the lower one-node solution is similar to the sphaleron of the Weinberg-Salam theory.

In the NBI theory β is the only dimensionful parameter giving a natural scale of length, i.e. theories with different values of β are equivalent up to rescaling. Setting $\beta = 1$ we obtain the equations of motion for the symmetrized trace NBI model

$$\frac{d}{dr}\left\{ \frac{w'}{2(V^2 - K^2)} \left(\frac{K^2\sqrt{1 + V^2}}{1 + K^2} - \frac{(2V^2 - K^2)\mathcal{A}}{\sqrt{V^2 - K^2}} \right) \right\} \tag{30}$$
$$= \frac{w V(K^2\mathcal{A} - V^2)}{(V^2 - K^2)\sqrt{1 + V^2}}.$$

For the ordinary trace model one has

$$\frac{d}{dr}\left\{ \frac{w'}{\sqrt{1 + V^2 + 2K^2}} \right\} = -\frac{w V}{\sqrt{1 + V^2 + 2K^2}}, \tag{31}$$

We are looking for the solutions satisfying the boundary conditions (25,26). For large r both equations reduce to that of the usual YM theory, so the solutions are not much different in the far zone. Near the origin the equations are different, more careful analysis reveals that the nature of stationary points associated with the origin is different for two versions of the theory.

A trivial solution to these equations (valid for both models) is an embedded abelian monopole $w = 0$. In the BI theory it has the finite energy. From the general analysis, as discussed in [14] for the ordinary trace, one finds that w can not have local minima for $0 < w < 1$, $w < -1$ and can not have local maxima for $-1 < w < 0$, $w > 1$. The same remains true for the symmetrized trace. Thus any solution which starts at the origin on the interval $-1 < w < 1$ must remain within the strip $-1 < w < 1$. Once w leaves the strip, it diverges in a finite distance. Regular solutions exist for a discrete sequence of b shown in the table I together with corresponding masses M_n for the first six n which is the number of zeroes of $w(r)$. The $n = 1$ solution is analogous to the sphaleron known in the Weinberg-Salam theory [10, 11], it is expected to have one decay mode. Higher odd-n solutions may be interpreted as excited sphalerons, they are expected to

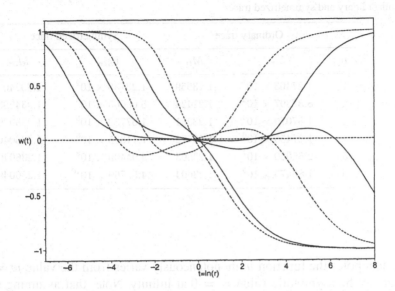

Figure 3. Sphaleron glueball solutions w_n for $n = 1, 2, 3$ in the symmetrized trace (solid line) and ordinary trace (dashed line) models

have n decay directions. Even-n solutions are topologically trivial, they can be regarded as sphaleronic excitation of the vacuum. Qualitatively picture is the same as for the ordinary trace [4], but the discrete values of b are rather different.

Numerical solutions for both models are shown in the figure 3. It is surprising that the solutions with the ordinary and the symmetrized trace are rather similar in spite of the substantial difference of the lagrangians. They have however somewhat different behavior near the origin: those with the symmetrized trace leave the vacuum value $w = 1$ faster and stay longer in the intermediate region where $w(r)$ is close to zero. In this region the magnetic charge is almost unscreened, so this is the particle core. Thus for all n solutions are more compact in the ordinary trace case. For both models the parameters b_n grow infinitely with increasing node number n. This means that there is no limiting solution as $n \to \infty$ contrary to the EYM case where such solutions do exist.

4. Magnetic monopoles and hybrid solutions

Magnetic monopoles are associated with the deformed D3-branes with non zero transverse coordinates X^m interpreted as Higgs scalars. The deformation can be thought of as caused by an open string attached to the brane. In the BPS limit the solutions are the same as for the quadratic YM theory [17, 18] Monopoles for the ordinary trace model were constructed by Grandi, Moreno and Schaposnik

TABLE I.   Values of b and M for first six glueball solutions in NBI models with ordinary and symmetrized traces

	Ordinary trace		Symmetrized trace	
n	b_{tr}	M_{tr}	b_{Str}	M_{Str}
1	1.27463×10^1	1.13559	1.23736×10^2	1.20240
2	8.87397×10^2	1.21424	5.05665×10^3	1.234583
3	1.87079×10^4	1.23281	1.67739×10^5	1.235979
4	1.27455×10^6	1.23572	7.11885×10^6	1.236046
5	2.65030×10^7	1.23603	4.94499×10^8	1.2360497
6	1.80475×10^9	1.23604	4.52769×10^{10}	1.2360497

[23]. For monopoles the function w monotoneously varies from the value $w = 1$ at the origin to the asymptotic value $w = 0$ at infinity. Note, that assuming the asymptotic value $w = 0$ for pure gauge NBI theory we will get only embedded abelian solution $w \equiv 0$. Our aim here is to show that, in addition, there are hybrid NBI-Higgs solutions for which the function $w(r)$ oscillates in the core region. In other words, starting from the vacuum $w = 1$ at the origin the function $w(r)$ tries to follow the sphaleronic behavior, but finally turns back to the monopole regime.

Adding to the NBI action the Higgs term $S = S_{NBI} + S_H$ where S_H is taken in the usual form

$$S_H = \frac{1}{8\pi} \int \left(D_\mu \phi^a D^\mu \phi^a - \frac{\lambda}{2} \left(\phi^a \phi^a - v^2 \right) \right), \tag{32}$$

one obtains the NBI-Higgs theory, containing, apart from β, the second parameter λ (without loss of generality we put the gauge coupling constant equal to unity). For spherically symmetric static purely magnetic configurations the YM ansatz remains the same, while for the Higgs field

$$\phi^a = \frac{H(r)}{r} n^a. \tag{33}$$

For simplicity we consider here the square root form of the NBI action (20). Performing an integration over spherical angles one obtains the energy functional (equal to minus action for static configurations)

$$E = 4\pi \int dr\, r^2 \left\{ 2\beta^2 (\mathcal{R} - 1) + \frac{1}{2r^2} \left((H' - \frac{H}{r})^2 + \frac{2}{r^2} H^2 w^2 \right) + V \right\}, \tag{34}$$

where

$$R = \sqrt{1 + \frac{1}{\beta^2 r^4}\left(r^2 w'^2 + \frac{1}{2}(w^2 - 1)^2\right)}, \quad V = \frac{\lambda}{4}\left(\frac{H^2}{r^2} - 1\right)^2. \tag{35}$$

Varying this functional one finds the equations of motion

$$r^2 w'' = w(RH^2 + w^2 - 1) + r^2 \frac{R'}{R}w', \tag{36}$$

$$r^2 H'' = 2Hw^2 - \lambda H(r^2 - H^2). \tag{37}$$

Boundary conditions at infinity for a solution with a unit magnetic charge read

$$\lim_{r \to \infty} w(r) = 0, \quad \lim_{r \to \infty} \frac{H(r)}{r} = 1, \tag{38}$$

while at the origin

$$w(0) = 1, \quad H(0) = 0. \tag{39}$$

Starting with (39) one can construct the following power series solution converging in a non-zero domain around the origin:

$$w = 1 - br^2 + \frac{\beta b^2 (22b^2 + \beta^2) + d^2 (6b^2 + \beta^2)^{\frac{3}{2}}}{10\beta (2b^2 + \beta^2)}r^4 + O(r^6) \tag{40}$$

$$H = dr^2 - \left(\frac{1}{10}\lambda d + \frac{2}{5}db\right)r^4 + O(r^6), \tag{41}$$

where b and d are free parameters. For $\beta \to \infty$ the theory reduces to the standard YMH-theory, admitting monopoles. In [23] it was shown that monopole solutions to the Eqs.(36, 37) continue to exist up to some limiting value β_{cr}.

Now we have to explain why one can expect to have also the hybrid solutions. Near the origin the Higgs field is close to zero, so the influence of the term $H^2 KR$ is negligible, and the YM field behaves like in the pure NBI case. As was argued in [4], NBI theories with different β are equivalent up to rescaling, and so for β large enough the solution starts forming just near the origin. But for larger r the role of Higgs is increased, so one can expect that some solutions can be trapped to the monopole asymptotic regime. More precisely, in the region of $r \approx 1/\sqrt{\beta}$, the function $w(r)$ is similar to the sphaleron solution of [4]: starting with $w = 1$ it passes through $w = 0$ and then tends to the value $w = 1$. After leaving this region the solution enters the region where it has properties of the NBI monopole and at $r \to \infty$ both field functions tend to their asymptotical values (38). The Higgs field $H(r)$ for these hybrid solutions behaves qualitatively in the same way as for the monopoles.

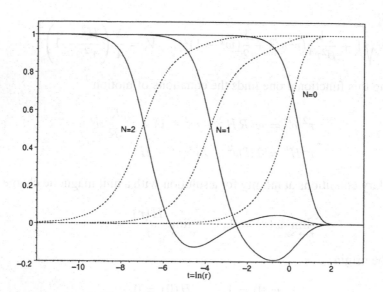

Figure 4. Magnetic monopole and first two hybrid solutions in the ordinary trace model for $\beta = 30$, $\lambda = 1/2$. Solid line — w, dashed line — H/r

To obtain hybrid solutions numerically we introduce the logarithmic variable $t = \ln(r)$ and apply a shooting strategy to find the values of parameters b and d ensuring the monopole asymptotic conditions (40-41) after several oscillations of w. As an initial guess for b one can take the (appropriately rescaled for given β) glueball values found in [4]. Another parameter d turns out to be weakly sensitive on β for β large enough. The resulting solutions for $n = 1, 2$ and $\lambda = 1/2$ are shown on Fig. 1,2 together with the ground state monopole ($n = 0$). The masses increase with n and converge rapidly to the mass of an embedded Abelian solution with frozen Higgs:

$$w(r) \equiv 0, \qquad H(r) \equiv r. \qquad (42)$$

Although this singular solution does not satisfy the boundary conditions (39) it has finite energy within the NBI-Higgs theory, which can be obtained by substituting the Eq. (42) into the Eq. (34):

$$E_{lim} = 2 \int \beta^2 (\mathcal{R} - 1) r^2 dr = \sqrt{\beta} \int \left(\sqrt{4 + \frac{2}{x^4}} - 2 \right) x^2 dx = 1.467338\sqrt{\beta}.$$

$$(43)$$

With decreasing β, the discrete values of the parameter b_n also decrease until relatively small values of β. Then, with β further decreasing both parameters b and d start growing until some critical value of $\beta_{cr\,n}$ is reached near which parameters

b_n and d_n tend to infinity and monopole solutions with given number of zeroes cease to exist. The lowest of these critical values is $\beta_{cr\,0} \approx 0.45$ for unexcited monopole solution. The excited solutions disappear at greater values of β. The mass of excited monopoles is well described by the formula (43), even for the lowest excited solution the difference with the exact numerical value is less then 4% for all values of β. The figure 4 shows the behavior of functions w, H for some intermediate value of β. Note, that at critical β all branches of monopole solutions (including unexcited branches) converge to the limiting Abelian solution (42) (with different rate).

The excited monopole solutions also exist in the Einstein-Yang-Mills-Higgs theory [24]. There the role of non-linear excitations is played by Bartnick-McKinnon gravitating sphalerons of EYM theory [12]. The phase diagram (regions of existence in parameter space) is somewhat different in our case, the details will be given elsewhere.

5. Non-commutative monopoles

Here we discuss another aspect of the D-brane picture of gauge theories, which is the direct subject of the present workshop. Recently it was discovered that gauge theories on noncommutative manifolds

$$[x_\mu, x_\nu] = i\theta_{\mu\nu} \tag{44}$$

are connected with the gauge theories on D-branes with the constant background Kalb-Ramond field B turned on [25]

$$B_{\mu\nu} = -\frac{\theta_{\mu\nu}}{(2\pi\alpha')^2}. \tag{45}$$

The relation between these two versions is non-local and is defined perturbatively through the Seiberg-Witten map [26] (for a more recent discussion see [27, 28]). Namely, the YM theory on a noncommutative four-dimensional space

$$\hat{S} = \text{Tr} \int \left(\frac{1}{4\hat{g}^2} \hat{F}_{\mu\nu} * \hat{F}^{\mu\nu} + ... \right) d^4x, \tag{46}$$

defined using the star-product

$$F(x) * G(x) = \exp\left(\frac{i\theta_{\mu\nu}}{2} \partial_\mu \partial'_\nu \right) F(x)G(x')|_{x'=x}, \tag{47}$$

and the D-brane theory with A_μ, $F_{\mu\nu}$ are related perturbatively via

$$\hat{A}_\mu = A_\mu - \frac{\theta^{\alpha\beta}}{4} \{A_\alpha, \partial_\beta A_\mu + F_{\beta\mu}\}_+ + O(\theta^2). \tag{48}$$

The issue of magnetic monopoles in both treatments of the non-commutative YM was discussed recently in a number of papers [29–32, 30]. It was argued that BPS-saturated monopoles exist in the non-commutative case as well. Apart from the BPS bound most of the previous discussion was perturbative in terms of the non-commutativity parameter $\theta_{\mu\nu}$.

Adding the constant B-field spoils the spherical symmetry of monopoles and therefore their non-perturbative treatment in the D-brane picture becomes rather complicated. At best one can construct an axially symmetric model using $B_{\mu\nu}$ as a Kalb-Ramond analog of the homogeneous magnetic (electric) field. Even in this case the NBI model is still too complicated both for Tr and Str versions. Here we give a non-perturbative monopole solution in the simplest case of the $U(1)$ gauge field with Abelian Higgs. As was shown by Gibbons [3], the system of BI $U(1)$ and Higgs fields possesses the boost symmetry (in the mixed space of coordinates and the field variables) which can be used as a solution generating technique to add a constant magnetic field to the pointlike magnetic monopole (resp. electric field to the electric BIon). Reinterpreted as the Kalb-Ramond field, this homogeneous field may be accounted for the parameter of non-commutativity.

We start with the DBI action

$$S_{DBI} = -\int d^4x \sqrt{-\det\left(\eta_{\mu\nu} + \partial_\mu y \partial_\nu y + F_{\mu\nu}\right)} \qquad (49)$$

with one external coordinate y (playing the role of the Higgs field) and introduce the magnetic potential χ

$$\mathbf{H} = -\nabla\chi, \qquad (50)$$

where $\mathbf{H}$ is the magnetic field strength — canonical conjugate to the magnetic induction $\mathbf{B}$:

$$\mathbf{H} = -\frac{\partial L}{\partial \mathbf{B}}. \qquad (51)$$

Performing the corresponding Legendre transformation we obtain the following hamiltonian functional

$$\mathcal{H} = \int d^3x \sqrt{1 - (\nabla\chi)^2 + (\nabla y)^2 + (\nabla\chi)^2(\nabla y)^2 - (\nabla\chi \cdot \nabla y)^2}, \qquad (52)$$

which can be interpreted as the volume of the three-dimensional hypersurface parametrized by coordinates x^i in the five-dimensional pseudoeuclidean space $\{x^i, y, \chi\}$ with the metric $\mathrm{diag}(+, +, +, +, -)$ (minus corresponds to χ). We use the symmetries of this functional to generate first the scalar charge from the monopole charge and then to generate a constant background field which will be then interpreted as the B field. So we start with the spherically symmetric configurations. The field equations are then reduced to

$$y'' = 2\frac{y'\left(\chi'^2 - y'^2 - 1\right)}{r}, \quad \chi'' = 2\frac{\chi'\left(\chi'^2 - y'^2 - 1\right)}{r}, \qquad (53)$$

where prime denotes the derivative with respect to the radial variable r. It is easy to see that two potentials should be proportional. Depending on which potential dominates, one can find three different types of behaviour:

1. The spacelike vector in the $\{y, \chi\}$ plane. By some rotation the magnetic field can be removed. This is the catenoidal solution [3]. Since it does not exist for all r, we will not consider it further.
2. The timelike vector in the $\{y, \chi\}$ plane. By a rotation it can be reduced to a $U(1)$ monopole without excitations of the transverse degrees of freedom. The potential for this particular solution (with unit charge) is

$$\chi_0(r) = \int^r \frac{dr}{\sqrt{1 + r^4}}, \qquad (54)$$

and could be written explicitly in terms of elliptic integrals.
3. The lightlike vector $y = \pm \chi$. This is the BPS monopole:

$$\chi_{BPS}(r) = \pm y_{BPS}(r) = \frac{1}{r}. \qquad (55)$$

To obtain the non-BPS monopole solution that also has a nonzero Higgs counterpart $y(r)$ one can simply perform a boost in the $\{\chi, y\}$ plane:

$$\chi(r) = \cosh \psi \, \chi_0(r) \qquad y(r) = \sinh \psi \, \chi_0(r). \qquad (56)$$

The next step is to perform a boost in the $\{\chi, z\}$ plane to generate the constant background magnetic field. To understand why this field may be equally interpreted as a B field one should notice that the field equations do not change if we replace $F_{\mu\nu}$ by $F_{\mu\nu} + B_{\mu\nu}$ with constant B.

So, if we denote $\chi = g(\rho, z)$, then after the second boost we obtain:

$$\cosh \phi \, g + \sinh \phi \, z = \cosh \psi \, \chi_0 \left(\sqrt{\rho^2 + (\cosh \phi \, z + \sinh \phi \, g)^2} \right), \qquad (57)$$

where $\rho = \sqrt{x^2 + y^2}$ and χ_0 is defined by the Eq.(54).

This nonlinear equation cannot be solved explicitly but it is simple to explore it numerically. The key point is to note that for a given g, ρ, z, using equations (54),(57), one can find the vector $\mathbf{F} + \mathbf{B}$ (magnetic *induction* plus B-field). Then the monopole field is obtained by subtracting the constant background. Note that, depending on the values of the boosts parameters ϕ and ψ, the solution can become double-valued. Let us consider this feature in more detail. For magnetic monopole without excitations of the transversal component the three-dimensional hypersurface $\chi_0(x, y, z)$ is spacelike everywhere except for the origin where it touches the lightcone. When we boost in the $\{\chi, y\}$ directions, the surface $\chi(x, y, z)$ acquires the timelike piece which can cause multivaluedness

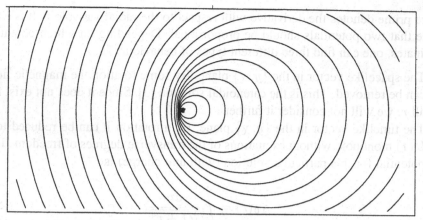

Figure 5.   Non-commutative $U(1)$ monopole: constant $|\mathbf{F}|$ curves

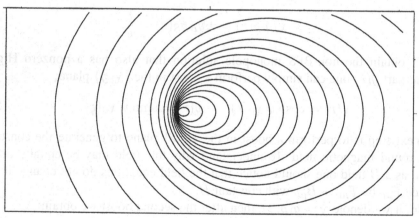

Figure 6.   Non-commutative $U(1)$ monopole: constant y curves

after boosting in the $\{\chi, z\}$ directions. (When treated as a hypersurface in the five-dimensional space $\{\mathbf{r}, \chi, y\}$ it remains of course spacelike). This effect is interpreted from the string theory point of view as tilting the D-brane, but from the point of view of 3-dimensional field theory this multivaluedness should be interpreted as a signal that no well defined solution exists. It is worth noting that for BPS solution such multivaluedness emerges for any value of the background field.

In the figures 5,6 the sections of level surfaces of constant y and constant $|\mathbf{B}|$ are shown. The full solution is axially symmetric and is obtained by rotating the pictures along the symmetry axis.

6. Discussion

We have discussed some new issues associated with the D-brane picture of gauge theories. Apart from giving a nice geometric framework, D-branes suggest a modification of dynamics of the YM field introducing the Born-Infeld type lagrangian. This latter breaks the conformal invariance of the YM equations removing the obstruction for existence of classical glueballs in the SU(2) theory in four dimensions. Topological reason for existence of such glueballs lies in the vacuum periodicity which holds equally in the ordinary YM theory and in the NBI theory, with an important difference that in the latter case the potential barriers between neighboring vacua have finite heights. Classical NBI glueballs (more precisely, half of them) are sphalerons mediating the topological transitions. We have found that they exist both for the ordinary trace and the symmetrized trace versions of the NBI theory with somewhat different core structure. We have also shown that in the NBI theory with the triplet Higgs one encounters, apart from the usual magnetic monopoles, the hybrid solutions which can be regarded as sphaleronic excitations of monopoles. Finally, adding the constant Kalb-Ramond field, one is able to account for non-commutative monopoles. We presented a new nonperturbative axisymmetric solution for the U(1) non-commutative monopole with Higgs.

Acknowledgements

One of the authors (DG) is grateful to the organizers of the Workshop for invitation and support and especially to Steven and Diana Duplij for a stimulating atmosphere during this meeting. He is also grateful to LAPTH (Annecy) for hospitality while the final version of this paper was completed. This work was supported in part by the RFBR grant 00-02-16306.

References

1. J. Polchinski, *String Theory*, Vol. I and II of *Cambridge Monograph on Mathematical Physics*, Cambridge University Press, 1998.
2. A. Giveon and D. Kutasov, *Brane dynamics and gauge theory*, Rev. Mod. Phys. **71** (1999), 983, hep-th/9802067.
3. G. W. Gibbons, *Born-Infeld particles and Dirichlet p-branes*, Nucl. Phys. **B514** (1998), 603–639, hep-th/9709027.
4. D. Gal'tsov and R. Kerner, *Classical glueballs in non-Abelian Born-Infeld theory*, Phys. Rev. Lett. **84** (2000), 5955–5958, hep-th/9910171.
5. V. V. Dyadichev and D. V. Gal'tsov, *Sphaleron glueballs in NBI theory with symmetrized trace*, Nucl. Phys. **B590** (2000), 504–518, hep-th/0006242.
6. S. Deser, *Absence of static solutions in source - free Yang-Mills theory*, Phys. Lett. **B64** (1976), 463.
7. H. Pagels, *Absence of periodic solutions to scale invariant classical field theories*, Phys. Lett. **B68** (1977), 466.

8. S. Coleman, *There are no classical glueballs*, Commun. Math. Phys. **55** (1977), 113.
9. G. Gibbons, *Self-gravitating Magnetic monopoles, Global Monopole and black holes*, Vol. 383 in *Lecture Notes In Phys.*, Springer-Verlag, Berlin, 1991.
10. R. Dashen, B. Hasslacher, and A. Neveu, *Nonperturbative methods and extended hadron models in field theory. 3. Four-dimensional nonabelian models*, Phys. Rev. **D10** (1974), 4138.
11. N. S. Manton, *Topology in the Weinberg-Salam theory*, Phys. Rev. **D28** (1983), 2019.
12. R. Bartnik and J. Mckinnon, *Particle - like solutions of the Einstein Yang-Mills equations*, Phys. Rev. Lett. **61** (1988), 141–144.
13. G. Lavrelashvili and D. Maison, *Dilatonic sphalerons and nonabelian black holes*, hep-th/9307159.
14. M. S. Volkov and D. V. Gal'tsov, *Gravitating non-Abelian solitons and black holes with Yang-Mills fields*, Phys. Rept. **319** (1999), 1, hep-th/9810070.
15. A. A. Tseytlin, *On non-abelian generalisation of the Born-Infeld action in string theory*, Nucl. Phys. **B501** (1997), 41–52, hep-th/9701125.
16. J. P. Gauntlett, J. Gomis, and P. K. Townsend, *BPS bounds for worldvolume branes*, JHEP **01** (1998), 003, hep-th/9711205.
17. D. Brecher, *BPS states of the non-Abelian Born-Infeld action*, Phys. Lett. **B442** (1998), 117–124, hep-th/9804180.
18. D. Brecher and M. J. Perry, *Bound states of D-branes and the non-Abelian Born-Infeld action*, Nucl. Phys. **B527** (1998), 121–141, hep-th/9801127.
19. J.-H. Park, *A study of a non-Abelian generalization of the Born-Infeld action*, Phys. Lett. **B458** (1999), 471–476, hep-th/9902081.
20. M. Zamaklar, *Geometry of the nonabelian DBI dyonic instanton*, Phys. Lett. **B493** (2000), 411–420, hep-th/0006090.
21. T. Hagiwara, *A nonabelian Born-Infeld lagrangian*, J. Phys. **A14** (1981), 3059.
22. D. V. Galtsov and M. S. Volkov, *Sphalerons in Einstein Yang-Mills theory*, Phys. Lett. **B273** (1991), 255–259.
23. N. Grandi, E. F. Moreno, and F. A. Schaposnik, *Monopoles in non-Abelian Dirac-Born-Infeld theory*, Phys. Rev. **D59** (1999), 125014, hep-th/9901073.
24. P. Breitenlohner, P. Forgacs, and D. Maison, *Gravitating monopole solutions. 2*, Nucl. Phys. **B442** (1995), 126–156, gr-qc/9412039.
25. A. Connes, M. R. Douglas, and A. Schwarz, *Noncommutative geometry and matrix theory: Compactification on tori*, JHEP **02** (1998), 003, hep-th/9711162.
26. N. Seiberg and E. Witten, *String theory and noncommutative geometry*, JHEP **09** (1999), 032, hep-th/9908142.
27. S. Terashima, *The non-Abelian Born-Infeld action and noncommutative gauge theory*, JHEP **07** (2000), 033, hep-th/0006058.
28. N. A. Nekrasov, *Trieste lectures on solitons in noncommutative gauge theories*, hep-th/0011095.
29. A. Hashimoto and K. Hashimoto, *Monopoles and dyons in non-commutative geometry*, JHEP **11** (1999), 005, hep-th/9909202.
30. S. Goto and H. Hata, *Noncommutative monopole at the second order in theta*, Phys. Rev. **D62** (2000), 085022, hep-th/0005101.
31. K. Hashimoto, H. Hata, and S. Moriyama, *Brane configuration from monopole solution in non- commutative super Yang-Mills theory*, JHEP **12** (1999), 021, hep-th/9910196.
32. S. Moriyama, *Noncommutative monopole from nonlinear monopole*, Phys. Lett. **B485** (2000), 278–284, hep-th/0003231.

QUANTUM DEFORMATIONS OF SPACE-TIME SUSY AND NONCOMMUTATIVE SUPERFIELD THEORY

P. KOSIŃSKI
Institute of Physics, University of Łódź,
ul. Pomorska 149/53 90–236 Łódź, Poland

JERZY LUKIERSKI *
Institute of Theoretical Physics, University of Wrocław
pl. M. Borna 9, 50-205 Wrocław, Poland

P. MAŚLANKA
Institute of Physics, University of Łódź,
ul. Pomorska 149/53 90–236 Łódź, Poland

Abstract. We review shortly present status of quantum deformations of Poincaré and conformal supersymmetries. After recalling the κ–deformation of D=4 Poincaré supersymmetries we describe the corresponding star product multiplication for chiral superfields. In order to describe the deformation of chiral vertices in momentum space the integration formula over κ–deformed chiral superspace is proposed.

1. Introduction

The noncommutative space–time coordinates were introduced as describing algebraically the quantum gravity corrections to commutative flat (Minkowski) background (see e.g. [1, 2]) as well as the modification of D–brane coordinates in the presence of external background tensor fields (e.g. $B_{\mu\nu}$ in $D = 10$ string theory; see [3]–[5]). We know well that both gravity and string theory have better properties (e.g. less divergent quantum perturbative expansions) after their supersymmetrization. It appears therefore reasonable, if not compelling, to consider the supersymmetric extensions of the noncommutative framework.

The generic relation for the noncommutative space–time generators $\hat{x}_\mu$

$$[\hat{x}_\mu, \hat{x}_\nu] = i\Theta_{\mu\nu}(\hat{x}) = i\left(\Theta_{\mu\nu} + \Theta^\rho_{\mu\nu}\hat{x}_\rho + \dots\right) \tag{1}$$

* lukier@ift.uni.wroc.pl

S. Duplij and J. Wess (eds.), Noncommutative Structures in Mathematics and Physics, 79–91.

has been usually considered for constant value of the commutator (1), i.e. for $\Theta_{\mu\nu}(\hat{x}) = \Theta_{\mu\nu}$. In such a case the multiplication of the fields $\phi_k(\hat{x})$ depending on the noncommutative (Minkowski) space–time coordinates can be represented by noncommutative Moyal $*$–product of classical fields $\phi_k(x)$ on standard Minkowski space

$$\phi_k(\hat{x})\phi_l(\hat{x}) \longleftrightarrow \phi_k(x) * \phi_l(x) = \phi_k(y)e^{\frac{i}{2}\Theta^{\mu\nu}\frac{\partial}{\partial y_\mu}\frac{\partial}{\partial y_\nu}}\phi_l(z)|_{x=y} \qquad (2)$$

It appears that the relation (1) with constant $\Theta_{\mu\nu}$ can be consistently supersymmetrized (see e.g. [6]–[9]) by supplementing the standard relations for the odd Grassmann superspace coordinates (further we choose $D = 4$ $N = 1$ SUSY and $\alpha, \beta = 1, 2$).

$$\{\theta_\alpha, \theta_\beta\} = \{\theta_\alpha, \bar{\theta}_{\dot{\beta}}\} = \{\bar{\theta}_{\dot{\alpha}}, \bar{\theta}_{\dot{\beta}}\} = 0 \qquad [\hat{x}_\mu, \theta_\alpha] = [\hat{x}_\mu, \bar{\theta}_{\dot{\alpha}}] = 0 \qquad (3)$$

Such a choice of superspace coordinates $(\hat{x}_\mu, \theta_\alpha, \theta_{\dot{\alpha}})$ implies that the supersymmetry transformations remain classical:

$$\hat{x}'_\mu = \hat{x}_\mu - i\left(\bar{\epsilon}\sigma_k\theta_\alpha - \bar{\theta}\sigma_k\epsilon\right)$$

$$\theta'^\alpha = \theta_\alpha + \epsilon_\alpha \qquad \bar{\theta}'^{\dot{\alpha}} = \bar{\theta}_{\dot{\alpha}} + \bar{\epsilon}_{\dot{\alpha}} \qquad (4)$$

i.e. the covariance requirements of deformed superspace formalism do not require the deformation of classical Poincaré supersymmetries[1].

Our aim here is to consider the case when the standard Poincaré supersymmetries can not be preserved. For this purpose we shall consider the case with linear Lie–algebraic commutator (1). Its supersymmetrization leads to the deformed superspace coordinates $\hat{z}_A = (\hat{x}_\mu, \hat{\theta}_\alpha, \bar{\hat{\theta}}_{\dot{\beta}})$ satisfying Lie superalgebra relation:

$$[\hat{z}_A, \hat{z}_B] = i\Theta^C_{AB}\hat{z}_C \qquad (5)$$

where Θ^C_{AB} satisfies graded Jacobi identity:

$$\Theta^D_{AB}\Theta^E_{CD} + \text{graded cycl. } (A, B, C) \qquad = 0 \qquad (6)$$

It appears that in such a case for some choices of the "structure constants" Θ^C_{AB} one can find the deformed quantum $D = 4$ Poincaré supergroup, which provide the relations (5) as describing the deformed translations and deformed supertranslations.

---

[1] It should be stresssed, however, that the introduction of constant tensor $\Theta_{\mu\nu}$ in (1) leads to breaking $(O(3,1) \rightarrow O(2) \times O(1,1))$ of $D = 4$ Lorentz symmetry. The way out is to consider $\Theta_{\mu\nu}$ as a constant field, with generator of Lorentz subalgebra containing contribution which rotates the $\Theta_{\mu\nu}$ components (see e.g. [10]). The relation (1) can be made covariant only for $D = 2$ ($\Theta_{\mu\nu} \equiv \epsilon_{\mu\nu}$ for $D = 2$); for 2+1 Euclidean case see [11]

The plan of the paper is following: In Sect. 2 we shall briefly review the considered in literature quantum deformations of Poincaré and conformal supersymmetries. The list of these deformations written in explicit form as Hopf algebras is quite short, and only the knowledge of large class of classical r-matrices shows that many quantum deformations should be still discovered. As the only nontrivial quantum deformation of $D = 4$ supersymmetry given in the literature is the so-called κ-deformation, obtained in 1993 [12]–[14].

In Sect. 3 we consider the Fourier supertransform of superfields in classical (undeformed) and κ-deformed form. We present also the integration formula over κ-deformed superspace, which provides the description in supermomentum space leading to the κ-deformed Feynmann superdiagrams.

In Sect. 4 we consider the κ-deformed superfield theory in chiral superspace. We introduce the $*$-product multiplication of κ-deformed superfields. It appears that there are two distinguished $*$-products, which both can be written in closed form: one described by standard supersymmetric extenion of CBA formula and other physical, providing the addition of fourmomenta and Grassmann momenta in terms of the coproduct formulae. In such a way we obtain the supersymmetric extension of two $*$-products, considered recently in [15].

In Sect. 5 we shall present some remarks and general diagram describing the deformation scheme of superfield theory.

2. Quantum Deformations of Space-Time Supersymmetries

There are two basic space-time symmetries in D dimensions:
- Conformal symmetries $O(D, 2)$, having another interpretation as anti-de-Sitter symmetries in $D + 1$ dimensions
- Poincaré symmetries $T^{D-1,1} \supset O(D - 1, 1)$.

i) Quantum deformations of conformal supersymmetries.

The conformal symmetries can be supersymmetrized without introducing tensorial central charges in $D = 1, 2, 3, 4$ and 6. One gets:

$$D = 1 : O(2,1) \longrightarrow OSp(N; 2|R) \quad \text{or} \quad SU(1, 1 : N)$$

$$D = 2 : O(2,2) = O(1,2) \otimes O(1,2) \longrightarrow OSp(M; 2|R) \otimes OSp(N; 2|R)$$

$$D = 3 \quad O(3,2) \longrightarrow OSp(N; 4|R)$$

$$D = 4 \quad O(4,2) \longrightarrow SU(2, 2; N)$$

$$D = 6 \quad O(6,2) \longrightarrow U_\alpha U(4; N|H)$$

All conformal supersymmetries listed above are described by simple Lie superalgebras. It is well–known that for every simple Lie superalgebra one can introduce the q–deformed Cartan–Chevaley basis describing quantum (Hopf–algebraic) Drinfeld–Jimbo deformation [16, 17]. These q–deformed relations have been explicitly written in physical basis of conformal superalgebra in different dimensions (see e.g. [18]). It is easy to see that the deformation parameter q appears as dimensionless.

It follows, however, that there is another class of deformations of conformal and superconformal symmetries, with dimensionfull parameter κ, playing the role of geometric fundamental mass. For $D = 1$ one can show that the Jordanian deformation of $SL(2; R) \simeq O(2,1)$ describes the κ–deformation of $D = 1$ conformal algebra [22]. This result can be extended supersymmetrically, with the following classical $\hat{r}$–matrix describing Jordanian deformation $U_\kappa(OSp(1;2|R))$ [23]

$$r = \frac{1}{\kappa} h \wedge e \overset{SUSY}{\Longrightarrow} r = \frac{1}{\kappa}(h \wedge e + Q^+ \wedge Q^+)$$

Jordanian deformation	Jordanian deformation
of $Sp(2; R) \simeq O(2; 1; R)$	of $OSp(1, 2; R)$
($D = 1$ conformal)	($D = 1$ superconformal) (7)

The $OSp(1; 2; R)$ Jordanian classical $\hat{r}$–matrix can be quantized by the twist method. Semi–closed form for the twist function has been obtained in [24].

It appears that one can extend the Jordanian deformations of $D = 1$ conformal algebra to $D > 1$; for $D = 3$ and $D = 4$ the extended Jordanian classical r–matrices were given in [22]. It should be also mentioned that the generalized Jordanian deformation of $D = 3$ conformal $O(3, 2)$ algebra has been obtained in full Hopf–algebraic form [25]. The extension of Jordanian deformation of $OSp(1, 2; R)$ for $D > 1$ superconformal algebras is not known even in its infinitesimal form given by classical r–matrices.

ii) Quantum deformations of Poincaré supersymmetries.

Contrary to DJ scheme for simple Lie (super)algebras it does not exist a systematic way of obtaining quantum deformations of non–semisimple Lie (super)algebras. A natural framework for the description of deformed semi–direct products, like quantum Poincaré algebra, are the noncocommutative bicrossproduct Hopf algebras (see e.g. [26]). It appears however that in the literature it has not been formulated any effective scheme describing these quantum bicrossproducts.

One explicit example of quantum deformation of $D = 4$ Poincaré superalgebra and its dual $D = 4$ Poincaré group in form of graded bicrossproduct Hopf algebra was given in [14]. By means of quantum contraction of q–deformed $N = 1$ anti–de–Sitter superalgebra $U_q(OSp(1|4))$ there was obtained in [12] the κ–deformed $D = 4$ Poincaré subalgebra $U_\kappa(\mathcal{P}_{4;1})$. Subsequently by nonlinear change of generators the quantum superalgebra $U_\kappa(\mathcal{P}_{4;1})$ was written in chiral bicrossproduct basis [13]. The κ–deformed Poincaré subalgebra is given by the deformation of the following graded cross–product [2]

$$p_{4;1} = \left(SL(2;C) \oplus (\overline{SL(2;C)} \raisebox{0.2ex}{$\supset$}\!\!\!\!+ \, \overline{T}_{0;2}\right) \ltimes T_{4;2} \tag{8}$$

where the generators of $SL(2;C)$ are given by two–spinor generators $M_{\alpha\beta} = \frac{1}{8}(\sigma^{\mu\nu})_{\alpha\beta} M_{\mu\nu}$, the generators of $(\overline{SL(2;C)}$ by $M_{\dot\alpha\dot\beta} = M^*_{\alpha\beta} = \frac{1}{8}\overline{\sigma}^{\mu\nu}_{\dot\alpha\dot\beta} M_{\mu\nu}$, $\overline{T}_{0;2}$ describes two antichiral supercharges $\overline{Q}_{\dot\alpha}$, and $T_{4;2}$ the graded Abelian superalgebra

$$T_{4;2}: \qquad [P_\mu, P_\nu] = [P_\mu, Q_\alpha] = \{Q_\alpha, Q_\beta\} = 0 \tag{9}$$

The relations (9) describe the algebra of generators of translations and supertranslations in chiral superspace. The algebra $(SL(2;c) \oplus (\overline{SL(2;c)} \raisebox{0.2ex}{$\supset$}\!\!\!\!+ \, \overline{T}_{0;2})$ has the form

$$sl(2;c): \qquad [M_{\alpha\beta}, M_{\gamma\delta}] = \epsilon_{\alpha\gamma} M_{\beta\delta} - \epsilon_{\beta\gamma} M_{\alpha\delta} \tag{10a}$$
$$+ c_{\beta\delta} M_{\alpha\gamma} - \epsilon_{\alpha\delta} M_{\beta\gamma}$$

$$\overline{sl(2;c)} \raisebox{0.2ex}{$\supset$}\!\!\!\!+ \, \overline{T}_{0;2}: \qquad [M_{\dot\alpha\dot\beta}, M_{\dot\gamma\dot\delta}] = \epsilon_{\dot\alpha\dot\gamma} M_{\dot\beta\dot\delta} - \epsilon_{\dot\beta\dot\gamma} M_{\dot\alpha\dot\delta}$$
$$+ \epsilon_{\dot\beta\dot\delta} M_{\dot\alpha\dot\gamma} - \epsilon_{\dot\alpha\dot\delta} M_{\dot\beta\dot\gamma}$$

$$[M_{\dot\alpha\dot\beta}, Q_{\dot\gamma}] = \epsilon_{\dot\alpha\dot\gamma} Q_{\dot\beta} - \epsilon_{\dot\beta\dot\gamma} Q_{\dot\alpha}$$
$$\{Q_{\dot\alpha}, Q_{\dot\beta}\} = 0 \tag{10b}$$

It should be observed that in the cross-product (8) the basic supersymmetry algebra $\{Q_\alpha, Q_{\dot\beta}\} = 2(\sigma^\mu p_\mu)_{\alpha\dot\beta}$ is the one belonging to the cross–relations.

The κ–deformed bicrossproduct is given by the formula

$$U_\kappa(p_{4;2}) = (SL(2;c) \oplus \overline{SL(2;c)} \raisebox{0.2ex}{$\supset$}\!\!\!\!+ \, \overline{T}_{0;2}) \bowtie T^\kappa_{4;2} \tag{11}$$

The relations (9) and (10a) remain valid but $T^\kappa_{4;2}$ describes now the Hopf algebra with deformed coproducts:

$$\Delta P_0 = P_0 \otimes 1 + 1 \otimes P_0$$
$$\Delta P_i = P_i \otimes e^{-\frac{P_0}{\kappa}} + 1 \otimes P_i$$
$$\Delta Q_\alpha = Q_\alpha \otimes e^{-\frac{P_0}{2\kappa}} + 1 \otimes Q_\alpha \tag{12}$$

The cross–relations are the following ($M_i = \frac{1}{2}\epsilon_{ijk} M_{jk}, N_i = M_{i0}$):

---

[2] In [13] for the crossproduct formula describing $D = 4$ superPoincaré algebra the following notation was used: $p_{4;1} = O(1,3;2) \ltimes T_{4,2}$. In the notation (8) proposed in present paper the extension of Lorentz algebra by odd generators is described more accurately.

$$[M_i, P_j] = i\epsilon_{ijk}P_k \qquad [M_i, P_0] = 0$$

$$[N_i, P_j] = i\delta_{ij}[\frac{\kappa}{2}(1 - e^{-\frac{2P_0}{\kappa}} + \frac{1}{2\kappa}\vec{P}^2) + \frac{1}{\kappa}P_iP_j]$$

$$[N_i, P_0] = iP_i \qquad\qquad (13)$$

and

$$[M_i, Q_\alpha] = -\frac{1}{2}(\sigma_i)_\alpha^\beta Q_\beta$$

$$[N_i, Q_\alpha] = \frac{1}{2}i e^{-\frac{P_0}{\kappa}}(\sigma_i)_\alpha^\beta Q_\beta + \frac{1}{2\kappa}\epsilon_{ijk}P_j(\sigma_k)_\alpha^\beta Q_\beta$$

$$\{Q_\alpha, Q_{\dot\beta}\} = 4\kappa\delta_{\alpha\dot\beta}\sinh\frac{P_0}{2\kappa} - 2e^{\frac{P_0}{2\kappa}}p_i(\sigma_i)_{\alpha\dot\beta} \qquad (14)$$

The notion of bicrossproduct (11) implies also the modification of primitive coproducts for $SL(2; c) \oplus \overline{SL(2; c)} \mathbin{\supset\!\!\!+} T_{0;2}$ generators. One gets:

$$\Delta M_i = M_i \otimes 1 + 1 \otimes M_i$$

$$\Delta N_i = N_l \otimes 1 + e^{-\frac{P_0}{\kappa}} \otimes N_i + \frac{1}{\kappa}\epsilon_{ijk} P_j \otimes M_k$$

$$\qquad - \frac{i}{4\kappa}(\sigma_i)_{\alpha\dot\beta} Q_\alpha \otimes e^{\frac{P_0}{\kappa}} Q_{\dot\beta}$$

$$\Delta Q_j = Q_{\dot\alpha} \otimes 1 + e^{\frac{P_0}{2\kappa}} \otimes Q_{\dot\alpha} \qquad (15)$$

It appears that the classical $N = 1$ $D = 4$ Poincaré superalgebra can be put as well in the form

$$p_{4;1} = (SL(2; c) \mathbin{\supset\!\!\!+} T_{0;2}) \oplus (\overline{SL(2; c)} \ltimes \overline{T}_{4;2}) \qquad (16)$$

where $\overline{T}_{4;2}^{\,0}$ describe the translation and supertranslation generators $(P_\kappa, Q_{\dot\alpha})$. Subsequently the κ–deformation of $D = 4$ $N = 1$ Poincaré superalgebra can be obtained by deforming (16) into graded bicrossproduct Hopf superalgebra

$$U_\kappa(p_{4;1}) = (SL(2; c) \mathbin{\supset\!\!\!+} T_{0;2}) \oplus (\overline{SL(2; c)} \mathbin{\bowtie} \overline{T}_{4;2}^{\,\kappa}) \qquad (17)$$

In order to describe the κ–deformed chiral superspace one should consider the Hopf superalgebra $\tilde{T}_{4;2}^{\,\kappa}$ obtained by dualization of the relations (9) and (12), and describing by functions $C(\hat{z}_A)$ on κ–deformed chiral superspace $\hat{z}_A = (\hat{z}_\mu, \hat{\theta}_\alpha)$, where $\widehat{z}_\mu$ denotes the complex space–time coordinates. One obtains the following set of relations:

$$[\widehat{z}_0, \widehat{z}_i] = \frac{i}{\kappa} \widehat{z}_i \qquad\qquad [\widehat{z}_i, \widehat{z}_j] = 0$$

$$[\widehat{z}_0, \widehat{\theta}_\alpha] = \frac{i}{2\kappa} \widehat{\theta}_\alpha \qquad [\widehat{z}_i, \widehat{\theta}_\alpha] = 0$$

$$\{\widehat{\theta}_\alpha, \widehat{\theta}_\beta\} = 0 \qquad\qquad\qquad (18a)$$

and the primitive coproducts:

$$\Delta \widehat{z}_\mu = \widehat{z}_\mu \otimes 1 + 1 \otimes \widehat{z}_\mu \qquad \Delta \widehat{\theta}_\alpha = \widehat{\theta}_\alpha \otimes 1 + 1 \otimes \widehat{\theta}_\alpha \qquad (18b)$$

The κ–deformed chiral superfield theory is obtained by cosidering suitably ordered superfields. In the following Section we shall consider the superFourier transform of deformed superfields and consider the κ–deformed chiral superfield theory.

3. Fourier Supertransforms and κ–deformed Berezin Integration

i) Fourier supertransform on classical superspace.

The superfields are defined as functions on superspace. Here we shall restrict ourselves to $D = 4$ chiral superspace $z_A = (z_\mu, \theta_\alpha)$ ($\mu = 0, 1, 2, 3; \alpha = 1, 2$) and to chiral superfields $\Phi(z, \theta)$.

The Fourier supertransform of the chiral superfield and its inverse take the form:

$$\Phi(x, \theta) = \frac{1}{(2\pi)^2} \int d^4p \, d^2\eta \, \widetilde{\Phi}(p, \eta) e^{i(px+\eta\theta)} \qquad (19a)$$

$$\widetilde{\Phi}(p, \eta) = \frac{1}{(2\pi)^2} \int d^4x \, d^2\theta \, \Phi(x, \theta) e^{-i(px+\eta\theta)} \qquad (19b)$$

The Fourier supertransforms were considered firstly in [29, 30]. It appears that the set of even and odd variables $(z_\mu, \theta_\alpha; p_\mu, \eta_\alpha)$ describes the superphase space, with Grassmann variables η_α describing "odd momenta". The Berezin integration rules are valid in both odd position and momentum sectors:

$$\int d^2\theta = \int d^2\theta \, \theta_\alpha = 0 \qquad \frac{1}{2} \int d^2\theta \, \theta_\alpha \theta^\alpha = 1 \qquad (20a)$$

$$\int d^2\eta = \int d^2\eta \, \eta_\alpha = 0 \qquad \frac{1}{2} \int d^2\theta \, \eta_\alpha \eta^\alpha = 1 \qquad (20b)$$

where $\eta^\alpha = \epsilon^{\alpha\beta}\eta_\beta$ and $\eta_\alpha\eta^\alpha = 2\eta_1\eta_2$. It is easy to see that $\theta^2 = \frac{1}{2}\theta_\alpha\theta^\alpha$ $\eta^2 = \frac{1}{2}\eta_\alpha\eta^\alpha$ play the role of Dirac deltas, because

$$\int d^2\theta \, \theta^2 \, \Phi(z,\theta) = \Phi(z,\theta) \, |_{\theta=0} \qquad (21a)$$

$$\int d^2\eta \, \eta^2 \, \tilde{\Phi}(p,\eta) = \tilde{\Phi}(p,\eta) \, |_{\eta=0} \qquad (21b)$$

The formulae (19a)–(19b) in component formalism

$$\Phi(z,\theta) = \Phi(z) + \Psi^\alpha(z)\theta_\alpha + F(z)\theta^2 \qquad (22a)$$

lead to

$$\tilde{\Phi}(p,\eta) = \tilde{F}(p) - \tilde{\Psi}^\nu(p)\eta_\nu - \tilde{\Phi}(p)\eta^2 \qquad (22b)$$

Let us consider for example the chiral vertex $\Phi^3(z,\theta)$, present in Wess–Zumino model. This vertex can be written in momentum superspace as follows:

$$\int d^4z \, d^2\theta \, \Phi^3(z,\theta) = \int d^4p_1 \dots d^4p_3 \, d^2\eta_1 \dots d^2\eta_3$$
$$\cdot \Phi(p_1,\eta_1) \, \Phi(p_2,\eta_2) \, \Phi(p_3,\eta_3)\delta^4(p_1+p_2+p_3)(\eta_1+\eta_2+\eta_3)^2 \quad (23)$$

We see therefore that in Feynmann superdiagrams the chiral vertex (23) will be represented by the product of Dirac deltas describing the conservation at the vertex of the fourmomenta as well as the Grassmann odd momenta.

ii) Fourier supertransform on κ–deformed superspace.

Following the formulae (18a)–(18b) we obtain the supersymmetric extension of of κ–deformed Minkowski space to κ–deformed superspace $\hat{x}_\mu \longrightarrow (\hat{x}_\mu, \hat{\theta}_\alpha)$. The ordered superexponential is defined as follows:

$$: e^{i(p_\mu \hat{z}^\mu + \eta_\alpha \hat{\theta}^\alpha)} := e^{-ip_0 \hat{z}_0} \, e^{i(\vec{p}\vec{z} + \eta^\alpha \hat{\theta}_\alpha)} \qquad (24)$$

where (p_μ, θ_α) satisfy the Abelian graded algebra)9), i.e.

$$[p_\mu, p_\nu] = [p_\mu, \eta_\alpha] = \{\eta_\alpha, \eta_\beta\} = 0 \qquad (25)$$

From the formulae (8) and (24)–(25) follows that:

$$: e^{i(p_\mu \hat{z}^\mu + \eta_\alpha \hat{\theta}^\alpha)} :: e^{i(p'_\mu \hat{z}^\mu + \eta'_\alpha \hat{\theta}^\alpha)} :=: e^{i\Delta^{(2)}_\mu(p,p')\hat{z}^\mu + \Delta^{(2)}_\alpha(\eta,\eta')\hat{\theta}^\alpha} : \qquad (26)$$

where

$$\Delta_0(p,p') = p_0 + p'_0$$
$$\Delta_i(p,p') = p_i + e^{-\frac{p_0}{\kappa}} p'_i$$

$$\Delta_\alpha(\eta, \eta') = \eta_\alpha + e^{-\frac{P_0}{2\kappa}}\eta'_\alpha \tag{27}$$

The κ–deformed Fourier supertransform can be defined as follows:

$$\Phi(\widehat{z}, \widehat{\theta}) := \frac{1}{(2\pi)^2} \int d^4p\, d^2\eta\, \widetilde{\Phi}_\kappa(p, \eta) : e^{i(p\widehat{z}+\eta\widehat{\theta})} : \tag{28}$$

If we define inverse Fourier supertransform

$$\widehat{\Phi}(p, \eta) = \frac{1}{(2\pi)^2} \int d^4\widehat{z}\, d^2\widehat{\theta}\, \Phi(\widehat{z}, \widehat{\theta}) : e^{-i(p\widehat{z}+\eta\widehat{\theta})} : \tag{29}$$

under the assumption that $(\widehat{\theta}^2 = \frac{1}{2}\widehat{\theta}_\alpha\widehat{\theta}^\alpha)$

$$\int d^2\widetilde{\theta}\, \widetilde{\theta}^2 = 1 \tag{30a}$$

or equivalently $(\eta^2 \equiv \frac{1}{2}\eta_\alpha\eta^\alpha)$

$$\frac{1}{(2\pi)^4} \iint d^4\widehat{z}\, d^2\widehat{\theta} : e^{i(p\widehat{z}+\eta\widehat{\theta})} := \delta^4(p) \cdot \eta^2 \tag{30b}$$

one gets

$$\widehat{\Phi}_\kappa(p, \eta) = e^{-\frac{4p_0}{\kappa}}\widetilde{\Phi}\left(e^{\frac{p_0}{\kappa}}\vec{p}, p_0, e^{\frac{p_0}{2\kappa}}\eta_\alpha\right) \tag{31}$$

For κ–deformed chiral fields one can consider their local powers, and perform the κ–deformed superspace integrals. One gets

$$\iint d^4\widehat{z}\, d^2\widehat{\theta} : \Phi(\widehat{z}, \widehat{\theta}) = \widehat{\Phi}(0, 0)$$

$$\iint d^4\widehat{z}\, d^2\widehat{\theta}\Phi^2(\widehat{z}, \widehat{\theta}) = \int d^4p_1\, d^4p_2\, d^2\eta_1\, d^2\eta_2 \tag{32a}$$

$$\widetilde{\Phi}_\kappa(p_1, \eta_1)\, \widetilde{\Phi}_\kappa(p_2, \eta_2)\, \delta(p_{01}+p_{02})\delta^{(3)}\left(\vec{p}_1 + e^{\frac{p_{01}}{\kappa}}\vec{p}_2\right)\left(\eta_1 + e^{\frac{p_{01}}{2\kappa}}\eta_2\right)^2$$

$$\iint d^4\widehat{z}\, d^2\widehat{\theta}\Phi^3(\widehat{z}, \widehat{\theta}) = \int \prod_{i=1}^{3} d^4p_i\, d^2\eta_i \cdot \widetilde{\Phi}_\kappa(p_i, \eta_i)$$

$$\cdot\, \delta(p_{01}+p_{02}+p_{03}) \cdot \delta^{(3)}\left(\vec{p}_1 + e^{\frac{p_{01}}{\kappa}}\vec{p}_2 + \frac{p_0+p_{02}}{\kappa}\vec{p}_3\right)$$

$$\cdot\, \left(\eta_1 + e^{\frac{p_{02}}{2\kappa}}\eta_2 + e^{\frac{p_{01}+p_{02}}{\kappa}}\eta_3\right)^2 \tag{32b}$$

The formulae (32a) can be used for the description of κ–deformed vertices in Wess–Zumino model for chiral superfields.

4. Star Product for κ–deformed Superfield Theory

In this section we shall extend the star product for the functions on κ–deformed Minkowski space given in [15] to the case of functions on κ–deformed chiral superspace, described by the relations (18a)–(18b).

The CBH $\star$–product formula for unordered exponentials takes the form

$$e^{ip_\mu z^\mu + \overline{\eta}_{\dot\alpha}\overline{\theta}^{\dot\alpha}} \cdot e^{ip'_\nu z^\nu + \overline{\eta}'_{\dot\beta}\overline{\theta}^{\dot\beta}} = e^{i\gamma_\mu(p,p')z^\mu + \overline{\sigma}_{\dot\alpha}(p,p',\overline{\eta},\overline{\eta}')\overline{\theta}^{\dot\alpha}} \tag{33}$$

where

$$\gamma_0 = p_0 + p'_0 \tag{34a}$$

$$\gamma_k = \frac{p_k e^{\frac{p'_0}{\kappa}} f\left(\frac{p_0}{\kappa}\right) + p'_k f\left(\frac{p'_0}{\kappa}\right)}{f\left(\frac{p_0+p'_0}{\kappa}\right)} \tag{34b}$$

$$\overline{\sigma}_{\dot\alpha} = \frac{\overline{\eta}_{\dot\alpha} e^{\frac{p'_0}{2\kappa}} f\left(\frac{p_0}{2\kappa}\right) + \overline{\eta}'_{\dot\alpha} f\left(\frac{p'_0}{2\kappa}\right)}{f\left(\frac{p_0+p'_0}{2\kappa}\right)} \tag{34c}$$

and $f(x) \equiv \frac{e^x-1}{x}$. The star product multiplication reproduces the formula (33).

$$e^{ip_\mu z^\mu + \overline{\eta}_{\dot\alpha}\overline{\theta}^{\dot\alpha}} \star e^{ip'_\nu z^\nu + \overline{\eta}'_{\dot\beta}\overline{\theta}^{\dot\beta}} = e^{i\gamma_\mu(p,p')z^\mu + \overline{\sigma}_{\dot\alpha}(p,p',\overline{\eta},\overline{\eta}')\overline{\theta}^{\dot\alpha}} \tag{35}$$

For arbitrary superfields $\phi(z,\theta)$ and $\chi(z,\theta)$ one gets

$$\phi(z,\theta) \star \chi(z,\theta) =$$

$$= \phi\left(\frac{1}{i}\frac{\partial}{\partial p_\mu}, \frac{\partial}{\partial \overline{\eta}_{\dot\alpha}}\right) \chi\left(\frac{1}{i}\frac{\partial}{\partial p'_\mu}, \frac{\partial}{\partial \overline{\eta}'_{\dot\alpha}}\right) e^{i\gamma_\mu(p,p')z^\mu + \overline{\sigma}_{\dot\alpha}(p,p',\overline{\eta},\overline{\eta}')\overline{\theta}^{\dot\alpha}} \Bigg|_{\substack{p=0 \\ p'=0 \\ \overline{\eta}=0 \\ \overline{\eta}'=0}} \tag{36}$$

or equivalently

$$\phi(z,\theta) \star \chi(z,\theta) = e^{iz^\mu\left(\gamma_\mu\left(\frac{\partial}{\partial y}, \frac{\partial}{\partial y'}\right) - \frac{\partial}{\partial y^\mu} - \frac{\partial}{\partial y'^\mu}\right) - \overline{\theta}^{\dot\alpha}\left(\overline{\sigma}^{\dot\alpha}\left(\frac{\partial}{\partial y}, \frac{\partial}{\partial y'}, \frac{\partial}{\partial \omega}, \frac{\partial}{\partial \omega'}\right) - \frac{\partial}{\partial \omega_{\dot\alpha}} - \frac{\partial}{\partial \omega'_{\dot\alpha}}\right)}$$
$$\cdot \phi(y,\omega)\chi(y',\omega')\Big|_{\substack{y=y'=z \\ \omega=\omega'=\overline{\theta}}} \tag{37}$$

In particular we get

$$z^i \star z^j = z^i z^j$$
$$z^0 \star z^i = z^0 z^i + \frac{i}{2\kappa} z^i$$
$$z^i \star z^0 = z^0 z^i - \frac{i}{2\kappa} z^i$$
$$z^i \star \overline{\theta}^{\dot\alpha} = z^i \overline{\theta}^{\dot\alpha}$$
$$\overline{\theta}^{\dot\alpha} \star z^i = z^i \overline{\theta}^{\dot\alpha}$$
$$z^0 \star \overline{\theta}^{\dot\alpha} = z^0 \overline{\theta}^{\dot\alpha} + \frac{i}{4\kappa} \overline{\theta}^{\dot\alpha}$$
$$\overline{\theta}^{\dot\alpha} \star z^0 = z^0 \overline{\theta}^{\dot\alpha} - \frac{i}{4\kappa} \overline{\theta}^{\dot\alpha}$$
$$\overline{\theta}^{\dot\alpha} \star \overline{\theta}^{\dot\beta} = \overline{\theta}^{\dot\alpha} \overline{\theta}^{\dot\beta} \tag{38}$$

Star product $\circledast$ corresponding to the multiplication of ordered exponentials (24) takes the form:

$$e^{ip_\mu z^\mu + \overline{\eta}_{\dot\alpha} \overline{\theta}^{\dot\alpha}} \; \circledast \; e^{ip'_\mu z^\mu + \overline{\eta}'_{\dot\alpha} \overline{\theta}^{\dot\alpha}}$$

$$= e^{i(p_0 + p'_0 z^0 + i(e^{\frac{p'_0}{\kappa}} p_\kappa + p'_\kappa) z^\kappa + (e^{\frac{p'_0}{2\kappa}} \overline{\eta}_{\dot\alpha} + \overline{\eta}'_{\dot\alpha}) \overline{\theta}^{\dot\alpha}} \tag{39}$$

The superalgebra (18a) of κ–deformed superspace is obtained from the following relations:

$$z^k \circledast \overline{\theta}^{\dot\alpha} = \overline{\theta}^{\dot\alpha} \circledast z^k = z^k \overline{\theta}^{\dot\alpha}$$
$$\overline{\theta}^{\dot\alpha} \circledast \overline{\theta}^{\dot\beta} = \overline{\theta}^{\dot\alpha} \overline{\theta}^{\dot\beta}$$
$$z^k \circledast z^i = z^k z^i$$
$$z^0 \circledast z^i = z^0 z^i$$
$$z^i \circledast z^0 = z^0 z^i - \frac{i}{\kappa} z^i$$
$$z^0 \circledast \overline{\theta}^{\dot\alpha} = z^0 \overline{\theta}^{\dot\alpha}$$
$$\overline{\theta}^{\dot\alpha} \circledast z^0 = z^0 \overline{\theta}^{\dot\alpha} - \frac{i}{2\kappa} \overline{\theta}^{\dot\alpha} \tag{40}$$

Similarly like in nonsupersymmetric case the star–product (39) is more physical because reproduces the composition law of even and odd momenta consistent with coalgebra structure.

5. Final Remarks

In this lecture we outlined present status of quantum deformations of space–time supersymmetries[3], and for the case of κ–deformation of $D = 4$ supersymmetries proposed the corresponding deformation of chiral superfield theory. It appears that only the κ–deformed chiral superspace generators describe a closed sub-algebra of κ–deformed $D = 4$ Poincaré group. At present it can be obtained the κ–deformation of superfield theory on real superspace can be obtained. The deformation of chiral superfield theory can be described by the following diagram:

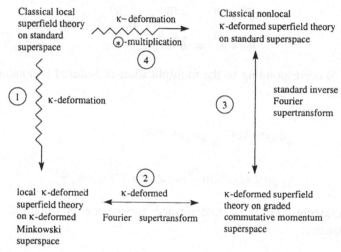

Figure 7. κ–deformation of local superfield theory

The star product $\circledast$ given by formula (39) (see ④ on Fig. 1) is selected by the choice of superFourier transform (28), with ordered Fourier exponential described by (24). Equivalently, the $\circledast$–product multiplication can be obtained by the following three consecutive steps:

i) Deformation of local superfield theory (see① on Fig. 1)
ii) κ–deformed superfield transform (28) (see② on Fig. 1)
iii) inverse classical Fourier transform (see③ on Fig. 1)

$$\Phi(z,\theta) = \frac{1}{(2\pi)^2} \int d^4p\, d^2\theta\, e^{-i(p_\mu z^\mu + \eta_\alpha \theta^\alpha)} \widetilde{\Phi}(p,\eta) \tag{41}$$

obtained in the limit $\kappa \to \infty$ from the inverse Fourier transform (29).

---

[3] We did not consider here however, the quantum deformations of infinite – parameter super-conformal symmetries in $1 + 1$ dimensions, described by superVirasoro algebras as well as affine $OSp(N; 2)$–superalgebras

Finally it should be observed that for the deformation (1) with constant $\widehat{\theta}_{\mu\nu}$ there were calculated some explicit corrections to physical processes, in particular for $D = 4$ QED [29]–[31]. We would like to stress that these calculations should be repeated for Lie algebraic deformations of space–time and superspace, in particular in the κ–deformed framework. The preliminary results in this direction has been obtained in [32, 33].

References

1. S. Dopplicher, K. Fredenhagen, J. Roberts, Phys. Lett. **B331**, 39 (1994); Comm. Math. Phys. **172**, 187 (1995)
2. L.J. Garay, Int. Journ. Mod. Phys. **A10**, 145 (1995)
3. Chong-Sun Chu, Pei-Ming Ho, hep–th/9812219; hep–th/9906192
4. N. Seiberg, E. Witten, JHEP 9909: 032 (1999)
5. J. Madore, S. Schraml, P. Schupp, J. Wess, hep—th/0001203
6. C. Chu, F. Zamorra, hep–th/9912153
7. S. Ferrara, M. Lledo, hep–th/0002084
8. S. Terashima, hep–th/0002119
9. A.A. Bichl, J.M. Grimstrup, H. Grosse, L. Popp, M. Schweda, R. Wulkenhaar, hep–th/0007050
10. S. Dopplicher, Ann. Inst. Henri Poinc. **64**, 543 (1996)
11. J. Lukierski, P. Stichel, W.J. Zakrzewski, Ann,. Phys. **261**, 224 (1997)
12. J. Lukierski, A. Nowicki, J. Sobczyk, J. Phys. **26A**, L1109 (1993)
13. P. Kosiński, J. Lukierski, P. Maślanka, J. Sobczyk, J. Phys. **27A**, 6827 (1994); ibid. **28A**, 2255 (1995)
14. P. Kosiński, J. Lukierski, P. Maślanka, J. Sobczyk, J. Math. Phys. **37**,3041 (1996)
15. P. Kosiński, J. Lukierski, P. Maślanka, hep–th/0009120
16. M. Chaichian, P.P. Kulish, Phys. Lett. **B234**, 72 (199)
17. S.M. Khoroshkin, V.N. Tolstoy, Comm. Math. Phys. **141**, 599 (1991)
18. J. Lukierski, A. Nowicki, Phys. Lett. B279, 299 (1992)
19. V. Dobrev, Journ. Phys. **A26**, 1317 (1993)
20. L. Dabrawski, V.K. Dobrev, R. Floreanini, V. Husain, Phys. Lett. **B302**, 215 (1993)
21. C. Juszczak, Journ. Phys. **A27**, 385 (1994)
22. J. Lukierski, P. Minnaert, M. Mozrzymas, Phys. Lett. **B371**, 215 (1996)
23. C. Juszczak, J. Sobczyk, math.QA/9809006
24. R. Celeghini, P. Kulish, q–alg/9712006,
25. F. Herranz, Journ. Phys. **A30**, 6123 (1997)
26. S. Majid, "Foundations of Quantum Group Theory" Cambridge Univ. Press, 1995
27. D. Leites, B.M. Zupnik "Multiple Processes at High Energies" (in Russian), Tashkent 1976, p. 3–26
28. F.A. Berezin, M.S. Marinov, Ann. Phys. **104**, 336 (1977)
29. I. Mocioiu, M. Pospelov, R. Roiban, Phys. Lett. **B489**, 390 (2000)
30. N. Chair, M.M. Sheikh–Jabbari, hep–th/0009037
31. M. Chaichian, M.M. Sheikh–Jabbari, A. Tureanu hep–th/0010175
32. J. Lukierski, H. Ruegg, W. Ruehl, Phys. Lett. **313**, 357 (1993)
33. L.C. Biedenharn, B. Mueller, M. Tarlini, Phys. Lett. **318**, 613 (1993)

THE HOWE DUALITY AND LIE SUPERALGEBRAS

DIMITRY LEITES[†][‡]
Department of Mathematics, University of Stockholm, Roslagsv. 101, Kräftriket hus 6, S-106 91, Stockholm, Sweden

IRINA SHCHEPOCHKINA[§]
Independent Univ. of Moscow, Bolshoj Vlasievsky per., 121002 Moscow, Russia

Abstract. Howe's duality is considered from a unifying point of view based on Lie superalgebras. New examples are offered. In particular, we construct several simplest spinor-oscillator representations and compute their highest weights for the "stringy" Lie superalgebras (i.e., Lie superalgebras of complex vector fields (or their nontrivial central extensions) on the supercircle $S^{1|n}$ and its two-sheeted cover associated with the Möbius bundle).

In our two lectures we briefly review, on the most elementary level, several results and problems unified by "Howe's duality". Details will be given elsewhere. The ground field in the lectures is $\mathbb{C}$.

1. Introduction

In his famous preprint [24] R. Howe gave an inspiring explanation of what can be "dug out" from H. Weyl's "wonderful and terrible" book [55], at least as far as invariant theory is concerned, from a certain unifying viewpoint. According to Howe, much is based on a remarkable correspondence between certain irreducible representations of Lie subalgebras Γ and Γ' of the Lie algebra $o(V)$ or $sp(V)$ provided Γ and Γ' are each other's "commutants", i.e., centralizers. This correspondence is known ever since as *Howe's correspondence* or *Howe's duality*. In [24] and subsequent papers Howe gave several examples of such a correspondence previously known, mostly, inadvertently. Let us remind

---

[‡] We gratefully acknowledge financial support of an NFR grant and RFBR grant 99-01-00245, respectively. D.L is thankful to B. Feigin, E. Poletaeva, V. Serganova and Xuan Peiqi for helpful discussions.

[†] `mleites@matematik.su.se`

[§] `ira@paramonova.mccme.ru`

S. Duplij and J. Wess (eds.), Noncommutative Structures in Mathematics and Physics, 93–111.
© 2001 *Kluwer Academic Publishers. Printed in the Netherlands.*

some of them (omitting important Jacquet-Langlands-Shimizu correspondence, S. Gelbart's contributions, etc.) :

1) decomposition of $o(V)$-module $S^{\cdot}(V)$ into spherical harmonics;

2) *Lefschetz decomposition* of $\mathfrak{sp}(V)$-module $\Lambda^{\cdot}(V)$ into primitive forms (sometimes this is called Hodge–Lépage decomposition);

3) a striking resemblance between *spinor representation* of $o(n)$ and *oscillator* (*Shale–Segal–Weil–metaplectic–...*) representation of $\mathfrak{sp}(2n)$.

As an aside Howe gives the "shortest possible" proof of the *Poincaré lemma*. (Recall that this lemma states that in any sufficiently small open star-shaped neighborhood of any point on any manifold any closed differential form is exact.) In this proof, Lie superalgebras, that lingered somewhere in the background in the previous discussion but were treated rather as a nuisance than help, are instrumental to reach the goal. This example shows also that the requirement of reductivity of Γ and Γ' to form a "dual pair" is extra. Elsewhere we will investigate what are the actual minimal restrictions on Γ and Γ' needed to reach one of the other problems usually solved by means of Howe duality: decompose the symmetric or exterior algebra of a module over $\Gamma \oplus \Gamma'$. Howe's manuscript was written at the time when supersymmetry theory was being conceived. By the time [24] was typed, the definition of what is nowadays called *superschemes* ([34]) was not yet rewritten in terms to match physical papers (language of points was needed; now we can recommend [5]) nor translated into English and, therefore, was unknown; the classification of simple finite dimensional Lie superalgebras over $\mathbb{C}$ had just been announced. This was, perhaps, the reason for a cautious tone with which Howe used Lie superalgebras, although he made transparent how important they might be for a lucid presentation of his ideas and explicitly stated so.

Since [25], the published version of [24], though put aside to stew for 12 years, underwent only censorial changes, we believe it is of interest to explore what do we gain by using Lie superalgebras from the very beginning (an elaboration of other aspects of this idea [4] are not published yet). Here we briefly elucidate some of Howe's results and notions and give several new examples of Howe's dual pairs. In the lectures we will review the known examples 1) – 3) mentioned above but consider them in an appropriate "super" setting, and add to them:

4) a refinement of the Lefschetz decomposition — J. Bernstein's decomposition ([2]) of the space $\Omega_{\hbar}^{\cdot}$ of "twisted" differential forms on a symplectic manifold with values in a line bundle with connection whose curvature form differs by a factor $\hbar$ from the canonical symplectic form;

5) a decomposition of the space of differential forms on a hyper-Kählerian manifold similar to the Lefschetz one ([53]) but with $\mathfrak{sp}(4)$ instead of $\mathfrak{sp}(2) = \mathfrak{sl}(2)$ and its refinement associated with the $\mathfrak{osp}(1|4)$.

6) Apart from general clarification of the scenery and new examples even in the old setting, i.e., on manifolds, the superalgebras introduced *ab ovo* make it manifest that there are at least two types of Howe's correspondence: the conven-

tional one and several "ghost" ones associated with quantization of the antibracket [40].

7) Obviously, if $\Gamma \oplus \Gamma'$ is a maximal subalgebra of $\mathfrak{osp}$, then (Γ, Γ') is an example of Howe dual pair. Section 6 gives some further examples, partly borrowed from [49], where more examples can be found.

We consider here only finite dimensional Lie superalgebras with the invariant theory in view. In another lecture (§§3,4) we consider spinor-oscillator representations in more detail. In these elementary talks we do not touch other interesting applications such as Capelli identities ([30],[43]), or prime characteristic ([47]). Of dozens of papers with examples of Howe's duality in infinite dimensional cases and still other examples, we draw attention of the reader to the following selected ones: [12], and various instances of bose-fermi correspondence, cf. [13] and [26]. Observe also that the Howe duality often manifests itself for q-deformed algebras, e.g., in Klimyk's talk at our conference, or [6]. To treat this q-Howe duality in a similar way, we first have to explicitly q-quantize Poisson superalgebras $\mathfrak{po}(2n|m)$ (for $mn = 0$ this is straightforward replacement of (super)commutators from [39] with q-(super)commutators.

2. The Poisson superalgebra $\mathfrak{g} = \mathfrak{po}(2n|m)$

2.1. Certain $\mathbb{Z}$-gradings of $\mathfrak{g}$. Recall that $\mathfrak{g}$ is the Lie superalgebra whose superspace is $\mathbb{C}[q, p, \Theta]$ and the bracket is the *Poisson bracket* $\{\cdot, \cdot\}_{P.b.}$ is given by the formula

$$\{f, g\}_{P.b.} = \sum_{i \le n} \left(\frac{\partial f}{\partial p_i} \frac{\partial g}{\partial q_i} - \frac{\partial f}{\partial q_i} \frac{\partial g}{\partial p_i} \right) - \\ (-1)^{p(f)} \sum_{j \le m} \frac{\partial f}{\partial \theta_j} \frac{\partial g}{\partial \theta_j} \text{ for } f, g \in \mathbb{C}[p, q, \Theta]. \tag{2.1}$$

Sometimes it is more convenient to redenote the Θ's and set

$$\xi_j = \tfrac{1}{\sqrt{2}}(\Theta_j - i\Theta_{r+j}); \quad \eta_j = \tfrac{1}{\sqrt{2}}(\Theta_j + i\Theta_{r+j})$$

for $j \le r = [m/2]$ (here $i^2 = -1$), $\theta = \Theta_{2r+1}$

and accordingly modify the bracket (if $m = 2r$, there is no term with θ):

$$\{f, g\}_{P.b.} = \sum_{i \le n} \left(\frac{\partial f}{\partial p_i} \frac{\partial g}{\partial q_i} - \frac{\partial f}{\partial q_i} \frac{\partial g}{\partial p_i} \right) - \\ (-1)^{p(f)} \left[\sum_{j \le m} \left(\frac{\partial f}{\partial \xi_j} \frac{\partial g}{\partial \eta_j} + \frac{\partial f}{\partial \eta_j} \frac{\partial g}{\partial \xi_j} \right) + \frac{\partial f}{\partial \theta} \frac{\partial g}{\partial \theta} \right].$$

Setting $\deg_{Lie} f = \deg f - 2$ for any monomial $f \in \mathbb{C}[p, q, \Theta]$, where $\deg p_i = \deg q_i = \deg \Theta_j = 1$ for all i, j, we obtain the *standard* $\mathbb{Z}$-grading of $\mathfrak{g}$:

degree of f	-2	-1	0	1	$\cdots$
f	1	p, q, θ	$f : \deg f = 2$	$f : \deg f = 3$	$\cdots$

Clearly, $\mathfrak{g} = \underset{i \geq -2}{\oplus} \mathfrak{g}_i$ with $\mathfrak{g}_0 \simeq \mathfrak{osp}(m|2n)$. Consider now another, "*rough*", grading of $\mathfrak{g}$. To this end, introduce: $Q = (q, \xi)$, $P = (p, \eta)$ and set

$$\deg Q_i = 0, \ \deg \theta = 1, \ \deg P_i = \begin{cases} 1 & \text{if } m = 2k \\ 2 & \text{if } m = 2k + 1. \end{cases} \qquad (*)$$

Remark. Physicists prefer to use half-integer values of deg for $m = 2k + 1$ by setting $\deg \theta = \frac{1}{2}$ and $\deg P_i = 1$ at all times.

The above grading $(*)$ of the polynomial algebra induces the following *rough grading* of the Lie superalgebra $\mathfrak{g}$. For $m = 2k$ just delete the columns of odd degrees and delete the degrees by 2:

$m = 2k + 1$:

degree	$\cdots$	2	1	0	-1	-2
elements	$\cdots$	$\mathbb{C}[Q]P^2$	$\mathbb{C}[Q]P\theta$	$\mathbb{C}[Q]P$	$\mathbb{C}[Q]\theta$	$\mathbb{C}[Q]$

2.2. Quantization. We call the nontrivial deformation $\mathcal{Q}$ of the Lie superalgebra $\mathfrak{po}(2n|m)$ *quantization* (for details see [40]). There are many ways to quantize $\mathfrak{g}$, but all of them are equivalent. Recall that we only consider $\mathfrak{g}$ whose elements are represented by polynomials; for functions of other types (say, Laurent polynomials) the uniqueness of quantization may be violated.

Consider the following quantization, so-called QP-quantization, given on linear terms by the formulas:

$$\mathcal{Q} : Q \mapsto \hat{Q}, \quad P \mapsto \hbar \frac{\partial}{\partial Q}, \qquad (*)$$

where $\hat{Q}$ is the operator of left multiplication by Q; an arbitrary monomial should be first rearranged so that the Q's stand first (normal form) and then apply $(*)$ term-wise.

The deformed Lie superalgebra $\mathcal{Q}(\mathfrak{po}(2n|2k))$ is the Lie superalgebra of differential operators with polynomial coefficients on $\mathbb{R}^{n|k}$. Actually, it is an analog of $\mathfrak{gl}(V)$. This is most clearly seen for $n = 0$. Indeed,

$$\mathcal{Q}(\mathfrak{po}(0|2k)) = \mathfrak{gl}(\Lambda^{\cdot}(\xi)) = \mathfrak{gl}(2^{k-1}|2^{k-1}).$$

In general, for $n \neq 0$, we have

$$\mathcal{Q}(\mathfrak{po}(2n|2k)) = \text{"}\mathfrak{gl}\text{"}(\mathcal{F}(Q)) = \mathfrak{diff}(\mathbb{R}^{n|k}).$$

For $m = 2k - 1$ we consider $\mathfrak{po}(0|2k - 1)$ as a subalgebra of $\mathfrak{po}(0|2k)$; the quantization sends $\mathfrak{po}(0|2k - 1)$ into $\mathfrak{q}(2^{k-1})$. For $n \neq 0$ the image of $\mathcal{Q}$ is an infinite dimensional analog of q, indeed (for $J = i(\theta + \frac{\partial}{\partial\theta})$ with $i^2 = -1$):

$$\mathcal{Q}(\mathfrak{po}(2n|2k-1)) = \mathfrak{qdiff}(\mathbb{R}^{n|k}) = \{D \in \mathfrak{diff}(\mathbb{R}^{n|k}) : [d, J] = 0\}.$$

2.3. Fock spaces and spinor-oscillator representations. The Lie superalgebras $\mathfrak{diff}(\mathbb{R}^{n|k})$ and $\mathfrak{qdiff}(\mathbb{R}^{n|k})$ have indescribably many irreducible representations even for $n = 0$. But one of the representations, the identity one, in the superspace of functions on $\mathbb{R}^{n|k}$, is the "smallest" one. Moreover, if we consider the superspace of $\mathfrak{diff}(\mathbb{R}^{n|k})$ or $\mathfrak{qdiff}(\mathbb{R}^{n|k})$ as the *associative* superalgebra (denoted $\mathrm{Diff}(\mathbb{R}^{n|k})$ or $\mathrm{QDiff}(\mathbb{R}^{n|k})$), this associative superalgebra has only *one* irreducible representation — the same identity one. This representation is called *the Fock space.*

As is known, the Lie superalgebras $\mathfrak{osp}(m|2n)$ are rigid for $(m|2n) \neq (4|2)$. Therefore, the through map

$$\mathfrak{h} \longrightarrow \mathfrak{g}_0 = \mathfrak{osp}(m|2n) \subset \mathfrak{g} = \mathfrak{po}(2n|m) \xrightarrow{\;\mathcal{Q}\;} \mathfrak{diff}(\mathbb{R}^{n|k})$$

sends any subsuperalgebra $\mathfrak{h}$ of $\mathfrak{osp}(m|2n)$ (for $(m|2n) \neq (4|2)$) into its isomorphic image. (One can also embed $\mathfrak{h}$ into $\mathfrak{diff}(\mathbb{R}^{n|k})$ directly.) The irreducible subspace of the Fock space which contains the constants is called the *spinor-oscillator representation* of $\mathfrak{h}$. In particular cases, for $m = 0$ or $n = 0$ this subspace turns into the usual *spinor* or *oscillator representation*, respectively. We have just given a unified description of them. (A more detailed description follows.)

2.4. Primitive alias harmonic elements. The elements of $\mathfrak{osp}(m|2n)$ (or its subalgebra $\mathfrak{h}$) act in the space of the spinor-oscillator representation by inhomogeneous differential operators of order ≤ 2 (order is just the filtration associated with the "rough" grading):

$m = 2k$:				$m = 2k + 1$:					
degree	-1	0	1	degree	-2	-1	0	1	2
elements	$\hat{P}^2$	$\hat{P}\hat{Q}$	$\hat{Q}^2$	elements	$\hat{P}^2$	$\hat{P}\hat{\theta}$	$\hat{P}\hat{Q}$	$\hat{Q}\hat{\theta}$	$\hat{Q}^2$

The elements from $(\mathbb{C}[Q])^{\hat{P}^2}$ for $m = 2k$ or $(\mathbb{C}[Q, \theta])^{\hat{P}\hat{\theta}}$ for $m = 2k + 1$ are called *primitive* or *harmonic* ones. More generally, let $\mathfrak{h} \subset \mathfrak{osp}(m|2n)$ be a $\mathbb{Z}$-graded Lie superalgebra embedded consistently with the rough grading of $\mathfrak{osp}(m|2n)$. Then the elements from $(\mathbb{C}[Q])^{\mathfrak{h}_{-1}}$ for $m = 2k$ or $(\mathbb{C}[Q, \theta])^{\mathfrak{h}_{-1}}$ for $m = 2k + 1$ will be called $\mathfrak{h}$-*primitive* or $\mathfrak{h}$-*harmonic.*

2.4.1. Nonstandard $\mathbb{Z}$-gradings of $\mathfrak{osp}(m|2n)$. It is well known that one simple Lie superalgebra can have several nonequivalent Cartan matrices and systems of Chevalley generators, cf. [20]. Accordingly, the corresponding divisions into *positive* and *negative* root vectors are distinct. The following problem arises: *How the passage to nonstandard gradings affects the highest weight of the spinor-oscillator representation defined in sec. 3?* (Cf. [44])

2.5. Examples of dual pairs. Two subalgebras Γ, Γ' of $\mathfrak{g}_0 = \mathfrak{osp}(m|2n)$ will be called a *dual pair* if one of them is the centralizer of the other in $\mathfrak{g}_0$.

If $\Gamma \oplus \Gamma'$ is a maximal subalgebra in $\mathfrak{g}_0$, then, clearly, Γ, Γ' is a dual pair. A generalization: consider a pair of mutual centralizers Γ, Γ' in $\mathfrak{gl}(V)$ and embed $\mathfrak{gl}(V)$ into $\mathfrak{osp}(V \oplus V^*)$. Then Γ, Γ' is a dual pair (in $\mathfrak{osp}(V \oplus V^*)$). For a number of such examples see [49]. Let us consider several of these examples in detail.

2.5.1. $\Gamma = \mathfrak{sp}(2n) = \mathfrak{sp}(W)$ and $\Gamma' = \mathfrak{sp}(2) = \mathfrak{sl}(2) = \mathfrak{sp}(V \oplus V^*)$. Clearly, $\mathfrak{h} = \Gamma \oplus \Gamma'$ is a maximal subalgebra in $\mathfrak{o}(W \otimes (V \oplus V^*))$. The Fock space is just $\Lambda^\cdot(W)$.

The following classical theorem and its analog 5.2 illustrate the importance of the above notions and constructions.

Theorem. *The Γ'-primitive elements of $\Lambda^\cdot(W)$ of each degree i constitute an irreducible Γ-module $P^i_{\mathfrak{sp}}$, $0 \leq i \leq n$.*

This action of Γ' in the superspace of differential forms on any symplectic manifold is well known: Γ' is generated (as a Lie algebra) by operators X_+ of left multiplication by the symplectic form ω and X_-, application of the bivector dual to ω.

2.5.2. $\Gamma = \mathfrak{o}(2n) = \mathfrak{o}(W)$ and $\Gamma' = \mathfrak{sp}(2) = \mathfrak{sl}(2) = \mathfrak{sp}(V \oplus V^*)$. Clearly, $\mathfrak{h} = \Gamma \oplus \Gamma'$ is a maximal subalgebra in $\mathfrak{sp}(W \otimes (V \oplus V^*))$. The Fock space is just $S^\cdot(W)$.

Theorem. *The Γ'-primitive elements of $S^\cdot(W)$ of each degree i constitute an irreducible Γ-module P^i_0, $i = 0, 1, \ldots$.*

This action of Γ' in the space of polynomial functions on any Riemann manifold is also well known: Γ' is generated (as a Lie algebra) by operators X_+ of left multiplication by the quadratic polynomial representing the metric g and X_- is the corresponding Laplace operator.

Clearly, a mixture of Examples 2.5.1 and 2.5.2 corresponding to symmetric or skew-symmetric forms on a supermanifold is also possible: *the space of Γ'-primitive elements of $S^\cdot(W)$ of each degree i is an irreducible Γ-module*, cf. [44] and Sergeev's papers [51], [52].

In [24], [25] the dual pairs had to satisfy one more condition: the through action of both Γ and Γ' on the identity $\mathfrak{g}_0$-module should be completely reducible. Even for the needs of the First Theorem of Invariant Theory this is too strong a requirement, cf. examples with complete irreducibility in [51, 52] with our last example, in which the complete reducibility of $\mathfrak{pe}(n)$ is violated. Investigation of the requiremets on Γ and Γ' needed for the First Theorem of Invariant Theory will be given elsewhere.

2.5.3. Bernstein's square root of the Lefschetz decomposition. Let L be the space of a (complex) line bundle over a connected symplectic manifold (M^{2n}, ω) with connection ∇ such that the curvature form of ∇ is equal to $\hbar\omega$ for some

$\hbar \in \mathbb{C}$. This $\hbar$ will be called a *twist*; the space of tensor fields of type ρ (here ρ : $\mathfrak{sp}(2n) \longrightarrow \mathfrak{gl}(U)$ is a representation which defines the space $\Gamma(M, U)$ of tensor fields with values in U), and twist $\hbar$ will be denoted by $T_\hbar(\rho)$. Let us naturally extend the action of X_+, X_- from the space Ω of differential forms on M onto the space $\Omega_\hbar$ of twisted differential forms using the isomorphism of *spaces* $T_\hbar(\rho) \simeq T(\rho) \otimes \Gamma(L)$, where $\Gamma(L) = \Omega_\hbar^0$ is the space of sections of the line bundle L, i.e., the space of twisted functions.

Namely, set $X_+ \mapsto X_+ \otimes 1$, etc. Let $D_+ = d + \alpha$ be the connection ∇ itself and $D_- = [X_-, D_+]$. On $\Omega_\hbar$, introduce a superspace structure setting $p(\varphi \otimes s) = \deg \varphi \pmod 2$, for $\varphi \in \Omega$, $s \in \Omega_\hbar^0$.

Theorem. ([2]) *On $\Omega_\hbar$, the operators D_+ and D_- generate an action of the Lie superalgebra $\mathfrak{osp}(1|2)$ commuting with the action of the group $\hat{G}$ of ∇-preserving automorphisms of the bundle L.*

Bernstein studied the $\hat{G}$-action, more exactly, the action of the Lie algebra $\mathfrak{po}(2n|0)$ corresponding to $\hat{G}$; we are interested in the part of this action only: in $\mathfrak{sp}(2n) = \mathfrak{po}(2n|0)_0$-action.

In Example 2.5.1 the space P^i consisted of differential forms with constant coefficients. Denote by $\mathcal{P}^i = P^i \otimes S^{\cdot}(V)$ the space of primitive forms with polynomial coefficients. The elements of the space $\sqrt{\mathcal{P}}_\hbar^i = \operatorname{Ker} D_- \cap \mathcal{P}_\hbar^i$ will be called ∇-*primitive forms* of degree i (and twist $\hbar$).

Bernstein showed that $\sqrt{\mathcal{P}}_\hbar^i$ is an irreducible $\mathfrak{g} = \mathfrak{po}(2n|0)$-module. It could be that over subalgebra $\mathfrak{g}_0$ the module $\sqrt{\mathcal{P}}_\hbar^i$ becomes reducible but the general theorem of Howe (which is true for $\mathfrak{osp}(1|2n)$) states that this is not the case, it remains irreducible. Shapovalov and Shmelev literally generalized Bernstein's result for $(2n|m)$-dimensional supermanifolds, see review [37]. In particular, Shapovalov, who considered $n = 0$, "took a square root of Laplacian and the metric".

2.5.4. Inspired by Bernstein's construction, let us similarly define a "square root" of the hyper-Kähler structure. Namely, on a hyper-Kählerean manifold (M, ω_1, ω_2) consider a line bundle L with two connections: ∇_1 and ∇_2, whose curvature forms are equal to $\hbar_1 \omega_1$ and $\hbar_2 \omega_2$ for some $\hbar_1, \hbar_2 \in \mathbb{C}$. The pair $\hbar = (\hbar_1, \hbar_2)$ will be called a *twist*; the space of tensor fields of type ρ and twist $\hbar$ will be denoted by $T_\hbar(\rho)$. Verbitsky [53] defined the action of $\mathfrak{sp}(4)$ in the space Ω of differential forms on M. Let us naturally extend the action of the generators $X_j^\pm$ for $j = 1, 2$ of of $\mathfrak{sp}(4)$ from Ω onto the space $\Omega_\hbar$ of twisted differential forms using the isomorphism $T_\hbar(\rho) \simeq T(\rho) \otimes \Gamma(L)$, where $\Gamma(L) = \Omega_\hbar^0$ is the space of sections of the line bundle L; here X_j^+ is the operator of multiplication by ω_j and X_j^- is the operator of convolution with the dual bivector.

Define the space of primitive i-forms (with constant coefficients) on the hyper-Kählerean manifold (M, ω_1, ω_2) by setting

$$P^i = \operatorname{Ker} X_1^- \cap \operatorname{Ker} X_2^- \cap \Omega^i. \qquad (HK)$$

According to the general theorem [25] this space is an irreducible $\mathfrak{sp}(2n; \mathbb{H})$-module.

Set $D_i^- = [X_i^-, D_i^+]$. The promised square root of the decomposition (HK) is the space

$$\mathcal{P}_\hbar^i = \operatorname{Ker} D_1^- \cap \operatorname{Ker} D_2^- \cap \Omega_\hbar^i. \qquad (\sqrt{\text{HK}})$$

The operators $D_i^\pm$, where $D_i^+ = \nabla_i$, generate $\mathfrak{osp}(1|4)$.

2.6. Further examples of dual pairs. The following subalgebras $\mathfrak{g}_1(V_1) \oplus \mathfrak{g}_2(V_2)$ are maximal in $\mathfrak{g}(V_1 \otimes V_2)$, hence, are dual pairs:

$\mathfrak{g}_1$	$\mathfrak{g}_2$	$\mathfrak{g}$
$\mathfrak{osp}(n_1|2m_1)$	$\mathfrak{osp}(n_2|2m_2)$	$\mathfrak{osp}(n_1 n_2 + 4m_1 m_2 | 2n_1 m_2 + 2n_2 m_1)$
$\mathfrak{o}(n)$	$\mathfrak{osp}(n_2|2m_2)$	$\mathfrak{osp}(nn_2|2nm_2), n \neq 2, 4$
$\mathfrak{sp}(2n)$	$\mathfrak{osp}(n_2|2m_2)$	$\mathfrak{osp}(2mn_2|4nm_2)$
$\mathfrak{pe}(n_1)$	$\mathfrak{pe}(n_2)$	$\mathfrak{osp}(2n_1 n_2 | 2n_1 n_2), n_1, n_2 > 2$
$\mathfrak{osp}(n_1|2m_1)$	$\mathfrak{pe}(n_2)$	$\mathfrak{pe}(n_1 n_2 + 2m_1 n_2)$ if $n_1 \neq 2m_1$
		$\mathfrak{spe}(n_1 n_2 + 2m_1 n_2)$ if $n_1 = 2m_1$
$\mathfrak{o}(n)$	$\mathfrak{pe}(m)$	$\mathfrak{pe}(nm)$
$\mathfrak{sp}(2n)$	$\mathfrak{pe}(m)$	$\mathfrak{pe}(2nm)$

In particular, on the superspace of polyvector fields, there is a natural $\mathfrak{pe}(n)$-module structure, and $\mathfrak{pe}(1)$, its dual partner in $\mathfrak{osp}(2n|2n)$, is spanned by the divergence operator Δ ("odd Laplacian"), called the *BRST operator* ([1]), the even operator of $\mathfrak{pe}(1)$ being $\deg_x - \deg_\theta$, where $\theta_i = \pi(\frac{\partial}{\partial x_i})$, π being the shift of parity operator.

For further examples of maximal subalgebras in $\mathfrak{gl}$ and $\mathfrak{q}$ see [49]. These subalgebras give rise to other new examples of Howe dual pairs. For the decomposition of the tensor algebra corresponding to some of these examples see [51, 52], some of the latter are further elucidated in [3]. Some further examples of Howe's duality, considered in a detailed version of our lectures, are: (1) over reals; (2) dual pairs in simple subalgebras of $\mathfrak{po}(2n|m)$ distinct from $\mathfrak{osp}(m|2n)$; in particular, (3) embeddings into $\mathfrak{po}(2n|m; r)$, the nonstandard regradings of the Poisson super-algebra, cf. [50]; (4) a "projective" version of the Howe duality associated with embeddings into the Lie superalgebra of Hamiltonian vector fields, the quotient of the Poisson superalgebra, in particular, the exceptional cases in dimension $(2|2)$, cf. [40]. It is also interesting to consider the prime characteristic and an "odd" Howe's duality obtained from quantization of the antibracket (the main objective of [4]), to say nothing of q-quantized versions of the above.

3. Generalities on spinor and spinor-like representations

3.1. The spinor and oscillator representations of Lie algebras. The importance of the spinor representation became clear very early. One of the reasons is the following. As is known from any textbook on representation theory, the fundamental representations $R(\varphi_1) = W$, $R(\varphi_2) = \Lambda^2(W)$, ..., $R(\varphi_{n-1}) = \Lambda^{n-1}(W)$ of $\mathfrak{sl}(W)$, where dim $W = n$ and φ_i is the highest weight of $\Lambda^i(W)$, are irreducible. Any finite dimensional irreducible $\mathfrak{sl}(n)$-module L^λ is completely determined by its highest weight $\lambda = \sum \lambda_i \varphi_i$ with $\lambda_i \in \mathbb{Z}_+$. The module L^λ can be realized as a submodule (or quotient) of $\otimes \left(R(\varphi_i)^{\otimes \lambda_i} \right)$.

Similarly, every irreducible $\mathfrak{gl}(n)$-module L^λ, where $\lambda = (\lambda_1, \dots \lambda_{n-1}; c)$ and c is the eigenvalue of the unit matrix, is realized in the space of tensors, perhaps, twisted with the help of c-densities, namely in the space $\otimes_i \left(R(\varphi_i)^{\otimes \lambda_i} \right) \otimes$ tr^c, where tr^c is the Lie algebraic version of the cth power of the determinant, i.e., infinitesimally, trace, given for any $c \in \mathbb{C}$ by the formula $X \mapsto c \cdot \mathrm{tr}(X)$ for any matrix $X \in \mathfrak{gl}(W)$. Thus, all the irreducible finite dimensional representations of $\mathfrak{sl}(W)$ are naturally realized in the space of tensors, i.e., in the subspaces or quotient spaces of the space $T_q^p = \underbrace{W \otimes \cdots \otimes W}_{p} \underbrace{\otimes W^* \otimes \cdots \otimes W^*}_{q}$, where W is the space of the identity representation. For $\mathfrak{gl}(W)$, we have to consider the space $T_q^p \otimes \mathrm{tr}^c$.

For $\mathfrak{sp}(W)$, the construction is similar, except the fundamental module $R(\varphi_i)$ is now a *part* of the module $\Lambda^i(\mathrm{id})$ consisting of the *primitive forms*.

For $\mathfrak{o}(W)$, the situation is totally different: not all fundamental representations can be realised as (parts of) the modules $\Lambda^i(\mathrm{id})$. The exceptional one (or two, for $\mathfrak{o}(2n)$) of them is called the *spinor representation*; for $\mathfrak{o}(W)$, where dim $W = 2n$, it is realized in the Grassmann algebra $E^\cdot(V)$ of a "half" of W, where $W = V \oplus V^*$ is a decomposition into the direct sum of subspaces isotropic with respect to the form preserved by $\mathfrak{o}(W)$. For dim $W = 2n+1$, it is realized in the Grassmann algebra $E^\cdot(V \oplus W_0)$, where $W = V \oplus V^* \oplus W_0$ and W_0 is the 1-dimensional space on which the orthogonal form is nondegenerate.

The quantization of the harmonic oscillator leads to an infinite dimensional analog of the spinor representation which after Howe we call *oscillator* representation of $\mathfrak{sp}(W)$. It is realized in $S^\cdot(V)$, where as above, V is a maximal isotropic subspace of W (with respect to the skew form preserved by $\mathfrak{sp}(W)$). The remarkable likeness of the spinor and oscillator representations was underlined in a theory of *dual Howe's pairs*, [23].

The importance of spinor-oscillator representations is different for distinct classes of Lie algebras and their representations. In the description of irreducible finite dimensional representations of classical matrix Lie algebras $\mathfrak{gl}(n)$, $\mathfrak{sl}(n)$ and $\mathfrak{sp}(2n)$ we can do without either spinor or oscillator representations. We can not do without spinor representation for $o(n)$, but a pessimist might say that spinor representation constitutes only $\frac{1}{n}$th of the building bricks. Our, optimistic, point of view identifies the spinor representations as one of the two possible types of the building bricks.

For the Witt algebra $\mathfrak{witt}$ and its central extension, the Virasoro algebra $\mathfrak{vir}$, *every* irreducible highest weight module is realized as a quotient of a spinor or, equivalently, oscillator representation, see [8], [10]. This miraculous equivalence is known in physics under the name of *bose-fermi correspondence*, see [18], [26].

For the list of generalizations of $\mathfrak{witt}$ and $\mathfrak{vir}$, i.e., simple (or close to simple) stringy Lie superalgebras or Lie superalgebras of vector fields on N-extended supercircles, often called by an unfortunate (as explained in [21]) name "super-conformal algebras", see [21]. The importance of spinor-oscillator representations diminishes as N grows, but for the most interesting — *distinguished* ([21]) — stringy superalgebras it is high, cf. [11], [46].

3.2. Semi-infinite cohomology. An example of applications of spinor-oscillator representations: semi-infinite (or BRST) cohomology of Lie superalgebras. These cohomology were introduced by Feigin first for Lie algebras ([9]); then he extended the definition to Lie superalgebras via another construction, equivalent to the first one for Lie algebras ([7]). For an elucidation of Feigin's construction see [14], [31] and [54]. Feigin rewrote in mathematical terms and generalized the constructions physicists used to determine the *critical dimensions* of string theories, i.e., the dimensions in which the quantization of the superstring is possible, see [42], [18]. These critical dimensions are the values of the central element (central charges) on the spinor-oscillator representation constructed from the adjoint representation; to this day not for every central element of all distinguished simple stringy superalgebras their values are computed on every spinor-oscillator representation, not even on the ones constructed from the adjoint representations.

4. The spinor-oscillator representations and Lie superalgebras

4.1. Spinor (Clifford–Weil–wedge– ...) and oscillator representations. As we saw in [40], $\mathfrak{po}(2n|m)_0 \cong \mathfrak{osp}(m|2n)$, the superspace of elements of degree 0 in the standard $\mathbb{Z}$-grading of $\mathfrak{po}(2n|m)$ or, which is the same, the superspace of quadratic elements in the representation by generating functions. At our first lecture we defined the *spinor-oscillator representation* as the through map (here $k = [\frac{m}{2}]$ and $\mathcal{Q}$ is the quantization)

$$\mathfrak{g} \longrightarrow \mathfrak{po}(2n|m) \overset{\mathcal{Q}}{\longrightarrow} \begin{cases} \mathfrak{diff}(n|k) & \text{if } m = 2k \\ \mathfrak{qdiff}(n|k) & \text{if } m = 2k - 1, \end{cases}$$

where $\mathrm{Im}(\mathfrak{g}) \subset \mathfrak{po}(2n|m)_0 = \mathfrak{osp}(m|2n)$. Actually, such requirement is too restrictive, we only need that the image of $\mathfrak{g}$ under embedding into $\mathfrak{po}(2n|m)$

remains rigid under quantization. So various simple subalgebras of $\mathfrak{po}(2n|m)$ will do as ambients of $\mathfrak{g}$.

This spinor-oscillator representation is called the *spinor representation* of $\mathfrak{g}$ if $n = 0$, or the *oscillator representation* if $m = 0$. We will denote this representation $\mathrm{Spin}(V)$ and set $\mathrm{Osc}(V) = \mathrm{Spin}(\Pi(V))$, where V is the standard representation of $\mathfrak{osp}(m|2n)$. In other words, if $\mathrm{Spin}(V)$ is a representation of $\mathfrak{osp}(m|2n)$, then $\mathrm{Osc}(V)$ is a representation of $\mathfrak{osp}(2n|m)$, so $\mathrm{Osc}(V)$ only exists for m even.

If V is a $\mathfrak{g}$-module without any bilinear form, but we still want to construct a spinor-oscillator representation of $\mathfrak{g}$, consider the module $W = V \oplus V^*$ (where in the infinite dimensional case we replace V^* with the *restricted* dual of V; roughly speaking, if $V = \mathbb{C}[x]$, then $V^* = \mathbb{C}[[\frac{\partial}{\partial x}]]$, whereas the restricted dual is $\mathbb{C}[\frac{\partial}{\partial x}]$) endowed with the form (for $v_1, w_1 \in V$, $v_2, w_2 \in V^*$) symmetric for the plus sign and skew-symmetric otherwise:

$$B((v_1, v_2), (w_1, w_2)) = v_2(w_1) \pm (-1)^{p(v_1)p(w_2)} w_2(v_1).$$

Now, in W, select a maximal isotropic subspace U (not necessarily V or V^*) and realize the spinor-oscillator representation of $\mathfrak{g}$ in the exterior algebra of U.

Observe that the classical descriptions of spinor representations differ from ours, see, e.g., [17], where the embedding of $\mathfrak{g}$ (in their case $\mathfrak{g} = \mathfrak{o}(n)$) into the quantized algebra (namely into $\mathcal{Q}(\mathfrak{po}(0|n-1))$) is considered, not into $\mathfrak{po}(0|m)$. The existence of this embedding is not so easy to see unless told, whereas our constructions are manifest and bring about the same result.

To illustrate our definitions and constructions, we realize the orthogonal Lie algebra $\mathfrak{o}(n)$ as the subalgebra in the Lie superalgebra $\mathfrak{po}(0|n)$.

Case $\mathfrak{o}(2k)$. Basis:

$$X_1^+ = \xi_2\eta_1, \quad \ldots, \quad X_{k-1}^+ = \xi_k\eta_{k-1}, \quad X_k^+ = \eta_k\eta_{k-1};$$
$$X_1^- = \xi_1\eta_2, \quad \ldots, \quad X_{k-1}^- = \xi_{k-1}\eta_k, \quad X_k^- = \xi_{k-1}\xi_k;$$
$$H_1 = \xi_1\eta_1 - \xi_2\eta_2, \quad \ldots, \quad H_{k-1} = \xi_{k-1}\eta_{k-1} - \xi_k\eta_k, \quad H_k = \xi_{k-1}\eta_{k-1} + \xi_k\eta_k.$$

For $R(\varphi_k)$ take the subspacespace functions $\mathbb{C}[\xi]_{ev}$ which contains the constants $\mathbb{C} \cdot \hat{1}$, where $\hat{1}$ is just the constant function 1; clearly, $\hat{1}$ is the vacuum vector.

Quantization (see above) sends: ξ_i into $\hat{\xi}_i$, and η_i into $\hbar\frac{\partial}{\partial\xi_i}$, so $X_i^\pm\hat{1} = 0$ for $i < k$, hence, $H_i\hat{1} = [X_i^+, X_i^-]\hat{1} = 0$ for $i < k$. Contrariwise,

$$H_k\hat{1} = [X_k^+, X_k^-]\hat{1} = [\partial_k\partial_{k-1}, \hat{\xi}_{k-1}\hat{\xi}_k]\hat{1} = \partial_k(-\hat{\xi}_{k-1}\partial_{k-1} + 1)\hat{\xi}_k\hat{1} = \hat{1}.$$

So we see that the spinor representation is indeed a fundamental one.

Case $\mathfrak{o}(2k + 1)$. Basis:

$$X_1^+ = \xi_2\eta_1, \quad \ldots, \qquad X_{k-1}^+ = \xi_k\eta_{k-1}, \qquad X_k^+ = \sqrt{2}\eta_k\theta;$$
$$X_1^- = \xi_1\eta_2, \quad \ldots, \qquad X_{k-1}^- = \xi_{k-1}\eta_k, \qquad X_k^- = \sqrt{2}\theta\xi_k;$$
$$H_1 = \xi_1\eta_1 - \xi_2\eta_2, \ldots, \; H_{k-1} = \xi_{k-1}\eta_{k-1} - \xi_k\eta_k, \quad H_k = 2\xi_k\eta_k.$$

For $R(\varphi_k)$ consider the space of even functions $\mathbb{C}[\xi_1, \ldots, \xi_k, \theta]_{\mathrm{ev}}$ and realize $\mathfrak{o}(2k + 1)$ so that $\xi_i \mapsto \hat{\xi}_i$, $\eta_i \mapsto \hbar\frac{\partial}{\partial\xi_i}$, $\theta \mapsto \hbar(\hat{\theta} + \frac{\partial}{\partial\theta})$. As above for $\mathfrak{o}(2k)$, set $\hbar = 1$.

Then, as above, $H_i v = [X_i^+, X_i^-]\hat{1} = 0$ for $i < k$, whereas

$$H_k\hat{1} = [X_k^+, X_k^-]\hat{1} = \frac{2}{2}\left(\partial_k(\hat{\theta} + \frac{\partial}{\partial\theta})^2\hat{\xi}_k + \hat{\xi}_k(\hat{\theta} + \frac{\partial}{\partial\theta})^2\partial_k\right)\hat{1} = \hat{1}.$$

So $\hat{1}$ is indeed the highest weight vector of the kth fundamental representation.

4.2. Stringy superalgebras. Case $\mathfrak{vir}$. For the basis of $\mathfrak{vir}$ take $e_i = t^{i+1}\frac{d}{dt}$, $i \in \mathbb{Z}$, and the central element z; let the bracket be

$$[e_i, e_j] = (j - i)e_{i+j} - \frac{1}{12}\delta_{ij}(i^3 - i)z. \qquad (*)$$

We advise the reader to refresh definitions of stringy superalgebras and various modules over them, see [21], where we also try to convince physicists not to use the term "superconformal algebra" (except, perhaps, for $\mathfrak{k}^L(1|1)$ and $\mathfrak{k}^M(1|1)$). In particular, recall that $\mathcal{F}_{\lambda,\mu} = \mathrm{Span}(\varphi_i = t^{\mu+i}(dt)^\lambda \mid i \in \mathbb{Z})$.

Statement. *The only instances when $\mathcal{F}_{\lambda,\mu}$ possesses an invariant symmetric nondegenerate bilinear form are the space of half-densities, $\sqrt{\mathrm{Vol}} = \mathcal{F}_{1/2,0}$, and its twisted version, $\mathcal{F}_{1/2,1/2}$ and in both cases the form is:*

$$(f\sqrt{dt}, g\sqrt{dt}) = \int fg \cdot dt;$$

the only instances when $\mathcal{F}_{\lambda,\mu}$ possesses an invariant skew-symmetric forms are the quotient space of functions modulo constants, $d\mathcal{F} = \mathcal{F}_{0,0}/\mathbb{C} \cdot 1$, and $\frac{1}{2}$-twisted functions, $\sqrt{t}\mathcal{F} = \mathcal{F}_{0,1/2}$ and in both cases the form is:

$$(f, g) = \int f \cdot dg.$$

Let $\partial_i = \frac{\partial}{\partial\varphi_i}$ (where $\varphi_i = t^{\mu+i}(dt)^\lambda$). Let $\mathrm{osc}(\sqrt{\mathrm{Vol}})$ be the $\mathfrak{vir}$-submodule of the *exterior* algebra on φ_i for $i < 0$ containing the constant $\hat{1}$. Since the generators

e_i of $\mathfrak{vir}$ acts on $\mathcal{F}_{\lambda,\mu}$ as (sums over $i \in \mathbb{Z}$)

$$e_1 = \sum(\mu + i + 2\lambda)\varphi_{i+1}\partial_i = \sum i\varphi_{i+1}\partial_i,$$
$$e_{-1} = \sum(\mu + i + 1)\varphi_i\partial_{i+1} = \sum(i+1)\varphi_i\partial_{i+1};$$
$$e_2 = \sum(\mu + i - \lambda)\varphi_{i+1}\partial_i = \sum i\varphi_{i+1}\partial_i,$$
$$e_{-2} = \sum(\mu + i + 3\lambda)\varphi_i\partial_{i+1} = \sum(i+1)\varphi_i\partial_{i+1},$$

and representing e_0 and z as brackets of $e_{\pm 1}$ and $e_{\pm 2}$ from $(*)$ we immediately deduce that the highest weights (c, h) of $\mathrm{osc}(\sqrt{\mathrm{Vol}})$ is $(-\frac{1}{3}, 0)$.

For the spinor representations $\mathrm{spin}(\sqrt{t}\mathcal{F})$ and $\mathrm{spin}(d\mathcal{F})$ (realized on the *symmetric* algebra of φ_i for $i < 0$) we similarly obtain that the highest weights (c, h) are $(\frac{1}{6}, \frac{1}{2})$ for $\mathrm{spin}(\sqrt{t}\mathcal{F})$ and $(-\frac{1}{6}, 0)$ for $\mathrm{spin}(d\mathcal{F})$.

Observe that the representations $\mathrm{spin}(\sqrt{t}\mathcal{F})$, $\mathrm{spin}(d\mathcal{F})$ and $\mathrm{osc}(\sqrt{\mathrm{Vol}})$ are constructed on a half of the generators used to construct $\mathrm{Spin}(\mathcal{F}_{\lambda;\mu})$.

4.3. The highest weights of the spinor representations of $\mathfrak{k}^L(1|n)$ and $\mathfrak{k}^M(1|n)$. In the following theorem we give the coordinates $(c, h; H_1, \dots)$ of the highest weight of the spinor representations $\mathrm{Spin}(\mathcal{F}_{\lambda;\mu})$ of the contact superalgebra $\mathfrak{k}^L(1|n)$ with respect to z (the central element), K_t, and, after semicolon, on the elements of Cartan subalgebra, respectively. For $\mathfrak{k}^M(1|n)$ we write $\tilde{h}; \tilde{H}_i$. (Observe that for $n > 4$ the Cartan subalgebra has more generators than just $H_1 = K_{\xi_1\eta_1}, \dots, H_k = K_{\xi_k\eta_k}$ which generate the Cartan subalgebra of $\mathfrak{k}(1|2k)$, the algebra of contact vector fields with polynomial coefficients.)

n	0	1	2	≥ 3
c	$12\lambda^2 - 12\lambda + 2$	$-12\lambda + 3$	6	0
h	$(\mu + 2\lambda)(\mu + 1)$	$\mu + 2\lambda$	$2\mu + 2\lambda + \nu$	$2^{n-1}(\mu + \lambda) + 2^{n-3}$
$\tilde{h}$	$-$	$2\mu + 3\lambda - \frac{1}{4}$	$2\mu + 2\lambda - \frac{1}{2}$	$2^{n-1}(\mu + \lambda)$

Theorem. *Let $(c, h; H_1, \dots)$ be the highest weight of the spinor representation $\mathrm{Spin}(\mathcal{F}_{\lambda;\mu})$ of $\mathfrak{k}^L(1|n)$. The highest weight of the oscillator representation $\mathrm{Osc}(\mathcal{F}_{\lambda;\mu}) = \mathrm{Spin}(\Pi(\mathcal{F}_{\lambda;\mu}))$ is $(-c, h; H_1, \dots)$ and similarly for $\mathfrak{k}^M(1|n)$.*

For $n \neq 2$, all the coordinates of the highest weight other than c, h vanish. For $n = 2$ the value of H on the highest weight vector from $\mathrm{Spin}(\mathcal{F}_{\lambda,\nu;\mu})$ is equal to ν.

The values of c and h (or $\tilde{h}$) on modules $\mathrm{Spin}(\mathcal{F}_{\lambda;\mu})$ are given in the above table.

Up to rescaling, these results are known for small n, see [29], [28] and refs.

Remark. For the contact superalgebras $\mathfrak{g}$ on the $1|n$-dimensional supercircle our choice of $\mathfrak{g}$-modules $V = \mathcal{F}_{\lambda;\mu}$ from which we constructed $\mathrm{Spin}(V \oplus V^*)$ is natural for small n: there are no other modules! For larger n it is only justified if we are interested in semi-infinite cohomology of $\mathfrak{g}$ and not in representation theory *per se*. For the superalgebras $\mathfrak{g}$ of series $\mathfrak{vect}$ and $\mathfrak{svect}$ the adjoint module $\mathfrak{g}$ is of the form $T(\mathrm{id}^*)$, i.e, it is either coinduced from multidimensional representation ($\mathfrak{vect}$), or is a submodule of such a coinduced module ($\mathfrak{svect}$). Spinor-oscillator representations of this type were not studied yet, cf. sec. 5.

4.4. Other spinor representations. 1) Among various Lie superalgebras for which it is interesting to study spinor-oscillator representations, the simple (or close to them) maximal subsuperalgebras of $\mathfrak{po}$ are most interesting. The list of such maximal subalgebras is being completed; various maximal subalgebras listed in [48] distinct from the sums of mutual centralizers also provide with spinor representations.

As an interesting example consider A. Sergeev's Lie superalgebra $\mathfrak{as}$, the nontrivial central extension of the Lie superalgebra $\mathfrak{spe}(4)$ preserving the odd bilinear form and the volume on the $(4|4)$-dimensional superspace, see [49, 50]. Namely, consider $\mathfrak{po}(0|6)$, the Lie superalgebra whose superspace is the Grassmann superalgebra $\Lambda(\xi, \eta)$ generated by $\xi_1, \xi_2, \xi_3, \eta_1, \eta_2, \eta_3$ and the bracket is the Poisson bracket. Recall also that the quotient of $\mathfrak{po}(0|6)$ modulo center is $\mathfrak{h}(0|6) = \mathrm{Span}(H_f \mid f \in \Lambda(\xi, \eta))$, where

$$H_f = (-1)^{p(f)} \sum \left(\frac{\partial f}{\partial \xi_j} \frac{\partial}{\partial \eta_j} + \frac{\partial f}{\partial \eta_j} \frac{\partial}{\partial \xi_j} \right).$$

Now, observe that $\mathfrak{spe}(4)$ can be embedded into $\mathfrak{h}(0|6)$. Indeed, setting $\deg \xi_i = \deg \eta_i = 1$ for all i we introduce a $\mathbb{Z}$-grading on $\Lambda(\xi, \eta)$ which, in turn, induces a $\mathbb{Z}$-grading on $\mathfrak{h}(0|6)$ of the form $\mathfrak{h}(0|6) = \underset{i \geq -1}{\oplus} \mathfrak{h}(0|6)_i$. Since $\mathfrak{sl}(4) \cong \mathfrak{o}(6)$, we can identify $\mathfrak{spe}(4)_0$ with $\mathfrak{h}(0|6)_0$.

It is not difficult to see that the elements of degree -1 in the standard gradings of $\mathfrak{spe}(4)$ and $\mathfrak{h}(0|6)$ constitute isomorphic $\mathfrak{sl}(4) \cong \mathfrak{o}(6)$-modules. It is subject to a direct verification that it is really possible to embed $\mathfrak{spe}(4)_1$ into $\mathfrak{h}(0|6)_1$.

A. Sergeev's extension $\mathfrak{as}$ is the result of the restriction onto $\mathfrak{spe}(4) \subset \mathfrak{h}(0|6)$ of the cocycle that turns $\mathfrak{h}(0|6)$ into $\mathfrak{po}(0|6)$. The quantization (with parameter λ) deforms $\mathfrak{po}(0|6)$ into $\mathfrak{gl}(\Lambda(\xi))$; the through maps $T_\lambda : \mathfrak{as} \longrightarrow \mathfrak{po}(0|6) \longrightarrow \mathfrak{gl}(\Lambda(\xi))$ are representations of $\mathfrak{as}$ in the $4|4$-dimensional modules Spin_λ. The explicit form of T_λ is as follows:

$$T_\lambda : \begin{pmatrix} a & b \\ c & -a^t \end{pmatrix} + d \cdot z \mapsto \begin{pmatrix} a & b - \lambda \tilde{c} \\ c & -a^t \end{pmatrix} + \lambda d \cdot 1_{4|4},$$

where $1_{4|4}$ is the unit matrix and $\tilde{c}_{ij} = c_{kl}$ for any skew-symmetric matrix $c_{ij} = E_{ij} - E_{ji}$ and any even permutation $(1234) \mapsto (ijkl)$. Clearly, T_λ is an irreducible representation for any λ and $T_\lambda \not\simeq T_\mu$ for $\lambda \neq \mu$.

2) Maximal subalgebras (for further examples see [48]) and a conjecture. Let V_1 be a linear superspace of dimension $(r|s)$; let $\Lambda(n)$ be the Grassmann

superalgebra with n odd generators $\xi_1, \ldots, \xi_n$ and $\mathfrak{vect}(0|n) = \mathfrak{der}\Lambda(n)$ the Lie superalgebra of vector fields on the $(0|n)$-dimensional supermanifold.

Let $\mathfrak{g} = \mathfrak{gl}(V_1) \otimes \Lambda(n) \ni \mathfrak{vect}(0|n)$ be the semidirect sum (the ideal at the open part of $\ni$) with the natural action of $\mathfrak{vect}(0|n)$ on the ideal $\mathfrak{gl}(V_1) \otimes \Lambda(n)$. The Lie superalgebra $\mathfrak{g}$ has a natural faithful representation ρ in the space $V = V_1 \otimes \Lambda(n)$ defined by the formulas

$$\rho(X \otimes \varphi)(v \otimes \psi) = (-1)^{p(\varphi)p(\psi)} Xv \otimes \varphi\psi,$$
$$\rho(D)(v \otimes \psi) = -(-1)^{p(D)p(v)} v \otimes D\psi$$

for any $X \in \mathfrak{gl}(V_1)$, $\varphi, \psi \in \Lambda(n)$, $v \in V_1$, $D \in \mathfrak{vect}(0|n)$. Let us identify the elements from $\mathfrak{g}$ with their images under ρ, so we consider $\mathfrak{g}$ embedded into $\mathfrak{gl}(V)$.

Theorem ([48]) 1) *The Lie superalgebra* $\mathfrak{gl}(V_1) \otimes \Lambda(n) \ni \mathfrak{vect}(0|n)$ *is maximal irreducible in* $\mathfrak{sl}(V_1 \otimes \Lambda(n))$ *unless* a) $\dim V_1 = (1,1)$ *or* b) $n = 1$ *and* $\dim V_1 = (1,0)$ *or* $(0,1)$ *or* $(r|s)$ *for* $r \neq s$.

2) *If* $\dim V_1 = (1,1)$, *then* $\mathfrak{gl}(1|1) \cong \Lambda(1) \ni \mathfrak{vect}(0|1)$, *so*

$$\mathfrak{gl}(V_1) \otimes \Lambda(n) \ni \mathfrak{vect}(0|n) \subset \Lambda(n+1) \ni \mathfrak{vect}(0|n+1)$$

and it is the bigger superalgebra which is maximal irreducible in $\mathfrak{sl}(V)$.

3) *If* $n = 1$ *and* $\dim V_1 = (r|s)$ *for* $r > s > 0$, *then* $\mathfrak{g}$ *is maximal irreducible in* $\mathfrak{gl}(V)$.

Conjecture. Suppose $r + s = 2^N$. Then, $\dim V$ coincides with $\dim \Lambda(W)$ for some space W. We suspect that this coincidence is not accidental but is occasioned by the spinor representations of the maximal subalgebras described above. The same applies to $\mathfrak{q}(V_1) \otimes \Lambda(n) \ni \mathfrak{vect}(0|n)$, a maximal irreducible subalgebra in $\mathfrak{q}(V_1 \otimes \Lambda(n))$.

4.5. Selected problems. 1) The spinor and oscillator representations are realized in the symmetric (perhaps, supersymmetric) algebra of the maximal isotropic (at least for $\mathfrak{g} = \mathfrak{sp}(2k)$ and $\mathfrak{o}(2k)$) subspace V of the identity $\mathfrak{g}$-module $\mathrm{id} = V \oplus V^*$. But one could have equally well started from another $\mathfrak{g}$-module. For an interesting study of spinor representations constructed from $W \neq \mathrm{id}$, see [45].

To consider in a way similar to sec. 2 contact stringy superalgebras $\mathfrak{g} = \mathfrak{k}^L(1|n)$ and $\mathfrak{k}^M(1|n)$, as well as other stringy superalgebras from the list [21], we have to replace $\mathcal{F}_{\lambda,\mu}$ with modules $\mathcal{T}_\mu(W)$ of (twisted) tensor fields on the supercircle and investigate how does the highest weight of $\hat{1} \in \mathrm{Osc}(\mathcal{T}_\mu(W))$ or $\hat{1} \in \mathrm{Spin}(\mathcal{T}_\mu(W))$ constructed from an arbitrary irreducible $\mathfrak{co}(n)$-module $W = V \oplus V^*$ depend on the highest weight of W. (It seems that the new and absolutely remarkable spinor-like representation Poletaeva recently constructed [46] is obtained in this way.)

To give the reader a feel of calculations, we consider here the simplest non-trivial case $\mathfrak{o}(3) = \mathfrak{sl}(2)$. The results may (and will) be used in calculations of

$\mathrm{Spin}(\mathcal{T}_\mu(W))$ for $\mathfrak{g} = \mathfrak{k}^L(1|n)$ and $\mathfrak{k}^M(1|n)$ for $n = 3, 4$. As is known, for every $N \in \mathbb{Z}_+$ there exists an irreducible $(N + 1)$-dimensional $\mathfrak{g}$-module with highest weight N. This module possesses a natural nondegenerate $\mathfrak{g}$-invariant bilinear form which is skew-symmetric for $N = 2k + 1$ and symmetric for $N = 2k$. The corresponding embeddings $\mathfrak{g} \longrightarrow \mathfrak{o}(2k + 1)$ and $\mathfrak{g} \longrightarrow \mathfrak{sp}(2k)$ are called *principal*, see [19] and references therein. Explicitly, the images of the Chevalley generators $X^\pm$ of $\mathfrak{sl}(2)$ are as follows: $X^- \mapsto \sum X_i^-$,

$$
X^+ \mapsto \begin{cases} N(N + 1)X_N^+ + \sum\limits_{1 \le i \le N-1} i(N + 1 - i)X_i^+ & \text{for } N = 2k + 1 \\ N^2 X_N^+ + \sum\limits_{1 \le i \le N-1} i(2N - i)X_i^+ & \text{for } N = 2k. \end{cases}
$$

From the commutation relations between X^+ and X^- we derive that only $X_N^\pm$ give a nontrivial contribution to the highest weight HW of the $\mathfrak{sl}(2)$-module $\mathrm{Spin}(L^N)$; we have:

$$
\mathrm{HW} = \begin{cases} N(N + 1) & \text{if } N = 2k + 1 \\ -\frac{1}{2}N^2 & \text{if } N = 2k. \end{cases}
$$

2) Observe, that the notion of spinor-oscillator representation can be broadened to embrace the subalgebras of the Lie superalgebra $\mathfrak{h}$ of Hamiltonian vector fields and their images under quantization; we call the through map the *projective spinor-oscillator representation*. Since the Lie superalgebra $\mathfrak{h}$ has more deformations than $\mathfrak{po}$ ([40]), and since the sets of maximal simple subalgebras of $\mathfrak{po}$ and $\mathfrak{h}$ are distinct, the set of examples of projective spinor-oscillator representations differs from that of spinor-oscillator representations.

References

1. Batalin I., Tyutin I., *Generalized Field–Antifield formalism*, In: Dobrushin R. et. al. (eds.) *Topics in Statistical and theoretical Physics* (F.A.Berezin memorial volume), Transactions of AMS, series 2, **177**, 1996, 23–43

2. Bernstein J., *The Lie superalgebra* $\mathfrak{osp}(1|2)$, *connections over symplectic manifolds and representations of Poisson algebras*. In: [36], 9/1987–13 *preprint*

3. Cheng S., Wang W., *Howe duality for Lie superalgebras*, math.RT/0008093; *Remarks on the Schur-Howe-Sergeev duality*, math.RT/0008109

4. Deligne P., Leites D., *Howe's duality and Lie superalgebras* unpublished notes, IAS, Princeton, 1989

5. Deligne P. et al (eds.) *Quantum fields and strings: a course for mathematicians*. Vol. 1, 2. Material from the Special Year on Quantum Field Theory held at the Institute for Advanced Study, Princeton, NJ, 1996–1997. AMS, Providence, RI; Institute for Advanced Study (IAS), Princeton, NJ, 1999. Vol. 1: xxii+723 pp.; Vol. 2: pp. i–xxiv and 727–1501

6. Ding J., Frenkel I., *Spinor and oscillator representations of quantum groups* Lie theory and geometry, 127–165, Progr. Math., **123**, Birkhuser Boston, Boston, MA, 1994.

7. Feigin B., Usp. Mat. Nauk **39**, 195–196 (English translation: Russian Math. Surveys **39**, 155–156).

8. Feigin B., Fuchs D., *Representations of the Virasoro algebra.* In: Vershik A., Zhelobenko D. (eds.) Representation of Lie groups and related topics, Adv. Stud. Contemp. Math., **7**, Gordon and Breach, New York, 1990, 465–554

9. Feigin B., Leites D., *New Lie superalgebras of string theories* In: Markov M. (ed.), Group–theoretical methods in physics (Zvenigorod, 1982), **v. 1**, Nauka, Moscow, 1983, 269–273 (Harwood Academic Publ., Chur, 1985, **Vol. 1–3**, 623–629)

10. B. Feigin, E. Frenkel, *Semi-infinite Weil complex and the Virasoro algebra*, Commun. Math. Phys. **137**, 1991, 617–639. Erratum: Commun. Math. Phys. **147**, 1992, 647–648

11. Feigin B., Semikhatov A., Sirota V., Tipunin I., *Resolutions and characters of irreducible representations of the $N = 2$ superconformal algebra.* Nuclear Phys. **B 536**, 1999, no. 3, 617–656

12. Feigin B. L., Semikhatov A. M., Tipunin I. Yu., *Equivalence between chain categories of representations of affine* sl(2) *and* $N = 2$ *superconformal algebras*, J. Math. Phys. **39** (1998), no. 7, 3865–3905

13. Frenkel, I. B. *Two constructions of affine Lie algebra representations and boson-fermion correspondence in quantum field theory* J. Funct. Anal. **44**, 1981, no. 3, 259–327

14. Frenkel I., H. Garland H., G. Zuckerman G., *Semi-infinite cohomology and string theory*, Proc. Natl. Acad. Sci. U.S.A. **83**, 1986, 8442–8446.

15. Gendenshtein L. E.; Krive I. V. *Supersymmetry in quantum mechanics*, Soviet Phys. Uspekhi **28** (1985), no. 8, 645–666 (1986)

16. Gomis J., Paris J., Samuel S., *Antibracket, antifields and gauge-theory quantization*, Phys. Rept. **259** (1995) 1–191

17. Goto M., Grosshans F., *Semisimple Lie algebras.* Lecture Notes in Pure and Applied Mathematics, **Vol. 38**, Marcel Dekker, Inc., New York-Basel, 1978. vii+480 pp.

18. Green M., Schwarz J., Witten E., *Superstring theory* **Vol. 1, 2**. Second edition. Cambridge Monographs on Mathematical Physics. Cambridge University Press, Cambridge-New York, 1988. x+470 pp.; xii+596 pp.

19. Grozman P., Leites D., *Lie superalgebras of supermatrices of complex size. Their generalizations and related integrable systems*, In: Vasilevsky N. et. al. (eds.) *Proc. Internatnl. Symp. Complex Analysis and related topics*, Mexico, 1996, Birkhauser Verlag, 1999, 73–105

20. Grozman P., Leites D., *Defining relations for classical Lie superalgebras with Cartan matrix*, hep-th 9702073; (Czechoslovak J. Phys., 2001, to appear)

21. Grozman P., Leites D., Shchepochkina I., *Lie superalgebras of string theories*, hep-th 9702120

22. Grozman P., Leites D., Poletaeva E., *Defining relations for simple Lie superalgebras. Lie superalgebras without Cartan matrix*, In: Ivanov E. et. al. (eds.) *Supersymmetries and Quantum Symmetries* (SQS'99, 27-31 July, 1999), Dubna, JINR, 2000, 387–396

23. Howe R., *Remarks on classical invariant theory*, Trans. Amer. Math. Soc., **313**, n. 2, 1989, 539–570; Erratum to: *"Remarks on classical invariant theory"*. Trans. Amer. Math. Soc. **318**, no. 2, 1990, 823

24. Howe R., *Remarks on classical invariant theory (preprint* ca 1973–75)

25. Howe R., *Remarks on classical invariant theory*, Trans. Amer. Math. Soc., **313**, n. 2, 1989, 539–570; Erratum to: *"Remarks on classical invariant theory"*. Trans. Amer. Math. Soc. **318**, no. 2, 1990, 823

26. Kac V., *Infinite Dimensional Lie Algebras.* 3rd ed. Cambridge Univ. Press, Cambridge, 1992

27. Kac V., Peterson, D., *Spin and wedge representations of infinite-dimensional Lie algebras and groups.* Proc. Nat. Acad. Sci. U.S.A. **78**, 1981, no. 6, part 1, 3308–3312

28. Kac V., Wakimoto M., *Unitarizable highest weight representations of the Virasoro, Neveu-Schwarz and Ramond algebras.* In: Conformal groups and related symmetries: physical results

and mathematical background (Clausthal-Zellerfeld, 1985), 345–371, Lecture Notes in Phys., **261**, Springer, Berlin-New York, 1986

29.   Kent, A., Riggs, H., *Determinant formulae for the $N = 4$ superconformal algebras*. Phys. Lett. **B** 198 (1987), no. 4, 491–496

30.   Kostant B., Sahi S., *Jordan algebras and Capelli identities* Invent. Math. **112** (1993), no. 3, 657–664; Brylinski R., Kostant B., *Minimal representations of E_6, E_7, and E_8 and the generalized Capelli identity*, Proc. Nat. Acad. Sci. U.S.A. **91** (1994), no. 7, 2469–247

31.   Kostant B.; Sternberg S., Symplectic reduction, BRS cohomology, and infinite-dimensional Clifford algebras. Ann. Physics, 176 (1987), no. 1, 49–113

32.   Kotchetkoff Yu. *Deformations of Lie superalgebras. VINITI Depositions*, Moscow 1985, # 384–85 (in Russian)

33.   Kotchetkoff Yu. *Déformations de superalgébres de Buttin et quantification*, C.R. Acad. Sci. Paris, ser. I, **299: 14**, 1984, 643–645

34.   Leites D., *Spectra of graded-commutative rings*, Uspekhi Mat. Nauk, **29**, n. 3, 1974, 209–210 (in Russian)

35.   Leites D., *New Lie superalgebras and mechanics*, Soviet Math. Doklady, **18**, n. 5, 1977, 1277–1280

36.   Leites D. (ed.), *Seminar on supermanifolds*, Reports of Stockholm University, **1–34**, 1987– 1990, 2100 pp. *preprint*

37.   Leites D., *Lie superalgebras*, JOSMAR, **30(6)**, 1985, 2481–2512; id., *Introduction to the supermanifold theory*, Russian Math. Surveys, **35**, n.1, 1980, 3–53

38.   Leites D., *Quantization. Supplement 3*, In: Berezin F., Shubin M. *Schrödinger equation*, Kluwer, Dordrecht, 1991, 483–522

39.   Leites D., Poletaeva E., *Defining relations for classical Lie algebras of polynomial vector fields*, Math. Scand., **81**, 1997, no. 1, 1998, 5–19; Grozman P., Leites D., Poletaeva E., *Defining relations for simple Lie superalgebras. Lie superalgebras without Cartan matrix*, In: Ivanov E. et. al. (eds.) *Supersymmetries and Quantum Symmetries* (SQS'99, 27-31 July, 1999), Dubna, JINR, 2000, 387–396

40.   Leites D., Shchepochkina I., *How to quantize antibracket* , Theor. and Math. Physics, to appear

41.   Leites D., Shchepochkina I., *How to quantize antibracket*, Theor. and Math. Physics, to appear

42.   Marinov, M. S. *Relativistic strings and dual models of the strong interactions*. Soviet Physics Uspekhi **20**, 1977, no. 3, 179–208.; translated from Uspehi Fiz. Nauk **121**, 1977, no. 3, 377–425 (Russian)

43.   Molev A., *Factorial supersymmetric Schur functions and super Capelli identities*. Kirillov's seminar on representation theory, Amer. Math. Soc. Transl. Ser. 2, **181**, Amer. Math. Soc., Providence, RI, 1998, 109–137; Molev A., Nazarov M., *Capelli identities for classical Lie algebras*, Math. Ann. **313** (1999), no. 2, 315–357

44.   Nishiyama K., *Oscillator representations of orthosymplectic algebras*, J. Alg., **129**, 1990, 231–262; Hayashi, T., *Q-analogue of Clifford and Weyl algebras–Spinor and oscillator representations of quantum envelopping algebras*, Comm. Math. Phys. **127**, 129–144 (1990)

45.   Panyushev D., *The exterior algebra and 'Spin' of an orthogonal g-module* math.AG/0001161

46.   Poletaeva E., *A spinor-like representation of the contact superconformal algebra $K'(4)$*, hep-th/0011100 (J. Math. Phys. 42, no. 1, 2001); *Semi-infinite cohomology and superconformal algebras*, Comptes Rendus de l'Académie des Sciences, **t. 326**, 1998, Série I, 533–538; *Superconformal algebras and Lie superalgebras of the Hodge theory*, preprint MPI 99–136.

47.   Rybnikov G. L., *Tensor invariants of the Lie algebra $sl_2(C[t])$ and fundamental representations of the Lie algebra $\widehat{sp}_{2n}$*. (Russian) Zap. Nauchn. Sem. Leningrad. Otdel. Mat. Inst. Steklov. (LOMI) 172 (1989), Differentsialnaya Geom. Gruppy Li i Mekh. **10**, 137–144; Rybnikov G. L., *Fermionic dual pairs of representations of loop groups*. (Russian) Functional

Anal. Appl. **26** (1992), no. 1, 61–62 Adamovich A. M., Rybnikov G. L., *Tilting modules for classical groups and Howe duality in positive characteristic* Transform. Groups **1** (1996), no. 1-2, 1–34

48. Shchepochkina I., *Maximal subalgebras of simple Lie superalgebras,* (hep-th 9702120)
49. Shchepochkina I., *The five exceptional simple Lie superalgebras of vector fields* (Russian) Funktsional. Anal. i Prilozhen. **33** (1999), no. 3,59–72, 96 translation in Funct. Anal. Appl. **33** (1999), no. 3, 208–219 (hep-th 9702121); id.., *The five exceptional simple Lie superalgebras of vector fields and their fourteen regradings.* Represent. Theory, (electronic) **3** (1999), 373–415
50. Shchepochkina I., Post G., *Explicit bracket in an exceptional simple Lie superalgebra,* Internat. J. Algebra Comput. **8** (1998), no. 4, 479–495 (physics/9703022)
51. Sergeev A., *An analog of the classical invariant theory for Lie superlagebras,* Functsional. Anal. i ego Prilozh., **26**, no.3, 1992, 88–90 (in Russian); an expanded version in: math.RT/9810113
52. Sergeev A., *Vector and covector invariants of Lie superalgebras,* Functsional. Anal. i ego Prilozh., **30**, no.3, 1996, 90–93 (in Russian); an expanded version in: math.RT/9904079
53. Verbitsky, M. *Action of the Lie algebra of* $SO(5)$ *on the cohomology of a hyper-Kähler manifold.* Functional Anal. Appl. **24** (1990), no. 3, 229–230
54. Voronov A., *Semi-infinite homological algebra,* Invent. math. **113**, 1993, 103–146
55. Weyl H., *The classical groups. Their invariants and representations.* Fifteenth printing. Princeton Landmarks in Mathematics. Princeton Paperbacks. Princeton University Press, Princeton, NJ, 1997. xiv+320 pp.

ENVELOPING ALGEBRA OF GL(3) AND ORTHOGONAL POLYNOMIALS

ALEXANDER SERGEEV [†][‡]

(Correspondence) Department of Mathematics, University of Stockholm, Roslagsv. 101, Kräftriket hus 6, S-106 91, Stockholm, Sweden; mleites@matematik.su.se (On leave of absence from Balakovo Institute of Technique of Technology and Control, Branch of Saratov Technical University, Balakovo, Saratov Region, Russia)

Abstract. Let A be an associative algebra over $\mathbb{C}$ and L an invariant linear functional on it (trace). Let ω be an involutive antiautomorphism of A such that $L(\omega(a)) = L(a)$ for any $a \in A$. Then A admits a symmetric invariant bilinear form $\langle a, b \rangle = L(a\omega(b))$. For $A = U(\mathfrak{sl}(2))/\mathfrak{m}$, where $\mathfrak{m}$ is any maximal ideal of $U(\mathfrak{sl}(2))$, Leites and I have constructed orthogonal basis whose elements turned out to be, essentially, Chebyshev and Hahn polynomials in one discrete variable.

Here I take $A = U(\mathfrak{gl}(3))/\mathfrak{m}$ for the maximal ideals $\mathfrak{m}$ which annihilate irreducible highest weight $\mathfrak{gl}(3)$-modules of particular form (generalizations of symmetric powers of the identity representation). In whis way we obtain multivariable analogs of Hahn polynomials. Clearly, one can similarly consider $\mathfrak{gl}(n)$ and $\mathfrak{gl}(m|n)$ instead of $\mathfrak{gl}(3)$ but the amount of calculations is appalling.

§1. Background

1.1. Lemma. *Let A be an associative algebra generated by a set X. Denote by $[X, A]$ the set of linear combinations of the form $\sum [x_i, a_i]$, where $x_i \in X$, $a_i \in A$. Then $[A, A] = [X, A]$.*

Proof. Let us apply the identity ([3], p.561)

$$[ab, c] = [a, bc] + [b, ca]. \tag{1.1.1}$$

Namely, let $a = x_1 \ldots x_n$; let us induct on n to prove that $[a, A] \subset [X, A]$. For $n = 1$ the statement is obvious. If $n > 1$, then $a = xa_1$, where $x \in X$ and due to (1.1.1) we have

$$[a, c] = [xa_1, c] = [x, a_1c] + [a_1, cx].$$

[‡] I am thankful to D. Leites for encouragement and help and to ESI, Vienna, for hospitality and support.

[†] `sergeev@bittu.org.ru`

S. Duplij and J. Wess (eds.), Noncommutative Structures in Mathematics and Physics, 113–124.
© 2001 *Kluwer Academic Publishers. Printed in the Netherlands.*

1.2. Lemma. *Let A be an associative algebra and $a \mapsto \omega(a)$ be its involutive antiautomorphism (transposition for $A = \text{Mat}(n)$). Let L be an invariant functional on A (like trace, i.e., $L([A, A]) = 0$) such that $L(\omega(a)) = L(a)$ for any $a \in A$. Define the bilinear form on A by setting*

$$\langle u, v \rangle = L(u\omega(v)) \text{ for any } u, v \in A. \tag{1.2.1}$$

Then
   i) $\langle u, v \rangle = \langle v, u \rangle$;
   ii) $\langle xu, v \rangle = \langle u, \omega(x)v \rangle$;
   iii) $\langle ux, v \rangle = \langle u, v\omega(x) \rangle$;
   iv) $\langle [x, u], v \rangle = \langle u, [\omega(x), v] \rangle$.
   Proof. (Clearly, iii) is similar to ii)).

   i) $\langle u, v \rangle = L(u\omega(v)) = L(\omega(u\omega(v))) = L(v\omega(u)) = \langle v, u \rangle$.

   ii) $\langle xu, v \rangle = L(xu\omega(v)) = L(u\omega(v)x) = L(u\omega(\omega(x)v)) = \langle u, \omega(x)v \rangle$.

   iv) $\langle [x, u], v \rangle = \langle xu, v \rangle - \langle ux, v \rangle$
        $\langle u, \omega(x)v \rangle - \langle u, v\omega(x) \rangle = \langle u, [\omega(x), v] \rangle$.

1.3. Traces and forms on $U(\mathfrak{g})$. Let $\mathfrak{g}$ be a finite dimensional Lie algebra, $Z(\mathfrak{g})$ the center of $(U(\mathfrak{g})$, W the Weyl group of $\mathfrak{g}$ and $\mathfrak{h}$ a Cartan subalgebra of $\mathfrak{g}$. The following statements are proved in [1].
   1.3.1. Proposition. i) $U(\mathfrak{g}) = Z(\mathfrak{g}) \oplus [U(\mathfrak{g}), U(\mathfrak{g})]$.
   ii) *Let $\sharp : Z(\mathfrak{g}) \oplus [U(\mathfrak{g}), U(\mathfrak{g})] \longrightarrow Z(\mathfrak{g})$ be the natural projection. Then*

$$(uv)^{\sharp} = (vu)^{\sharp} \text{ and } (zv)^{\sharp} = z(v)^{\sharp} \text{ for any } u, v \in U(\mathfrak{g}) \text{ and } z \in Z(\mathfrak{g}).$$

   iii) $U(\mathfrak{g}) = S(\mathfrak{h})^{W} \oplus [U(\mathfrak{g}), U(\mathfrak{g})]$.
   iv) *Let λ be the highest weight of the irreducible finite dimensional $\mathfrak{g}$-module L^{λ} and φ the Harish-Chandra homomorphism. Then*

$$\varphi(u^{\sharp})(\lambda) = \frac{\text{tr}(u|_{L^{\lambda}})}{\dim L^{\lambda}}.$$

1.3.2. On $U(\mathfrak{g})$, define a form with values in $Z(\mathfrak{g})$ by setting

$$\langle u, v \rangle = (u\omega(v))^{\sharp}, \tag{$*$}$$

where ω is the Chevalley involution in $U(\mathfrak{g})$.
   Lemma *The form $(*)$ is nondegenerate on $U(\mathfrak{g})$.*
   Proof. Let $\langle u, v \rangle = 0$ for any $v \in U(\mathfrak{g})$. By Proposition 1.3.1

$$\text{tr}(u\omega(v)) = \varphi((u\omega(v))^{\sharp})(\lambda) \cdot \dim L(\lambda) = \varphi(\langle u, v \rangle)(\lambda) \cdot \dim L(\lambda) = 0;$$

hence, $u = 0$ on $L(\lambda)$ for any irreducible finite dimensional $L(\lambda)$, and, therefore, $u = 0$ in $U(\mathfrak{g})$.

1.3.3. Lemma. *For any $\lambda \in \mathfrak{h}^*$ define a $\mathbb{C}$-valued form on $U(\mathfrak{g})$ by setting*

$$\langle u, v \rangle_\lambda = \varphi(\langle u, v \rangle)(\lambda).$$

The kernel of this form is a maximal ideal in $U(\mathfrak{g})$.

Proof. The form $\langle \cdot, \cdot \rangle_\lambda$ arises from a linear functional $L(u) = \varphi(u^\sharp)(\lambda)$; hence, by Lemma 1.2 its kernel is a twosided ideal I in $U(\mathfrak{g})$. On $A = U(\mathfrak{g})/I$, the form induced is nondegenerate. If $z \in Z(\mathfrak{g})$, then

$$\langle z, v \rangle_\lambda = L(z\omega(v)) = L(z)L(\omega(v));$$

hence, $z - L(z) \in I$. Therefore, the only $\mathfrak{g}$-invariant elements in A are those from Span (1).

Let J be a twosided nontrivial ($\neq A, 0$) ideal in A and $J = \bigoplus_\mu J^\mu$ be the decomposition into irreducible finite dimensional $\mathfrak{g}$-modules (with respect to the adjoint representation). Since $J \neq A$, it follows that $J^0 = 0$. Hence, $L(J) = 0$ and $\langle J, A \rangle$. Thus, $J = 0$.

1.4. Gelfand–Tsetlin basis and transvector algebras. (For recapitulation on transvector algebras see [7].)

Let E_{ij} be the matrix units. In $\mathfrak{gl}(3)$, we fix the subalgebra $\mathfrak{gl}(2)$ embedded into the left upper corner and let $\mathfrak{h}$ denote the Cartan subalgebra of $\mathfrak{gl}(3) = $ Span $(E_{ii} : i = 1, 2, 3)$.

There is a one-to-one correspondence between finite dimensional irreducible representations of $\mathfrak{gl}(3)$ and the sets

$$(\lambda_1, \lambda_2, \lambda_3) \text{ such that } \lambda_1 - \lambda_2, \lambda_2 - \lambda_3 \in \mathbb{Z}_+.$$

Such sets are called highest weights of the corresponding irreducible representation whose space is denoted L^λ. With each such λ we associate a Gelfand–Tsetlin diagram Λ:

$$\lambda_{31} \qquad \lambda_{32} \qquad \lambda_{33}$$
$$\lambda_{21} \qquad \lambda_{22} \qquad\qquad (1.4.1)$$
$$\lambda_{11}$$

where the upper line coincides with λ and where "betweenness" conditions hold:

$$\lambda_{k,i} - \lambda_{k-1,i} \in \mathbb{Z}_+; \quad \lambda_{k-1,i} - \lambda_{k,i+1} \in \mathbb{Z}_+ \text{ for any } i = 1, 2; \ k = 2, 3. \quad (1.4.2)$$

Set

$$z_{21} = E_{21}, \ z_{12} = E_{12}; \ z_{13} = E_{13}, \ z_{32} = E_{32};$$
$$z_{31} = (E_{11} - E_{22} + 2)E_{31} + E_{21}E_{32}, \qquad (1.4.3)$$
$$z_{23} = (E_{11} - E_{22} + 2)E_{23} - E_{21}E_{13}.$$

Set $(L^\lambda)^+ = \mathrm{Span}\,(u : u \in L^\lambda, E_{12}u = 0)$.

1.4.1. Theorem. (see [4]) *Let v be a nonzero highest weight vector in L^λ, and Λ a Gelfand–Tsetlin diagram. Set*

$$v_\Lambda = z_{21}^{\lambda_{21}-\lambda_{11}} z_{31}^{\lambda_{31}-\lambda_{21}} z_{32}^{\lambda_{32}-\lambda_{22}} v$$

and let $l_{ki} = \lambda_{ki} - i + 1$. Then

i) *The vectors v_Λ parametrized by Gelfand–Tsetlin diagrams form a basis in L^λ.*

ii) *The $\mathfrak{gl}(3)$-action on vectors v_Λ is given by the following formulas*

$$E_{11}v_\Lambda = \lambda_{11}v_\Lambda;$$

$$E_{22}v_\Lambda = (\lambda_{21} + \lambda_{22} - \lambda_{11})v_\Lambda;$$

$$E_{33}v_\Lambda = (\sum_{i=1}^{3} \lambda_{3i} - \sum_{j=1}^{2} \lambda_{2j})v_\Lambda;$$

$$E_{12}v_\Lambda = -(l_{11} - l_{21})(l_{11} - l_{22})v_{\Lambda+\delta_{11}};$$

$$E_{21}v_\Lambda = v_{\Lambda-\delta_{11}};$$

$$E_{23}v_\Lambda = -\frac{(l_{21} - l_{31})(l_{21} - l_{32})(l_{21} - l_{33})}{(l_{21} - l_{22})}v_{\Lambda+\delta_{11}} - \frac{(l_{22} - l_{31})(l_{22} - l_{32})(l_{22} - l_{33})}{(l_{22} - l_{21})}v_{\Lambda+\delta_{22}};$$

$$E_{32}v_\Lambda = \frac{(l_{21} - l_{11})}{(l_{21} - l_{22})}v_{\Lambda-\delta_{11}} + \frac{(l_{22} - l_{11})}{(l_{22} - l_{21})}v_{\Lambda-\delta_{22}},$$

where $\Lambda \pm \delta_{ki}$ is obtained from Λ by replacing λ_{ki} with $\lambda_{ki} \pm 1$ and we assume that $v_\Lambda = 0$ if Λ does not satisfy conditions on GTs-diagrams.

iii) *The vectors v_Λ corresponding to the GTs-diagrams with $\lambda_{21} = \lambda_{11}$ form a basis of $(L^\lambda)^+$.*

§2. Formulations of main results

2.1. Modules $S^\alpha(V)$. Let $\mathfrak{g} = \mathfrak{gl}(3)$ be the Lie algebra of 3×3 matrices over $\mathbb{C}$. For any $\alpha \in \mathbb{C}$ denote by $S^\alpha(V)$ the irreducible $\mathfrak{g}$-module with highest weight $(\alpha, 0, 0)$.

If $\alpha \in \mathbb{Z}_+$, then $S^\alpha(V)$ is the usual α-th symmetric power of the identity $\mathfrak{g}$-module V. Namely:

$$S^\alpha(V) = \mathrm{Span}\,(x_1^{k_1}x_2^{k_2}x_3^{k_3} : k_1 + k_2 + k_3 = \alpha;\ k_1, k_2, k_3 \in \mathbb{Z}_+).$$

For $\alpha \notin \mathbb{Z}_+$ we have (like in semi-infinite cohomology of Lie superalgebras)

$$S^\alpha(V) = \mathrm{Span}\,(x_1^{k_1}x_2^{k_2}x_3^{k_3} : k_1 + k_2 + k_3 = \alpha;\ k_2, k_3 \in \mathbb{Z}_+).$$

Remark. The expression x^k for $k \in \mathbb{C}$ is understood as a formal one, satisfying $\frac{\partial x^k}{\partial x} = kx^{k-1}$.

On $S^\alpha(V)$ the $\mathfrak{g} = \mathfrak{gl}(3)$-action is given by $E_{ij} \mapsto x_i \frac{\partial}{\partial x_j}$.

2.2. Theorem. i) $S^\alpha(V)$ *is an irreducible $\mathfrak{g}$-module for any α.*

ii) *The kernel J^α of the corresponding to $S^\alpha(V)$ representation of $U(\mathfrak{g})$ is a maximal ideal if $\alpha \notin \mathbb{Z}_{<0}$.*

Set $\mathfrak{A}^\alpha = U(\mathfrak{g})/J^\alpha$ and let θ be the highest weight of the adjoint representation of $\mathfrak{g}$. Now consider $\mathfrak{A}^\alpha$ as $\mathfrak{g}$-module with respect to the adjoint representation.

iii) $\mathfrak{A}^\alpha = \overset{\infty}{\underset{k=0}{\oplus}} L^{k\theta}$ *if $\alpha \notin \mathbb{Z}_{\geq 0}$.*

iv) $\mathfrak{A}^\alpha = \overset{\alpha}{\underset{k=0}{\oplus}} L^{k\theta}$ *if $\alpha \in \mathbb{Z}_{\geq 0}$.*

v) *The form $\langle u, v \rangle_\alpha = \varphi(u\omega(v)^\sharp)(\alpha, 0, 0)$ is nondegenerate on $\mathfrak{A}^\alpha$ for $\alpha \notin \mathbb{Z}_{<0}$.*

2.3. Let $\mathfrak{h} = \mathrm{Span}\,(E_{11}, E_{22}, E_{33})$ be Cartan subalgebra in $\mathfrak{g}$ and $\varepsilon_1, \varepsilon_2, \varepsilon_3$ the dual basis of $\mathfrak{h}^*$. Let $Q = \{\sum k_i \varepsilon_i : \sum k_i = 0\}$ be the root lattice of $\mathfrak{g}$. For any $\mu \in Q$ define

$$(\mathfrak{A}^\alpha)_\mu = \{u \in \mathfrak{A}^\alpha : [h, u] = \mu(h)u \quad \text{for any} \quad h \in \mathfrak{h}\}. \tag{2.3.1}$$

Clearly, $\mathfrak{A}^\alpha$ is Q-graded:

$$\mathfrak{A}^\alpha = \underset{\mu \in Q}{\oplus} (\mathfrak{A}^\alpha)_\mu.$$

Theorem 2.5 below shows that $(\mathfrak{A}^\alpha)_\mu = Ru_\mu$, where $u_\mu \in \mathfrak{A}^\alpha$ is defined uniquely up to a constant factor and $R = \mathbb{C}[E_{11}, E_{22}, E_{33}]/(E_{11} + E_{22} + E_{33} - \alpha)$.

Denote by $(\mathfrak{A}^\alpha)^+$ the subalgebra of $\mathfrak{g}i$ consisting of vectors highest with respect to the fixed $\mathfrak{gl}(2)$:

$$(\mathfrak{A}^\alpha)^+ = \{u \in \mathfrak{A}^\alpha : [E_{12}, u] = 0\}. \tag{2.3.2}$$

The algebra $(\mathfrak{A}^\alpha)^+$ also admits Q-grading:

$$(\mathfrak{A}^\alpha)^+ = \underset{\nu \in Q}{\oplus} (\mathfrak{A}^\alpha)^+_\nu. \tag{2.3.3}$$

Denote: $Q^+ = \{\nu \in Q : (\mathfrak{A}^\alpha)^+_\nu \neq 0\}$.

Theorem 2.4 below shows that $(\mathfrak{A}^\alpha)^+_\nu = \mathbb{C}[E_{33}]u^+_\nu$, where $\nu \in Q^+$. For $f, g \in \mathbb{C}[E_{33}]$ and $\nu \in Q^+$ set

$$\langle f, g \rangle^+_\nu = \langle fu^+_\nu, gu^+_\nu \rangle_\alpha. \tag{2.3.4}$$

For $f, g \in R$ and $\mu \in Q$ set

$$\langle f, g \rangle_\mu = \langle fu_\mu, gu_\mu \rangle_\alpha. \tag{2.3.5}$$

For $k \geq 0$ and $\nu \in Q^+$ set

$$f_{k,\nu}(E_{33})u_\nu = \begin{cases} (\mathrm{ad}\ z_{31})^k(u_{\nu+k(\varepsilon_1-\varepsilon_3)}) & \text{for } \nu(E_{33}) \leq 0 \\ (\mathrm{ad}\ z_{23})^k(u_{\nu+k(\varepsilon_3-\varepsilon_2)}) & \text{for } \nu(E_{33}) \geq 0 \end{cases} \qquad \begin{matrix} (2.3.6) \\ (2.3.7) \end{matrix}$$

For $k, l \geq 0$ and $\nu \in Q^+$ set

$$f^\nu_{l,k}(E_{11}, E_{22}, E_{33})u_\nu =$$
$$\begin{cases} (\mathrm{ad}\ z_{21})^l(\mathrm{ad}\ z_{31})^k(u_{\nu+k(\varepsilon_1-\varepsilon_3)+l(\varepsilon_1-\varepsilon_2)}) & \text{for } \nu(E_{33}) \leq 0 \\ (\mathrm{ad}\ z_{21})^l(\mathrm{ad}\ z_{23})^k(u_{\nu+k(\varepsilon_3-\varepsilon_2)+l(\varepsilon_1-\varepsilon_2)}) & \text{for } \nu(E_{33}) \geq 0 \end{cases} \qquad \begin{matrix} (2.3.8) \\ (2.3.9) \end{matrix}$$

2.4. Theorem. 0) $(\mathfrak{A}^\alpha)^+_\nu = \mathbb{C}[E_{33}]u^+_\nu$, where u_ν is determined uniquely up to a constant factor.

1) $\langle (\mathfrak{A}^\alpha)^+_\nu, \mathfrak{A}^\alpha)^+_\nu \rangle_\alpha = 0$ for $\nu \neq \mu$.

2) The polynomials $f_{k,\nu}(E_{33})$ are orthogonal relative $\langle \cdot, \cdot \rangle^+_\nu$.

3) The polynomials $f_{k,\nu}(E_{33})$ satisfy the difference equation

$$(E_{33} - \nu(E_{33}) + 1)(E_{33} + \nu(E_{11}) - \alpha)\Delta f - E_{33}(E_{33} + \nu(E_{22}) - \alpha - 2)\nabla f =$$
$$k(k + 2\nu(E_{11}) + 2)f \quad \text{if } \nu(E_{33}) < 0;$$
$$(E_{33} + 1)(E_{33} + \nu(E_{11}) - \alpha)\Delta f - (E_{33} - \nu(E_{33}))(E_{33} + \nu(E_{22}) - \alpha - 2)\nabla f =$$
$$k(k - 2\nu(E_{11}) + 2)f \quad \text{if } \nu(E_{33}) \geq 0.$$

4) Explilcitely, $f_{k,\nu}(E_{33})$ is of the form

$$f_{k,\nu}(E_{33}) = \text{const} \times {}_3F_2 \left(\begin{matrix} -k, & k + 2\nu(E_{11}) + 2, & -E_{33} \\ & 1 - \nu(E_{33}), & \nu(E_{11}) - \alpha \end{matrix} \,\middle|\, 1 \right),$$

where

$${}_3F_2 \left(\begin{matrix} \alpha_1, \alpha_2, \alpha_3 \\ \beta_1, \beta_2 \end{matrix} \,\middle|\, z \right) = \sum_{i=0}^\infty \frac{(\alpha_1)_i(\alpha_2)_i(\alpha_3)_i}{(\beta_1)_i(\beta_2)_i} \frac{z^i}{i!}$$

is a generalized hypergeometric function, $(\alpha)_0 = 1$ and $(\alpha)_i = \alpha(\alpha+1)\ldots(\alpha+i-1)$ for $i > 0$.

2.5. Theorem. 0) $(\mathfrak{A}^\alpha)_\nu = \mathbb{C}[E_{11}, E_{22}, E_{33}]u_\nu$, where u_ν is determined uniquely up to a constant factor.

1) $\langle (\mathfrak{A}^\alpha)_\nu, \mathfrak{A}^\alpha)_\nu \rangle_\alpha = 0$ for $\nu \neq \mu$.

2) The polynomials $f^\nu_{l,k}(E_{11}, E_{22}, E_{33})$ form an orthogonal basis of R relative $\langle \cdot, \cdot \rangle_\nu$.

3) The polynomials $w(f_{l,k})(E_{11}, E_{22}, E_{33})$ for $w \in W$ form an orthogonal basis of R relative $\langle \cdot, \cdot \rangle_{w(\nu)}$ provided polynomials $f_{l,k}(E_{11}, E_{22}, E_{33})$ form an orthogonal basis of R relative $\langle \cdot, \cdot \rangle_\nu$.

4) *The polynomials $f^\nu_{l,k}(E_{11}, E_{22}, E_{33})$ for $\nu \in Q^+$ and $\nu(E_{33}) \le 0$ satisfy the system of two difference equations (where $H_1 = E_{11} - E_{22}$, $H_2 = E_{33}$)*

$$[f(H_1 + 2, H_2) - f(H_1, H_2)] \cdot \tfrac{1}{4}(H_1 - H_2 + \alpha + 1)(H_1 + H_2 - \alpha) -$$
$$[f(H_1, H_2) - f(H_1 - 2, H_2)] \cdot \tfrac{1}{4}(H_1 - H_2 + \alpha - \nu(E_{11}))(H_1 + H_2 - \alpha - 1 +$$
$$\nu(E_{22})) = [l^2 + l(\nu(E_{11}) + \nu(E_{22}) + 1)) + \nu(E_{22}) - \nu(E_{11})]f;$$

$$[2\alpha - \nu(H_2)(\alpha + 2 + \nu(H_2)) + H_2(2\alpha + 1 + 2\nu(H_2)) - 2H_2^2]f(H_1, H_2) -$$
$$\tfrac{1}{2}(H_2 + 1 - \nu(H_2))(H_1 - H_2 + \alpha - 2\nu(E_{11}))f(H_1 - 1, H_2 + 1) -$$
$$\tfrac{1}{2}H_2(H_1 - H_2 + \alpha + 2)f(H_1 + 1, H_2 + 1) -$$
$$\tfrac{1}{2}(H_2 + 1 - \nu(H_2))(\alpha - H_1 - H_2)f(H_1 + 1, H_2 + 1) -$$
$$\tfrac{1}{2}H_2(\alpha - H_1 - H_2 + 2 - 2\nu(E_{22}))f(H_1 - 1, H_2 - 1) =$$
$$[2k^2 + 4kl + 4k(1 + \nu(E_{11})) + 2l(1 + \nu(E_{11}) - \nu(E_{22})) +$$
$$\nu(E_{11})^2 - \nu(E_{22})^2 + 4\nu(E_{11})]f(H_1, H_2).$$

§3. Proof of Theorem 2.2

i) The module $S^\alpha(V)$ is irreducible if and only if it has no vacuum vectors (i.e, vectors annihilated by E_{12} and E_{23}. This is subject to a direct verification.

   ii) Follows from Exercise 858 of Ch. 8 of [1].

   iii) Let A_3 be the Weyl algebra (i.e., it is generated by the p_i and q_i for $i = 1, 2, 3$ satisfying

$$p_i p_j - p_j p_i = q_i q_j - q_j q_i = 0; \quad p_i q_j - q_j p_i = -\delta_{ij}. \tag{3.1}$$

Setting $E_{ij} \mapsto p_i q_j$ we see that the homomorphism $\varphi : U(\mathfrak{g}) \longrightarrow \text{End}(S^\alpha(V))$ factors through A_3 and A_3 acts on $S^\alpha(V)$ so that $p_i \mapsto x_i$ and $q_i \mapsto \frac{\partial}{\partial x_i}$. Let us describe the image of φ. To this end, on A_3, introcude a grading by setting

$$\deg p_i = 1 \quad \deg q_i = -1 \text{ for } i = 1, 2, 3. \tag{3.2}$$

Now it is clear that Im φ is the algebra B_3 of elements of degree 0.

   To describe highest weight elements in B_3, it suffices to describe same in $S^k(V) \otimes S^k(V^*)$. Let us identify $S^k(V) \otimes S^k(V^*)$ with End $(S^k(V))$, let $u \in$ End $(S^k(V))$ commutes with the action of E_{12} and E_{23} on $S^k(V)$. But then u is uniquely determined by its value on the lowest weight vector $x_3^k \in S^k(V)$; moreover, $E_{12}x_3^k = 0$. Hence,

$$u(x_3^k) = a_0 x_3^k + \sum_{i=0}^{k} a_i x_1^i x_3^{k-i},$$

so

$$u(x_3^k) = \frac{1}{k}a_0\left(\sum_{i=0}^{k} x_i\frac{\partial}{\partial x_i}\right)x_3^k + \sum_{i=0}^{k} \frac{(k-i)!}{k!}a_i\left(x_1\frac{\partial}{\partial x_3}\right)^i x_3^k.$$

This shows that the algebra of highest weight vectors in B_3 is generated by p_1q_3 and $z = p_1q_1 + p_2q_2 + p_3q_3$. If $\alpha \notin \mathbb{Z}_{\geq 0}$, then $\mathfrak{A}_\alpha$ is the quotient of B_3 modulo $(z - \alpha)$. This proves iii).

iv) In this case $\mathfrak{A}_\alpha = \mathrm{End}\,(S^k(V))$ and the proof follows from the arguments at the end of the above paragraph.

v) By 1.3.3 the kernel of $\langle\cdot,\cdot\rangle_\alpha$ in $U(\mathfrak{g})$ is a maximal ideal. But $\mathfrak{A}_\alpha = U(\mathfrak{g})/J^\alpha$, where J^α is maximal due to i). So J^α coincides with the kernel of $\langle\cdot,\cdot\rangle_\alpha$ in $U(\mathfrak{g})$ and the form is nondegenerate on $\mathfrak{A}_\alpha$.

§4. Proof of Theorem 2.4

0) Direct computations show that the set of elements from A_3 commuting with E_{12} is a subalgebra generated by p_1, q_2, p_3, q_3 and $z = p_1q_1 + p_2q_2 + p_3q_3$. So this algebra is the linear span of the elements of the form

$$u = p_1^{k_1} q_2^{k_2} p_3^{k_3} q_3^{k_4} z^{k_5}.$$

If $u \in B_3$, then $k_1 + k_3 = k_2 + k_4$, so

$$u = \begin{cases} p_1^{k_1} q_2^{k_2} p_3^{k_3-k_4} p_3^{k_4} q_3^{k_4} z^{k_5} & \text{if } k_3 \geq k_4 \\ p_1^{k_1} q_2^{k_2} p_3^{k_3} q_3^{k_3} q_3^{k_4-k_3} z^{k_5} & \text{if } k_3 \leq k_4. \end{cases} \qquad (4.1)$$

Hence, setting for $\nu = \sum k_i\varepsilon_i$ such that $\sum k_i = 0$, $k_1 \geq 0$ and $k_2 \leq 0$

$$u_\nu^+ = \begin{cases} p_1^{k_1} q_2^{-k_2} p_3^{k_3} & \text{if } k_3 \geq 0 \\ p_1^{k_1} q_2^{-k_2} q_3^{-k_3} & \text{if } k_3 \leq 0 \end{cases} \qquad (4.2)$$

we obtain the statement desired.

1) Let $u \in (\mathfrak{A}_\alpha^+)_\mu$, $v \in (\mathfrak{A}_\alpha^+)_\nu$, and $h \in \mathfrak{h}$. Then by heading iv) of Lemma 1.2 we obtain:

$$\langle[h,u],v\rangle = \mu(h)\langle u,v\rangle = \langle u,[h,v]\rangle = \nu(h)\langle u,v\rangle.$$

So $\langle u,v\rangle = 0$ if $\mu \neq \nu$.

2) Let $\nu(E_{33}) \leq 0$. We have:

$$f_{k,\nu}u_\nu^+ = (\mathrm{ad}\,z_{31})^k(u_{\nu+k(\varepsilon_1-\varepsilon_3)}^+) = \mathrm{ad}\,z_{31}(\mathrm{ad}\,z_{31})^{k-1}(u_{\nu+(k-1)(\varepsilon_1-\varepsilon_3)+(\varepsilon_1-\varepsilon_3)}^+) =$$
$$(\mathrm{ad}\,z_{31})f_{k-1,\nu+(\varepsilon_1-\varepsilon_3)}u_{\nu+(\varepsilon_1-\varepsilon_3)}^+.$$

Direct verification shows that (here $h = E_{33}$)

$$(\text{ad } z_{31})(fu_\nu^+) = \{(E_{33} - \nu(E_{33}))[(E_{33} - \alpha)\nu(E_{11}) + (\nu(E_{11}) - 1)\nu(E_{11})]f(h) - E_{33}[(E_{33} - \alpha)\nu(E_{11}) - (\nu(E_{22}) + 2)\nu(E_{11})]f(h-1)\}u_{\nu-(\varepsilon_1-\varepsilon_3)}.$$

$$(4.3)$$

It easily follows from Lemma 1.2 that for any $z \in U(\mathfrak{g})$ we have

$$\langle(\text{ad } z)(u), v\rangle = \langle u, (\text{ad } \omega(z))(v)\rangle,$$

but

$$\omega(z_{31}) = \omega((E_{11} - E_{22} + 2)E_{31} + E_{21}E_{32}) = E_{13}(E_{11} - E_{22} + 2) + E_{23}E_{12}.$$

Since fu_ν is a highest weight vector with respect to the fixed $\mathfrak{gl}(2)$, it follows that

$$(\text{ad } \omega(z_{31}))(fu_\nu) = (\text{ad } (E_{13}(E_{11} - E_{22} + 2)))(fu_\nu) =$$
$$(\nu(E_{11}) - \nu(E_{22}) + 2)\Delta f \cdot u_{\nu+(\varepsilon_1-\varepsilon_3)}.$$

Now, let us induct on k. For $k = 0$ the statement is obvious. For $k > 0$ and $\deg g < k$ we have

$$\langle f_{k,\nu}, g\rangle_\nu = \langle f_{k,\nu}u_\nu^+, gu_\nu^+\rangle =$$
$$\langle f_{k-1,\nu+(\varepsilon_1-\varepsilon_3)}u_{\nu+(\varepsilon_1-\varepsilon_3)}^+, (\text{ad } (\omega(z_{31})))(gu_\nu^+)\rangle =$$
$$\langle f_{k-1,\nu+(\varepsilon_1-\varepsilon_3)}, (\nu(E_{11}) - \nu(E_{22}) + 2)\Delta g\rangle_{\nu+(\varepsilon_1-\varepsilon_3)} = 0$$

by inductive hypothesis.

The case $\nu(E_{33}) \geq 0$ is similar.

3) Observe that $z = E_{13}E_{31} + E_{23}E_{32}$ belongs to the centralizer of $\mathfrak{gl}(2)$ in $U(\mathfrak{g})$. Let $\nu(E_{33}) \leq 0$. Then $u_\nu^+ = p_1^{k_1}q_2^{k_2}q_3^{k_3}$ as in (4.1.2). Having applied ad z

to fu_ν^+ we obtain:

$$(\text{ad } z)(fu_\nu^+) = E_{13}E_{31}fu_\nu^+ + fu_\nu^+E_{31}E_{13} - E_{13}fu_\nu^+E_{31} - E_{31}fu_\nu^+E_{13}+$$
$$E_{23}E_{32}fu_\nu^+ + fu_\nu^+E_{32}E_{23} - E_{23}fu_\nu^+E_{32} - E_{32}fu_\nu^+E_{23} =$$
$$E_{11}(E_{33}+1)fu_\nu^+ + fu_\nu^+E_{33}(E_{11}+1)-$$
$$f(E_{33}+1)u_\nu^+E_{11}(E_{33}+1) - f(E_{33}-1)E_{33}(E_{11}+1)u_\nu^+ + E_{22}(E_{33}+1)fu_\nu^+ +$$
$$fu_\nu^+E_{33}(E_{22}+1) - f(E_{33}+1)E_{22}u_\nu^+(E_{33}+1) - f(E_{33}-1)E_{33}u_\nu^+(E_{22}+1) =$$
$$(E_{11}+E_{22})(E_{33}+1)fu_\nu^+ +$$
$$(E_{33}-\nu(E_{33}))(E_{11}+1-\nu(E_{11})+E_{22}+1-\nu(E_{22}))fu_\nu^+ -$$
$$f(E_{33}+1)\cdot(E_{33}+1-\nu(E_{33}))(E_{11}+E_{22}-\nu(E_{11}))u_\nu^+ -$$
$$f(E_{33}-1)E_{33}(E_{11}+E_{22}-\nu(E_{22})+2)u_\nu^+ =$$
$$[f(E_{33}+1)\cdot(E_{33}+1-\nu(E_{33}))(E_{33}-\alpha+\nu(E_{11}))+$$
$$f(E_{33}-1)E_{33}(E_{33}+\nu(E_{22})-\alpha-2)-$$
$$(E_{33}-\alpha)(E_{33}+1)f - (E_{33}-\nu(E_{33}))(E_{33}+\nu(E_{11})+\nu(E_{22})-\alpha-2)f]u_\nu^+.$$

This gives us the right hand side of the first equation of heading 3).

Since ad z commutes with the $\mathfrak{gl}(2)$-action and preserves the degree of polynomial f, it follows that $(\text{ad } z)(fu_\nu) = c\cdot(fu_\nu)$. Counting the constant factor, we arrive to the first equation of heading 3).

The proof of the second equation is similar.

§5. Proof of Theorem 2.5

0) Recall that B_3 is the subalgebra of A_3 of the elements of degree 0 relative grading (3.2).

For $k \in \mathbb{Z}$ set $r_i^k = \begin{cases} = p_i^k & \text{if } k \geq 0 \\ , q_i^{-k} & \text{if } k \leq 0 \end{cases}$. For $\gamma = \sum k_i\varepsilon_i$, where $\sum k_i = 0$, set

$$u_\gamma = r_1^{k_1}r_2^{k_2}r_3^{k_3}.$$

Clearly, B_3 is the linear span of the elements of the form

$$p_1^{m_1}q_1^{l_1}p_2^{m_2}q_2^{l_2}p_3^{m_3}q_3^{l_3}, \quad \text{where } m_1 + m_2 + m_3 = l_1 + l_2 + l_3.$$

It is also clear that each such element can be represented in the form

$$f(E_{11},E_{22},E_{33})r_1^{k_1}r_2^{k_2}r_3^{k_3}.$$

This completes the proof of heading 0).

1) Proof is similar to that from sec. 4.2.

2) Let $\nu(E_{33}) \leq 0$. By setting $H_1 = E_{11} - E_{22}$, $H_2 = E_{22} - E_{33}$ we identify $R = \mathbb{C}[E_{11}, E_{22}, E_{33}]/(E_{11} + E_{22}, +E_{33} - \alpha)$ with $\mathbb{C}[H_1, H_2]$. Let Λ is a Gelfand–Tsetlin diagram of the following form:

$$\nu(E_{11}) + k + l \qquad\qquad 0 \qquad\qquad -(\nu(E_{11}) + k + l)$$
$$\nu(E_{11}) + l \qquad\qquad -(\nu(E_{22}) + l)$$
$$\nu(E_{11})$$

From the explicit formula for $f_{k,l}^{\nu}$ we derive that

$$f_{k,l}^{\nu} u_{\nu} = v_{\Lambda}. \tag{5.1}$$

Now, consider the following operators from the maximal commutative subalgebra of $U(\mathfrak{g})$:

$$E_{11}, E_{22},$$
$$\Omega_2 = E_{11}^2 + E_{22}^2 + E_{11} - E_{22} + 2E_{21}E_{12}, \tag{5.2}$$
$$\Omega_3 = E_{11}^2 + E_{22}^2 + E_{33}^2 + E_{11} - E_{22} + E_{11} - E_{33} + E_{22} - E_{33} +$$
$$2E_{21}E_{12} + 2E_{31}E_{13} + 2E_{32}E_{23}.$$

Then we have:

$$E_{11}v_{\Lambda} = \nu(E_{11})v_{\Lambda}; \quad E_{22}v_{\Lambda} = -\nu(E_{22})v_{\Lambda};$$
$$\Omega_2 v_{\Lambda} = [2l^2 + 2l(\nu(E_{11}) + \nu(E_{22}) + 1) + \tag{5.3}$$
$$\nu(E_{11})^2 + \nu(E_{22})^2]v_{\Lambda};$$
$$\Omega_3 v_{\Lambda} = 2(\nu(E_{11}) + k + l)(\nu(E_{11}) + k + l + 2]v_{\Lambda}.$$

It is easy to check that the operators (5.2) satisfy

$$\omega(E_{11}) = E_{11}; \quad \omega(E_{22}) = E_{22}; \quad \omega(\Omega_2) = \Omega_2; \quad \omega(\Omega_3) = \Omega_3$$

and, therefore, they are selfadjoint relative the form $\langle \cdot, \cdot \rangle$. Formula (5.3) makes it manifest that operators (5.2) separate the vectors v_{Λ}, hence, these vectors are pairwise orthogonal. Moreover, it is easy to see that $f_{k,l}^{\nu}$ is of the form

$$f_{k,l}^{\nu} = H_1^l H_2^k + \cdots,$$

where the dots designate the summands of degrees $\leq k + l$ of the form $H_1^a H_2^b$, where $(a, b) < (l, k)$ with respect to the lexicographic ordering. Thus, the $f_{l,k}^{\nu}$ constitute a basis of $\mathbb{C}[H_1, H_2]$.

3) The statement follows from the fact that the Weyl group acts on $\mathfrak{A}_\alpha$ and preserves the form $\langle \cdot, \cdot \rangle$.

4) Since the polynomials $f_{l,k}^\nu u_\nu$ are elements of a Gelfand–Tsetlin basis, they are eigenvectors for Ω_2 and Ω_3 with respect to the adjoint action of $\mathfrak{g} = \mathfrak{gl}(3)$ on $\mathfrak{A}_\alpha$. As we have shown in sec 5.2, we have

$$\Omega_2 f_{l,k}^\nu u_\nu = [2l^2 + 2l(\nu(E_{11}) + \nu(E_{22}) + 1) +$$
$$\nu(E_{11})^2 + \nu(E_{22})^2] f_{l,k}^\nu u_\nu;$$
$$\Omega_3 f_{l,k}^\nu u_\nu = 2(\nu(E_{11}) + k + l)(\nu(E_{11}) + k + l + 2] f_{l,k}^\nu u_\nu.$$

To derive the corresponding equations, we have to explicitly compute the actions of Ω_2 and Ω_3 on $f u_\nu$. These straightforward computations imply the second equation.

References

1. Dixmier J. *Algèbres envellopentes*, Gautier-Villars, Paris, 1974; *Enveloping algebras*, AMS, 1996

2. Leites D., Sergeev A., *Orthogonal polynomials of discrete variable and Lie algebras of complex size matrices*. Theor. and Math. Physics, **123** 2000, no.2, 582–609

3. Montgomery S., *Constructing simple Lie superalgebras from associative graded algebras*. J. Algebra **195** (1997), no. 2, 558–579

4. Molev A.I., *Yangians and transvector algebras*. Math.RT/9902060

5. Nikiforov A. F., Suslov S. K., Uvarov V. B. *Classical orthogonal polynomials of a discrete variable*. Translated from the Russian. Springer Series in Computational Physics. Springer-Verlag, Berlin, 1991. xvi+374 pp.

6. Pinczon G., *The enveloping algebra of the Lie superalgebra* $\mathfrak{osp}(1|2)$. J. Algebra, **132** (1990), 219–242

7. Zhelobenko D., *Predstavleniya reduktivnykh algebr Li*. (Russian) [Representations of reductive Lie algebras] Nauka, Moscow, 1994. 352 pp.; Zhelobenko D., Shtern A., *Predstavleniya grupp Li*. (Russian) [Representations of Lie groups] Spravochnaya Matematicheskaya Biblioteka. [Mathematical Reference Library], Nauka, Moscow, 1983. 360 pp.

NONINVERTIBILITY, SEMISUPERMANIFOLDS AND CATEGORIES REGULARIZATION

STEVEN DUPLIJ * †
Kharkov National University, Kharkov 61001, Ukraine

WŁADYSŁAW MARCINEK
Institute of Theoretical Physics, University of Wrocław, Pl. Maxa Borna 9, 50-204 Wrocław, Poland

Abstract. The categories with noninvertible morphisms are studied analogously to the semisupermanifolds with noninvertible transition functions. The concepts of regular n-cycles, obstruction and the regularization procedure are introduced and investigated. It is shown that the regularization of a category with noninvertible morphisms and obstruction form a 2-category. The generalization of some related structures to the regular case is given.

1. Introduction

In the supermanifold noninvertible generalization approach [1–3] we study here the obstructed cocycle conditions in the category theory framework and extend them to such structures as categories, functions, (co-) algebras, (co-) modules etc. This approach is connected with the higher regularity concept [4] and reconsidering the role of identities [5]. The introduced category regularization together with obstruction form a 2-category. Similar abstract structure generalizations were considered in topological QFT [6, 7], for n-categories [8–10], near-group categories [11, 12] (with noninvertible elements) and weak Hopf algebras [13, 14] in which the counit does not satisfy $\varepsilon(ab) = \varepsilon(a)\varepsilon(b)$ or satisfy first order (in our classification) regularity conditions [15, 16]. We first show how to deal with noninvertibility in the supermanifold theory [17, 18] and then apply this approach to more general structures.

* steven.a.duplij@univer.kharkov.ua

† WWW: http://gluon.physik.uni-kl.de/~duplij

S. Duplij and J. Wess (eds.), Noncommutative Structures in Mathematics and Physics, 125–139.
© 2001 *Kluwer Academic Publishers. Printed in the Netherlands.*

2. Supermanifolds and semisupermanifolds

In the supermanifold theory [17–19] the phenomenon of noninvertibility obviously arises from odd nilpotent elements and zero divisors of Grassmann algebras (also in the infinite dimensional case [20]). Despite the invertibility question is quite natural, the answer is not so simple and in some cases can be nontrivial, e.g. in some superalgebras one can introduce invertible analog of an odd symbol [21], or construct elements without number part which are not nilpotent even topologically [22]. Several guesses concerning inner noninvertibility inherent in the supermanifold theory were made before, e.g. "...there may be no inverse projection[0] at all" [23], "...a general SRS needs not have a body[0] " [24], or "...a body[0] may not even exist in the most extreme examples" [25]. It were also considered pure odd supermanifolds [26, 27] which give an important counterexample to the Coleman-Mandula theorem "...and provides us with a new, missed so far, version of the Poincaré supergroup" [28], exotic supermanifolds with nilpotent even coordinates [29] and supergravity with noninvertible vierbein [30]. Some problems with odd directions and therefore connected with noninvertibility in either event are described in [31, 32], and a perspective list of supermanifold problems was stated by D. Leites in [33].

The patch definition of a supermanifold $\mathfrak{M}_0$ in most cases differs from the patch definition of an ordinary manifold [34, 35] by "super-" terminology only and is well-known [36]. Let $\bigcup_\alpha \{U_\alpha, \varphi_\alpha\}$ is an atlas of a *supermanifold* $\mathfrak{M}_0$, then its gluing transition functions $\Phi_{\alpha\beta} = \varphi_\alpha \circ \varphi_\beta^{-1}$ satisfy the cocycle conditions

$$\Phi_{\alpha\beta}^{-1} = \Phi_{\beta\alpha}, \qquad \Phi_{\alpha\beta} \circ \Phi_{\beta\gamma} \circ \Phi_{\gamma\alpha} = 1_{\alpha\alpha} \tag{1}$$

on overlaps $U_\alpha \cap U_\beta$ and on triple overlaps $U_\alpha \cap U_\beta \cap U_\gamma$ respectively, where $1_{\alpha\alpha} \overset{\text{def}}{=} id\,(U_\alpha)$. To obtain a patch definition of an object analogous to supermanifold we try to weaken demand of invertibility of coordinate maps φ_α. Consider a generalized superspace $\mathfrak{M}$ covered by open sets U_α as $\mathfrak{M} = \bigcup_\alpha U_\alpha$. We assume here that the maps $\varphi_\alpha : U_\alpha \to V_\alpha \subset \mathbb{R}^{n|m}$ are not all homeomorphisms, i.e. among them there are noninvertible maps[1].

Definition 1. A *semisupermanifold* is a noninvertibly generalized superspace $\mathfrak{M}$ represented as a semiatlas $\mathfrak{M} = \bigcup_\alpha \{U_\alpha, \varphi_\alpha\}$ with invertible and noninvertible coordinate maps $\varphi_\alpha : U_\alpha \to V_\alpha \subset \mathbb{R}^{n|m}$.

We do not concretize here the details, how the invertibility appears here, but instead we will describe it by some general relations between semitransition

---

[0] number part.

[1] Under $\mathbb{R}^{n|m}$ we imply some its noninvertible generalization [3].

functions and other objects. We The noninvertibly extended gluing *semitransition functions* of a semisupermanifold are defined by the equations

$$\Phi_{\alpha\beta} \circ \varphi_\beta = \varphi_\alpha, \qquad \Phi_{\beta\alpha} \circ \varphi_\alpha = \varphi_\beta \tag{2}$$

instead of $\Phi_{\alpha\beta} = \varphi_\alpha \circ \varphi_\beta^{-1}$, which obviously extends the class of functions to non-invertible ones. Then we assume that instead of (1) the semitransition functions $\Phi_{\alpha\beta}$ of a semisupermanifold $\mathfrak{M}$ satisfy the following relations

$$\Phi_{\alpha\beta} \circ \Phi_{\beta\alpha} \circ \Phi_{\alpha\beta} = \Phi_{\alpha\beta} \tag{3}$$

on $U_\alpha \cap U_\beta$ overlaps (invertibility is extended to regularity) and

$$\Phi_{\alpha\beta} \circ \Phi_{\beta\gamma} \circ \Phi_{\gamma\alpha} \circ \Phi_{\alpha\beta} = \Phi_{\alpha\beta}, \tag{4}$$
$$\Phi_{\beta\gamma} \circ \Phi_{\gamma\alpha} \circ \Phi_{\alpha\beta} \circ \Phi_{\beta\gamma} = \Phi_{\beta\gamma}, \tag{5}$$
$$\Phi_{\gamma\alpha} \circ \Phi_{\alpha\beta} \circ \Phi_{\beta\gamma} \circ \Phi_{\gamma\alpha} = \Phi_{\gamma\alpha} \tag{6}$$

on triple overlaps $U_\alpha \cap U_\beta \cap U_\gamma$ and

$$\Phi_{\alpha\beta} \circ \Phi_{\beta\gamma} \circ \Phi_{\gamma\rho} \circ \Phi_{\rho\alpha} \circ \Phi_{\alpha\beta} = \Phi_{\alpha\beta}, \tag{7}$$
$$\Phi_{\beta\gamma} \circ \Phi_{\gamma\rho} \circ \Phi_{\rho\alpha} \circ \Phi_{\alpha\beta} \circ \Phi_{\beta\gamma} = \Phi_{\beta\gamma}, \tag{8}$$
$$\Phi_{\gamma\rho} \circ \Phi_{\rho\alpha} \circ \Phi_{\alpha\beta} \circ \Phi_{\beta\gamma} \circ \Phi_{\gamma\rho} = \Phi_{\gamma\rho}, \tag{9}$$
$$\Phi_{\rho\alpha} \circ \Phi_{\alpha\beta} \circ \Phi_{\beta\gamma} \circ \Phi_{\gamma\rho} \circ \Phi_{\rho\alpha} = \Phi_{\rho\alpha} \tag{10}$$

on $U_\alpha \cap U_\beta \cap U_\gamma \cap U_\rho$. We can write similar cycle relations to infinity and call them *tower relations* which satisfy identically in the standard invertible case [36].

REMARK 1. In any actions with noninvertible functions $\Phi_{\alpha\beta}$ we are not allowed to cancel by them, because the semigroup of $\Phi_{\alpha\beta}$'s is a semigroup without cancellation, and we are forced to exploit the corresponding semigroup methods [37, 38].

Conjecture 2. *The functions $\Phi_{\alpha\beta}$ satisfying the relations (3)–(10) can be viewed as some noninvertible generalization of the transition functions as cocycles in the corresponding Čech cohomology of coverings [39, 40].*

3. Obstructedness and additional orientation on semisupermanifolds

The semisupermanifolds defined above belong to a class of so called obstructed semisupermanifolds [1, 3] in the following sense. Let us rewrite relations (1) as the infinite series

$$n = 1: \ \Phi_{\alpha\alpha} = 1_{\alpha\alpha}, \tag{11}$$

$$n = 2: \quad \Phi_{\alpha\beta} \circ \Phi_{\beta\alpha} = 1_{\alpha\alpha}, \tag{12}$$

$$n = 3: \quad \Phi_{\alpha\beta} \circ \Phi_{\beta\gamma} \circ \Phi_{\gamma\alpha} = 1_{\alpha\alpha}, \tag{13}$$

$$n = 4: \quad \Phi_{\alpha\beta} \circ \Phi_{\beta\gamma} \circ \Phi_{\gamma\delta} \circ \Phi_{\delta\alpha} = 1_{\alpha\alpha} \tag{14}$$

$$\ldots \quad \ldots$$

Definition 3. A semisupermanifold is called *obstructed*, if some of the cocycle conditions (11)–(14) are broken.

It can happen that starting from some $n = n_m$ all higher cocycle conditions hold valid.

Definition 4. *Obstructedness degree* of a semisupermanifold is a maximal n_m for which the cocycle conditions (11)–(14) are broken. If all of them hold valid, then $n_m \overset{\text{def}}{=} 0$.

Obviously, that ordinary manifolds [35] (with invertible transition functions) have vanishing obstructedness, and the obstructedness degree for them is equal to zero, i.e. $n_m = 0$.

REMARK 2. The obstructed semisupermanifolds may have nonvanishing ordinary obstruction which can be calculated extending the standard methods [17] to the noninvertible case.

Therefore, using the obstructedness degree n_m, we have possibility to classify semisupermanifolds properly. Moreover, the pure soul supernumbers do not contain unity. Obviously that obstructed semisupermanifolds cannot have identity semitransition functions.

The orientation of ordinary manifolds is determined by the Jacobian sign of transition functions $\Phi_{\alpha\beta}$ written in terms of local coordinates on $U_\alpha \cap U_\beta$ overlaps [34, 35]. Since this sign belong to $\mathbb{Z}_2$, there exist two orientations on U_α. Two overlapping charts are *consistently oriented* (or *orientation preserving*) if $\Phi_{\alpha\beta}$ has positive Jacobian, and a manifold is *orientable* if it can be covered by such charts, thus there are two kinds of manifolds: orientable and nonorientable [35]. In supersymmetric case the role of Jacobian plays Berezinian [17] which has a "sign" belonging to $\mathbb{Z}_2 \oplus \mathbb{Z}_2$, and so there are four orientations on U_α and five corresponding kinds of supermanifold orientability [41, 42].

Definition 5. In case a nonvanishing Berezinian of $\Phi_{\alpha\beta}$ is nilpotent (and so has no definite sign in the previous sense) there exists additional *nilpotent orientation* on U_α of a semisupermanifold.

A degree of nilpotency of Berezinian allows us to classify semisupermanifolds having nilpotent orientability (see e.g. [43, 44]).

4. Higher regularity and obstruction

The above constructions have the general importance for *any* set of noninvertible mappings. The extension of $n = 2$ cocycle given by (3) can be viewed as some analogy with regular [45] or pseudoinverse [46] elements in semigroups or generalized inverses in matrix theory [47], category theory [48] and theory of generalized inverses of morphisms [49]. The relations (4)–(10) and with other n can be considered as noninvertible analogue of regularity for higher cocycles. Therefore, by analogy with (3)–(10) it is natural to formulate the general

Definition 6. An noninvertible mapping $\Phi_{\alpha\beta}$ is n-regular, if it satisfies on

overlaps $\overbrace{U_\alpha \cap U_\beta \cap \ldots \cap U_\rho}^{n}$ to the following conditions

$$\overbrace{\Phi_{\alpha\beta} \circ \Phi_{\beta\gamma} \circ \ldots \circ \Phi_{\rho\alpha} \circ \Phi_{\alpha\beta}}^{n+1} = \Phi_{\alpha\beta} + perm. \tag{15}$$

The formula (3) describes 3-regular mappings, the relations (4)–(6) correspond to 4-regular ones, and (7)–(10) give 5-regular mappings. Obviously that 3-regularity coincides with the ordinary regularity.

Let us consider a series of the selfmaps $e_{\alpha\alpha}^{(n)} : U_\alpha \to U_\alpha$ of a semisupermanifold defined as

$$e_{\alpha\alpha}^{(1)} = \Phi_{\alpha\alpha}, \tag{16}$$

$$e_{\alpha\alpha}^{(2)} = \Phi_{\alpha\beta} \circ \Phi_{\beta\alpha}, \tag{17}$$

$$e_{\alpha\alpha}^{(3)} = \Phi_{\alpha\beta} \circ \Phi_{\beta\gamma} \circ \Phi_{\gamma\alpha}, \tag{18}$$

$$e_{\alpha\alpha}^{(4)} = \Phi_{\alpha\beta} \circ \Phi_{\beta\gamma} \circ \Phi_{\gamma\delta} \circ \Phi_{\delta\alpha} \tag{19}$$

$$\ldots \ldots$$

We will call $e_{\alpha\alpha}^{(n)}$'s *tower identities (or obstruction of U_α)*. From (11)–(14) it follows that for ordinary supermanifolds obstruction coincide with the usual identity map

$$e_{\alpha\alpha}^{(n),ordinary} = 1_{\alpha\alpha}. \tag{20}$$

So the obstructedness degree can be treated as a maximal $n = n_m$ for which tower identities differ from the identity, i.e. (20) is broken. The obstruction gives the numerical measure of distinction of a semisupermanifold from an ordinary supermanifold. When morphisms are noninvertible (a semisupermanifold has a nonvanishing obstructedness), we cannot "return to the same point", because in general $e_{\alpha\alpha}^{(n)} \neq 1_{\alpha\alpha}$, and we have to consider "nonclosed" diagrams due to the fact that the relation $e_{\alpha\alpha}^{(n)} \circ \Phi_{\alpha\beta} = \Phi_{\alpha\beta}$ is noncancellative now (see REMARK 1).

Summarizing the above statements we propose the following intuitively consistent changing of the standard diagram technique as applied to noninvertible morphisms. In every case we get a new arrow which corresponds to the additional multiplier, and so for $n = 2$ we obtain

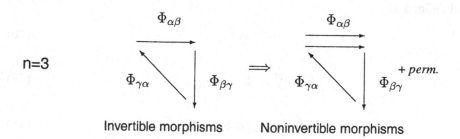

which describes the transition from (12) to (3) and presents the ordinary regularity condition for morphisms [48, 49]. The most intriguing semicommutative diagram is the triangle one

$$n=3$$

Invertible morphisms   Noninvertible morphisms

which generalizes the cocycle condition (1).

The higher n-regular semicommutative diagrams can be considered in the framework of generalized categories [9, 12, 50] in the following way.

5. Categories and 2-categories

There is an algebraic approach to the formalism considered in previous sections based on the category theory [5, 4]. A category $\mathcal{C}$ contains a collection $\mathcal{C}_0$ of objects and a collection $\hom(\mathcal{C})$ of arrows (morphisms) (see e.g. [51]). The collection hom $(\mathcal{C})$ is the union of mutually disjoint sets $\hom_{\mathcal{C}}(X, Y)$ of arrows $X \xrightarrow{f} Y$ from X to Y defined for every pair of objects $X, Y \in \mathcal{C}$. It may happen that for a pair $X, Y \in \mathcal{C}$ the set $\hom_{\mathcal{C}}(X, Y)$ is empty. The associative composition of morphisms is also defined. By an equivalence in $\mathcal{C}$ we mean a class of morphisms $\hom'(\mathcal{C}) = \bigcup_{X,Y \in (C_0)} \hom_{\mathcal{C}}'(X, Y)$ where $\hom_{\mathcal{C}}'(X, Y)$ is a subset of $\hom_{\mathcal{C}}(X, Y)$. Two objects X, Y of the category $\mathcal{C}$ is equivalent if and only if there is an morphism $X \xrightarrow{s} Y$ in $\hom_{\mathcal{C}}'(X, Y)$ such that

$$s^{-1} \circ s = id_X, \quad s \circ s^{-1} = id_Y \tag{21}$$

Let $X = (X_1, \cdots, X_n)$ be a sequence of objects of $\mathcal{C}$. Our category can contains a class of *noninvertible* morphisms [48, 4]. A (strict) 2-category $\mathcal{C}$ consists of a collection C_0 of objects as 0-cells and two collections of morphisms: C_1 and C_2 called 1-cells and 2-cells, respectively [52]. For every pair of objects $X, Y \in C_0$ there is a category $\mathcal{C}(X, Y)$ whose objects are 1-cell $f : X \to Y$ in C_1 and whose morphisms are 2-cells. For a pair of 1-cells $f, g \in C_1$ there is a 2-cell $s : f \to g$ in C_2. For every three objects $X, Y, Z \in C_0$ there is a bifunctor

$$c : \{\mathcal{C}(X, Y) \times \mathcal{C}(Y, Z) \longrightarrow \mathcal{C}(X, Z)\} \tag{22}$$

which is called a composition of 1-cells. There is an identity 1-cell $id_X \in \mathcal{C}(X, X)$ which acts trivially on $\mathcal{C}(X, Y)$ or $\mathcal{C}(Y, X)$. There is also 2-cell id_{id_X} which acts trivially on 2-cells.

Let $\mathcal{C}$ be a category with equivalence. Then one can see that collection of all equivalence classes of objects of $\mathcal{C}$ forms a 2-category $\mathbf{C}(\mathcal{C})$. These classes are 0-cells of $\mathbf{C}(\mathcal{C})$, 1-cells are classes of morphisms of $\mathcal{C}$. and 2-cells are maps between these classes. Observe that 1-cells of $\mathbf{C}(\mathcal{C})$ can be represented by morphisms of the underlying category $\mathcal{C}$, but such representation is not unique. One equivalence class can be represented by several equivalent morphisms. One can define 2-morphisms on equivalence classes, and $\mathbf{C}(\mathcal{C})$ becomes a 2-category. If the category $\mathcal{C}$ is equipped with certain additional structures, then one can transform them into $\mathbf{C}(\mathcal{C})$. If for instance $\mathcal{C}$ is monoidal category with product $\otimes : \mathcal{C} \times \mathcal{C} \longrightarrow \mathcal{C}$, then $\mathbf{C}(\mathcal{C})$ becomes the so-called semistrict monoidal 2-category. This means that the product $\otimes$ (under some natural conditions) is defined for all cells of the 2-category $\mathbf{C}(\mathcal{C})$. In the case of braided categories one can obtain the semistrict braided monoidal category [52]. Algebras, coalgebras, modules and comodules can be also included in this procedure. We apply such method to regularize categories with noninvertible morphisms and obstruction [5, 4].

6. Categories and regularization

Let $\mathcal{C}$ be a category with invertible and noninvertible morphisms [5] and equivalence. The equivalence in $\mathcal{C}$ is here defined as the class of invertible morphisms in the category $\mathcal{C}$.

Definition 7. A sequence of morphisms

$$X_1 \xrightarrow{f_1} X_2 \xrightarrow{f_2} \cdots \xrightarrow{f_{n-1}} X_n \xrightarrow{f_n} X_1 \tag{23}$$

such that there is an (endo-)morphism $e_{X_1}^{(3)} : X_1 \longrightarrow X_1$ defined uniquely by the following equation

$$e_{X_1}^{(n)} := f_n \circ \cdots \circ f_2 \circ f_1 \tag{24}$$

and subjects to the relation $f_1 \circ f_n \circ \cdots \circ f_2 \circ f_1 = f_1$ is said to be a *regular n-cycle* on C and it is denoted by $f = (f_1, \ldots f_n)$.

The (endo-)morphisms $e_{X_i}^{(n)} : X_i \longrightarrow X_i$ corresponding for $i = 2, \ldots, n$ are defined by a suitable cyclic permutation of above sequence.

Definition 8. The morphism $e_X^{(n)}$ is said to be an obstruction of X. The mapping $e^{(n)} : X \in C_0 \to e_X^{(n)} \in \hom(X, X)$ is called a regular n-cycle obstruction structure on C.

If

$$X_1 \xrightarrow{g_1} X_2' \xrightarrow{g_2} \cdots \xrightarrow{g_{n-1}} X_n' \xrightarrow{g_n} X_1$$

is an another n-tuple of morphisms such that $e_{X_1}^{(n)} : g_n \circ \cdots \circ g_2 \circ g_1$, then we assume that X_i' is equivalent to X_i, for $i = 2, \ldots, n$.

Definition 9. A map $s : f \Rightarrow g$ which sends the object X_i into equivalent object X_i' and morphism f_i into g_i is said to be obstruction n-cycle equivalence.

We have the diagram

$$
\begin{array}{ccc}
& X_2 \xrightarrow{f_2} \cdots \xrightarrow{f_{n-1}} X_n & \\
{\scriptstyle f_1}\nearrow & & \searrow{\scriptstyle f_n} \\
X_1 & \Downarrow s & X_1 \\
{\scriptstyle g_1}\searrow & & \nearrow{\scriptstyle g_n} \\
& X_2' \xrightarrow{g_2} \cdots \xrightarrow{g_{n-1}} X_n' &
\end{array}
\tag{25}
$$

Lemma 10. *There is a one to one correspondence between equivalence classes of regular n-cycles and regular n-cycle obstruction structures.*

If $f = (f_1, \ldots f_n)$ is a class of regular n-cycles, then there is the corresponding regular n-cycle obstruction structure $e : X \in C_0 \to e_X \in \hom(X, X)$ such that the relation (24) holds true. Let $e^{(n)} : X \in C_0 \to e_X^{(n)} \in \hom(X, X)$ be a regular n-cycle obstruction in C.

Definition 11. A morphism $\alpha : X \longrightarrow Y$ of the category $\mathcal{C}$ such that

$$\alpha \circ e_X^{(n)} = e_Y^{(n)} \circ \alpha \tag{26}$$

is said to be a regular n-cycle obstruction morphism from X to Y.

It follows from (23) that the morphism α is in fact a sequence of morphism $\alpha := (\alpha_1, \dots, \alpha_n)$ such that the diagram

$$\begin{array}{ccccccccc}
X_1 & \xrightarrow{f_1} & X_2 & \xrightarrow{f_2} & \cdots & \xrightarrow{f_{n-1}} & X_n & \xrightarrow{f_n} & X_1 \\
\alpha_1 \downarrow & & \downarrow & & & & \downarrow & & \downarrow \alpha_1 \\
Y_1 & \xrightarrow{g_1} & Y_2 & \xrightarrow{g_2} & \cdots & \xrightarrow{g_{n-1}} & Y_n & \xrightarrow{g_n} & Y_1
\end{array} \tag{27}$$

is commutative.

Definition 12. A collection of all equivalence classes of objects $\mathcal{C}_0$ with obstruction structures $e^{(n)} : X \in \mathcal{C}_0 \to e_X^{(n)} \in \hom(X, X)$ is denoted by $\mathfrak{Reg}_n(\mathcal{C})$ and called an obstruction n-cycle regularization of $\mathcal{C}$. The class of all regular n-cycle morphisms from X to Y is denote by $\mathfrak{Reg}_n(\mathcal{C})(X, Y)$.

Corollary 13. It follows from the Lemma 10 that the map $s : \alpha \longrightarrow \beta$ which sends an arbitrary regular n-cycle morphisms $\alpha \in \mathfrak{Reg}_n(\mathcal{C})(X, X')$ into a regular n-cycle morphisms $\beta \in \mathfrak{Reg}_n(\mathcal{C})(X, X')$ is a regular obstruction n-cycle equivalence.

One can define 2-morphisms and an associative composition of 2-morphisms such that $\mathfrak{Reg}_n(\mathcal{C})(X, Y)$ becomes a category for every two objects $X, Y \in \mathcal{C}_0$. If $\alpha : X \longrightarrow Y$ and $\beta : Y \longrightarrow Z$ are two n-cycle morphisms, then the composition $\beta \circ \alpha : X \to Z$ is also a n-cycle morphism. In this way we obtain the composition as bifunctors

$$c^{\mathfrak{Reg}_n} := \{\mathfrak{Reg}_n(\mathcal{C})(X, Y) \times \mathfrak{Reg}_n(\mathcal{C})(Y, Z) \longrightarrow \mathfrak{Reg}_n(\mathcal{C})(X, Z)\} \tag{28}$$

We summarize our considerations in the following lemma:

Lemma 14. The class $\mathfrak{Reg}_n(\mathcal{C})$ forms a (strict) 2-category whose 0-cells are equivalence classes of objects of $\mathcal{C}$ with obstructions, whose 1-cells are regular n-cycle obstruction morphisms, and whose 2-cells are regular obstruction n-cycle 2-morphisms.

7. Regularization of monoidal categories functions and Yang-Baxter equation

Let $C = C(I, \otimes)$ be a monoidal category, where I is the unit object and $\otimes$: $C \times C \longrightarrow C$ is the monoidal product [53, 54]. If the following relation

$$e_X^{(n)} \otimes e_Y^{(n)} = e_{X \otimes Y}^{(n)}. \tag{29}$$

holds true, then we have

Proposition 15. *The monoidal product of two regular n-cycles $X_1, \ldots, X_n$ and $Y_1, \ldots, Y_n$ with obstruction $e_{X_1}^{(n)}$, and $e_Y^{(n)}$, respectively, is the regular n-cycle*

$$X_1 \otimes Y_1, \otimes \cdots \otimes X_n \otimes Y_n$$

with the obstruction $e_{X \otimes Y}^{(n)}$.

One can see that in this case $\Re eg_n(C)$ is the so-called semistrict monoidal category [52].

Let C and D be two monoidal categories and let $\Re eg_n(C)$, $\Re eg_n(D)$ be their regularization 2-categories. We can introduce the notion of regular 2-functions, pseudonatural transformations and modifications. All definitions do not changed, but the preservation of the identity can be replaced by the requirement of preservation of obstruction morphisms $e_X^{(n)}$ and the invertibility is replaced by regularity. If, for instance, there is a regular 2-functor $\mathcal{F} : \Re eg_n(C) \longrightarrow \Re eg_n(C)$, then in addition to the standard definition [51] we have the following relation

$$\mathcal{F}(e_X) = e_{\mathcal{F}(X)}. \tag{30}$$

In the same manner we can "regularize" pseudo-natural transformations and modifications [50]. Let $\Re eg_n(C)$ be a semistrict monoidal 2-category. A pseudo-natural transformations $B = \{B_{X,X'} : X \otimes X' \to X' \otimes X\}$ and two regular modifications $B_{X \otimes Y, Z}$, $B_{X, Y \otimes Z}$ such that

$$
\begin{array}{ccc}
& B_{X \otimes Y, Z} & \\
X \otimes Y \otimes Z & \longrightarrow & Y \otimes Z \otimes X \\
B_{X,Y} \otimes e_Z \searrow & & \nearrow e_Y \otimes B_{X,Z} \\
& Y \otimes X \otimes Z &
\end{array}
\tag{31}
$$

and

$$
\begin{array}{ccc}
& B_{X, Y \otimes Z} & \\
X \otimes Y \otimes Z & \longrightarrow & Z \otimes X \otimes Y \\
e_X \otimes B_{Y,Z} \searrow & & \nearrow B_{X,Z} \otimes e_Y \\
& X \otimes Z \otimes Y &
\end{array}
\tag{32}
$$

and

$$B_{X,X'} \circ e_{X \otimes X'} = e_{X' \otimes X} \circ B_{X,X'}, \tag{33}$$

are said to be a regular n-cycle braiding. Obviously, these operations must satisfying all conditions of [52] with two changes indicated at the beginning of this section. Then the 2-category $\mathfrak{Reg}_n(\mathcal{C})$ is called a semistrict regular n-cycle braided monoidal category. This allows us to obtain here the following regular n-cycle Yang–Baxter equation [5, 4]

$$\mathbf{B}^{(1)}_{Y,Z,X} \circ \mathbf{B}^{(2)}_{Y,X,Z} \circ \mathbf{B}^{(1)}_{X,Y,Z} = \mathbf{B}^{(2)}_{Z,X,Y} \circ \mathbf{B}^{(1)}_{X,Z,Y} \circ \mathbf{B}^{(2)}_{X,Y,Z}, \tag{34}$$

where the notation

$$\mathbf{B}^{(1)}_{X,Y,Z} = B_{X,Y} \otimes e_Z, \quad \mathbf{B}^{(2)}_{X,Y,Z} = e_X \otimes B_{Y,Z}$$

has been used and the obstruction e_X is exploited instead of the identity Id_X. Solutions of the regular n-cycle Yang–Baxter equation (34) can be found by application of the endomorphism semigroup methods used in [55, 16].

8. Regularization of algebras, coalgebras, modules and comodules

Let $(\mathcal{C})$ be a monoidal category and $\mathfrak{Reg}_n(\mathcal{C})$ be its regularization . It is known that an associative algebra in the category $\mathcal{C}$ is an object $\mathcal{A}$ of this category such that there is an associative multiplication $m : \mathcal{A} \otimes \mathcal{A} \to \mathcal{A}$ which is also a morphism of this category. If the multiplication is in addition a regular n-cycle morphism, then the algebra $\mathcal{A}$ is said to be a regular n-cycle algebra. This means that we have the relation

$$m \circ (e_{\mathcal{A}} \otimes e_{\mathcal{A}}) = e_{\mathcal{A}} \circ m. \tag{35}$$

Obviously such multiplication not need to be unique. Denote by $\mathfrak{Reg}_n(\mathcal{C})(\mathcal{A} \otimes \mathcal{A}, \mathcal{A})$ a class of all such multiplications. We can see that a regular n-cycle 2-morphisms $s : m \Rightarrow n$ which send the multiplication m into a new one n should be an algebra homomorphism. One can define regular n-cycle coalgebra or bialgebra in a similar way. A comultiplication $\triangle : \mathcal{A} \longrightarrow \mathcal{A} \otimes \mathcal{A}$ can be regularized according to the relation

$$\triangle \circ e_{\mathcal{A}} = (e_{\mathcal{A}} \otimes e_{\mathcal{A}}) \circ \triangle. \tag{36}$$

In this case we obtain a class $\mathfrak{Reg}_n(\mathcal{C})(\mathcal{A}, \mathcal{A} \otimes \mathcal{A})$ of comultiplications.

Let $_{\mathcal{A}}\mathcal{C}$ be a category of all left $\mathcal{A}$ -modules, where $\mathcal{A}$ is a bialgebra. For the regularization $\mathfrak{Reg}_n(_{\mathcal{A}}\mathcal{C})$ of the $\mathcal{A}$–module action $\rho_M : \mathcal{A} \otimes M \longrightarrow M$ we use the following formula

$$\rho_M \circ (e_{\mathcal{A}} \otimes e_M) = e_M \circ \rho_M, \tag{37}$$

where $\rho_M : \mathcal{A} \otimes M \longrightarrow M$ is the left module action of $\mathcal{A}$ on M. The class of all such module actions is denoted by $\mathfrak{Reg}_n({}_\mathcal{A}\mathcal{C})(\mathcal{A} \otimes M, M)$. The monoidal operation in this category is given as the following tensor product of $\mathcal{A}$-modules

$$\rho_{M \otimes N} := (id_M \otimes \tau \otimes id_N) \circ (\rho_M \otimes \rho_N) \circ (\triangle \otimes id_{M \otimes N}), \qquad (38)$$

where $\tau : \mathcal{A} \otimes M \to M \otimes \mathcal{A}$ is the twist, i. e. $\tau(a \otimes m) := m \otimes a$ for every $a \in \mathcal{A}, m \in M$.

Lemma 16. *For the tensor product of module actions we have the following formula*

$$\rho_{M \otimes N} \circ (\mathbf{e}_\mathcal{A} \otimes \mathbf{e}_{M \otimes N}) = \mathbf{e}_{M \otimes N} \circ \rho_{M \otimes N}. \qquad (39)$$

This lemma means that the tensor product of two module actions satisfy our regularity condition if and only if these two actions also satisfy the regularity condition (37).

Observe that there is also a category $\mathcal{C}^\mathcal{A}$ of right $\mathcal{A}$-comodules, where $\mathcal{A}$ is an algebra. We can regularize this category in the following way. For the coaction we have

$$\rho \circ \mathbf{e}_\mathcal{A} = (\mathbf{e}_M \otimes \mathbf{e}_\mathcal{A}) \circ \rho_M, \qquad (40)$$

and

$$\rho_{M \otimes N} := (id_M \otimes m_\mathcal{A}) \circ (id_M \otimes \tau \otimes id_N) \circ (\rho_M \otimes \rho_N), \qquad (41)$$

where $\tau : M \otimes N \to N \otimes M$ is the twist, $m_\mathcal{A} : \mathcal{A} \otimes \mathcal{A} \to \mathcal{A}$ is the multiplication in $\mathcal{A}$.

Conclusions

Thus noninvertible extension of many abstract structures can be done in common general way: by introduction of the obstructions (or n-cycles) e which are analogs of units of the invertible case. In search of possible analogies we observe that "ln e" can play the role of first "fundamental group" for "space" of categories and vanishes for invertible morphisms, while its difference from "zero" can be treated as nontrivial "noninvertible topology" of such "space". We also note that "nil-" extension of supermanifolds – semisupermanifolds [56, 3] – can be compared with the "meta-" extension of supermanifolds– metamanifolds [57–59] – to find their complementarity or additivity and possibly for further generalizations simultaneously in both ways.

Acknowledgments. One of the authors (S.D.) would like to thank Andrzej Borowiec, Friedemann Brandt, Dimitry Leites, Jerzy Lukierski and Volodymyr

Lyubashenko for valuable discussions and Fang Li for fruitful correspondence and rare reprints. The NATO financial support is greatly acknowledged.

References

1. S. Duplij, *Semisupermanifolds and semigroups*, Krok, Kharkov, 2000.
2. S. Duplij, *Semigroup methods in supersymmetric theories of elementary particles*, Habilitation Thesis, Kharkov State University, math-ph/9910045, Kharkov, 1999.
3. S. Duplij, *On semi-supermanifolds*, Pure Math. Appl. **9** (1998), 283–310.
4. S. Duplij and W. Marcinek, *On higher regularity and monoidal categories*, Kharkov State University Journal (Vestnik KSU), ser. Nuclei, Particles and Fields **481** (2000), 27–30.
5. S. Duplij, *Higher regularity properties of mappings and morphisms*, Univ. Wrocław *preprint*, IFT UWr 931/00, math-ph/0005033, Wrocław, 2000, 12 p.
6. J. C. Baez and J. Dolan, *Higher-dimensional algebra and topological quantum field theory*, J. Math. Phys. **36** (1995), 6073–6105.
7. L. Crane and D. Yetter, *On algebraic structures implicit in topological quantum field theories*, Kansas State Univ. *preprint*, hep-th/9412025, Manhattan, 1994, 13 p.
8. J. C. Baez, *An introduction to n-categories*, Univ. California *preprint*, q-alg/9705009, Riverside, 1997, 34 p.
9. J. C. Baez and J. Dolan, *Higher-dimensional algebra III: n-categories and the algebra of opetopes*, Univ. California *preprint*, math/9702014, Riverside, 1997, 60 p.
10. J. C. Baez and J. Dolan, *From finite sets to Feynman diagrams*, Univ. California *preprint*, math.QA/0004133, Riverside, 2000, 30 p.
11. J. Siehler, *Braided near-group categories*, Virginia Tech. Univ. *preprint*, math.QA/0011037, Blacksburg, 2000, 8 p.
12. P. Greenberg and V. Sergiescu, *An acyclic extension of the braid group*, Comm. Math. Helv. **62** (1991), 185–239.
13. G. Bohm, F. Nill, and K. Szlachanyi, *Weak Hopf algebras I: Integral theory and C^*-structure*, Inst. Theor. Phys. FU *preprint*, math.QA/9805104, Berlin, 1998, 40 p.
14. F. Nill, *Axioms for weak bialgebras*, Inst. Theor. Phys. FU *preprint*, math.QA/9805104, Berlin, 1998, 48 p.
15. F. Li, *Weaker structures of Hopf algebras and singular solutions of Yang-Baxter equation*, Zhejiang Univ. *preprint*, Hangzhou, 2000, 6 p.
16. F. Li, *Weak Hopf algebras and new solutions of Yang-Baxter equation*, J. Algebra **208** (1998), 72–100.
17. F. A. Berezin, *Introduction to Superanalysis*, Reidel, Dordrecht, 1987.
18. D. Leites, *Supermanifold Theory*, Math. Methods Sci. Invest., Petrozavodsk, 1983.
19. C. Bartocci, U. Bruzzo, and D. Hernandez-Ruiperez, *The Geometry of Supermanifolds*, Kluwer, Dordrecht, 1991.
20. V. D. Ivashchuk, *Invertibility of elements in infinite-dimensional Grassmann-Banach algebras*, Theor. Math. Phys. **84** (1990), 13–22.
21. D. Leites and X. Peiqi, *Supersymmetry of the Schrödinger and Korteweg-de Vries operators*, Univ. Stockholm *preprint*, hep-th/9710045, Stockholm, 1997, 15 p.
22. V. Pestov, *Ground algebras for superanalysis*, Rep. Math. Phys. **29** (1991), 275–287.
23. I. B. Penkov, *D-modules on supermanifolds*, Inv. Math. **71** (1981), 501–512.
24. L. Crane and J. M. Rabin, *Super Riemann surfaces: uniformization and Teichmüller theory*, Comm. Math. Phys. **113** (1988), 601–623.

25. P. Bryant, *Supermanifolds, supersymmetry and Berezin integration*, in Complex Differential Geometry and Supermanifolds in Strings and Fields, (P. J. M. Bongaarts and R. Martini, eds.), Springer-Verlag, Berlin, 1988, pp. 150–167.

26. J. M. Rabin, *Manifold and supermanifold: Global aspects of supermanifold theory*, in Topological Properties and Global Structure of Space and Time, (P. G. Bergmann and V. De Sabbata, eds.), Plenum Press, New York, 1985, pp. 169–176.

27. J. M. Rabin, *Berezin integration on general fermionic supermanifolds*, Commun. Math. Phys. **103** (1986), 431–445.

28. P. Grozman, D. Leites, *An unconventional supergravity*, Talk at the Kiev NATO ARW, September 2000, *this volume*.

29. A. Konechny and A. Schwarz, *On $(k \oplus l \mid q)$-dimensional supermanifolds*, Univ. of California preprint, hep-th/9706003, Davis, 1997, 19 p.

30. N. Dragon, H. Günter, and U. Theis, *Supergravity with a noninvertible vierbein*, Univ. Hannover preprint, ITP-UH-21/97, hep-th/9707238, Hannover, 1997, 8 p.

31. P. Bryant, *Global properties of supermanifolds and their bodies*, Math. Proc. Cambridge Phil. Soc. **107** (1990), 501–523.

32. R. Catenacci, C. Reina, and P. Teofilatto, *On the body of supermanifolds*, J. Math. Phys. **26** (1985), 671–674.

33. D. Leites, *Selected problems of supermanifold theory*, Duke Math. J. **54** (1987), 649–656.

34. A. A. Kosinski, *Differential Manifolds*, Academic Press, Boston, 1993.

35. S. Lang, *Differential and Riemannian Manifolds*, Springer-Verlag, Berlin, 1995.

36. B. S. De Witt, *Supermanifolds*, 2nd edition, Cambridge Univ. Press, Cambridge, 1992.

37. A. Császár and E. Thümmel, *Multiplicative semigroups of continuous mappings*, Acta Math. Hung. **56** (1990), 189–204.

38. K. D. Magill, *Homomorphisms of semigroups of continuous selfmaps*, Bull. Alld. Math. Soc. **2** (1987), 1–36.

39. S. MacLane, *Homology*, Springer-Verlag, Berlin, 1967.

40. R. M. Switzer, *Algebraic Topology—Homotopy and Homology*, Springer-Verlag, Berlin, 1975.

41. V. V. Minachin, *Berezinians in substitution structures*, Func. Anal. Appl. **22** (1988), 90–91.

42. V. N. Shander, *Orientation of supermanifolds*, Func. Anal. Appl. **22** (1988), 91–92.

43. S. Duplij, *Some abstract properties of semigroups appearing in superconformal theories*, Semigroup Forum **54** (1997), 253–260.

44. S. Duplij, *On superconformal-like transformations and their nonlinear realization*, in Supersymmetries and Quantum Symmetries, (J. Wess and E. A. Ivanov, eds.), Springer-Verlag, Heidelberg, 1998, pp. 243–251.

45. A. H. Cliford, *The fundamental representation of a regular semigroup*, Semigroup Forum **10** (1975/76), 84–92.

46. W. D. Munn and R. Penrose, *Pseudoinverses in semigroups*, Math. Proc. Cambridge Phil. Soc. **57** (1961), 247–250.

47. C. R. Rao and S. K. Mitra, *Generalized Inverse of Matrices and its Application*, Wiley, New York, 1971.

48. D. L. Davis and D. W. Robinson, *Generalized inverses of morphisms*, Linear Algebra Appl. **5** (1972), 329–338.

49. M. Z. Nashed, *Generalized Inverses and Applications*, Academic Press, New York, 1976.

50. L. Breen, *Braided n-categories and Σ-structures*, Univ. Paris preprint, math.CT/9810045, Paris, 1998, 25 p.

51. S. MacLane, *Categories for the Working Mathematician*, Springer-Verlag, Berlin, 1971.

52. J. C. Baez and M. Neuchl, *Higher-dimensional algebra I: Braided monoidal 2-categories*, Univ. California preprint, q-alg/9511013, Riverside, 1995, 51 p.

53. D. N. Yetter, *Quantum groups and representations of monoidal categories*, Math. Proc. Camb. Phil. Soc. **108** (1990), 261–290.

54. A. Joyal and R. Street, *Braided monoidal categories*, Macquarie University preprint, Mathematics Reports 86008, North Ryde, New South Wales, 1986, 45 p.

55. F. Li, *Solutions of Yang-Baxter equation in an endomorphism semigroup and quasi-(co)braided almost bialgebras*, Zhejiang Univ. preprint, Hangzhou, 1999, 17 p.

56. S. Duplij, *Noninvertibility and "semi-" analogs of (super) manifolds, fiber bundles and homotopies*, Univ. Kaiserslautern preprint, KL-TH-96/10, q-alg/9609022, Kaiserslautern, 1996, 30 p.

57. D. Leites and V. Serganova, *Metasymmetry and Volichenko algebras*, Phys.Lett. **B252** (1990), 91–98.

58. D. Leites and V. Serganova, *Simple Volichenko algebras and symmetries wider than supersymmetry*, Univ. Stockholm preprint, Stockholm, 2000, 38 p.

59. D. Leites and V. Serganova, *Symmetries wider than supersymmetry*, Talk at the Kiev NATO ARW, September 2000, *this volume*.

AN OVERVIEW OF NEW SUPERSYMMETRIC GAUGE THEORIES
WITH 2-FORM GAUGE POTENTIALS

FRIEDEMANN BRANDT *

*Max-Planck-Institut für Mathematik in den Naturwissenschaften,
Inselstraße 22-26, D-04103 Leipzig, Germany*
After October 1, 2000:
Max-Planck-Institut für Gravitationsphysik, Albert-Einstein-Institut, Am Mühlenberg 1, D-14476 Golm, Germany

Abstract. An overview of new 4d supersymmetric gauge theories with 2-form gauge potentials constructed by various authors during the past five years is given. The key rôle of three particular types of interaction vertices is emphasized. These vertices are used to develop a connecting perspective on the new models and to distinguish between them. One example is presented in detail to illustrate characteristic features of the models. A new result on couplings of 2-form gauge potentials to Chern-Simons forms is presented.

1. Introduction

During the past five years, several new 4d supersymmetric gauge theories have been constructed by various authors [1]–[13]. Common to all these models is the presence of 2-form gauge potentials and a complicated (nonpolynomial) structure of interactions and symmetry transformations (gauge symmetries, supersymmetry). The initial motivation to construct such models came from string theory and focused the attention first on the vector-tensor (VT) multiplet [14, 15] of N=2 supersymmetry. Namely, in N=2 supersymmetric 4d heterotic string vacua, the dilaton is believed to reside in a VT multiplet (see, e.g., section 3 of the review [16]). In order to couple this multiplet to N=2 supergravity, its so-called central charge must be gauged and this leads inevitably to the structures characteristic of the new models (cf. remarks at the end of section 3). Only two of the works [1]–[13] are not devoted to the VT multiplet: in [11] a rather general class of new supersymmetric gauge theories with 2-form gauge fields is constructed, and [13] deals with the double tensor (TT) multiplet of N=2 supersymmetry and its

---

* fbrandt@aei-potsdam.mpg.de

S. Duplij and J. Wess (eds.), Noncommutative Structures in Mathematics and Physics, 141–152.
© 2001 Kluwer Academic Publishers. Printed in the Netherlands.

couplings to vector and hyper multiplets. The TT multiplet is believed to be the dilaton-multiplet of N=2 supersymmetric type IIB superstring vacua [16] and thus it should play there a rôle analogous to the VT multiplet in heterotic vacua.

The purpose of this contribution is to give an overview of the new models and to emphasize the key rôle of three types of cubic interaction vertices in these models. To this end, first a brief excursion to consistent interactions of p-form gauge potentials in general is made in section 2. This will also show how the new models fit in the recent classification [17–19] of interactions between p-form gauge potentials. The three particular types of interaction vertices are identified and discussed in some detail in section 3, including a new result on couplings of 2-form gauge potentials to Chern-Simons forms. Then these vertices and the supersymmetry multiplet structure are used to characterize the various models and to distinguish between them. In section 4, an explicit example is treated in detail to illustrate characteristic features of the new models. The example is an N=2 supersymmetric model found in [13], coupling the TT multiplet mentioned above to two N=2 vector multiplets. Section 5 contains a selection of open problems and possible future developments.

2. Interactions of p-form gauge potentials

Gauge invariance restricts the possible interactions of p-form gauge fields quite severely. In the simplest case, the gauge transformation of a p-form gauge potential $A = (1/p!)dx^{\mu_1} \wedge \cdots \wedge dx^{\mu_p} A_{\mu_1 \ldots \mu_p}$ is a natural generalization of the gauge transformation of the electromagnetic gauge field:

$$\delta^{(0)}_{\text{gauge}} A = d\omega \Leftrightarrow \delta^{(0)}_{\text{gauge}} A_{\mu_1 \ldots \mu_p} = p \partial_{[\mu_1} \omega_{\mu_2 \ldots \mu_p]} , \tag{1}$$

where $\omega_{\mu_1 \ldots \mu_{p-1}}$ are arbitrary gauge parameter fields. Analogously to the electromagnetic case, corresponding gauge invariant field strengths are thus

$$F = dA \Leftrightarrow F_{\mu_0 \ldots \mu_p} = (p+1) \partial_{[\mu_0} A_{\mu_1 \ldots \mu_p]} , \tag{2}$$

and the standard Lagrangian for a set of free p-form gauge fields is a linear combination of Maxwell-type kinetic terms $F_{\mu_0 \ldots \mu_p} F^{\mu_0 \ldots \mu_p}$.

A systematic investigation of the possible interaction vertices which can be added consistently to such a free Lagrangian $L^{(0)}$ was carried out by Henneaux and Knaepen [17–19]. They studied consistent deformations of the free Lagrangian $L^{(0)}$ and of the gauge transformations $\delta^{(0)}_{\text{gauge}}$,

$$L = L^{(0)} + g^\alpha V^{(1)}_\alpha + g^\alpha g^\beta V^{(2)}_{\alpha\beta} + \ldots \tag{3}$$

$$\delta_{\text{gauge}} = \delta^{(0)}_{\text{gauge}} + g^\alpha \delta^{(1)}_{\text{gauge}\,\alpha} + g^\alpha g^\beta \delta^{(2)}_{\text{gauge}\,\alpha\beta} + \ldots , \tag{4}$$

where g^α are continuous coupling constants (deformation parameters), such that the deformed Lagrangian L is invariant under the deformed gauge transformations

δ_{gauge} modulo a total derivative,

$$\delta_{\text{gauge}} L = \partial_\mu K^\mu. \tag{5}$$

To first order in the coupling constants, (5) requires that the $V_\alpha^{(1)}$ be $\delta_{\text{gauge}}^{(0)}$-invariant on-shell in the free theory modulo a total derivative. Furthermore, without loss of generality, one may neglect all $V_\alpha^{(1)}$ which vanish on-shell in the free theory modulo a total derivative because they can be removed by field redefinitions (such vertices are therefore called trivial ones). Henneaux and Knaepen found the following result for the remaining first-order vertices:

Category 1: Vertices that are $\delta_{\text{gauge}}^{(0)}$-invariant off-shell modulo a total derivative and therefore do not modify the gauge transformations to first order. There are two types of such vertices (modulo total derivatives). Those of the first type depend on p-form gauge fields only via the field strengths $F_{\mu_0 \ldots \mu_p}$ and their derivatives. Of course, there are infinitely many vertices of this type. Those of the second type are vertices of the Chern-Simons type

$$A \wedge F \wedge \cdots \wedge F \tag{6}$$

where the F's may have different form-degrees and all form-degrees must sum up to the spacetime dimension. These vertices are $\delta_{\text{gauge}}^{(0)}$-invariant only modulo a total derivative.

Category 2: Vertices that are $\delta_{\text{gauge}}^{(0)}$-invariant only on-shell in the free theory modulo a total derivative. These vertices are of particular interest because they are accompanied by deformations of the gauge transformations. A remarkable result is that, when ordinary gauge fields (1-form gauge potentials) are absent, all these vertices can be brought to the following form (modulo trivial vertices and vertices of category 1):

$$A \wedge F \wedge \cdots \wedge F \wedge \underbrace{{}^*F \wedge \cdots \wedge {}^*F}_{\text{at least one } {}^*F} \tag{7}$$

where ${}^*F$ denotes the Hodge dual of F and there must be at least one ${}^*F$ because otherwise the vertex would be of the Chern-Simons type (6). Again, the F's may have different form-degrees and all form-degrees must sum up to the spacetime dimension. Therefore there are only finitely many vertices (7) for a finite number of p-form gauge fields. The first order deformations of the gauge transformations which correspond to a vertex (7) take the form

$$\delta_{\text{gauge}}^{(1)} A = \omega \wedge F \wedge \cdots \wedge F \wedge {}^*F \wedge \cdots \wedge {}^*F \tag{8}$$

where one of the ${}^*F$'s that occurs in (7) is omitted (for instance, when (7) contains only one ${}^*F$, then (8) contains no ${}^*F$). When 1-form gauge potentials are present, (7) still gives nontrivial first-order vertices of category 2, but then there may be additional vertices of category 2 which cannot be brought to the form (7). In particular, when at least three 1-form gauge potentials are present, there are Yang-Mills cubic vertices which differ from (7) because they contain two 'naked' gauge potentials instead of only one (the structure of Yang-Mills cubic vertices is $A \wedge A \wedge {}^*F$ where the A's are 1-form gauge potentials and F is a 2-form field strength).

In four-dimensional spacetime there are three different types of cubic vertices (7) involving 1-form gauge potentials A_1, 2-form potentials A_2 and corresponding field strengths $F_2 = dA_1$ and $F_3 = dA_2$:

$$A_2 \wedge {}^*F_3 \wedge {}^*F_3 \tag{9}$$

$$A_1 \wedge {}^*F_2 \wedge {}^*F_3 \tag{10}$$

$$A_1 \wedge F_2 \wedge {}^*F_3 . \tag{11}$$

These are the vertices mentioned in the introduction.

3. Overview of the new models

In accordance with commonly used nomenclature (which is actually somewhat unfair, see remarks at the end of this section), the vertices (9), (10) and (11) will be referred to as "Freedman-Townsend" (FT), "Henneaux-Knaepen" (HK) and "Chapline-Manton" (CM) vertices, respectively. Each of the new supersymmetric models reviewed here contains at least one of these vertices. We label 1-form potentials and 2-form potentials by indices $a = 1, 2, \ldots$ and $i = 1, 2, \ldots$ respectively, and denote their component fields by A_μ^a and $B_{\mu\nu}^i = -B_{\nu\mu}^i$. The field strengths of A_μ^a are denoted by $F_{\mu\nu}^a = \partial_\mu A_\nu^a - \partial_\nu A_\mu^a$, the Hodge-dualized field strengths of $B_{\mu\nu}^i$ by $H^{i\mu} = \frac{1}{2}\varepsilon^{\mu\nu\rho\sigma}\partial_\nu B_{\rho\sigma}^i$. The vertices (9), (10) and (11) read explicitly, using a suitable normalization,

FT vertices: $\qquad \dfrac{1}{4} f_{ijk} H_\mu^i H_\nu^j B_{\rho\sigma}^k \, \varepsilon^{\mu\nu\rho\sigma} \qquad\qquad$ (12)

HK vertices: $\qquad T_{iab} H_\mu^i F^{a\mu\nu} A_\nu^b \qquad\qquad$ (13)

CM vertices: $\qquad \dfrac{1}{2} S_{iab} H_\mu^i F_{\nu\rho}^a A_\sigma^b \, \varepsilon^{\mu\nu\rho\sigma} \qquad\qquad$ (14)

where the f_{ijk}, T_{iab} and S_{iab} are constant coefficients, with

$$f_{ijk} = -f_{jik} \quad , \quad S_{iab} = S_{iba} .$$

[$S_{iab} = S_{iba}$ can be imposed without loss of generality because $S_{i[ab]}$ can be removed from the vertices (14) by subtracting trivial vertices.] These coefficients are subject to conditions imposed by (5) at second order in the coupling constants (deformation parameters). Viewing T_{iab} and S_{iab} as the entries of matrices T_i and S_i, these conditions read

$$f_{ijl}f_{klm} + f_{jkl}f_{ilm} + f_{kil}f_{jlm} = 0 \tag{15}$$

$$[T_i, T_j] = f_{ijk}T_k \tag{16}$$

$$(S_iT_j - S_jT_i) + (S_iT_j - S_jT_i)^\top = f_{ijk}S_k . \tag{17}$$

To derive these conditions, it was assumed that the zeroth order Lagrangian is $L^{(0)} = -(1/2)H^{\mu i}H^i_\mu - (1/4)F^a_{\mu\nu}F^{\mu\nu a}$, and that (9), (10) and (11) are the only vertices of category 2 with non-vanishing coefficients (vertices of category 1 do not modify these conditions, but switching on other vertices of category 2 might cause modifications or lead to additional conditions).

(15) and (16) were already found in [17] and require that the f_{ijk} be structure constants of a Lie algebra and that the T_i be representation matrices of that Lie algebra, respectively. (17) was not derived in a previous work, to my knowledge. It requires that the symmetric parts of the matrices $2(S_iT_j - S_jT_i)$ be equal to $f_{ijk}S_k$. This is fulfilled, for instance, if $S_i = NT_i + T_i^\top N$ where N is an arbitrary symmetric matrix (i.e., $S_{iab} = N_{ac}T_{icb} + N_{bc}T_{ica}$ with $N_{ab} = N_{ba}$), but there are other solutions as well.

The corresponding first order deformations of the gauge transformations are

$$\delta^{(1)}_{\text{gauge}} B^i_{\mu\nu} = -f_{ijk}(H^j_\mu\omega^k_\nu - H^j_\nu\omega^k_\mu) - \frac{1}{2}\varepsilon_{\mu\nu\rho\sigma}T_{iab}F^{\rho\sigma a}\omega^b + S_{iab}F^a_{\mu\nu}\omega^b$$

$$\delta^{(1)}_{\text{gauge}} A^a_\mu = -T_{iab}H^i_\mu\omega^b. \tag{18}$$

The following table gives an overview of the new supersymmetric models. The vertices discussed above are used to distinguish between the various models. In addition the number of supersymmetries (N=1 or N=2 supersymmetry) and the supersymmetry multiplets are given. In the case of N=1 supersymmetry, T and V stand for tensor multiplets (also called linear multiplets) and vector multiplets respectively. In the case of N=2 supersymmetry, VT, TT and V stand for vector-tensor multiplets, double-tensor multiplets and vector multiplets respectively.

susy	multiplets	interactions	papers
N=2	VT , V	HK , CM	[1, 2, 7, 8, 10]
N=2	VT , V	CM	[3–5, 9]
N=2	VT	CM	[6]
N=1	T , V	FT , HK , CM	[11]
N=2	VT	HK	[12]
N=2	TT , V	FT , HK	[13]

Of course, this table characterizes the various models only very roughly. The example in the next section is to illustrate characteristic features of these models. It is beyond the scope of this paper to review the various models in greater detail but I would like to add at least a few remarks: (a) Among all these models only those in [7] are locally supersymmetric, the other ones are globally supersymmetric. (b) The works on the VT multiplet overlap in part because some of these works rederive models which had already been found by means of other methods in previous works. (c) Models in the same row of the table may of course still differ. For instance, CM vertices in two models with the same multiplet content may contain different Chern-Simons forms (in the literature, this has led to a distinction between "linear" and "nonlinear" VT multiplets [2]). Different CM couplings correspond to different solutions to Eq. (17). Of course, analogous statements apply to the FT and HK vertices. (d) Some of the models in [11] possess extended ($N \geq 2$) supersymmetry. For instance, it has been pointed out in [12] that the model constructed there can be obtained from [11]. However, it is not clear how to sieve out systematically those models in [11] which have extended supersymmetry.

Finally a few comments on the history may be in order. Models with FT interactions were constructed already by Ogievetsky and Polubarinov [20] a long time before the work by Freedman and Townsend [21]. CM interactions have a long history too. It seems that they appeared first in the early 80's [22–24] and, again, the work by Chapline and Manton was not the first one with such interactions. CM interactions attracted particular attention because of their crucial rôle in the Green-Schwarz anomaly cancellation mechanism [25] (the anomaly cancellation is made possible by the deformation of the gauge transformations associated with CM vertices, see section 2).

HK interactions (in four-dimensional spacetime) were discovered much later. However, the first models with such interactions were not found by Henneaux and Knaepen. Rather, it seems that HK interactions occurred for the first time in [1] where the central charge of the VT multiplet was gauged. The connection of that gauging to HK vertices is the following. Gauging the central charge (e.g., via the Noether method) gives rise to a vertex $V_\mu j^\mu$ where V_μ is a 1-form gauge field and j^μ is the Noether current corresponding to the central charge symmetry. That Noether current is $j^\mu = H_\nu F^{\nu\mu}$, and thus the vertex $V_\mu j^\mu$ is a HK vertex. Combined FT and HK interactions, and the relation to Lie algebras, were found afterwards by Henneaux and Knaepen [17]. It seems that the first and so far only work with models containing simultaneously FT, HK and CM vertices is [11].

4. Example

The example is an N=2 supersymmetric model coupling one TT multiplet to two V multiplets and involves HK vertices but no FT or CM vertices. A TT multiplet contains two 2-form gauge potentials $B^i_{\mu\nu}$ ($i = 1, 2$), two real scalar fields a^i and two Weyl fermions χ and ψ. Each V multiplet contains a 1-form gauge potential A_μ, a complex scalar field ϕ and two Weyl fermions λ^i. The V multiplets are labeled by the index $a = 1, 2$. This field content is supplemented with auxiliary fields h^i_μ which are embedded in the TT multiplet. These auxiliary fields allow one to construct the model in a compact polynomial form. In fact, it would be very cumbersome to construct the model without these auxiliary fields because of the complicated nonpolynomial structure which arises then, see below. Note that, in contrast to other supersymmetric models, the auxiliary fields do not lead to an off-shell closed supersymmetry algebra. On the contrary, the auxiliary fields make the supersymmetry algebra even "more open" (a formulation of the TT multiplet with an off-shell closed supersymmetry algebra is not known).

	bosons			Weyl-fermions	
TT	$B^i_{\mu\nu}$	a^i	(h^i_μ)	χ	ψ
V^a	A^a_μ	ϕ^a		λ^{ai}	

Thanks to the inclusion of the auxiliary fields, the Lagrangian takes the following simple form (using conventions as [26] adapted to the Minkowski metric diag$(1, -1, -1, -1)$),

$$L = \partial_\mu a^i \partial^\mu a^i + h^i_\mu h^{\mu i} + 2h^i_\mu H^{\mu i} - i\chi\partial\bar\chi - i\psi\partial\bar\psi$$
$$-\frac{1}{4}\hat F^a_{\mu\nu}\hat F^{a\mu\nu} + \frac{1}{2}\hat D_\mu\phi^a \hat D^\mu\bar\phi^a - 2i\lambda^{ia}\hat D\bar\lambda^{ia} \qquad (19)$$

where

$$\hat F^a_{\mu\nu} = \hat D_\mu A^a_\nu - \hat D_\nu A^a_\mu = \partial_\mu A^a_\nu + g^i h^i_\mu \varepsilon^{ab} A^b_\nu - (\mu \leftrightarrow \nu)$$
$$\hat D_\mu\phi^a = \partial_\mu\phi^a + g^i h^i_\mu \varepsilon^{ab}\phi^b$$
$$\hat D\bar\lambda^{ia} = \sigma^\mu(\partial_\mu\bar\lambda^{ia} + g^i h^i_\mu \varepsilon^{ab}\bar\lambda^{ib}).$$

The g^i are real coupling constants (deformation parameters). Note that $\hat{D}_\mu$ has the form of a covariant derivative even though the auxiliary fields cannot be viewed as gauge fields (in fact, they substitute for field strengths, as the equations of motion give $h_\mu^i = -H_\mu^i + \dots$). The auxiliary fields also simplify the structure of the gauge and supersymmetry transformations considerably. The gauge transformations read

$$\delta_{\text{gauge}} A_\mu^a = \hat{D}_\mu \omega^a = \partial_\mu \omega^a + g^i h_\mu^i \varepsilon^{ab} \omega^b$$

$$\delta_{\text{gauge}} B_{\mu\nu}^i = \frac{1}{4} g^i \omega^a \varepsilon^{ab} \varepsilon_{\mu\nu\rho\sigma} \hat{F}^{b\rho\sigma} + \partial_\mu \omega_\nu^i - \partial_\nu \omega_\mu^i$$

$$\delta_{\text{gauge}} = 0 \quad \text{on other fields}$$

where ω^a and ω_μ^i are the gauge parameter fields associated with A_μ^a and $B_{\mu\nu}^i$ respectively. The supersymmetry transformations read, with constant anticommuting Weyl-spinors ξ^i as transformation parameters,

$$\delta_{\text{susy}} A_\mu^a = \varepsilon^{ij} \xi^i \sigma_\mu \bar{\lambda}^{ja} - \xi^i \Gamma^i \varepsilon^{ab} A_\mu^b + \text{c.c.}$$

$$\delta_{\text{susy}} \phi^a = 2\xi^i \lambda^{ia} - (\xi^i \Gamma^i + \bar{\xi}^i \bar{\Gamma}^i) \varepsilon^{ab} \phi^b$$

$$\delta_{\text{susy}} \lambda^{ia} = \frac{i}{2} (\varepsilon^{ij} \xi^j \sigma^{\mu\nu} \hat{F}_{\mu\nu}^a - \bar{\xi}^i \bar{\sigma}^\mu \hat{D}_\mu \phi^a) - (\xi^j \Gamma^j + \bar{\xi}^j \bar{\Gamma}^j) \varepsilon^{ab} \lambda^{ib}$$

$$\delta_{\text{susy}} B_{\mu\nu}^i = -\varepsilon^{ij} \xi^j \sigma_{\mu\nu} \chi + \xi^i \sigma_{\mu\nu} \psi$$

$$+ i g^i \varepsilon^{ab} (\bar{\phi}^a \xi^j \sigma_{\mu\nu} \lambda^{jb} + \varepsilon^{jk} A_{[\mu}^a \xi^j \sigma_{\nu]} \bar{\lambda}^{kb}) + \text{c.c.}$$

$$\delta_{\text{susy}} a^i = \frac{1}{2} (\xi^i \chi - \varepsilon^{ij} \xi^j \psi) + \text{c.c.}$$

$$\delta_{\text{susy}} \chi = -\bar{\xi}^i \bar{\sigma}^\mu (\varepsilon^{ij} h_\mu^j + i \partial_\mu a^i)$$

$$\delta_{\text{susy}} \psi = -\bar{\xi}^i \bar{\sigma}^\mu (h_\mu^i + i \varepsilon^{ij} \partial_\mu a^j)$$

$$\delta_{\text{susy}} h_\mu^i = \frac{i}{2} \partial_\mu (\xi^i \psi - \varepsilon^{ij} \xi^j \chi) + \text{c.c.}$$

where

$$\Gamma^i = \frac{i}{2} g^j (\varepsilon^{ij} \chi + \delta^{ij} \psi).$$

The commutator algebra of the supersymmetry and gauge transformations is rather complicated off-shell but on-shell it is quite simple,

$$[\delta_{\text{susy}}, \delta'_{\text{susy}}] \approx \delta_{\text{translation}} + \delta_{\text{gauge}} \tag{20}$$

$$[\delta_{\text{susy}}, \delta_{\text{gauge}}] \approx \delta'_{\text{gauge}} \tag{21}$$

$$[\delta_{\text{gauge}}, \delta'_{\text{gauge}}] \approx 0, \tag{22}$$

where $\approx$ is equality on-shell. (20) is the standard N=2 supersymmetry algebra on-shell (modulo gauge transformations), with vanishing central charge. I remark that the gauge transformations which appear on the right hand side of (20) involve explicitly the spacetime coordinates, see [13] and [27] for details and comments on this point. (21) illustrates a feature typical of many of the new models, namely that gauge and supersymmetry transformations do not commute (not even on-shell). Explicitly, the gauge parameter fields $\omega^{a\prime}$ and $\omega_\mu^{i\prime}$ of δ'_{gauge} on the right hand side of (21) read

$$\omega^{a\prime} = (\xi^i \Gamma^i + \bar{\xi}^i \bar{\Gamma}^i)\varepsilon^{ab}\omega^b$$

$$\omega_\mu^{i\prime} = -\frac{i}{2} g^i \varepsilon^{ab} \varepsilon^{jk} \omega^a (\xi^j \sigma_\mu \bar{\lambda}^{kb} - \lambda^{kb}\sigma_\mu \bar{\xi}^j)$$

where the ξ's and ω's are supersymmetry parameters and gauge parameter fields of δ_{susy} and δ_{gauge} on the left hand side of (21). According to (22), the gauge transformations commute on-shell which is also typical of the new models [note: the algebra of the gauge transformations is not related to the Lie algebra underlying Eqs. (15) through (17)!].

Let me finally discuss the nonpolynomial structure which arises when one eliminates the auxiliary fields. The Lagrangian (19) contains the auxiliary fields at most quadratically,

$$L = -\frac{1}{4}F_{\mu\nu}^a F^{a\mu\nu} + \partial_\mu a^i \partial^\mu a^i + \frac{1}{2}\partial_\mu \phi^a \partial^\mu \bar{\phi}^a$$
$$-i\chi\partial\bar{\chi} - i\psi\partial\bar{\psi} - 2i\lambda^{ia}\partial\bar{\lambda}^{ia} + 2h_\mu^i \mathcal{H}^{\mu i} + h_\mu^i K^{\mu i,\nu j} h_\nu^j$$

where

$$\mathcal{H}^{\mu i} = H^{\mu i} - g^i \varepsilon^{ab}(\frac{1}{2}F^{a\mu\nu}A_\nu^b + \frac{1}{4}\phi^a \overset{\leftrightarrow}{\partial^\mu} \bar{\phi}^b + i\lambda^{ja}\sigma^\mu \bar{\lambda}^{jb})$$

$$K^{\mu i,\nu j} = \eta^{\mu\nu}\delta^{ij} + \frac{1}{2}g^i g^j [\eta^{\mu\nu}(\phi^a \bar{\phi}^a - A_\rho^a A^{a\rho}) + A^{a\mu}A^{a\nu}]$$

The auxiliary fields can be eliminated by solving their algebraic equations of motion. The solution is

$$h_\mu^i = -(K^{-1})_{\mu i,\nu j} \mathcal{H}^{\nu j}, \tag{23}$$

where K^{-1} is the inverse of the field dependent matrix K, $(K^{-1})_{\mu i,\rho k}K^{\rho k,\nu j} = \delta_\mu^\nu \delta_i^j$. Note that K does not involve derivatives of the fields and therefore K^{-1} is nonpolynomial in the fields but still local. Hence, using (23), the Lagrangian, gauge and supersymmetry transformations become nonpolynomial but remain strictly local. Expanding the resulting Lagrangian in the coupling constants, one finds at first order HK vertices as well as vertices of category 1 which complete the HK vertices such that the sum is supersymmetric on-shell in the free theory modulo a total derivative,

$$
\begin{aligned}
L &= -\frac{1}{4}F^a_{\mu\nu}F^{a\mu\nu} + \partial_\mu a^i \partial^\mu a^i + \frac{1}{2}\partial_\mu \phi^a \partial^\mu \bar{\phi}^a \\
&\quad -i\chi\partial\bar{\chi} - i\psi\partial\bar{\psi} - 2i\lambda^{ia}\partial\bar{\lambda}^{ia} - \mathcal{H}^{\mu i}(K^{-1})_{\mu i,\nu j}\mathcal{H}^{\nu j} \\
&= L^{(0)} + \underbrace{g^i\varepsilon^{ab}H^i_\mu F^{a\mu\nu}A^b_\nu}_{\text{HK vertices}} + \underbrace{g^i\varepsilon^{ab}H^i_\mu(\frac{1}{2}\phi^a \overset{\leftrightarrow}{\partial^\mu} \bar{\phi}^b + 2i\lambda^{ja}\sigma^\mu\bar{\lambda}^{jb})}_{\substack{\text{category 1 vertices}\\ \text{(susy completion of HK vertices)}}} + \ldots(24)
\end{aligned}
$$

It was mentioned already that nonpolynomial structures as in this example are typical of the new gauge theories. They cannot be avoided in models with FT or HK vertices because they are necessary consequences of these vertices, already in the non-supersymmetric case. The use of appropriate auxiliary fields that simplify the construction is an almost indispensable tool for constructing complicated models of this type, especially supersymmetric ones. The finding of such auxiliary fields and their embedding in supersymmetry multiplets is in general a nontrivial and subtle ingredient of the construction. In contrast, models which contain CM vertices but no FT or HK vertices are simpler and the issue of auxiliary fields is less involved. In particular, such models are not necessarily nonpolynomial although supersymmetry often enforces a nonpolynomial dependence on scalar fields even in such models.

5. Comments

The following is a selection of open problems which may point to possible further developments in the field:

(i) In my opinion, the rôle of the matter fields (scalar fields, fermions) in the new supersymmetric models has not been fully understood yet. In particular, the relation of scalar fields to the underlying geometry (Lie algebra) is somewhat mysterious. A better understanding of this issue might be a key to a deeper understanding of the supersymmetry structure of the models and to a more systematic construction of such models.

(ii) Systematic classifications of the possible consistent and supersymmetric interactions involving p-form gauge potentials, analogous to the classification [17–19] of non-supersymmetric interactions, are largely missing. An exception is the classification of the lowest dimensional interaction vertices involving a TT multiplet in [13]. Supersymmetry supplements (5) with the additional requirement $\delta_{\text{susy}}L = \partial_\mu M^\mu$ where δ_{susy} are the deformed supersymmetry transformations. This restricts the possible interactions as compared to the non-supersymmetric case, and relates coefficients of various interaction terms. A typical example is (24) where the coefficients of the HK vertices are related to coefficients of interaction vertices of category 1. In fact, supersymmetry can even completely forbid interactions which would be allowed if supersymmetry were not imposed. An example is the absence of N=2 supersymmetric CM couplings of the TT multiplet [13]. Furthermore, it depends on the supersymmetry multiplet structure which interactions are possible. For instance, it was just mentioned that there are no N=2 supersymmetric CM couplings involving the TT multiplet, whereas such couplings do exist for the VT multiplet (cf. table in section 3). Such results could be relevant in the context of string theory when comparing properties of different superstring vacua.

(iii) Locally supersymmetric models with FT or HK couplings are almost completely missing so far. In fact, the only exception is the work [7] where N=2 supergravity models with VT multiplets were constructed. The construction of locally supersymmetric extensions of some of the other models could be of interest in the string theory context. In particular this applies to supergravity models with the TT multiplet because of the conjectured importance of this multiplet to type IIB superstring vacua (cf. introduction).

(iv) Recall that FT, HK and CM vertices are special cases of vertices (7). Non-supersymmetric models in spacetime dimensions > 4 with such vertices have been constructed already [17, 28]. Analogous globally or locally supersymmetric models in higher spacetime dimensions have not been constructed so far. In fact it seems that the only vertices (7) which have been used in supersymmetric models in spacetime dimensions > 4 so far are the familiar CM vertices (14). For instance, these vertices occur in 10-dimensional supergravity in connection with the Green-Schwarz anomaly cancellation mechanism (cf. remarks at the end of section 3).

References

1. P. Claus, B. de Wit, M. Faux, B. Kleijn, R. Siebelink and P. Termonia, *The vector-tensor supermultiplet with gauged central charge*, Phys. Lett. B 373 (1996) 81-88, hep-th/9512143.

2. P. Claus, B. de Wit, M. Faux and P. Termonia, *Chern-Simons couplings and inequivalent vector-tensor multiplets*, Nucl. Phys. B 491 (1997) 201-220, hep-th/9612203.

3. R. Grimm, M. Hasler and C. Herrmann, *The N=2 vector-tensor multiplet, central charge superspace, and Chern-Simons couplings*, Int. J. Mod. Phys. A 13 (1998) 1805-1816, hep-th/9706108.

4. N. Dragon, S.M. Kuzenko and U. Theis, *The vector-tensor multiplet in harmonic superspace*, Eur. Phys. J. C 4 (1998) 717-721, hep-th/9706169.

5. I. Buchbinder, A. Hindawi and B.A. Ovrut, *A two-form formulation of the vector-tensor multiplet in central charge superspace*, Phys. Lett. B 413 (1997) 79-88, hep-th/9706216.

6. N. Dragon and S.M. Kuzenko, *Self-interacting vector-tensor multiplet*, Phys. Lett. B 420 (1998) 64-68, hep-th/9709088.

7. P. Claus, B. de Wit, M. Faux, B. Kleijn, R. Siebelink and P. Termonia, *N=2 supergravity Lagrangians with vector-tensor multiplets*, Nucl. Phys. B 512 (1998) 148-178, hep-th/9710212.

8. N. Dragon and U. Theis, *Gauging the central charge*, hep-th/9711025;
The linear vector-tensor multiplet with gauged central charge, Phys. Lett. B 446 (1999) 314-320, hep-th/9805199.

9. E. Ivanov and E. Sokatchev, *On non-linear superfield versions of the vector-tensor multiplet*, Phys. Lett. B 429 (1998) 35-47, hep-th/9711038.

10. N. Dragon, E. Ivanov, S.M. Kuzenko, E. Sokatchev and U. Theis, *N=2 rigid supersymmetry with gauged central charge*, Nucl. Phys. B 538 (1999) 411-450, hep-th/9805152 .

11. F. Brandt and U. Theis, *D=4, N=1 supersymmetric Henneaux-Knaepen models*, Nucl. Phys. B 550 (1999) 495-510, hep-th/9811180.

12. U. Theis, *New N=2 supersymmetric vector-tensor interaction*, Phys. Lett. B 486 (2000) 443-447, hep-th/0005044.

13. F. Brandt, *New N=2 supersymmetric gauge theories: the double tensor multiplet and its interactions*, Nucl. Phys. B 587 (2000) 543-567, hep-th/0005086.

14. M.F. Sohnius, K.S. Stelle and P.C. West, *Off-mass-shell formulation of extended supersymmetric gauge theories*, Phys. Lett. B 92 (1980) 123-127.

15. B. de Wit, V. Kaplunovsky, J. Louis and D. Lüst, *Perturbative couplings of vector multiplets in N=2 heterotic string vacua*, Nucl. Phys. B 451 (1995) 53-95, hep-th/9504006.

16. J. Louis and K. Förger, *Holomorphic couplings in string theory*, Nucl. Phys. Proc. Suppl. 55 B (1997) 33-64, hep-th/9611184.

17. M. Henneaux and B. Knaepen, *All consistent interactions for exterior form gauge fields*, Phys. Rev. D 56 (1997) 6076-6080, hep-th/9706119.

18. M. Henneaux and B. Knaepen, *The Wess-Zumino consistency condition for p-form gauge theories*, Nucl. Phys. B 548 (1999) 491-526, hep-th/9812140.

19. M. Henneaux and B. Knaepen, *A theorem on first-order interaction vertices for free p-form gauge fields*, Int. J. Mod. Phys. A 15 (2000) 3535-3548, hep-th/9912052.

20. V.I. Ogievetsky and I.V. Polubarinov, *The notoph and its possible interactions*, Yad. Fiz. 4 (1966) 216-223, Sov. J. Nucl. Phys. 4 (1967) 156-161.

21. D.Z. Freedman and P.K. Townsend, *Antisymmetric tensor gauge theories and nonlinear σ-models*, Nucl. Phys. B 177 (1981) 282-296.

22. H. Nicolai and P.K. Townsend, *N=3 supersymmetry multiplets with vanishing trace anomaly: building blocks of the N>3 supergravities*, Phys. Lett. B 98 (1981) 257-260.

23. E. Bergshoeff, M. de Roo, B. de Wit and P. van Nieuwenhuizen, *Ten-dimensional Maxwell-Einstein supergravity, its currents, and the issue of its auxiliary fields*, Nucl. Phys. B 195 (1982) 97-136.

24. G.F. Chapline and N.S. Manton, *Unification of Yang-Mills theory and supergravity in ten dimensions*, Phys. Lett. B 120 (1983) 105-109.

25. M.B. Green and J.H. Schwarz, *Anomaly cancellations in supersymmetric D=10 gauge theory and superstring theory*, Phys. Lett. B 149 (1984) 117-122.

26. J. Wess and J. Bagger, *Supersymmetry and supergravity*, Princeton Series in Physics (Princeton Univ. Press, Princeton 1992).

27. F. Brandt, *Hidden symmetries of supersymmetric p-form gauge theories*, Phys. Lett. B (to appear), hep-th/0009133.

28. F. Brandt, J. Simón and U. Theis, *Exotic gauge theories from tensor calculus*, Class. Quant. Grav. 17 (2000) 1627-1636, hep-th/9910177.

SUPERSYMMETRIC R^4 ACTIONS AND QUANTUM CORRECTIONS TO SUPERSPACE TORSION CONSTRAINTS

KASPER PEETERS[1], PIERRE VANHOVE[1] and ANDERS WESTERBERG[2] *

[1] *CMS/DAMTP, Wilberforce Road, Cambridge CB3 0WA, United Kingdom*

[2] *NORDITA, Blegdamsvej 17, DK-2100 Copenhagen Ø, Denmark*

Abstract. We present the supersymmetrisation of the anomaly-related R^4 term in eleven dimensions and show that it induces no non-trivial modifications to the on-shell supertranslation algebra and the superspace torsion constraints before inclusion of gauge-field terms.[1]

1. Higher-derivative corrections and supersymmetry

The low-energy supergravity limits of superstring theory and D-brane effective actions receive infinite sets of correction terms, proportional to increasing powers of $\alpha' = l_s^2$ and induced by superstring theory massless and massive modes. At present, eleven-dimensional supergravity lacks a corresponding microscopic underpinning that could similarly justify the presence of higher-derivative corrections to the classical Cremmer-Julia-Scherk action [1]. Nevertheless, some corrections of this kind are calculable from unitarity arguments and super-Ward identities in the massless sector of the theory [2] or by anomaly cancellation arguments [3, 4].

Supersymmetry puts severe constraints on higher-derivative corrections. For example, it forbids the appearance of certain corrections (like, e.g., R^3 corrections to supergravity effective actions [5]), and groups terms into various invariants [6–9]. The structure of the invariants that contain anomaly-cancelling terms is of great

---

[1] Based on talks given by K.P. at the SPG meeting, Cambridge, February 2000, by A.W. at the Nordic Network Meeting, Copenhagen, May 2000, and by P.V. at the Fradkin Memorial Conference, Moscow, June 2000, and at the ARW Conference, Kiev, September 2000.

* k.peeters,p.vanhove@damtp.cam.ac.uk, a.westerberg@nordita.dk

S. Duplij and J. Wess (eds.), Noncommutative Structures in Mathematics and Physics, 153–159.
© 2001 *Kluwer Academic Publishers. Printed in the Netherlands.*

importance due to the quantum nature of the anomaly-cancellation mechanism and is the main concern of this note.

Higher-derivative additions to the supergravity actions are in general compatible with supersymmetry only if the transformation rules for the fields also receive higher-derivative corrections:

$$\left(\delta_0 + \sum_n (\alpha')^n \delta_n\right)\left(S_0 + \sum_n (\alpha')^n S_n\right) = 0. \tag{1}$$

As a consequence, the field-dependent structure coefficients on the right-hand side of the supersymmetry algebra,

$$[\delta_1^{\text{susy}}, \delta_2^{\text{susy}}] = \delta^{\text{translation}} + \delta^{\text{susy}} + \delta^{\text{gauge}} + \delta^{\text{Lorentz}}, \tag{2}$$

will be modified as well. When the theory is formulated in superspace the structure of the algebra is related to the structure of the tangent bundle, the link being provided by the constraints on the superspace torsion. In particular, corrections to the parameters modify the superspace constraints. However, since some corrections are reabsorbable by suitable rotations of the tangent bundle basis, not all corrections are physical.

We report here on the supersymmetrization of the anomaly-related terms $(\alpha')^2 B \wedge F^4$ for super-Maxwell theory coupled to $N=1$ supergravity in ten dimensions and $(\alpha'_M)^3 C \wedge t_8 R^4$ (where $(\alpha'_M)^3 = 4\pi (l_P)^6$) in eleven dimensions performed in [10]. In both cases, these superinvariants do not imply any modifications to the superspace constraints. We present here only the more salients aspect of the analysis and refer to the article [10] for computational and bibliographical details.

Our main motivation to look for non-trivial corrections to superspace constraints comes from the link between these constraints and the kappa symmetry of M-branes [11–13] and D-branes [14–16]. Classical kappa invariance of the M- and D-brane world-volume actions — a key requirement for these objects to be supersymmetric — imposes the on-shell constraints on the background superspace supergravity fields, among them the superspace torsion. For this reason, any non-trivial modification to the constraints is expected to require new terms in the world-volume actions for the branes in order for kappa symmetry to be preserved.

2. Construction of an abelian F^4 superinvariant in D=10

As a first step in our analysis of the implications of higher-derivative corrections to the supersymmetry algebra, we discuss the construction of the abelian $(\alpha')^2(t_8 F^4 - B \wedge F^4)$ for $N=1$ super-Maxwell theory coupled to gravity in ten dimensions.

The field content of the on-shell super-Maxwell theory comprises an abelian vector A_μ and a negative-chirality Majorana-Weyl spinor χ. Since we are interested in local supersymmetry invariance we have to take into account also the interactions with the zehnbein $e_\mu{}^r$, the negative-chirality Majorana-Weyl gravitino ψ_μ and the two-form $B_{\mu\nu}$ from the supergravity multiplet. The classical action (leaving out the gravitational sector)

$$S_{F^2} = \int d^{10}x\, e\left[-\frac{1}{4}F_{\mu\nu}F^{\mu\nu} - 8\bar{\chi}\,\slashed{D}(\omega)\chi + 2\bar{\chi}\Gamma^\mu\Gamma^{\nu\rho}\psi_\mu F_{\nu\rho}\right] \quad (3)$$

is invariant under the local supersymmetry transformations

$$\delta A_\mu = -4\,\bar{\epsilon}\Gamma_\mu\chi, \qquad \delta\chi = \frac{1}{8}\Gamma^{\mu\nu}\epsilon F_{\mu\nu}, \quad (4)$$

For local supersymmetry we have to consider the transformations of the supergravity multiplet fields as well (neglecting terms proportional to the two-form $B_{\mu\nu}$ and the corresponding field strength, $H_{\mu\nu\rho}$):

$$\delta e_\mu{}^r = 2\bar{\epsilon}\Gamma^r\psi_\mu, \quad \delta\psi_\mu = D_\mu(\omega)\epsilon + \cdots, \quad \delta B_{\mu\nu} = \frac{1}{\sqrt{2}}\bar{\epsilon}\Gamma_{[\mu}\psi_{\nu]}. \quad (5)$$

The F^4 action invariant under the local supersymmetry transformations listed above is [17, 18, 10]:

$$S_{F^4} = \frac{(\alpha')^2}{32}\int d^{10}x\Big[\tfrac{1}{6}e\,t_8^{(r)}F_{r_1r_2}\cdots F_{r_7r_8} + \tfrac{1}{3\sqrt{2}}\epsilon_{10}^{(r)}B_{r_1r_2}F_{r_3r_4}\cdots F_{r_9r_{10}}$$

$$-\tfrac{32}{5}e\,t_8^{(r)}\eta_{r_2r_3}(\bar{\chi}\Gamma_{r_1}D_{r_4}(\omega)\chi)F_{r_5r_6}F_{r_7r_8} + \tfrac{12\cdot32}{5}e(\bar{\chi}\Gamma_{r_1}D_{r_2}(\omega)\chi)F^{r_1m}F_m{}^{r_2}$$

$$-\tfrac{16}{5!}\epsilon_{10}^{(r)}(\bar{\chi}\Gamma_{r_1\cdots r_4}\Gamma_{r_5}D_{r_6}(\omega)\chi)F_{r_7r_8}F_{r_9r_{10}} + \tfrac{16}{3}e\,t_8^{(r)}(\bar{\psi}_{r_1}\Gamma_{r_2}\chi)F_{r_3r_4}F_{r_5r_6}F_{r_7r_8}$$

$$+\tfrac{8}{3}e\,(\bar{\psi}_m\Gamma^{mr_1\cdots r_6}\chi)F_{r_1r_2}\cdots F_{r_5r_6}\Big]. \quad (6)$$

Note that our string-amplitude based analysis has allowed us to group also the fermionic terms using the well-known t_8 tensor. The local supersymmetry invariance of the combined action $S_{F^2} + S_{F^4}$ requires that the supersymmetry transformations be modified according to ($F^2 := F^{mn}F_{nm}$)

$$\delta A_\mu = -4\,\bar{\epsilon}\Gamma_\mu\chi - (\alpha')^2\Big[\frac{1}{4}(\bar{\epsilon}\Gamma_\mu\chi)F^2$$

$$- (\bar{\epsilon}\Gamma^m\chi)F_{m\mu}^2 - \frac{1}{8}(\bar{\epsilon}\Gamma^{r_1\cdots r_4}{}_\mu\chi)F_{r_1r_2}F_{r_3r_4}\Big],$$

$$\delta\chi = \frac{1}{8}\Gamma^{\mu\nu}\epsilon F_{\mu\nu} + \frac{1}{768}(\alpha')^2\Big[t_8^{(r)}\Gamma_{r_7r_8}\epsilon - \Gamma^{r_1\cdots r_6}\epsilon\Big]F_{r_1r_2}F_{r_3r_4}F_{r_5r_6}. \quad (7)$$

It can be verified that the structure of the supersymmetry algebra is not modified by the order-$(\alpha')^2$ corrections [17, 18, 10]:

$$[\delta_{\epsilon_1}^{(\alpha')^0} + \delta_{\epsilon_1}^{(\alpha')^2}, \delta_{\epsilon_2}^{(\alpha')^0} + \delta_{\epsilon_2}^{(\alpha')^2}]A_\mu = [\delta_{\epsilon_1}^{(\alpha')^0}, \delta_{\epsilon_2}^{(\alpha')^0}]A_\mu + \mathcal{O}((\alpha')^4). \quad (8)$$

Consequently, the structure of the superspace torsion constraints will be the same as for the classical theory to this order. This observation is related to the fact that it is possible to supersymmetrise the Dirac-Born-Infeld actions while imposing only the classical constraints [19].

3. Construction of the $C \wedge R^4$ superinvariant in D=11

Noticing the close parallel between the classical supersymmetry transformations for the super-Maxwell and the supergravity fields

$$
\delta\chi \quad = \tfrac{1}{8}\Gamma^{\mu\nu}\epsilon\, F_{\mu\nu}\,, \qquad \delta\psi_{rs} \quad = \frac{1}{8}\Gamma^{\mu\nu}\epsilon\, R_{\mu\nu rs} + \cdots\,, \tag{9}
$$

$$
\delta F_{\mu\nu} = -8D_{[\mu}(\bar{\epsilon}\Gamma_{\nu]}\chi)\,, \qquad \delta R_{\mu\nu}{}^{rs} = -8D_{[\mu}(\bar{\epsilon}\Gamma_{\nu]}\psi^{rs})
$$

$$
+4D_{[\mu}(\bar{\epsilon}\Gamma_{\nu]}\psi^{rs} + 2\,\bar{\epsilon}\Gamma^{[r}\psi^{s]}{}_{\nu]}) + \cdots\,,
$$

it is tempting to make the following substitution in the super-Maxwell action:

$$
F_{r_1 r_2} \to R_{r_1 r_2 s_1 s_2}\,, \quad \chi \to \psi_{s_1 s_2}\,, \quad D_r\chi \to D_r\psi_{s_1 s_2}\,. \tag{10}
$$

Unfortunately, the difference in structure between the equations of motion for the gauge potential and the spin connection implies that the previous mapping does not commute with supersymmetry, as can be seen by the presence of the second line in the supersymmetry transformation of the Riemann tensor above. Another crucial difference between the super-Maxwell and supergravity cases is that, when subtracting all the lowest-order equations of motions, it is necessary to make the following substitution for the Riemann tensor:

$$
R_{mn}{}^{pq} \to W_{mn}{}^{pq} - \frac{16}{d-2}\delta_{[m}{}^{[p}(\bar{\psi}_{|r|}\Gamma^{|r|}\psi_{n]}{}^{q]} - \bar{\psi}^{|r|}\Gamma^{q]}\psi_{n]r})\,. \tag{11}
$$

Taking all these facts into account, as well as the information from string-amplitude analysis that the extra s-type indices in (10) should be contracted with an additional $t_8^{(s)}$ tensor, we arrive at the following M-theory $C \wedge R^4$ invariant after lifting to eleven dimensions [10]:

$$(\alpha'_M)^{-3}\mathcal{L}_{\Gamma[0]} = + \frac{1}{192} e t_8^{(r)} t_8^{(s)} W_{r_1 r_2 s_1 s_2} \cdots W_{r_7 r_8 s_7 s_8} \tag{12}$$

$$+ \frac{1}{(48)^2} \varepsilon^{t_1 t_2 t_3 r_1 \cdots r_8} t_8^{(s)} C_{t_1 t_2 t_3} W_{r_1 r_2 s_1 s_2} \cdots W_{r_7 r_8 s_7 s_8} \, ,$$

$$(\alpha'_M)^{-3}\mathcal{L}_{\Gamma[1]} = - 4 e t_8^{(s)} (\bar\psi_{s_1 s_2} \Gamma_{r_1} D_{r_2} \psi_{s_3 s_4}) W_{r_1 r_3 s_5 s_6} W_{r_3 r_2 s_7 s_8}$$

$$- \frac{1}{4} e t_8^{(s)} (\bar\psi_{r_1} \Gamma_{r_2} \psi_{s_7 s_8}) W_{r_1 r_2 s_1 s_2} W_{mn s_3 s_4} W_{nm s_5 s_6}$$

$$- e t_8^{(s)} (\bar\psi_{r_1} \Gamma_{r_2} \psi_{s_7 s_8}) W_{r_1 m s_1 s_2} W_{mn s_3 s_4} W_{nr_2 s_5 s_6}$$

$$+ e t_8^{(s)} (\bar\psi_{r_1} \Gamma_{s_7} \psi_{r_2 s_8}) W_{r_1 r_2 s_1 s_2} W_{mn s_3 s_4} W_{nm s_5 s_6}$$

$$- 4 e t_8^{(s)} (\bar\psi_{r_1} \Gamma_{s_7} \psi_{r_2 s_8}) W_{r_1 m s_1 s_2} W_{mn s_3 s_4} W_{nr_2 s_5 s_6}$$

$$+ \frac{2}{9} e t_8^{(s)} (\bar\psi_m \Gamma_n \psi_{m s_8}) W_{pq s_1 s_2} W_{qp s_3 s_4} W_{n s_7 s_5 s_6}$$

$$- \frac{8}{9} e t_8^{(s)} (\bar\psi_m \Gamma_n \psi_{m s_8}) W_{np s_1 s_2} W_{pq s_3 s_4} W_{q s_7 s_5 s_6} \, ,$$

$$(\alpha'_M)^{-3}\mathcal{L}_{\Gamma[3]} = + 2 e t_8^{(s)} (\bar\psi_{s_5 s_6} \Gamma_{r_1 r_2 r_3} D_{r_4} \psi_{s_7 s_8}) W_{r_1 r_2 s_1 s_2} W_{r_3 r_4 s_3 s_4}$$

$$- \frac{1}{8} e t_8^{(s)} (\bar\psi_m \Gamma^{m r_1 r_2} \psi_{s_7 s_8}) W_{r_1 r_2 s_1 s_2} W_{pn s_3 s_4} W_{np s_5 s_6}$$

$$+ \frac{1}{2} e t_8^{(s)} (\bar\psi_m \Gamma^{m r_1 r_2} \psi_{s_7 s_8}) W_{r_1 p s_1 s_2} W_{pn s_3 s_4} W_{nr_2 s_5 s_6}$$

$$+ e t_8^{(s)} (\bar\psi_m \Gamma^{r_1 r_2 r_3} \psi_{s_7 s_8}) W_{r_1 r_2 s_1 s_2} W_{mn s_3 s_4} W_{nr_3 s_5 s_6} \, ,$$

$$(\alpha'_M)^{-3}\mathcal{L}_{\Gamma[5]} = + \frac{1}{8} e t_8^{(s)} (\bar\psi^{r_6} \Gamma^{r_1 \cdots r_5} \psi_{s_7 s_8}) W_{r_1 r_2 s_1 s_2} W_{r_3 r_4 s_3 s_4} W_{r_5 r_6 s_5 s_6} \, ,$$

$$(\alpha'_M)^{-3}\mathcal{L}_{\Gamma[7]} = + \frac{1}{48} e t_8^{(s)} (\bar\psi_m \Gamma_{m r_1 \cdots r_6} \psi_{s_7 s_8}) W_{r_1 r_2 s_1 s_2} W_{r_3 r_4 s_3 s_4} W_{r_5 r_6 s_5 s_6} \, .$$

Even if the elfbein supersymmetry transformation rule receives $(\alpha'_M)^3$ modifications, by computing the closure of the supersymmetry algebra (2), we find [10] that the translation parameter does *not* receive corrections that cannot be absorbed by field redefinitions.

4. Superspace approach

It can be argued that in the completely general Ansatz for the dimension zero torsion constraint

$$T_{ab}{}^r = (C\Gamma^{r_1})_{ab} X^r{}_{r_1} + \frac{1}{2!} (C\Gamma^{r_1 r_2})_{ab} X^r{}_{r_1 r_2} + \frac{1}{5!} (C\Gamma^{r_1 \cdots r_5})_{ab} X^r{}_{r_1 \cdots r_5},$$

(13)

the coefficient $X^r{}_{r_1}$ can be set equal to $\delta^r{}_{r_1}$, and all fully antisymmetric tensors contained in $X^r{}_{r_1 r_2}$ and $X^r{}_{r_1 \cdots r_5}$ to zero by a choice of tangent bundle basis (see, e.g., [21]). This leaves as the only candidates for non-trivial M-theory corrections the SO(1,10) representations **429** and **4290** of the $\Gamma_{[2]}$ and $\Gamma_{[5]}$ coefficients, respectively. Therefore, from the component analysis of the previous section we conclude that the higher-order invariant (12) does not induce any modifications to the torsion constraint (13).

Howe showed in [20], that imposing *only* the constraint

$$T_{ab}{}^r = (C\Gamma^r)_{ab}$$

(14)

on the dimension-zero component of the superspace torsion, the classical, on-shell, eleven-dimension supergravity theory of [1] follows *without* having to introduce a four-form superfield. An analysis of the superspace Bianchi identities for this superfield would necessitate a more complete analysis of the R^4 invariant (12) with the inclusion of higher powers of the four-form field strength.

In this context, let us also mention that in parallel with our component-space based approach to uncover the superspace underlying M-theory, a complementary line of attack based on an analysis of the superspace Bianchi identities has been initiated by Cederwall et al. in [21].

Acknowledgements

K.P. and P.V. are supported by PPARC grant PPA/G/S/1998/00613. P.V. thanks NATO for partial support.

References

1. E. Cremmer, B. Julia and J. Scherk, *Supergravity Theory in Eleven Dimensions*, Phys. Lett. **B76** (1978) 409.
2. Z. Bern, L. Dixon, D.C. Dunbar, M. Perelstein and J.S. Rozowski, *On the relationship between Yang-Mills theory and gravity and its implication for ultraviolet divergences* Nucl. Phys. **B530** (1998) 401, hep-th/9802162.
3. M.J. Duff, J.T. Liu and R. Minasian, *Eleven Dimensional Origin of String/String Duality: A One Loop Test*, Nucl. Phys. **B452** (1995) 261, hep-th/9506126.

4. C. Vafa and E. Witten, *A One-Loop Test of String Duality*, Nucl. Phys. **B447** (1995) 261, hep-th/9505053.
5. M.T. Grisaru, *Two-loop renormalizability of Supergravity*, Phys. Lett. **66B** (1977) 75.
6. E. A. Bergshoeff and M. de Roo, *The Quartic Effective Action of the Heterotic String and Supersymmetry*, Nucl. Phys. **B328** (1989) 439.
7. M. de Roo, H. Suelmann and A. Wiedemann, *The Supersymmetric Effective Action of the Heterotic String in Ten Dimensions*, Nucl. Phys. **B405** (1993) 326, hep-th/9210099.
8. H. Suelmann, *String effective actions and supersymmetry*, PhD Thesis Groningen University, 1994.
9. M.B. Green and S. Sethi, *Supersymmetry Constraint on Type IIB Supergravity*, Phys.Rev. **D59** (1999) 046006, hep-th/9808061.
10. K. Peeters, P. Vanhove and A. Westerberg, *Supersymmetric Higher-Derivative Actions and their Associated Superalgebras from String Theory*, DAMTP-2000-82, to appear.
11. E. Bergshoeff, E. Sezgin, and P.K. Townsend, *Properties of the eleven-dimensional supermembrane theory*, Ann. Phys. **185** (1988) 330.
12. I. Bandos, K. Lechner, A. Nurmagambetov, P. Pasti, D. Sorokin and M. Tonin, *Covariant action for the superfive-brane of M theory*, Phys. Rev. Lett. **78** (1997) 4332, hep-th/9701037.
13. M. Aganagic, J. Park, C. Popescu and J.H. Schwarz, *World volume action of the M-theory five-brane*, Nucl. Phys. **B496** (1997) 191, hep-th/9701166.
14. M. Cederwall, A. von Gussich, B.E.W. Nilsson, P. Sundell and A. Westerberg, *The Dirichlet super-p-branes in ten-dimensional type IIA and type IIB supergravity*, Nucl. Phys. **B490** (1997) 179, hep-th/9611159.
15. E. Bergshoeff and P.K. Townsend, *Super D-branes* Nucl. Phys. **B490** (1997) 145, hep-th/9611173.
16. M. Aganagic, C. Popescu and J.H. Schwarz, *Gauge-invariant and gauge-fixed D-brane actions*, Nucl. Phys. **B495** (1997) 99, hep-th/9612080.
17. R.R. Metsaev and M.A. Rakhmanov, *Fermionic terms in the open superstring effective action*, Phys. Lett. **B193** (1987) 202.
18. E. Bergshoeff, M. Rakowski and E. Sezgin, *Higher derivative super Yang-Mills theories* Phys. Lett. **B185** (1987) 371.
19. J. Bagger and Al. Galperin, *A new Goldstone multiplet for partially broken supersymmetry*, Phys. Rev **D55** (1997) 1091, hep-th/9608177.
20. P.S. Howe, *Weyl Superspace*, Phys.Lett. **B415** (1997) 149, hep-th/9707184.
21. M. Cederwall, U. Gran, M. Nielsen and B.E.W. Nilsson, *Manifestly Supersymmetric M-Theory*, hep-th/0007035.

MASSIVE SUPERPARTICLE WITH SPINORIAL CENTRAL CHARGES

S. FEDORUK
Ukrainian Engineering-Pedagogical Academy, 61003 Kharkiv, 16 Universitetska Str., Ukraine

V. G. ZIMA
Kharkiv National University, 61077 Kharkiv, 4 Svobody Sq., Ukraine

Abstract. We construct the manifestly Lorenz-invariant formulation of the $N = 1$ $D = 4$ massive superparticle with spinorial central charges. The model possesses from one to three κ-symmetries. The local transformations of κ-symmetry are written out. The using of index spinor for construction of the tensorial central charges is considered. The equivalence at the classical level between the massive $D = 4$ superparticle with one κ-symmetry and the massive $D = 4$ spinning particle is obtained.

1. Introduction

Recently it became clear that some interesting supersymmetric theories admit besides scalar central charges which are presented in conventional $D = 4$ Poincare supersymmetry [1, 2] also nonscalar central charges: tensorial [3]-[7] or spinor [8, 9] ones. Although the tensorial central charges in the supersymmetry algebra are usually associated with topological contributions of the extended objects it is attractive to consider the pure superparticle models having symmetry of this kind. Such models were firstly obtained in massless case [10] for $D = 4$ with two or three local κ-symmetries.

Central charge is a quantity which is inert with respect to SUSY but transforms under internal or Lorentz groups.

We construct the model of the massive $D = 4$ nonextended superparticle with spinorial central charges possessing one or two local κ–symmetries [1]. In particular in such a way we obtain the superparticle with a single κ-symmetry which is equivalent to the usual spinning (spin $1/2$) particle [12, 13] in the spinorial central charge background.

---

[1] The Lagrangian of the massive superparticle with vector central charge and with two κ–symmetries has been presented already in [11]

S. Duplij and J. Wess (eds.), Noncommutative Structures in Mathematics and Physics, 161–180.
© 2001 *Kluwer Academic Publishers. Printed in the Netherlands.*

Generalized central extension of $N = 1$ 4-dimensional supersymmetry algebra

$$\{Q, \overline{Q}\} = 2(\gamma^\mu) P_\mu \tag{1}$$

with Majorana supercharges $Q^+ = Q$, energy–momentum vector P and γ–matrices in Majorana representation, so that $C = \gamma_0 = C^{-1}$ and as in any representation we have $C^T = -C$, can be written in the form

$$\{Q, Q\} = 2\mathcal{Z} \tag{2}$$

where $\mathcal{Z}^T = \mathcal{Z}$ is the most general symmetric matrix of Abelian generalized central charges with a total of ten real entries. In decomposition of this matrix on the basis defined by products of γ–matrices we have tensorial central charges as coefficients

$$\mathcal{Z}C = (\gamma^\mu)\mathcal{P}_\mu + \frac{i}{2}(\gamma^{\mu\nu})Z_{\mu\nu} \tag{3}$$

where $\mathcal{P}$ is (in general) a linear combination of the energy–momentum vector P and a "string charge". Six real charges $Z_{\mu\nu} = -Z_{\nu\mu}$ are related to the symmetric complex Weyl spin–tensor $Z_{\alpha\beta} = Z_{\beta\alpha}$ by the relation

$$Z_{\mu\nu} = \frac{1}{2}(\bar{Z}_{\dot\alpha\dot\beta}\tilde\sigma_{\mu\nu}^{\dot\alpha\dot\beta} - Z_{\alpha\beta}\sigma_{\mu\nu}^{\alpha\beta}). \tag{4}$$

The spin–tensors $Z_{\alpha\beta}$ and $\bar{Z}_{\dot\alpha\dot\beta} = \overline{(Z_{\alpha\beta})}$ represent the self–dual and anti–self–dual parts of the central charge matrix. The tensorial central charges commute with four–momentum and transform as components of a tensor under the Lorentz group transformations.

There are two types of model with central charges. Some of them have exact SUSY due to presence of special tensorial central charge coordinates which transform together with Grassmannian spinor θ and space–time vector x. Their derivative with respect to development parameter τ absorbs the tensorial part in SUSY variation of product $\theta\dot\theta$ in complete analogy with absorption of the vector part in the variation by the space–time vector x under ordinary SUSY transformations. Other models have no similar coordinates and their SUSY is reached only on the mass shell. Here we examine namely the second type model which is obtained by adding to coordinates of nonextended massive superparticle certain dynamical even spinor ζ. This spinor parameterizes [15] in the rest frame of particle the compact group manifold of quantum–theory rotation group $SU(2)$.

In this paper we use the $D = 4$ spinor conventions of [2]. Majorana and Weyl odd spinors are denoted by the same literal. One can easy identifies the meaning of a denotation viewing its nearest encirclement. Bispinor expressions with Majorana spinors are written, as a rule, in conventional form which makes obvious a transition to Weyl spinors.

2. Action and its symmetries

2.1. SUPERPARTICLE LAGRANGIAN

Let us take for superparticle Lagrangian the expression

$$
\begin{aligned}
L &= L_{\text{super}} + L_{\text{SCC}} \\
&\equiv p\dot{x} + i\bar{\theta}\mathcal{Z}C\dot{\theta} - \frac{e}{2}(p^2 + m^2) + L_{\text{SCC}} \\
&\equiv p\dot{x} + i\mathcal{P}_{\alpha\dot{\beta}}(\theta^\alpha\dot{\bar{\theta}}^{\dot{\beta}} - \dot{\theta}^\alpha\bar{\theta}^{\dot{\beta}}) \\
&\quad + i\mathcal{Z}_{\alpha\beta}\theta^\alpha\dot{\theta}^\beta + i\bar{\mathcal{Z}}_{\dot{\alpha}\dot{\beta}}\bar{\theta}^{\dot{\alpha}}\dot{\bar{\theta}}^{\dot{\beta}} - \frac{e}{2}(p^2 + m^2) + L_{\text{SCC}}\,.
\end{aligned}
\tag{5}
$$

Here e is Lagrange multiplier for mass constraint $p^2 + m^2 \approx 0$. Let us take spinorial central charge Lagrangian L_{SCC}, i.e. a part of Lagrangian (5) containing kinetic term for commuting spinor coordinates ζ^α, $\bar{\zeta}^{\dot{\alpha}} = \overline{(\zeta^\alpha)}$ and generating constraint on these variables, in the form

$$
L_{\text{SCC}} = \dot{\zeta}v + \bar{v}\dot{\bar{\zeta}} - \lambda(\zeta\hat{p}\bar{\zeta} - j)\,.
\tag{6}
$$

where v is canonical conjugate momentum for ζ and λ is Lagrange multiplier for "spin constraint"

$$
r - j \equiv \zeta\hat{p}\bar{\zeta} - j \approx 0 \qquad (r \equiv \zeta\hat{p}\bar{\zeta})\,.
\tag{7}
$$

This constraint gives us at $j \neq 0$ the completeness condition

$$
r\delta_\alpha^\beta = \zeta_\alpha(\bar{\zeta}\tilde{p})^\beta + (\hat{p}\bar{\zeta})_\alpha\zeta^\beta \qquad \text{and} \quad \text{c. c.}
\tag{8}
$$

for spinors ζ, $\hat{p}\bar{\zeta}$. Here matrices $\hat{p}$ and $\tilde{p}$ are the contractions of the space–time momentum p and σ–matrices with lower and upper spinor indices, respectively. Similar to j numerical constant plays the role of "classical spin" in the index spinor formalism [14, 11, 15] which is attractive in task of particle spin description with commuting spinors. In what follows it is important that some equations of motion following from the Lagrangian (5) read as

$$
\dot{\zeta} = 0, \quad \dot{p} = 0\,.
\tag{9}
$$

We can construct the vector and tensor central charges $\mathcal{Z}$ in the Lagrangian (5) with the spinors ζ, $\hat{p}\bar{\zeta}$ without derivatives in τ. So on shell, due to (9), these quantities will be constants. Such constructions for central charges $\mathcal{Z}$ in terms of spinorial ones do not modify the equations (9) and are specified below.

2.2. SUPERSYMMETRY

SUSY transformations can be viewed as global translations in odd spinor coordinates θ accompanied translations in space–time vector x and possibly some scalar or tensorial even central charge coordinates y which leave invariant certain differential 1–forms. Mentioned forms covariantly transform under global Lorentz and internal groups and are invariant under space–time translations. These forms are sums of corresponding even coordinate differentials and terms which are bilinear in odd spinor coordinates and their differentials. Using such forms one can construct theories with exact off–shell SUSY but in absence of some central charge coordinates and corresponding differential forms it is possibly to reach SUSY only on shell. In the case under consideration we have unique fundamental vectorial superform

$$\omega^\mu \equiv \dot\omega^\mu \, d\tau = dx + id\bar\theta\gamma^\mu\theta = dx + i\theta\sigma^\mu d\bar\theta - id\theta\sigma^\mu\bar\theta \qquad (10)$$

and usual SUSY transformations are

$$\delta x^\mu = -i\bar\theta\gamma^\mu\delta\theta = +i\theta\sigma^\mu\delta\bar\theta - i\delta\theta\sigma^\mu\bar\theta \qquad (11)$$

with constant $\delta\theta$. Supercharges can be obtained as coefficients at the derivative $(\delta\theta)^\cdot$ in the integrand of the local variation of the action in Hamiltonian form. The variation of the Lagrangian (5) is

$$\delta L = i\mathcal{P}(\theta\sigma(\delta\bar\theta)^\cdot - (\delta\theta)^\cdot\sigma\bar\theta) - 2i(Z_{\alpha\beta}(\delta\theta^\alpha)^\cdot\theta^\beta + \bar Z_{\dot\alpha\dot\beta}(\delta\bar\theta^{\dot\alpha})^\cdot\bar\theta^{\dot\beta}) -$$
$$i(\dot Z_{\alpha\beta}\delta\theta^\alpha\theta^\beta + \dot{\bar Z}_{\dot\alpha\dot\beta}\delta\bar\theta^{\dot\alpha}\bar\theta^{\dot\beta}) + (iZ_{\alpha\beta}\delta\theta^\alpha\theta^\beta + i\bar Z_{\dot\alpha\dot\beta}\delta\bar\theta^{\dot\alpha}\bar\theta^{\dot\beta})^\cdot +$$
$$\mathcal{P}(\delta\omega)^\cdot - (p-\mathcal{P})^\cdot\delta x + ((p-\mathcal{P})\delta x)^\cdot . \qquad (12)$$

We see that $\delta L = 0$ if $(\delta\theta)^\cdot = 0$ up to surface terms in absence of tensorial central charge coordinates. Equations of motion (9) are twice used for this conclusion. In the first place we use these equations to change multiplier p at $\dot x$ by $\mathcal{P}$ and as consequence to collect variations of x and θ at vectorial part of Z in variation of superform (10). In the second place we use equations (9) to represent variation with tensorial part of Z as total derivative. Constancy of $\delta\theta$ is used as well. The price for the presence of the supersymmetry is the infinite number of the spin states in the spectrum. At the restriction of the bosonic spinor sector to the index spinor one [14, 11, 15] the number of the states in spectrum becomes finite but the supersymmetry disappears. But in both cases the models possess local κ-symmetries.

In coordinate representation for odd variables one obtains as generators of SUSY transformations

$$Q = \frac{\partial}{\partial\theta} + Z\theta . \qquad (13)$$

In terms of Weyl spinor we have

$$Q_\alpha = \frac{\partial}{\partial\theta^\alpha} + (\hat{P}\bar{\theta})_\alpha + \theta^\beta Z_{\beta\alpha}, \tag{14}$$

$$\bar{Q}_{\dot\alpha} = \frac{\partial}{\partial\bar{\theta}^{\dot\alpha}} + (\theta\hat{P})_{\dot\alpha} + \bar{Z}_{\dot\alpha\dot\beta}\bar{\theta}^{\dot\beta}. \tag{15}$$

So generators of SUSY contain "anomalous" extra pieces with central charges. The algebra (2) of SUSY generators

$$\{Q_\alpha, Q_\beta\} = 2Z_{\alpha\beta}, \quad \{Q_\alpha, \bar{Q}_{\dot\beta}\} = 2P_{\alpha\dot\beta} \tag{16}$$

is the $N = 1$ $D = 4$ SUSY algebra extended by tensorial central charges.

One can introduce terms with derivatives of central charge coordinates y to the multipliers at central charges in the Lagrangian (5). Then the model becomes SUSY invariant not only quasi-invariant.

2.3. κ–SYMMETRY

Grassmannian constraints of the model (5) are

$$d_\theta \equiv -ip_\theta - Z\theta \approx 0. \tag{17}$$

In terms of Weyl spinor we have

$$d_{\theta\alpha} \equiv -ip_{\theta\alpha} - (\hat{P}\bar{\theta})_\alpha - \theta^\beta Z_{\beta\alpha} \approx 0, \tag{18}$$

$$\bar{d}_{\theta\dot\alpha} \equiv -i\bar{p}_{\theta\dot\alpha} - (\theta\hat{P})_{\dot\alpha} - \bar{Z}_{\dot\alpha\dot\beta}\bar{\theta}^{\dot\beta} \approx 0. \tag{19}$$

Poisson brackets algebra of constraints (17) is

$$\{d_\theta, d_\theta\} = 2iZ. \tag{20}$$

In terms of Weyl spinors it is

$$\{d_{\theta\alpha}, d_{\theta\beta}\} = 2iZ_{\alpha\beta}, \quad \{\bar{d}_{\theta\dot\alpha}, \bar{d}_{\theta\dot\beta}\} = 2i\bar{Z}_{\dot\alpha\dot\beta},$$

$$\{d_{\theta\alpha}, \bar{d}_{\theta\dot\beta}\} = 2iP_{\alpha\dot\beta}. \tag{21}$$

Let us analyze all possibilities of different numbers of κ–symmetries. The number of κ–symmetries is defined rank of Poisson bracket matrix for the fermionic constraints. In considered case

$$\det Z = (P^2)^2 - P_{\alpha\dot\alpha}P_{\beta\dot\beta}Z^{\alpha\beta}\bar{Z}^{\dot\alpha\dot\beta} + \frac{1}{4}Z^{\alpha\beta}Z_{\alpha\beta}\bar{Z}_{\dot\alpha\dot\beta}\bar{Z}^{\dot\alpha\dot\beta}. \tag{22}$$

The characteristic polynomial, which is obtained by substitution $p^0 \rightarrow p^0 - \lambda$ in (22), has the form

$$\lambda^2(\lambda^2 - 4p^0\lambda + A) + 2B\lambda + \det \mathcal{Z}$$

where

$$A = 4(p^0)^2 - 2\mathcal{P}^2 - \sigma^0_{\alpha\dot{\alpha}}\sigma^0_{\beta\dot{\beta}}Z^{\alpha\beta}\bar{Z}^{\dot{\alpha}\dot{\beta}},$$

$$B = 2p^0\mathcal{P}^2 - \sigma^0_{\alpha\dot{\alpha}}\mathcal{P}_{\beta\dot{\beta}}Z^{\alpha\beta}\bar{Z}^{\dot{\alpha}\dot{\beta}}.$$

Thus if $A = 0$, $B = 0$, $\det \mathcal{Z} = 0$ we have three fermionic first class constraints and superparticle model with $3/4$ conserved SUSY. In case $B = 0$, $\det \mathcal{Z} = 0$ but $A \neq 0$ two eigenvalues λ among four ones are zero and superparticle model conserves $1/2$ SUSY. Only in case $\det \mathcal{Z} = 0$ but $A \neq 0$, $B \neq 0$ we have system with $1/4$ conserved SUSY.

Noted that some superparticle model associated with superalgebra with tensorial central charges was considered in [16]. Bosonic constraints of the model [16] are generalized mass shell condition

$$\mathcal{Z}C\mathcal{Z} = 0 \tag{23}$$

which in Weyl spinor notation reads

$$Z_\alpha{}^\beta Z_\beta{}^\gamma = \mathcal{P}^2\delta_\alpha{}^\gamma, \quad \bar{Z}^{\dot{\alpha}}{}_{\dot{\beta}}\bar{Z}^{\dot{\beta}}{}_{\dot{\gamma}} = \mathcal{P}^2\delta^{\dot{\alpha}}{}_{\dot{\gamma}}, \quad Z_\alpha{}^\beta\mathcal{P}_{\beta\dot{\alpha}} + \mathcal{P}_{\alpha\dot{\beta}}\bar{Z}^{\dot{\beta}}{}_{\dot{\alpha}} = 0. \tag{24}$$

It is easy to see that the model [16] preserves two supersymmetries or more. Preserving of one supersymmetry is not possible in that model. From (24) we have $B = 0$, $\det \mathcal{P} = 0$ and thus necessarily two eigenvalues λ among four ones are zero. Thus the condition (24) are too much strong to have system with $1/4$ conserved SUSY.

3. Equivalence between massive spinning particle and superparticle with one κ–symmetry

3.1. SPINNING PARTICLE IN THE PSEUDOCLASSICAL APPROACH

In the pseudoclassical approach the Lagrangian of spinning particle has the following form [12, 13]

$$L_{1/2} = p^\mu\dot{x}_\mu + \frac{i}{2}(\psi^\mu\dot{\psi}_\mu + \psi_5\dot{\psi}_5) - \frac{e}{2}(p^2 + m^2) - i\chi(p\psi + m\psi_5). \tag{25}$$

The spin variables in this description are the Grassmannian (pseudo)vector ψ_μ and the Grassmannian (pseudo)scalar ψ_5. Besides mass constraint $T \equiv p^2 + m^2 \approx 0$

in Hamiltonian formalism the physical sector of the model is subjected to the Grassmannian constraints from which one Dirac constraint

$$D \equiv p^\mu \psi_\mu + m\psi_5 \approx 0 \qquad (26)$$

plays the role of the first class constraint and five self-conjugacy condition for the Grassmannian variables

$$g^\mu \equiv p_\psi^\mu - \frac{i}{2}\psi^\mu \approx 0, \qquad g_5 \equiv p_{\psi 5} - \frac{i}{2}\psi_5 \approx 0 \qquad (27)$$

are the second class constraints. Thus the number of physical odd degrees of freedom in the model (25) is [number of $(\psi_\mu, \psi_5, p_{\psi\mu}, p_{\psi 5})$] – [number of the second class constraints (g_μ, g_5)] – 2[number of the first class constraint (D)] = 3.

The usual model of the massive CBS superparticle [22] with Grassmannian spinor coordinates θ^α, $\bar{\theta}^{\dot\alpha}$ has only the fermionic spinor constraints

$$d_{\theta\alpha} \equiv -ip_{\theta\alpha} - (\hat{p}\bar{\theta})_\alpha \approx 0, \qquad \bar{d}_{\theta\dot\alpha} \equiv -i\bar{p}_{\theta\dot\alpha} - (\theta\hat{p})_{\dot\alpha} \approx 0$$

which all are the second class constraints. Here the number of the physical odd degrees of freedom is [number of $(\theta^\alpha, \bar{\theta}^{\dot\alpha}, p_{\theta\alpha}, \bar{p}_{\theta\dot\alpha})$] – [number of $(d_\theta, \bar{d}_\theta)$] = 4. In order to obtain desired three physical fermionic degrees of freedom it is necessary that from fermionic four spinor constraints three constraints are of the second class whereas one constraint should be of the first class. Such situation with nonsymmetric separation of the fermionic constraints into the ones of first and second class has been proposed in massless superparticle models [10] as well as in the massive particle case [23]. Precisely the situation with one first class fermionic constraint has been presented in [23] in the construction of $N = 4 \to N = 1$ PBGS in $d = 1$. The relation between that model and our one will be given below. Thus in the massive case the equivalence of spinning particle and superparticle with tensorial central charges with one κ-symmetry is expected. Let us note that in massless case [24, 25] the spinning particle is equivalent, at least on classical level, to the usual CBS superparticle without any central charges. This fact of identifying the local fermionic invariances of spinning particle and κ-symmetries of superparticle is essential for superfield formulation of massless superparticle theory [24, 25] and consequent generalizations on superbranes [26].

Accounting above mentioned preliminary arguments for the possible relation between massive spinning particle and massive superparticle with tensorial central charges we take the following way for construction of the superparticle model. We shall realize the covariant transition, under preservation of the physical content, from the model of the massive spinning particle to the system with Grassmannian spinor variables. As result of this procedure we arrive at model of the $N = 1$ $D = 4$ massive superparticle with tensorial central charges possessing one gauge fermionic invariance (κ-symmetry).

Covariant transition from the Grassmannian vector ψ_μ and scalar ψ_5 to the Grassmannian spinors θ^α, $\bar\theta^{\dot\alpha}$ requires using of the commuting spinor variables ζ^α, $\bar\zeta^{\dot\alpha}$.

The total system which we consider as initial under transition to Grassmannian spinors is in fact the sum of the two sectors coupled through the space-time momentum. One of these sectors is the usual massive spinning particle with Lagrangian (25) whereas the second is the sector of the bosonic spinor with Lagrangian (6). Thus the Lagrangian of the initial system has the following form

$$L = L_{1/2} + L_{\mathrm{SCC}}$$

$$= p\dot{x} + \frac{i}{2}(\psi\dot\psi + \psi_5\dot\psi_5) - \frac{e}{2}(p^2 + m^2) - i\chi(p\psi + m\psi_5)$$

$$+ \dot\zeta v + \bar{v}\dot{\bar\zeta} - \lambda(\zeta\hat{p}\bar\zeta - j). \tag{28}$$

As result of the constraint $\zeta\hat{p}\bar\zeta = j$ the sign of the constant j defines the sign of the energy. In following we consider the positive energy sector where $j > 0$.

3.2. CONVERSION OF SPINNING PARTICLE TO SUPERPARTICLE WITH TENSORIAL CENTRAL CHARGES

The conversion of spinning particle model described by the Grassmannian variables ψ_μ, ψ_5 to the model with the Grassmannian spinor variables θ^α, $\bar\theta^{\dot\alpha}$ is realized by the general resolution [11] of the form

$$\psi_\mu = r^{-1/2}(\theta\sigma_\mu\tilde{p}\zeta + \bar\zeta\tilde{p}\sigma_\mu\bar\theta) - m\rho\zeta\sigma_\mu\bar\zeta, \tag{29}$$

$$\psi_5 = r^{-1/2}m(\zeta\theta + \bar\theta\bar\zeta) + r\rho + \tilde\psi_5. \tag{30}$$

The initial Grassmannian variables ψ_μ, ψ_5 (5 variables) are expressed in terms of two Grassmannian scalars ρ, $\tilde\psi_5$ and three components of spinor θ. Just for projections of $\psi_\mu \equiv -\frac{1}{2}\tilde\sigma_\mu{}^{\dot\alpha\alpha}\hat\psi_{\alpha\dot\alpha}$ in the basis formed by spinors ζ^α, $(\bar\zeta\tilde{p})^\alpha$ we have

$$\zeta\hat\psi\bar\zeta = 2r^{1/2}(\zeta\theta + \bar\theta\bar\zeta), \quad \bar\zeta\tilde{p}\hat\psi\tilde{p}\bar\zeta = 2mr^2\rho, \tag{31}$$

$$\zeta\hat\psi\tilde{p}\bar\zeta = 2r^{1/2}(\zeta\tilde{p}\bar\theta), \quad \bar\zeta\tilde{p}\hat\psi\bar\zeta = 2r^{1/2}(\theta\tilde{p}\bar\zeta), \tag{32}$$

where $\hat\psi = \psi^\mu\sigma_\mu$. The fourth component of the spinor

$$\phi = i(\theta\zeta - \bar\zeta\bar\theta) \tag{33}$$

does not participate in the expression for ψ-variables. The inversion of (29), (30) and (33) looks as follows

$$\theta_\alpha = \frac{1}{4}r^{-3/2}\left[(\zeta\hat\psi\bar\zeta)(\hat{p}\bar\zeta)_\alpha + 2(\bar\zeta\tilde{p}\hat\psi\bar\zeta)\zeta_\alpha\right] + \frac{i}{2}r^{-1}\phi(\hat{p}\bar\zeta)_\alpha,$$

$$\bar{\theta}_{\dot{\alpha}} = \frac{1}{4} r^{-3/2} \left[(\zeta \hat{\psi} \bar{\zeta})(\zeta \hat{p})_{\dot{\alpha}} + 2(\zeta \tilde{\psi} \hat{p} \zeta) \bar{\zeta}_{\dot{\alpha}} \right] - \frac{i}{2} r^{-1} \phi(\zeta \hat{p})_{\dot{\alpha}},$$

$$\rho = \frac{1}{2m} r^{-2} (\bar{\zeta} \hat{p} \hat{\psi} \hat{p} \zeta),$$

$$\tilde{\psi}_5 = \frac{1}{m} (p^\mu \psi_\mu + m \psi_5) - (2mr)^{-1} (\zeta \hat{\psi} \bar{\zeta})(p^2 + m^2).$$

In the new variables the Dirac constraint takes a simple form. On mass shell $p^2 + m^2 = 0$ we have

$$D = p\psi + m\tilde{\psi}_5 = m\tilde{\psi}_5 \approx 0. \tag{34}$$

Moreover, we can extract from the new variables a pure gauge degree of freedom for fermionic local symmetry of the spinning particle [12, 13] (world-line supersymmetry)

$$\delta \chi = \dot{\epsilon}, \quad \delta e = -2i\epsilon \chi, \quad \delta \psi_\mu = -\epsilon p_\mu, \quad \delta \psi_5 = -\epsilon m, \quad \delta x_\mu = i\epsilon \psi_\mu.$$

In the new variables this transformation takes the form

$$\delta \theta_\alpha = -\frac{1}{4} \epsilon r^{-1/2} (\hat{p} \bar{\zeta})_\alpha, \quad \delta \bar{\theta}_{\dot{\alpha}} = -\frac{1}{4} \epsilon r^{-1/2} (\zeta \hat{p})_{\dot{\alpha}},$$

$$\delta \rho = -\frac{1}{2} \epsilon m r^{-1}, \quad \delta \tilde{\psi}_5 = -\frac{1}{2m} \epsilon (p^2 + m^2) \approx 0.$$

Thus, the only transformed are the variable ρ and one component of spinor θ

$$\delta(\theta \zeta + \bar{\zeta} \bar{\theta}) = \frac{1}{2} \epsilon r^{1/2}.$$

Subsequently the combination $\rho + mr^{-3/2}(\theta \zeta + \bar{\zeta} \bar{\theta})$ of this component θ and ρ is invariant under the gauge transformations, $\delta[\rho + mr^{-3/2}(\theta \zeta + \bar{\zeta} \bar{\theta})] = 0$, whereas the variable

$$\rho - mr^{-3/2}(\theta \zeta + \bar{\zeta} \bar{\theta}) \tag{35}$$

is the pure gauge degree of freedom, $\delta[\rho - mr^{-3/2}(\theta \zeta + \bar{\zeta} \bar{\theta})] = -mr^{-1}\epsilon$.

Accounting the equation of motion for bosonic spinor $\zeta = 0$ and substituting the resolving expressions (29), (30) for ψ_μ, ψ_5 in the Lagrangian (28) we arrive at the Lagrangian

$$\begin{aligned}
L = &\, p(\dot{x} - i\theta\sigma\dot{\bar{\theta}} + i\dot{\theta}\sigma\bar{\theta}) - im^2 r^{-1}(\theta\zeta\bar{\zeta}\dot{\bar{\theta}} - \dot{\theta}\zeta\bar{\zeta}\bar{\theta}) \\
&+ \frac{i}{2} r^2 \left[\rho + mr^{-3/2}(\theta\zeta + \bar{\zeta}\bar{\theta}) \right] \left[\dot{\rho} + mr^{-3/2}(\dot{\theta}\zeta + \bar{\zeta}\dot{\bar{\theta}}) \right] \\
&+ \frac{i}{2} r \left[\rho - mr^{-3/2}(\theta\zeta + \bar{\zeta}\bar{\theta}) \right] \dot{\tilde{\psi}}_5 + \frac{i}{2} r \tilde{\psi}_5 \left[\dot{\rho} - mr^{-3/2}(\dot{\theta}\zeta + \bar{\zeta}\dot{\bar{\theta}}) \right] \\
&+ \frac{i}{2} \tilde{\psi}_5 \dot{\tilde{\psi}}_5 - im\chi\tilde{\psi}_5 - \frac{e}{2}(p^2 + m^2) \\
&+ \dot{\zeta}v + \bar{v}\dot{\bar{\zeta}} - \lambda(\zeta\hat{p}\bar{\zeta} - j).
\end{aligned} \tag{36}$$

It should be stressed that the equation $\dot\zeta = 0$ for bosonic spinor, which has been used for derivation of the Lagrangian (36), is reproduced by the same Lagrangian (36). As we see from the Lagrangian, the gauge variable (35) is the corresponding conjugate variable for $\tilde\psi_5$ which generates the local transformations. The simpler gauge fixing condition for it

$$\rho - mr^{-3/2}(\theta\zeta + \bar\zeta\bar\theta) = 0$$

gives us the possibility to resolve the scalar ρ in term of spinor projection $(\theta\zeta + \bar\zeta\bar\theta)$. We take the more general condition of this type

$$\rho - mr^{-3/2}(\theta\zeta + \bar\zeta\bar\theta) = 2(k-1)mr^{-3/2}(\theta\zeta + \bar\zeta\bar\theta) \tag{37}$$

which is the gauge fixing condition at all k except $k = 0$. At $k = 0$ (37) is reduced to the condition on gauge invariant variable

$$\rho + mr^{-3/2}(\theta\zeta + \bar\zeta\bar\theta) = 0$$

and of course it is not a gauge fixing.

Substituting in the Lagrangian (36) the constraint condition $\tilde\psi_5 = 0$ (the equation of motion for the Lagrange multiplier χ) and the expression

$$\rho = (2k-1)mr^{-3/2}(\theta\zeta + \bar\zeta\bar\theta) \tag{38}$$

(following from the gauge fixing condition (37)) we obtain the Lagrangian

$$L = \rho\dot\omega_\theta + iZ_{\alpha\beta}\theta^\alpha\dot\theta^\beta + i\bar{Z}_{\dot\alpha\dot\beta}\bar\theta^{\dot\alpha}\dot{\bar\theta}^{\dot\beta} + iZ_{\alpha\dot\beta}(\theta^\alpha\dot{\bar\theta}^{\dot\beta} - \dot\theta^\alpha\bar\theta^{\dot\beta}) - \frac{e}{2}(p^2 + m^2)$$
$$+ \dot\zeta v + \bar{v}\dot{\bar\zeta} - \lambda(\zeta\hat{p}\bar\zeta - j). \tag{39}$$

In this expression $\omega_\theta \equiv \dot\omega_\theta\, d\tau = dx - id\theta\sigma\bar\theta + i\theta\sigma d\bar\theta$ is the usual $N = 1$ superinvariant ω-form. The quantities $Z_{\alpha\beta} = Z_{\beta\alpha}$, $\bar{Z}_{\dot\alpha\dot\beta} = \overline{(Z_{\alpha\beta})}$ and $Z_{\alpha\dot\beta} = \overline{(Z_{\beta\dot\alpha})}$ are expressed in terms of bosonic spinor ζ (for similar formula see [10])

$$Z_{\alpha\beta} = 2k^2m^2j^{-1}\zeta_\alpha\zeta_\beta, \qquad Z_{\alpha\dot\beta} = (2k^2 - 1)m^2j^{-1}\zeta_\alpha\bar\zeta_{\dot\beta}. \tag{40}$$

$Z_{\alpha\beta}$ and $\bar{Z}_{\dot\alpha\dot\beta}$ are tensor central charges (types $(1,0)$ and $(0,1)$) and $Z_{\alpha\dot\beta}$ is vector one (type $(1/2, 1/2)$) for the $D = 4$ $N = 1$ supersymmetry algebra [17]-[21].

The same result is obtained if we consider the connection of the systems (28) and (5) in the Hamiltonian formalism. Precisely there is the canonical transformation which connect the models with each other. Now in order to make equal the number of Grassmannian variables in the models we introduce pure gauge variable ϕ in the initial model of the spinning particle. Its pure gauge nature is achieved by the presence of the first class constraint

$$p_\phi \approx 0 \tag{41}$$

in the initial model. So in the canonical transformation we imply that the term $p_\phi \dot{\phi} - \mu p_\phi$ is added to the Lagrangian (28). Here μ is Lagrange multiplier. The resolution of ϕ in terms of the spinors is given by the expression (33).

As the generating function of the canonical transformation from system with coordinates ψ_μ, ψ_5, ϕ, x^μ, ζ^α, $\bar{\zeta}^{\dot\alpha}$ to the system with coordinates θ^α, $\bar{\theta}^{\dot\alpha}$, ρ, $\tilde{\psi}_5$, x'^μ, ζ'^α, $\bar{\zeta}'^{\dot\alpha}$ we take

$$F = -p_\psi^\mu \psi_\mu(p_\mu, \zeta, \theta, \rho) - p_{\psi 5}\psi_5(\zeta, \theta, \rho, \tilde{\psi}_5) - p_\phi \phi(\zeta, \theta)$$
$$+ \zeta^\alpha v'_\alpha + \bar{v}'_{\dot\alpha}\bar{\zeta}^{\dot\alpha} - p^\mu x'_\mu. \tag{42}$$

Here the expressions for old variables in term of new ones from the right hand side of the equations (29), (30), (33) have been used. That construction of the generating function (42) reproduces, by definition of the canonical transformation, the resolution (29), (30), (33) of the initial Grassmannian coordinates in spinors $\psi_\mu = -\partial_l F/\partial p_\psi^\mu$, $\psi_5 = -\partial_l F/\partial p_{\psi 5}$, $\phi = -\partial_l F/\partial p_\phi$ and leaves invariable bosonic spinor coordinates $\zeta'^\alpha = \partial F/\partial v'_\alpha = \zeta^\alpha$, $\bar{\zeta}'^{\dot\alpha} = \partial F/\partial \bar{v}'_{\dot\alpha} = \bar{\zeta}^{\dot\alpha}$ and the momentum vector $p'_\mu = -\partial F/\partial x'^\mu = p_\mu$. The expression of new Grassmannian momenta in terms of initial ones are

$$p_{\theta\alpha} = -\partial_r F/\partial\theta^\alpha = r^{-1/2}(\sigma_\mu \bar{p}\zeta)_\alpha p_\psi^\mu - mr^{-1/2}\zeta_\alpha p_{\psi 5} + i\zeta_\alpha p_\phi,$$

$$\bar{p}_{\theta\dot\alpha} = -\partial_r F/\partial\bar{\theta}^{\dot\alpha} = r^{-1/2}(\bar{\zeta}\bar{p}\sigma_\mu)_{\dot\alpha} p_\psi^\mu - mr^{-1/2}\bar{\zeta}_{\dot\alpha} p_{\psi 5} - i\bar{\zeta}_{\dot\alpha} p_\phi,$$

$$p_\rho = -\partial_r F/\partial\rho = -m(\zeta\sigma_\mu\bar{\zeta})p_\psi^\mu + rp_{\psi 5}, \quad p_{\tilde{\psi}5} = -\partial_r F/\partial\tilde{\psi}_5 = p_{\psi 5}.$$

The expressions of the initial bosonic spinor momenta $v_\alpha = \partial F/\partial\zeta^\alpha$, $\bar{v}_{\dot\alpha} = \partial F/\partial\bar{\zeta}^{\dot\alpha}$ and space-time coordinate $x_\mu = -\partial F/\partial p^\mu$ in terms of the new phase space coordinates contain besides corresponding new phase variables the additional terms depending on the new Grassmannian phase space variables. These terms arise because of the dependence of the resolution expressions (29), (30), (33) on ζ, $\bar{\zeta}$ and p. Here we do not need the expressions for v', $\bar{v}'$ and x' in the explicit form due to independence of all constraints on these phase variables.

Now we eliminate the variables $\tilde{\psi}_5$, $p_{\tilde{\psi}5}$ by means of the Dirac constraint (26) and gauge fixing condition for Dirac constraint

$$p_{\tilde{\psi}5} - i(k-1)mr^{-1/2}\left[\theta\zeta + \bar{\zeta}\bar{\theta}\right] \approx 0 \tag{43}$$

at $k \neq 0$ [2]. After fulfillment of the additional canonical transformation $p_\rho \to p_{\rho'} = p_\rho - ikmr^{1/2}\left[\theta\zeta + \bar{\zeta}\bar{\theta}\right]$, which leads to resolving form $p_{\rho'} \approx 0$ of one

---

[2] The diagonalized Dirac constraint $D' \equiv D - ip^\mu g_\mu - img_5 = -i[p_\mu(p_\psi^\mu + \frac{i}{2}\psi^\mu) + m(p_{\psi 5} + \frac{1}{2}\psi_5)] \approx 0$ has in new variables the form $D' = \frac{1}{4}r^{-1/2}\left[\bar{\zeta}\bar{p}p\zeta + \bar{p}_\theta \bar{p}\zeta\right] - \frac{1}{2}mr^{-1}p_\rho + \frac{1}{2}m\tilde{\psi}_5 \approx 0$. The Poisson bracket of the condition (43) and D' is equal to $(km)/2$, i.e. at $k = 0$ the condition (43) does not fix the gauge for the Dirac constraint.

Fermi-constraint from (27), we eliminate the variables ρ, p_ρ with the help of two from five second class Fermi-constraints (27). Because of the resolving form of the constraints with respect to eliminated variables, $\tilde{\psi}_5 \approx 0$ and $p_{\rho'} \approx 0$, the Dirac brackets for remaining variables are the same as their Poisson brackets. After that the remaining Grassmannian constraints take the following form

$$\bar{\zeta}\bar{p}p_\theta - \bar{p}_\theta\bar{p}\zeta \approx 0, \tag{44}$$

$$[\bar{\zeta}\bar{p}p_\theta + \bar{p}_\theta\bar{p}\zeta] - 4ik^2m^2\left[\theta\zeta + \bar{\zeta}\bar{\theta}\right] \approx 0, \tag{45}$$

$$\zeta\left[-ip_\theta - \hat{p}\bar{\theta}\right] \approx 0, \qquad \left[-i\bar{p}_\theta - \theta\hat{p}\right]\bar{\zeta} \approx 0 \tag{46}$$

which are the same as the projections on spinors ζ, $\hat{p}\bar{\zeta}$ of the Grassmannian spinor constraints

$$d_{\theta\alpha} \equiv -ip_{\theta\alpha} - (\hat{p}\bar{\theta})_\alpha - \theta^\beta Z_{\beta\alpha} - Z_{\alpha\dot\beta}\bar{\theta}^{\dot\beta} \approx 0, \tag{47}$$

$$\bar{d}_{\theta\dot\alpha} \equiv -i\bar{p}_{\theta\dot\alpha} - (\theta\hat{p})_{\dot\alpha} - \bar{Z}_{\dot\alpha\dot\beta}\bar{\theta}^{\dot\beta} - \theta^\beta Z_{\beta\dot\alpha} \approx 0 \tag{48}$$

with quantities $Z_{\alpha\beta}$, $Z_{\alpha\dot\beta}$ defined in (40). From invariance of the variables ζ^α, $\bar{\zeta}^{\dot\alpha}$, p_μ under the canonical transformation, all bosonic constraints, i.e. $p^2 + m^2 \approx 0$ and $\zeta\hat{p}\bar{\zeta} - j \approx 0$, are not changed. The system with remaining variables and the constraints is described by the above mentioned Lagrangian (5). The Lagrangian (5) reproduces accurately this set of the constraints and nothing else.

Thus we establish that the model described by Lagrangian $L = L_{1/2} + L_{b.s.}$ is equivalent physically to the model with Lagrangian $L = L_{super} + L_{b.s.}$ at classical level. Here $L_{1/2}$ is the Lagrangian (25) of the massive spinning particle (spin 1/2) whereas L_{super} is Lagrangian of the massive $N = 1$ superparticle with tensorial central charges (40)

$$L_{super} = p\dot{\omega}_\theta + iZ_{\alpha\beta}\theta^\alpha\dot{\theta}^\beta + i\bar{Z}_{\dot\alpha\dot\beta}\bar{\theta}^{\dot\alpha}\dot{\bar{\theta}}^{\dot\beta} + iZ_{\alpha\dot\beta}(\theta^\alpha\dot{\bar{\theta}}^{\dot\beta} - \dot{\theta}^\alpha\bar{\theta}^{\dot\beta}) - \frac{e}{2}(p^2 + m^2). \tag{49}$$

Lagrangians $L_{b.s.}$ of the bosonic spinor in the both equivalent models are quite identical.

It should be noted that the value of constant k in the formula (40) for central charges of the superparticle is nonzero, $k \neq 0$, in the case of its equivalence to the spinning particle. But in general the value $k = 0$ is not forbidden in model of superparticle with central charges. Next we consider the cases both with $k \neq 0$ and $k = 0$. As we see below at $k \neq 0$ and $k = 0$ we have superparticle models with one and two κ-symmetries respectively.

3.3. ANALYSIS ON LEVEL OF PHYSICAL DEGREES OF FREEDOM

Alternative way for a proof of classical equivalence of the massive spin $1/2$ particle (25) and the massive superparticle with central charges (49), at $k \neq 0$, possessing one κ-symmetry is the reduction of both models to physical degrees of freedom [27]. In the examining positive energy sector after choice of gauge $\psi_- = \psi_0 - \psi_5 = 0$ for Dirac constraint and exclusion of $\psi_+ = \psi_0 + \psi_5$ by means of the constraint condition we obtain for the physical odd degrees of freedom of spinning particle [28, 27] the Lagrangian in the form of $L^{(\text{ph})}_{1/2,\text{Gr}} = \frac{i}{2}\vec{\psi}\dot{\vec{\psi}}$. On the other hand the Grassmannian part of the superparticle Lagrangian L_{super} takes the form

$$L^{(\text{ph})}_{\text{super,Gr}} = i\bar{q}\dot{q} - iq\dot{\bar{q}} + 2k^2 i\eta\dot{\eta}$$

after using of the variables

$$\eta = mr^{-1/2}(\theta\zeta + \bar{\zeta}\bar{\theta}), \qquad \sigma = -imr^{-1/2}(\theta\zeta - \bar{\zeta}\bar{\theta}), \tag{50}$$

$$q = r^{-1/2}(\theta\hat{p}\bar{\zeta}), \qquad \bar{q} = r^{-1/2}(\zeta\hat{p}\bar{\theta}). \tag{51}$$

Setting

$$q = (\psi_1 + i\psi_2)/2, \qquad \bar{q} = (\psi_1 - i\psi_2)/2, \qquad \eta = \psi_3/2k$$

we obtain exactly the same Grassmannian part of the Lagrangian

$$L^{(\text{ph})}_{\text{super,Gr}} = L^{(\text{ph})}_{1/2,\text{Gr}} = \frac{i}{2}\vec{\psi}\dot{\vec{\psi}}. \tag{52}$$

Such Lagrangian for the physical odd variables comes out also from work [23] in non-Lorentz covariant Grassmannian sector $N = 4 \rightarrow N = 1$ PBGS. In first order formalism the target space action of this work has the Lagrangian

$$L = \vec{P}\dot{\vec{\Pi}} - P^0\Pi^0 + \frac{e}{2}(P^{02} - \vec{P}^2 - 1) - \Theta\dot{\Theta} - \vec{\Psi}\dot{\vec{\Psi}} \tag{53}$$

where $\Pi^0 = \dot{X}^0 + \Theta\dot{\Theta} + \vec{\Psi}\dot{\vec{\Psi}}, \vec{\Pi} = \dot{\vec{Y}} - \Theta\dot{\vec{\Psi}} + \Theta\dot{\vec{\Psi}}$ (we remain here the notations of [23]). In accounting the last expressions, the Lagrangian (53) takes the form

$$L = \vec{P}\dot{\vec{Y}} - P^0\dot{X}^0 + \frac{e}{2}(P^{02} - \vec{P}^2 - 1)$$

$$-(P^0 + 1)\left[\vec{\Psi} - \frac{1}{P^0 + 1}\vec{P}\Theta\right]\left[\dot{\vec{\Psi}} - \frac{1}{P^0 + 1}\vec{P}\dot{\Theta}\right].$$

After using of the variables

$$\vec{\psi} = \sqrt{2}(P^0 + 1)^{1/2}\left[\vec{\Psi} - \frac{1}{P^0 + 1}\vec{P}\Theta\right]$$

we obtain exactly the Lagrangian (52) for Grassmannian variables.

3.4. SUPERPARTICLE WITH INDEX SPINOR

In order to analyze the properties of the obtained massive superparticle with tensorial central charges let us consider the model of spinning particle with index spinor [14, 11, 15] as additional bosonic coordinates. It is naturally because we have used for bosonic spinor the relation $\zeta \hat{p} \bar{\zeta} - j \approx 0$ which is inherent in the index spinor approach. In the Hamiltonian formalism the index spinor sector is restricted by the spinor self-conjugacy conditions

$$d_\zeta \equiv i p_\zeta - \hat{p} \bar{\zeta} \approx 0, \qquad \bar{d}_\zeta \equiv -i \hat{p}_\zeta - \zeta \hat{p} \approx 0 \qquad (54)$$

which are the second class constraints in the massive case. It is achieved in above model (28) by the substitution $v = -i \hat{p} \bar{\zeta}$, $\bar{v} = i \zeta \hat{p}$. Then $L_{\text{b.s.}}$ (6) takes the form of the index spinor Lagrangian [14]

$$L_{\text{index}} = -i \dot{\zeta} \hat{p} \bar{\zeta} + i \zeta \hat{p} \dot{\bar{\zeta}} - \lambda(\zeta \hat{p} \bar{\zeta} - j). \qquad (55)$$

The constraint $\zeta \hat{p} \bar{\zeta} - j \approx 0$ included in the Lagrangian generates in Hamiltonian formalism the spin constraint

$$\frac{i}{2} (\zeta p_\zeta - \bar{p}_\zeta \bar{\zeta}) - j \approx 0 \qquad (56)$$

which together with second class constraints (54) leads [14] to the particle state of the single spin associated with given sector of index spinor. Spin of the particle in the quantum spectrum is the value of the constant j renormalized by ordering constants (thus j can be named "classical spin").

The realization of the previously considered canonical transformation to the model with Lagrangian $L' = L_{1/2} + L_{\text{index}}$, i.e. L_{index} instead $L_{\text{b.s.}}$ in (28), leads to the Lagrangian

$$
\begin{aligned}
L' = {} & p\dot{\omega} + i Z_{\alpha\beta} \theta^\alpha \dot{\theta}^\beta + i \bar{Z}_{\dot{\alpha}\dot{\beta}} \bar{\theta}^{\dot{\alpha}} \dot{\bar{\theta}}^{\dot{\beta}} + i Z_{\alpha\dot{\beta}} (\theta^\alpha \dot{\bar{\theta}}^{\dot{\beta}} - \dot{\theta}^\alpha \bar{\theta}^{\dot{\beta}}) \\
& + i Y_{\alpha\beta} \zeta^\alpha \dot{\zeta}^\beta + i \bar{Y}_{\dot{\alpha}\dot{\beta}} \bar{\zeta}^{\dot{\alpha}} \dot{\bar{\zeta}}^{\dot{\beta}} + i Y_{\alpha\dot{\beta}} (\zeta^\alpha \dot{\bar{\zeta}}^{\dot{\beta}} + \dot{\zeta}^\alpha \bar{\zeta}^{\dot{\beta}}) \\
& - i N (\dot{\zeta} \hat{p} \bar{\zeta} - \zeta \hat{p} \dot{\bar{\zeta}}) \\
& - \frac{e}{2} (p^2 + m^2) - \lambda(\zeta \hat{p} \bar{\zeta} - j). \qquad (57)
\end{aligned}
$$

Here the form $\omega \equiv \dot{\omega} \, d\tau = dx - i d\zeta \sigma \bar{\zeta} + i \zeta \sigma d\bar{\zeta} - i d\theta \sigma \bar{\theta} + i \theta \sigma d\bar{\theta}$ is invariant with respect to the transformations of the usual $N = 1$ supersymmetry with Grassmannian spinor parameter and "bosonic supersymmetry" with c-number spinor parameter [14, 11, 15]. The central charges $Z_{\alpha\beta}$, $Z_{\alpha\dot{\beta}}$ have the same form (40).

So the kinetic terms of the space-time coordinate and Grassmannian spinor in L' (57) are identical to the corresponding terms in L (5) and hence the algebras of the fermionic constraints in both models are identical. But the kinetic terms of the index spinor in Lagrangian L' are different from the kinetic terms of the bosonic spinor in Lagrangian L by additional terms with quantities

$$Y_{\alpha\beta} = 2k(k-2)m^2 j^{-1}\theta_\alpha\theta_\beta, \qquad \bar{Y}_{\dot\alpha\dot\beta} = -\overline{(Y_{\alpha\beta})},$$

$$Y_{\alpha\dot\beta} = -(2k^2 - 4k + 1)m^2 j^{-1}\theta_\alpha\bar\theta_{\dot\beta} \tag{58}$$

which can be regarded as the central charges of the "bosonic SUSY" as well as

$$N \equiv j^{-1}\left[(\theta\hat{p}\bar\theta) + 2(2k-1)m^2 j^{-1}(\theta\zeta)(\bar\zeta\bar\theta)\right]. \tag{59}$$

The appearance of these extra terms is the result of modification of index spinor momenta $p_\zeta, \bar{p}_\zeta$ under the canonical transformation and, as consequence, the modification of the spin constraint (56) and bosonic spinor constraints (54) expressed by new variables.

Specific peculiarity of the model (57) with index spinor is an interconnection between usual fermionic supersymmetry and "bosonic one" and at present its meaning is not yet quite clear. Some duality appears in the invariance under permutation of Grassmannian and bosonic spinors both ω-form and certain terms with central charges of different types.

4. Gauge symmetries of massive superparticle with tensorial central charges

For local transformation of the Grassmannian spinor

$$\delta\theta^\alpha = i\kappa(\bar\zeta\bar{p})^\alpha, \qquad \delta\bar\theta^{\dot\alpha} = -i\bar\kappa(\bar{p}\zeta)^{\dot\alpha} \tag{60}$$

and standard Siegel transformation [29, 30] of the space-time coordinate

$$\delta x_\mu = -i\theta\sigma_\mu\delta\bar\theta + i\delta\theta\sigma_\mu\bar\theta \tag{61}$$

with local complex Grassmannian parameter $\kappa(\tau)$ the variation of the Lagrangians up to a total derivative is

$$\delta L = -2k^2 m^2(\theta\zeta + \bar\zeta\bar\theta)(\kappa - \bar\kappa)^{\cdot} + 2k^2 m^2(\theta\zeta + \bar\zeta\bar\theta)^{\cdot}(\kappa - \bar\kappa)$$
$$- 4km^2 j^{-1}[(\theta\hat{p}\bar\zeta)\zeta\dot\zeta + (\zeta\hat{p}\bar\theta)\dot{\bar\zeta}\zeta](\kappa - \bar\kappa). \tag{62}$$

As we see, $\delta L = 0$ for real $\kappa = \bar\kappa$ at arbitrary values of constant k. But at $k = 0$ we have $\delta L = 0$ for arbitrary complex parameter κ. Thus at $k \neq 0$ when the tensor central charge $Z_{\alpha\beta}$ is present the models have one κ-symmetry with real

Grassmannian parameter $\kappa = \bar{\kappa}$. But at $k = 0$ when there is only the vector central charge $Z_{\alpha\dot\beta}$ we have two κ-symmetries with complex Grassmannian parameter κ.

A first class constraint is associated to each gauge symmetry in Hamiltonian formalism. As is already noted our systems are described by the fermionic constraints (covariant derivatives) (47), (48). Their Poisson brackets algebra is

$$\{d_{\theta\alpha}, d_{\theta\beta}\} = 2iZ_{\alpha\beta}, \quad \{\bar{d}_{\theta\dot\alpha}, \bar{d}_{\theta\dot\beta}\} = 2i\bar{Z}_{\dot\alpha\dot\beta},$$

$$\{d_{\theta\alpha}, \bar{d}_{\theta\dot\beta}\} = 2i\left(p_{\alpha\dot\beta} + Z_{\alpha\dot\beta}\right) \tag{63}$$

with central charges (40). Covariant separation of the fermionic first and second class constraints is achieved by the projection on the spinors ζ_α, $(\hat{p}\bar{\zeta})_\alpha$. Let us put

$$\chi_\theta \equiv \zeta d_\theta = -i\zeta p_\theta - \zeta\hat{p}\bar{\theta} \approx 0, \quad \bar{\chi}_\theta \equiv \bar{d}_\theta\bar{\zeta} = -i\bar{p}_\theta\bar{\zeta} - \theta\hat{p}\bar{\zeta} \approx 0, \tag{64}$$

$$g_\theta \equiv \bar{\zeta}\tilde{p}d_\theta + \bar{d}_\theta\tilde{p}\zeta = -i(\bar{\zeta}\tilde{p}p_\theta + \bar{p}_\theta\tilde{p}\zeta) - 4k^2m^2(\theta\zeta + \bar{\zeta}\bar{\theta}) \approx 0, \tag{65}$$

$$f_\theta \equiv i(\bar{\zeta}\tilde{p}d_\theta - \bar{d}_\theta\tilde{p}\zeta) = \bar{\zeta}\tilde{p}p_\theta - \bar{p}_\theta\tilde{p}\zeta \approx 0. \tag{66}$$

The nonzero Poisson brackets of these projections are

$$\{\chi_\theta, \bar{\chi}_\theta\} = 2ij, \quad \{g_\theta, g_\theta\} = 16k^2m^2ij. \tag{67}$$

Thus the constraints χ_θ, $\bar{\chi}_\theta$ are always the second class constraints whereas the constraint f_θ is always the first class constraint generating one κ-symmetry with local parameter $(\kappa + \bar{\kappa})$ on variable $(\theta\zeta - \bar{\zeta}\bar{\theta})$, $\{f_\theta, \theta\zeta - \bar{\zeta}\bar{\theta}\} = 2r$, $\delta(\theta\zeta - \bar{\zeta}\bar{\theta}) = ir(\kappa + \bar{\kappa})$. The constraint g_θ is the second class constraint at $k \neq 0$. But at $k = 0$ the constraint g_θ becomes the first class constraint and generates additional κ-symmetry with local parameter $i(\kappa - \bar{\kappa})$ on variable $(\theta\zeta + \bar{\zeta}\bar{\theta})$, $\{g_\theta, \theta\zeta + \bar{\zeta}\bar{\theta}\} = -2ir$, $\delta(\theta\zeta + \bar{\zeta}\bar{\theta}) = ir(\kappa - \bar{\kappa})$.

Thus we obtain the models of the $D = 4$ $N = 1$ massive superparticle with tensorial central charges possessing one or two Siegel κ-symmetries. In the language of the brane theories these models correspond to the BPS superbrane configurations preserving $1/4$ or $1/2$ of supersymmetry (see [21] and references therein).

It should be noted that constant k in the construction of the superparticle appears in the gauge fixing condition under transition from the spinning particle. Therefore at all $k \neq 0$ the superparticle has quite similar systems of the constraints and the same number of physical degrees of freedom. The models at all $k \neq 0$ are equivalent. Under transformations which can be considered as canonical transformations

$$\theta^\alpha \to \theta^\alpha + br^{-1}(\theta\zeta + \bar{\zeta}\bar{\theta})(\bar{\zeta}\tilde{p})^\alpha, \quad \bar{\theta}^{\dot\alpha} \to \bar{\theta}^{\dot\alpha} + br^{-1}(\theta\zeta + \bar{\zeta}\bar{\theta})(\tilde{p}\zeta)^{\dot\alpha} \tag{68}$$

where b is real number the Lagrangian L (or L') transforms into the same Lagrangian with ak in place of k where $a \equiv 1 + 2b$. As final result at level of the free superparticle we have two substantially different models of the massive superparticle with tensorial central charges. First of them at $k = 1/\sqrt{2}$ has only tensor central charge $Z_{\alpha\beta}$ and possesses one κ-symmetry. Second model at $k = 0$ has only vector central charge $Z_{\alpha\dot\beta}$ and possesses two κ-symmetries.

5. Quantum spectrum of the models

In process of the construction it is established the equivalence at classical level between the massive $D = 4\ N = 1$ superparticle with one κ-symmetry and the massive $D = 4\ n = 1$ spinning particle. But they may lead to distinct quantum theories [27]. Below we establish that the spinning particle and superparticle with tensorial central charges, which have index spinor as additional one, have identical state spectrum. By analogy with results in paper [12–14] the first operator quantization of the spinning particle with index spinor described by Lagrangian $L_{1/2} + L_{\text{index}}$ is immediate. Wave function in the model is defined by Dirac spinor with (anti)holomorphic dependence in index spinor of homogeneity degree $2J$ where J is the classical spin j renormalized by the ordering constant. Writing Dirac spinor in terms of Weyl spinors as $\begin{pmatrix} \psi \\ \chi \end{pmatrix}$, in according to analysis carried out in [14] we have in holomorphic case two multispinor fields $\psi_{\alpha_1...\alpha_{2J}\beta}$ and $\chi_{\alpha_1...\alpha_{2J}\dot\beta}$ which are symmetrical in $2J$ indices αs. Here β and $\dot\beta$ correspond to bispinor index. These fields are connected with each other by Dirac equation

$$\begin{pmatrix} 0 & \tilde{p} \\ \hat{p} & 0 \end{pmatrix} \begin{pmatrix} \psi \\ \chi \end{pmatrix} = m \begin{pmatrix} \psi \\ \chi \end{pmatrix} \tag{69}$$

(quantum counterpart of the Dirac constraint (26)). Comparison with superparticle model is more immediate if we take the field $\chi_{\alpha_1...\alpha_{2J}\dot\beta}$ as basic one. But the field $\psi_{\alpha_1...\alpha_{2J-1}\alpha_{2J}\beta} = \phi_{(\alpha_1...\alpha_{2J}\beta)} + \phi_{(\alpha_1...\alpha_{2J-1}}\epsilon_{\alpha_{2J})\beta}$ exhibits simply that two spins $J \pm \frac{1}{2}$ are presented in spectrum at fixed J as it should be when one adds spin J which is given by index spinor and spin $\frac{1}{2}$ which corresponds to the Grassmannian variables ψ_μ, ψ_5 of the pseudoclassical mechanics under quantization.

The quantization of the superparticle (57) is suitable to carry out in variables (50), (51) in term of which the fermionic constraints (64)-(66) take the extremely simple forms

$$ip_q + \bar{q} \approx 0, \quad i\bar{p}_q + q \approx 0,$$

$$ip_\eta + 2k^2\eta \approx 0, \tag{70}$$

$$p_\sigma \approx 0.$$

We gauging out the variable σ, the introduce the Dirac brackets for taking into account of the fermionic second class constraints and the represent the remaining fermionic variables q, $\bar{q}$, η (in fact $\vec{\psi}$) by means of the usual Pauli σ-matrices. Thus the wave function of this problem has two components depending appropriately on index spinor and space-time variables. The quantization of the bosonic spinor sector shows certain difference with [14]. Additional term of the form $q\bar{q}$ in spin constraint (56) arising due to interaction of bosonic and fermionic sectors leads to different homogeneity degrees (which correspond to different representations of Lorentz group) for two components of wave function. Bosonic spinor constraints (54) ((anti)homogeneity conditions) acquire the additional terms both with $q\bar{q}$ and also $q\eta$ (or $\bar{q}\eta$). These last terms, which are proportional σ_+ (or σ_-), $\sigma_\pm \equiv (\sigma_1 \pm i\sigma_2)/2$ in matrix realization of odd variables, connect two components of wave function. As result the irreducible $(2J+1)$-component spinor field $\phi_{\alpha_1\ldots\alpha_{2J+1}}$, in term of which one component of wave function is determined, is expressed by Dirac equation

$$p_{\gamma\beta}\chi_{\alpha_1\ldots\alpha_{2J}}{}^{\dot{\beta}} = m\phi_{\alpha_1\ldots\alpha_{2J}\gamma} \qquad (71)$$

via field $\chi_{\alpha_1\ldots\alpha_{2J}\dot{\beta}}$ which determines second component of wave function. This last field $\chi_{\alpha_1\ldots\alpha_{2J}\dot{\beta}}$ can be identified with basic field of the spinning particle spectrum.

In case of models (28) and (5), when there is not present the truncation of bosonic spinor sector to the index one because of absence of bosonic spinor constraints, the quantum equivalence apparently remains too. One can expect it from the quite identity of bosonic sectors of the models (28) and (5) and identifying of physical fermionic degrees of freedom which has been demonstrated in Sec. 2.

In case of the Lagrangian (5) one can include vector central charge Z_μ into vector of space-time momentum by the shift $p_\mu \to p_\mu + Z_\mu$ after taking into account the bosonic spinor equation of motion $\zeta = 0$. Therefore at $k = 0$, when there is vector central charge only, it disappears completely from the action and superparticle model reduces in fact to massless case. Unlike this in the particle model (57) with index bosonic spinor at $k = 0$ the redefinition of momentum does not exclude vector central charge due to accompanying modification of bosonic spinor and spin constraints. In this case the wave function contains two usual spin-tensor fields $\phi_{\alpha_1\ldots\alpha_{2J\pm1}}$, satisfying massive Klein-Gordon equation and disconnected with each other because of missing terms with $q\eta$ in bosonic spinor constraints.

6. Conclusion

In this work we presented the manifestly Lorentz-invariant formulation of the $D = 4$ $N = 1$ free massive superparticle with tensorial central charges. The tensorial central charges are construct by commuting bosonic spinor and also by index spinor. The model possesses in general one or two κ-symmetries. In particular case the model contains a real parameter k and at $k \neq 0$ it has one κ-symmetry while at $k = 0$ the number of κ-symmetries is two. The local transformations of κ-symmetry are written out. It is obtained the equivalence at classical level between the massive $D = 4$ superparticle with one κ-symmetry and the massive $D = 4$ spinning particle.

Acknowledgments. We would like to thank I.A.Bandos, E.A.Ivanov, S.O.Krivonos, J.Lukierski, A.Yu.Nurmagambetov, D.P.Sorokin, A.A.Zheltukhin for interest to the work and for many useful discussions. The authors are grateful to I.A.Bandos, V.P.Berezovoj, A.Yu.Nurmagambetov, D.P.Sorokin for the hospitality at the NSC Kharkov Institute of Physics and Technology. This work was partially supported by research grant of the Ministry of Education and Science of Ukraine.

References

1. R.Haag, J.Lopuszanski and M.Sonius, Nucl. Phys. **B78**(1995)257
2. J.Wess and J.Bagger, *Supersymmetry and Supergravity*, 1983 (Princeton: Princeton University Press)
3. J. van Holten and A. van Proyen, J. Phys. **A15**(1982)3763
4. P.K.Townsend, *P-brane democracy*, in *Particles, Strings and Cosmology*, eds. J.Bagger, G.Domokos, A.Falk and S.Kovesi-Domokos (World Scientific 1996), hep-th/9507048; *Four lectures on M-theory*, hep-th/9612121; *F-theory from its superalgebra*, hep-th/9712004
5. I.Bars, Phys. Lett. **B 373**(1996)68; Phys. Rev. **D 54**(1996)2503
6. Y.Eisenberg and S.Solomon, Phys. Lett. **B220**(1989)562
7. S.F.Hewson and M.J.Perry, Nucl. Phys. **B492**(1997)249
   S.F.Hewson, Nucl. Phys. **B 501**(1997)445; *An approach to F-theory*, hep-th/9712017
8. P.d'Auria and P.Fré, Nucl. Phys. **B201**(1982)101
9. E.Sezgin, Phys. Lett. **B392**(1997)323
10. I.Bandos and J.Lukierski, Mod. Phys. Lett. **A 14**(1999)1257;
    I.Bandos, J.Lukierski and D.Sorokin, Phys. Rev. **D 61**(2000)045002
11. V.G.Zima and S.O.Fedoruk, Class. and Quantum Grav. **16**(1999)3653
12. F.A.Berezin and M.S.Marinov, JETP Lett. **21**(1975)678; Ann. Phys. **107**(1977)336
    A.Barducci, R.Casalbuoni and L.Lusanna, Nuovo Cimento A **35**(1976)377;
    L.Brink, S.Deser, B.Zumino, DiVecchia and P.S.Howe, Phys. Lett. **B 64**(1976)435;
    L.Brink, DiVecchia and P.S.Howe, Nucl. Phys. **B 118**(1977)76
13. V.D.Gershun and V.I.Tkach, JETP Lett. **29**(1979)320;
    P.S.Howe, S.Penati, M.Pernici and P.Townsend, Phys. Lett. **B 215**(1988)555
14. V.G.Zima and S.O.Fedoruk, JETP Lett. **61**(1995)251
15. V.G.Zima and S.O.Fedoruk, Phys. of Atomic Nucl. **63**(2000)617
16. I.Rudychev and E.Sezgin, *Superparticles, p-Form Coordinates and the BPS Condition*, hep-th/9711128;
17. J.A.de Azcarraga, J.P.Gauntlett, J.H.Izquerdo and P.K.Townsend, Phys. Rev. Lett. **63**(1989)2443
18. D.P.Sorokin and P.K.Townsend, Phys. Lett. **B 412**(1997)265
19. H.Hammer, Nucl. Phys. **B 521**(1998)503
20. S.Ferrara and M.Porrati, Phys. Lett. **B 423**(1998)255, **B 458**(1999)43

21. J.P.Gauntlett and C.M.Hull, JHEP **0001**(2000)004;
     J.P.Gauntlett, G.W.Gibbons, C.M.Hull and P.K.Townsend, *BPS states of* $D = 4\ N = 1$
     supersymmetry, hep-th/0001024
22. R.Casalbuoni, Phys. Lett. **B 62**(1976)49; Nuovo Cimento **A 33**(1976)389;
     L.Brink and J.H.Schwarz, Phys. Lett. **B 100**(1981)310
23. F.Delduc, E.Ivanov and S.Krivonos, Nucl. Phys. **B 576**(2000)196
24. D.V.Volkov and A.A.Zheltukhin, Lett. Math. Phys. **17**(1989)141
25. D.P.Sorokin, V.I.Tkach and D.V.Volkov, Mod. Phys. Lett. **A4**(1989)901;
     D.P.Sorokin, V.I.Tkach, D.V.Volkov and A.A.Zheltukhin, Phys. Lett. **B 216**(1989)302
26. I.A.Bandos, M.Cederwall, D.P.Sorokin and D.V.Volkov, Mod. Phys. Lett. **A9**(1994)2987;
     I.A.Bandos, D.P.Sorokin, M.Tonin, P.Pasti and D.V.Volkov, Nucl. Phys. **B 446**(1995)79;
     I.A.Bandos, D.P.Sorokin and D.V.Volkov, Phys. Lett. **B 352**(1995)269
     D.P.Sorokin, Phys. Rep. **329**(2000)1 (and refs. therein)
27. P.K.Townsend, Phys. Lett. **B 261**(1991)65
28. J.P.Gauntlett, J.Gomis and P.K.Townsend, Phys. Lett. **B 248**(1990)288
29. J.A.de Azcarraga and J.Lukierski, Phys. Lett. **B 113**(1982)170
30. W.Siegel, Phys. Lett. **B 128**(1983)397

ROTATING SUPER BLACK HOLE AS SPINNING PARTICLE

ALEXANDER BURINSKII *

NSI, Russian Academy of Sciences, Moscow, Russia

1. Introduction

The Kerr rotating black hole solution displays some remarkable features indicating a relation to the structure of the spinning elementary particles. In particular, in the 1969 Carter [1] observed, that if three parameters of the Kerr-Newman metric are adopted to be ($\hbar$=c=1) $e^2 \approx 1/137$, $m \approx 10^{-22}$, $a \approx 10^{22}$, $ma = 1/2$, then one obtains a model for the four parameters of the electron: charge, mass, spin and magnetic moment, and the gyromagnetic ratio is automatically the same as that of the Dirac electron. Investigations along this line [2–6] allowed to find out stringy structures in the real and complex Kerr geometry and to put forward a conjecture on the baglike structure of the source of the Kerr-Newman solution. The earlier investigations [2, 13, 5] showed that this source represents a rigid rotator (a relativistic disk) built of an exotic matter with superconducting properties. Since 1992 black holes have paid attention of string theory. In 1992 the Kerr solution was generalized by Sen to low energy string theory [7], and it was shown [17] that near the Kerr singular ring the Kerr-Sen solution acquires a metric similar to the field around a heterotic string. The point of view has appeared that black holes can be treated as elementary particles [8]. On the other hand, a description of a spinning particle based only on the bosonic fields cannot be complete, and involving fermionic degrees of freedom is required. Therefore, the spinning particle must be based on a super-Kerr-Newman black hole solution [18] representing a natural combination of the Kerr spinning particle and superparticle model. Angular momentum L of spinning particles is very high $\mid a \mid = L/m \geq m$, and the horizons of the Kerr metric disappear. There appears a naked ring-like singularity which has to be regularized being replaced by a smooth matter source. In this review we consider a source representing a rotating superconducting bag with a smooth domain wall boundary described by a supersymmetric version of

* bur@ibrae.ac.ru

S. Duplij and J. Wess (eds.), Noncommutative Structures in Mathematics and Physics, 181–193.

the $U(I) \times U'(I)$ field model [23]. In fact, this model of the Kerr-Newman source represents a generalization of the Witten superconducting string model [16] for the superconducting baglike sources [6].

2. Complex source of Kerr geometry and its stringy interpretation

The Kerr-Newman solution can be represented in the Kerr-Schild form

$$g_{\mu\nu} = \eta_{\mu\nu} + 2he_\mu^3 e_\nu^3, \tag{1}$$

where $\eta_{\mu\nu}$ is metric of an auxiliary Minkowski space $\eta_{\mu\nu} = diag(-1, 1, 1, 1)$, and h is a scalar function. Vector field e^3 is null, $e_\mu^3 e^{3\mu} = 0$, and tangent to PNC (principal null congruence) of the Kerr geometry. The Kerr PNC is twisting i.e. corresponding to a vortex of a null radiation. [1] One of the main peculiarities of the Kerr geometry is singular ring representing a branch line of the Kerr space on the 'positive' ($r > 0$) and 'negative'($r < 0$) sheets which are divided by the disk $r = 0$ spanned by this ring. The Kerr singular ring is exhibited as a pole of the function $h(r, \theta) = \frac{mr - e^2/2}{r^2 + a^2 \cos^2 \theta}$, where r and θ are the oblate spheroidal coordinates. The Kerr PNC is in-going on the 'negative' sheet of space, it crosses the disk $r = 0$ and turns into out-going one on the 'positive' sheet. Appearance of the Kerr singular ring on the real space-time can also be observed in the Coulomb solution $f = e/\tilde{r}$ when its point-like source is shifted in complex region $(x_0, y_0, z_0) \to (0, 0, ia)$. Radial distance $\tilde{r}$ becomes complex in this case and can be expressed as $\tilde{r} = r + ia \cos \theta$ (Appel, 1987 !). Similarly, the source of Kerr-Newman solution can be considered from complex point of view as a "particle" propagating along a complex world-line [9, 12] parametrized by complex time.

The objects described by the complex world-lines occupy an intermediate position between particles and strings. Like the strings they form the two-dimensional surfaces or the world-sheets in the space-time. It was shown that the complex Kerr source may be considered as a complex hyperbolic string which requires an orbifold-like structure of the world-sheet. In many respects this string is similar to the 'mysterious' $N = 2$ string of superstring theory shedding a light on the puzzle of its physical interpretation. As we have already mentioned, there is one more stringy structure in the Kerr geometry connected with the Kerr singular ring. In fact the both these stringy structures are different exhibitions of some membrane-like source. This source has a complex interpretation alongside with some real image in the form of a rotating bubble which will be discussed further.

The Kerr PNC may be obtained from the complex source by a retarded-time construction. The rays of PNC are the tracks of null planes of the complex light cones emanated from the complex world line [11, 12]. The complex light cone

---

[1] Besides, the Kerr PNC is geodesic and shear free, it represents a bundle of twistors and can be described by the Kerr theorem [10, 11, 9, 12].

with the vertex at some point x_0 of the complex world line $x_0^\mu(\tau)$: $(x_\mu - x_{0\mu})(x^\mu - x_0^\mu) = 0$, can be split into two families of null planes: "left" planes spanned by null vectors e^1 and e^3, and "right" planes spanned by null vectors e^2 and e^3. The Kerr PNC arises as the real slice of the family of the "left" null planes of the complex light cones which vertices lie on the straight complex world line $x_0(\tau)$.

Only the cones lying on the strip $|Im\tau| \le |a|$ have a real slice. Therefore, the ends of the resulting complex string are open. To satisfy the complex boundary conditions, an orbifold-like structure of the worldsheet must be introduced [9, 12], which is closely connected with the above mentioned Kerr's twosheetedness.

3. Super-Kerr-Newman geometry

A supergeneralization of the Kerr-Newman solution can be obtained as a natural combination of the Kerr spinning particle and superparticle [18]. In fact, the complex structure of the Kerr geometry suggests the way of its supergeneralization.

Note, that any exact solution of the Einstein gravity is indeed a trivial solution of supergravity field equations. The supergauge freedom allows one to turn any gravity solution into a form containing spin-3/2 field ψ_i satisfying the supergravity field equations. However, since this spin-3/2 field can be gauged away by the reverse transformation, such supersolutions have to be considered as *trivial*. The hint how to avoid this triviality problem follows from the complex structure of the Kerr geometry. In fact, from the complex point of view the Schwarzschild and Kerr geometries are equivalent and connected by a *trivial* complex shift.

The *non-trivial* twisting structure of the Kerr geometry arises as a result of the complex *shift of the real slice* concerning the center of the solution [11, 9]. Similarly, it is possible to turn a *trivial* super black hole solution into a *non-trivial*. The *trivial supershift* can be represented as a replacement of the complex world line by a superworldline $X_0^\mu(\tau) = x_0^\mu(\tau) - i\theta\sigma^\mu\bar\zeta + i\zeta\sigma^\mu\bar\theta$, parametrized by Grassmann coordinates ζ, $\bar\zeta$, or as a corresponding coordinate replacement in the Kerr solution

$$x'^\mu = x^\mu + i\theta\sigma^\mu\bar\zeta - i\zeta\sigma^\mu\bar\theta; \qquad \theta' = \theta + \zeta, \qquad \bar\theta' = \bar\theta + \bar\zeta, \qquad (2)$$

Assuming that coordinates x^i before the supershift were the usual c-number coordinates one sees that coordinates acquire nilpotent Grassmann contributions after supertranslations. Therefore, there appears a natural splitting of the space-time coordinates on the c-number 'body'-part and a nilpotent part - the so called 'soul'. The 'body' subspace of superspace, or B-slice, is a submanifold where the nilpotent part is equal to zero, and it is a natural analogue to the real slice of the complex case.

Reproducing the real slice procedure of the Kerr geometry in superspace one has to use the replacements:

a/ complex world line $\rightarrow$ superworldline,

b/ complex light cone → superlightcone,
c/ real slice → body slice.

Performing the body-slice procedure to superlightcone constraints

$$s^2 = [x_\mu - X_{0\mu}(\tau)][x^\mu - X_0^\mu(\tau)] = 0, \qquad (3)$$

one selects the body and nilpotent parts of this equation and obtains three equations. The first one is the discussed above real slice condition of the complex Kerr geometry claiming that complex light cones can reach the real slice. The nilpotent part of (3) yields two B-slice conditions

$$[x^\mu - x_0^\mu(\tau)](\theta\sigma_\mu\bar\zeta - \zeta\sigma_\mu\bar\theta) = 0; \qquad (4)$$

$$(\theta\sigma\bar\zeta - \zeta\sigma\bar\theta)^2 = 0. \qquad (5)$$

These equations can be resolved by representing the complex light cone equation via the commuting two-component spinors Ψ and $\tilde\Psi$: $x_\mu = x_{0\mu} + \Psi\sigma_\mu\tilde\Psi$. "Right" (or "left") null planes of the complex light cone can be obtained keeping Ψ constant and varying $\tilde\Psi$ (or keeping $\tilde\Psi$ constant and varying Ψ.) As a result we obtain the equations $\tilde\Psi\bar\theta = 0$, $\tilde\Psi\bar\zeta = 0$, which in turn are conditions of proportionality of the commuting spinors $\tilde\Psi(x)$ determining the PNC of the Kerr geometry and anticommuting spinors $\bar\theta$ and ζ, these conditions providing the left null superplanes of the supercones to reach B-slice. It also leads to $\theta\bar\theta = \zeta\bar\zeta = 0$, and equation (5) is satisfied automatically.

Thus, as a consequence of the B-slice and superlightcone constraints we obtain a non-linear submanifold of superspace $\theta = \theta(x)$, $\bar\theta = \bar\theta(x)$. The original four-dimensional supersymmetry is broken, and the initial supergauge freedom which allowed to turn the super geometry into trivial one is lost. Nevertheless, there is a residual supersymmetry based on free Grassmann parameters θ^1, $\bar\theta^1$.

The above B-slice constraints yield in fact the non-linear realization of broken supersymmetry introduced by Volkov and Akulov [20, 21] and considered in N=1 supergravity by Deser and Zumino [19]. It is assumed that this construction is similar to the Higgs mechanism of the usual gauge theories and $\zeta^\alpha(x)$, $\bar\zeta^{\dot\alpha}(x)$ represent Goldstone fermion which can be eaten by appropriate local supertransformation $\epsilon(x)$ with a corresponding redefinition of the tetrad and spin-3/2 field. Complex character of supertranslations in the Kerr case demands to use in this scheme the N=2 supergravity. We omit here details referring to [18] and mention only that in the resulting exact solution the torsion and Grassmann contributions to tetrad cancel, and metric takes the exact Kerr-Newman form. However there are the extra wave fermionic fields on the bosonic Kerr-Newman background propagating along the Kerr PNC and concentrating near the Kerr singularity. Solution contains also an extra axial singularity which is coupled topologically with singular ring threading it.

4. Baglike source of the Kerr-Newman solution

The above consideration of super-Kerr-Newman solution is based on the massless fields providing description of the rotating super-black-hole. It could be the end of story since the source of a rotating black hole is hidden behind the horizons.

However, the value of angular momentum for spinning particles is very high regarding the mass parameter and the horizons disappear uncovering the Kerr singular ring. To get a regularized solution the massless fields of the black hole solution have to get a mass in the core region forming a matter source removing the Kerr singularity and twosheetedness of the Kerr space.[2]

Obtaining a regular Kerr source represents an old problem. In the first disk-like model given by Israel [2] a truncation of the negative sheet was used. As a result there appeared a source distribution on the surface of the disk $r = 0$. Analyzing the resulting stress-energy tensor Hamity showed [13] that this disk has to be in a rigid relativistic rotation and built of an exotic matter having zero energy density and negative pressure. In the development of this model given by López [5] the truncation is placed at the coordinate surface $r = r_e = \frac{e^2}{2m}$ (where $h = 0$), and the region $r < r_e$ is replaced by Minkowski space. As a result the source takes the form of the highly oblate and infinitely thin elliptic shell of the Compton radius $a = \frac{1}{2m}$ and of the thickness of the classical Dirac electron radius r_e. For small angular momentum the source takes the form of the Dirac electron model, a charged sphere of the classical size r_e. The fields out of the shell have the exact Kerr-Newman form. Interior of the shell is flat. The shell is charged and rotating, and built of a superconducting matter. In corotating space one sees that matter has a negative pressure and zero energy density.

The López source represents a bubble with an infinitely thin domain wall boundary. In the paper [6] an attempt was undertaken to get the source of the Kerr-Newman solution with a smooth matter distribution. Retaining the metric in the Kerr-Schild form (1) and the form and properties of the Kerr PNC, it was assumed that function $h(r, \theta)$ takes a more general form $h = \frac{f(r)}{r^2 + a^2 \cos^2 \theta}$, where the function $f(r)$ is continuous and takes the usual Kerr-Newman form $f_{KN}(r) = mr - e^2/2$ in the external region. In the same time, in a neighborhood of the Kerr disk $r \leq r_0$ (the core region) including the Kerr singularity, the function $f(r)$ has to satisfy some conditions of regularity to provide finiteness of the metric and the stress-energy tensor of source.

It was shown that this regularity is achieved for the function $f(r) \sim r^n$ with $n \geq 4$. In the case $n = 4$, $f(r) = f_0(r) = \alpha r^4$, (in the nonrotational case $a = 0$) space-time has a constant curvature in the core and generated by a homogeneous matter distribution with energy density $\rho = \frac{1}{8\pi} 6\alpha$. Therefore, assuming that matter in the core has a homogenous distribution one can estimate

---

[2] This problem is actual for black hole physics, too. See for example [22] and references therein.

the boundary of the core region r_0 as a point of intersection of $f_0(r)$ and $f_{KN}(r)$. Regularity of the stress-tensor demands continuity of the function $f(r)$ up to first derivative, therefore, the resulting smooth function $f(r)$ must be interpolating between functions $f_0(r)$ and $f_{KN}(r)$ near the boundary of the core $r \approx r_0$.

Let us now mention that general metric (1) can be expressed via orthonormal tetrad as follows [6] $g_{\mu\nu} = m_\mu m_\nu + n_\mu n_\nu + l_\mu l_\nu - u_\mu u_\nu$, and the corresponding stress-energy tensor of the source (following from the Einstein equations) may be represented in the form $T_{\mu\nu}^{(af)} = (8\pi)^{-1}[(D+2G)g_{\mu\nu} - (D+4G)(l_\mu l_\nu - u_\mu u_\nu)]$, where u_μ is the unit time-like four-vector, l_μ is the unit vector in radial direction, and n_μ, m_μ are two more space-like vectors. Here

$$D = -f''/(r^2 + a^2 \cos^2 \theta), \tag{6}$$

$$G = (f'r - f)/(r^2 + a^2 \cos^2 \theta)^2, \tag{7}$$

and the Boyer-Lindquist coordinates t, r, θ, ϕ are used.

Like to the results for singular (infinitely thin) shell-like source [13, 5], the stress-energy tensor can be diagonalized in a comoving coordinate system showing that the source represents a relativistic rotating disk. However, in this case, the disk is separated into ellipsoidal layers each of which rotates rigidly with its own angular velocity $\omega(r) = a/(a^2 + r^2)$. In the comoving coordinate system the tensor $T_{\mu\nu}$ takes the form

$$T_{\mu\nu} = \frac{1}{8\pi} \begin{pmatrix} 2G & 0 & 0 & 0 \\ 0 & -2G & 0 & 0 \\ 0 & 0 & 2G+D & 0 \\ 0 & 0 & 0 & 2G+D \end{pmatrix}, \tag{8}$$

that corresponds to energy density $\rho = \frac{1}{8\pi} 2G$, radial pressure $p_{rad} = -\frac{1}{8\pi} 2G$, and tangential pressure $p_{tan} = \frac{1}{8\pi}(D + 2G)$.

Setting $a = 0$ for the non-rotating case, we obtain $\Sigma = r^2$, the surfaces $r = const.$ are spheres and we have spherical symmetry for all the above relations. The region described by $f(r) = f_0(r)$ is the region of constant value of the scalar curvature invariant $R = 2D = -2f_0''/r^2 = -24\alpha$, and of a constant value of energy density. If we assume that the region of a constant curvature is closely extended to the boundary of source r_0 which is determined as a root of the equation

$$f_0(r_0) = f_{KN}(r_0), \tag{9}$$

then, smoothness of the $f(r)$ in a small neighborhood of r_0, say $|r - r_0| < \delta$, implies a smooth interpolation for the derivative of the function $f(r)$ between $f_0'(r)|_{r=r_0-\delta}$ and $f_{KN}'(r)|_{r=r_0+\delta}$. Such a smooth interpolation on a small distance δ shall lead to a shock-like increase of the second derivative $f''(r)$ by $r \approx r_0$.

In charged case for $\alpha \leq 0$ (AdS internal geometry of core) there exists only one positive root r_0, and second derivative of the smooth function $f''(r)$ is positive near this point. Therefore, there appears an extra tangential stress near r_0 caused by the term $D = -f''(r)/(r^2 + a^2 \cos^2 \theta)|_{r=r_0}$ in the expression (8). It can be interpreted as the appearance of an effective shell (or a domain wall) confining the charged ball-like source with a geometry of a constant curvature inside the ball. The case $\alpha = 0$ represents the bubble with a flat interior which has in the limit $\delta \to 0$ an infinitely thin shell. It corresponds to the López model.

The internal geometry of the ball is de Sitter one for $\alpha > 0$, anti de Sitter one for $\alpha < 0$ and flat one for $\alpha = 0$.

Let us consider peculiarities of the rotating Kerr source. In this case the surfaces $r = const.$ are ellipsoids described by the equation $\frac{x^2 + y^2}{r^2 + a^2} + \frac{z^2}{r^2} = 1$. Energy density inside the core will be constant only in the equatorial plane $\cos \theta = 0$. Therefore, the Kerr singularity is regularized and the curvature is constant in string-like region $r < r_0$ and $\theta = \pi/2$ near the former Kerr singular ring. The ratio $\frac{stress|_{\theta=0}}{stress|_{\theta=\pi/2}} < (r_e/a)^4 = e^8 < 10^{-8}$ shows a strong increase of the stress near the string-like boundary of the disk.

5. Field model: From superconducting strings to superconducting bags

The known models of the bags and cosmic bubbles with smooth domain wall boundaries are based on the Higgs scalar field ϕ with a Lagrange density of the form $L = -\frac{1}{2}\partial_\mu\phi\partial^\mu\phi - \frac{\lambda^2}{8}(\phi^2 - \eta^2)^2$ leading to the kink planar solution (the wall is placed in xy-plane at $z = 0$) $\phi(z) = \eta \tanh(z/\delta)$, where $\delta = \frac{2}{\lambda\eta}$ is the wall thickness. The kink solution describes two topologically distinct vacua $< \phi >= \pm\eta$ separated by the domain wall.

The stress–energy tensor of the domain wall is $T_\mu^\nu = \frac{\lambda^2\eta^4}{4} \cosh^{-4}(z/\delta)diag(1, 1, 1, 0)$, indicating a surface stress within the plane of the wall which is equal to the energy density. When applied to the spherical bags or cosmic bubbles [27, 28], the thin wall approximation is usually assumed $\delta \ll r_0$, and a spherical domain wall separates a false vacuum inside the ball $(r < r_0) < \phi >_{in}= -\eta$ from a true outer vacuum $< \phi >_{out}= \eta$.

In the gauge string models, the Abelian Higgs field provides confinement of the magnetic vortex lines in superconductor. Similarly, in the models of superconducting bags, the gauge Yang-Mills or quark fields are confined in a bubble (or cavity) in superconducting QCD-vacuum.

A direct application of the Higgs model for modelling superconducting properties of the Kerr source is impossible since the Kerr source has to contain the external long range Kerr-Newman electromagnetic field, while in the models of strings and bags the situation is quite opposite: vacuum is superconducting in external region and electromagnetic field acquires a mass there from Higgs field turning into a short range field. An exclusion represents the $U(I) \times \tilde{U}(I)$ cosmic string model given by Vilenkin-Shellard and Witten [15, 16] which represents a doubling of the usual Abelian Higgs model. The model contains two sectors, say A and B, with two Higgs fields ϕ_A and ϕ_B, and two gauge fields A_μ and B_μ yielding two sorts of superconductivity A and B. It can be adapted to the bag-like source in such a manner that the gauge field A_μ of the A sector has to describe a long-range electromagnetic field in outer region of the bag while the chiral scalar field of this sector ϕ_A has to form a superconducting core inside the bag which must be unpenetrable for A_μ field.

The sector B of the model has to describe the opposite situation. The chiral field ϕ_B must lead to a B-superconductivity in outer region confining the gauge field B_μ inside the bag.

The corresponding Lagrangian of the Witten $U(I) \times \tilde{U}(I)$ field model is given by [16]

$$L = -(D^\mu \phi_A)(\overline{D_\mu \phi_A}) - (\tilde{D}^\mu \phi_B)(\overline{\tilde{D}_\mu \phi_B}) - \frac{1}{4}F_A^{\mu\nu}F_{A\mu\nu} - \frac{1}{4}F_B^{\mu\nu}F_{B\mu\nu} - V, \tag{10}$$

where $F_{A\mu\nu} = \partial_\mu A_\nu - \partial_\nu A_\mu$ and $F_{B\mu\nu} = \partial_\mu B_\nu - \partial_\nu B_\mu$ are field stress tensors, and the potential has the form

$$V = \lambda(\bar{\phi}_B \phi_B - \eta^2)^2 + f(\bar{\phi}_B \phi_B - \eta^2)\bar{\phi}_A \phi_A + m^2 \bar{\phi}_A \phi_A + \mu(\bar{\phi}_A \phi_A)^2. \tag{11}$$

Two Abelian gauge fields A_μ and B_μ interact separately with two complex scalar fields ϕ_B and ϕ_A so that the covariant derivative $D_\mu \phi_A = (\partial_\mu + ieA_\mu)\phi_A$ is associated with A sector, and covariant derivative $\tilde{D}_\mu \phi_B = (\partial_\mu + igB_\mu)\phi_B$ is associated with B sector. The model fully retains the properties of the usual bag models which are described by B sector providing confinement of B_μ gauge field inside bag, and it acquires the long range electromagnetic field A_μ in the outer-to-the-bag region described by sector A. The A and B sectors are almost independent interacting only through the potential term for scalar fields. This interaction has to provide synchronized phase transitions from superconducting B-phase inside the bag to superconducting A-phase in the outer region. The synchronization of this transition occurs explicitly in a supersymmetric version of this model given by Morris [23].

5.1. SUPERSYMMETRIC MORRIS MODEL

In Morris model, the main part of Lagrangian of the bosonic sector is similar to the Witten field model. However, model has to contain an extra scalar field Z providing synchronization of the phase transitions in A and B sectors. [3]

---

[3] In fact the Morris model contains five complex chiral fields $\phi_i = \{Z, \phi_-, \phi_+, \sigma_-, \sigma_+\}$. However, the following identification of the fields is assumed $\phi = \phi_+$; $\bar{\phi} = \phi_-$ and $\sigma = \sigma_+$; $\bar{\sigma} = \sigma_-$. In previous notations $\phi \sim \phi_A$ and $\sigma \sim \phi_B$.

The effective Lagrangian of the Morris model has the form

$$L = -2(D^\mu\phi)\overline{(D_\mu\phi)} - 2(\tilde{D}^\mu\sigma)(\overline{\tilde{D}_\mu}\sigma) - \partial^\mu Z \partial_\mu \bar{Z}$$
$$-\frac{1}{4}F^{\mu\nu}F_{\mu\nu} - \frac{1}{4}F_B^{\mu\nu}F_{B\mu\nu} - V(\sigma, \phi, Z), \qquad (12)$$

where the potential V is determined through the superpotential W as

$$V = \sum_{i=1}^{5}|W_i|^2 = 2|\partial W/\partial\phi|^2 + 2|\partial W/\partial\sigma|^2 + |\partial W/\partial Z|^2. \qquad (13)$$

The following superpotential, yielding the gauge invariance and renormalizability of the model, was suggested [4]

$$W = \lambda Z(\sigma\bar{\sigma} - \eta^2) + (cZ + m)\phi\bar{\phi}, \qquad (14)$$

where the parameters λ, c, m, and η are real positive quantities.

The resulting scalar potential V is then given by

$$V = \lambda^2(\bar{\sigma}\sigma - \eta^2)^2 + 2\lambda c(\bar{\sigma}\sigma - \eta^2)\phi\bar{\phi} + c^2(\bar{\phi}\phi)^2 + \qquad (15)$$
$$2\lambda^2\bar{Z}Z\bar{\sigma}\sigma + 2(c\bar{Z} + m)(cZ + m)\bar{\phi}\phi.$$

5.1.1. *Supersymmetric vacua*

From (13) one sees that the supersymmetric vacuum states, corresponding to the lowest value of the potential, are determined by the conditions

$$F_\sigma = -\partial\bar{W}/\partial\bar{\sigma} = 0; \qquad (16)$$
$$F_\phi = -\partial\bar{W}/\partial\bar{\phi} = 0; \qquad (17)$$
$$F_Z = -\partial\bar{W}/\partial\bar{Z} = 0, \qquad (18)$$

and yield $V = 0$. These equations lead to two supersymmetric vacuum states:

$$I) \quad Z = 0; \quad \phi = 0; \quad |\sigma| = \eta; \quad W = 0; \qquad (19)$$

and

$$II) \quad Z = -m/c; \quad \sigma = 0; \quad |\phi| = \eta\sqrt{\lambda/c}; \quad W = \lambda m\eta^2/c. \qquad (20)$$

We shall take the state I for external region of the bag, and the state II as a state inside the bag.

The treatment of the gauge field A_μ and B_μ in B is similar in many respects because of the symmetry between A and B sectors allowing one to consider the

---

[4] Superpotential is homomorphic function of $\{Z, \phi, \bar{\phi}, \sigma, \bar{\sigma}\}$.

state $\Sigma = \eta$ in outer region as superconducting one in respect to the gauge field B_μ. Field B_μ acquires the mass $m_B = g\eta$ in outer region, and the $\tilde{U}(I)$ gauge symmetry is broken, which provides confinement of the B_μ field inside the bag. The bag can also be filled by quantum excitations of fermionic, or non Abelian fields. The interior space of the Kerr bag is regularized in this model since the Kerr singularity and twofoldedness are suppressed by function $f = f_0(r)$. However, a strong increase of the fields near the former Kerr singularity can be retained leading to the appearance of traveling waves along the boundary of the disk.

5.2. SUPERSYMMETRIC BUBBLE BASED ON THE MORRIS FIELD MODEL

It is shown in [6] that in the planar thin wall approximation, and by neglecting the gauge fields there is a supersymmetric BPS-saturated domain wall solution interpolating between supersymmetric vacua I) and II). This domain wall displays the usual structure of stress-energy tensor with a tangential stress. The non-zero components of the stress-energy tensor take the form

$$T_{00} = -T_{xx} = -T_{yy} = \frac{1}{2}[\delta_{ij}(\Phi^i{}_{,z})(\Phi^j{}_{,z}) + V]; \tag{21}$$

$$T_{zz} = \frac{1}{2}[\delta_{ij}(\Phi^i{}_{,z})(\Phi^j{}_{,z}) - V], \tag{22}$$

where $\Phi_i = \{Z, \phi_-, \phi_+, \sigma_-, \sigma_+\}$. One can estimate the mass and energy of a bubble formed by such a domain wall in global supersymmetry setting vacuum I) as external one and vacuum II) as an internal vacuum. Using the Tolman relation $M = \int dx^3 \sqrt{-g}(-T_0^0 + T_1^1 + T_2^2 + T_3^3)$, replacing coordinate z on radial coordinate r, and integrating over sphere one obtains

$$M_{bubble} = -4\pi \int V(r)r^2 dr = -4\pi \int (\Phi^i{}_{,r})^2 r^2 dr. \tag{23}$$

The resulting effective mass is negative, which is caused by gravitational contribution of the tangential stress. The repulsive gravitational field was obtained in many singular and smooth models of domain walls [32, 25, 30, 31]. One should note, that similar gravitational contribution to the mass caused by interior of the bag will be $M_{gr.int} = \int Dr^2 dr = -\frac{2}{3}\Lambda r_0^3$. It depends on the sign of curvature inside the bag and will be negative in de Sitter case and positive in AdS one.

The total energy of a uncharged bubble forming from the supersymmetric BPS saturated domain wall is

$$E_{0bubble} = E_{wall} = 4\pi \int_0^\infty \rho r^2 dr \approx 4\pi r_0^2 \epsilon_{min}, \tag{24}$$

where r_0 is radius of the bubble, and $\epsilon_{min} = W(0) - W(\infty) = \lambda m \eta^2/c$. Corresponding total mass following from the Tolman relation will be negative

$M_{0bubble} = -E_{wall} \approx -4\pi r_0^2 \epsilon_{min}$. It is the known fact showing that the uncharged bubbles are unstable and form the time-dependent states [30, 31].

For charged bubbles there are extra positive terms: contribution caused by the energy and mass of the external electromagnetic field $E_{e.m.} = M_{e.m.} = \frac{e^2}{2r_0}$, and contribution to mass caused by gravitational field of the external electromagnetic field (determined by Tolman relation for the external e.m. field) $M_{grav.e.m.} = E_{e.m.} = \frac{e^2}{2r_0}$. As a result the total energy for charged bubble is

$$E_{tot.bubble} = E_{wall} + E_{e.m.} = 4\pi r_0^2 \epsilon_{min} + \frac{e^2}{2r_0}, \qquad (25)$$

and the total mass will be

$$M_{tot.bubble} = M_{0bubble} + M_{e.m.} + M_{grav.e.m.} = \qquad (26)$$

$$-E_{wall} + 2E_{e.m.} = -4\pi r_0^2 \epsilon_{min} + \frac{e^2}{r_0}. \qquad (27)$$

Minimum of the total energy is achieved by $r_0 = (\frac{e^2}{16\pi\epsilon_{min}})^{1/3}$, which yields the following expressions for total mass and energy of the stationary state

$$M_{tot}^* = E_{tot}^* = \frac{3e^2}{4r_0}. \qquad (28)$$

One sees that the resulting total mass of charged bubble is positive, however, due to negative contribution of $M_{0bubble}$ it can be lower than BPS energy bound of the domain wall forming this bubble. This remarkable property of the bubble models ('ultra-extreme' states for the Type I domain walls in [30]) allows one to overcome BPS bound [33] and opens the way to get the ratio $m^2 \ll e^2$ which is necessary for particle-like models.

5.3. BAGLIKE SOURCE IN SUPERGRAVITY

In supergravity the scalar potential has a more complicate form [21, 30, 31, 29]

$$V_{sg} = e^{k^2 K}(K^{i\bar{j}} D_i W \overline{D_j W} - 3k^2 W \bar{W}), \qquad (29)$$

where K is Kähler potential $K^{i\bar{j}} = \frac{\partial^2 K}{\partial \Phi_i \partial \Phi_j}$, and $k^2 = 8\pi G_N$, G_N is the Newton constant. In the small kW limit, this expression turns into potential of global susy. In this approximation, the above treatment of the charged domain wall bubble will be valid in supergravity. The preserving supersymmetry vacuum state has to satisfy the condition $D_i W \equiv W_i + k^2 K_i W = 0$. This condition is satisfied for the internal vacuum state II) only in the limit $k^2 \to 0$ since $W = \lambda m \eta^2/c$ inside the bag, and $D_i W \approx k^2 K_i W$ there. In the order k^2 the vacuum state II) does not preserve supersymmetry. There appears also an extra contribution to stress-energy tensor having the leading term

$$T_{\mu\nu} = 3(k^2/8\pi)e^{k^2 K}|W|^2 g_{\mu\nu},\tag{30}$$

and yielding the negative cosmological constant $\Lambda = -3k^4 e^{k^2 K}|W|^2$ and to anti-de Sitter space-time for the bag interior. General expression for cosmological constant inside the bag has the form

$$\Lambda = k^4 e^{k^2 K} \sum_i \{k^2|K_i W|^2 - 3|W|^2\}.\tag{31}$$

It yields AdS vacuum if $k^2|K_i W|^2 - 3|W|^2 < 0$.

In the same time the vacuum state I) in external region has $W = 0$ and $\Lambda = 0$, and it preserves supersymmetry for strong chiral fields.

6. Conclusion

A regularized source of the Kerr-Newman solution is considered having the structure of a rotating bag with AdS interior and a smooth domain wall boundary. It is shown that the Witten superconducting string model can be generalized and adapted forming a charged superconducting bag with AdS interior and a long range external gauge field which is necessary for description of charged black holes. Since 1968 a successive accumulation of evidences is observed relating the structure of Kerr geometry with physics of elementary particles.

Acknowledgments. We would like to thank organizers of this Workshop for kind invitation and financial support.

References

1. B. Carter, *Global structure of the Kerr family of gravitational fields*, Phys. Rev. **174**(1968) 1559.
2. W. Israel, *Source of Kerr metric*, Phys. Rev. **D2** (1970) 641.
3. A. Burinskii, *Microgeons with spin*, Sov. Phys. JETP **39**(1974) 193.
4. D. Ivanenko and A. Burinskii, *Gravitational strings in the models of spinning elementary particles*, Izv. VUZ Fiz. **5**(1975)135.
5. C. A. López, *Extended model of the electron in general relativity*, Phys. Rev. **D30** (1984) 313.
6. A. Burinskii, *Supersymmetric superconducting bag as a core of Kerr spinning particle*, e-print hep-th/0008129.
7. A. Sen, *Rotating charged black hole solution in heterotic string theory* , Phys.Rev.Lett., **69**(1992)1006-1009.
8. A. Sen, *Extremal black holes and elementary string states*, Modern Phys. Lett. **A 10**(1995)2081.

C. Holzhey and F. Wilczek, *Black holes as elementary particles*, Nucl. Phys. **B380**(1992)447, hep-th/9202014

9. A. Burinskii, *String - like structures in complex Kerr geometry*, in Relativity Today, (R.P. Kerr and Z. Perjes eds,), Academiai Kiado, Budapest 1994, pp. 149-158, gr-qc/9303003.
————, *Complex string as source of Kerr geometry*, in Espec. Space Explorations, **9** (**C2**)(1995) 60, Moscow, Belka, hep-th/9503094.

10. G.C. Debney, R.P. Kerr, A. Schild, *Solutions of the Einstein and Einstein-Maxwell equations.*, J.Math.Phys. **10**(1969)1842.

11. A. Burinskii, R.P. Kerr and Z. Perjes, *Nonstationary Kerr Congruences*, e-print gr-qc/9501012.

12. A. Burinskii, *The Kerr geometry, complex world lines and hyperbolic strings*, Phys.Lett. **A** **185**(1994)441.

13. V. Hamity, *Interior of Kerr metric*, Phys. Let. **A56**(1976)77.

14. C.A. López, *Material and electromagnetic sources of the Kerr-Newman geometry*, Nuovo Cimento **B 76**(1983)9;

15. A. Vilenkin and E.P.S. Shellard, *Cosmic Strings and Other Topological Defects* (Cambrige University Press, 1994)

16. E. Witten, *Superconducting strings*, Nucl.Phys., **B249**(1985)557.

17. A. Burinskii, *Some properties of the Kerr solution to low energy string theory*, Phys.Rev. **D** **52**(1995)5826, hep-th/9504139.

18. A. Burinskii, *Kerr spinning particle, strings and superparticle models*, Phys.Rev. **D 57** (1998)2392, hep-th/9704102. ————, *Super-Kerr-Newman solution to broken N=2 supergravity*, Class. Quantum Grav.**16**(1999)3497, hep-th/9903032; ————, *Non-trivial supergeneralization of the Kerr-Newman solution*, Ann.Phys.(Leipzig), **9** (2000) Spec.Issue 34, hep-th/9910045.

19. S. Deser and B. Zumino, *Broken supersymmetry and supergravity*, Phys. Rev. Lett. **38**(1977) 1433.

20. D.V. Volkov and V.P. Akulov, *Possible universal neutrino interaction*, JETP Lett. **16**(1972)367.

21. J. Wess and J. Bagger, *Supersymmetry and supergravity*, Princeton, New Jersey 1983.

22. I. Dymnikova, *Vacuum nonsingular black hole*, Gen.Rel.Grav. **24**(1992)235; E.Elizalde and S.R. Hildebrandt, *Regular quantum interiors for black holes*, gr-qc/0007030

23. J.R. Morris, *Supersymmetry and gauge invariance constraints in a $U(I) \times U(I)'$-Higgs superconducting cosmic string model*, Phys.Rev.**D** **53**(1996)2078, hep-ph/9511293.

24. C.A. López, *Internal structure of a classical spinning electron*, Gen.Rel.Grav.**24** (1992)2851.

25. C.A. López, *Dynamics of charged bubbles in general relativity and models of particles* Phys.Rev. **D 38**(1988) 3662.

26. M. Gürses and F.Gürsey, *Lorentz covariant treatment of the Kerr-Schild geometry*, Journ. Math.Phys. **16**(1975)2385.

27. A. L. Macpherson and B.A. Campbell, *Biased discrete supersymmetry breaking and Fermi balls*, Phys.Let.**B**(1995)205.

28. J.R. Morris and D. Bazeia, *Supersymmetry breaking and Fermi balls*, Phys.Rev. **D** **54**(1996)5217.

29. J.R. Morris, *Cosmic strings in supergravity*, Phys.Rev.**D** **56**(1997)2378. hep-ph/9706302

30. M. Cvetič and H. Soleng, *Supergravity domain walls* , Phys.Rept.**282**(1997)159, hep-th/9604090

31. M. Cvetič, S. Griffies and S.J. Rey, *Static domain walls in N=1 supergravity*, Nucl.Phys. **B** **381**(1992)301-328, hep-th/9206004.

32. J.Ipser and P. Sikivie, *Gravitationally repulsive domain walls*, Phys.Rev. **D 30**(1983) 712-719.

33. G.W. Gibbons and C.M. Hull, *A Bogomolny bound for general relativity and solitons in N=2 supergravity*, Phys. Lett. **B 109**(1982)190.

CLASSIFYING N-EXTENDED 1-DIMENSIONAL SUPER SYSTEMS

FRANCESCO TOPPAN *
CBPF, DCP, Rua Dr. Xavier Sigaud 150,
cep 22290-180 Rio de Janeiro (RJ), Brazil

1. Introduction

In this talk I will report some results obtained in a joint collaboration with A. Pashnev, concerning the classification of the irreducible representations of the N-extended Supersymmetry in 1 dimension and which find applications to the construction of Supersymmetric Quantum Mechanical Systems [1].

This mathematical problem finds immediate application to the theory of dimensionally (to one temporal dimension) supersymmetric $4d$ theories, which gets 4 times the number of supersymmetries of the original models (the $N = 8$ supergravity being e.g. associated with the a $N = 32$ Supersymmetric Quantum Mechanical theory). Due to a lack of superfield formalism for $N > 4$, only partial results are known [2] and [3].

More recently, Supersymmetric and Superconformal Quantum Mechanics have been applied in describing e.g. the low-energy effective dynamics of a certain class of black holes, for testing the AdS/CFT correspondence in the case of AdS_2, in investigating the light-cone dynamics of supersymmetric theories.

In this report of the work with Pashnev, two main results will be presented. At first a peculiar property of supersymmetry in one dimension is exhibited, namely that any finite dimensional multiplet containing d bosons and d fermions in different spin states are put into classes of equivalence individuated by irreducible multiplets of just two spin states, where all bosons and all fermions are grouped in the same spin. Later it is shown that all irreducible multiplets of this kind are in one-to-one correspondence with the classification of real-valued Clifford Γ matrices of Weyl type.

This classification refines (in the case of "non-Euclidean" supersymmetry, see below) the results obtained in [4] and [5]. Another reference where some aspects

* toppan@cbpf.br

195

S. Duplij and J. Wess (eds.), Noncommutative Structures in Mathematics and Physics, 195–201.
© 2001 *Kluwer Academic Publishers. Printed in the Netherlands.*

of the theory of the representation of 1-dimensional supersymmetry are discussed is given by [6].

The mathematical problem we are investigating can be stated as follows, finding the irreducible representation of the supersymmetry algebra

$$\{Q_i, Q_j\} = \omega_{ij} H, \tag{1}$$

where Q_i, $i = 1, 2, \cdots, N$ are supercharges and

$$H = -i\frac{\partial}{\partial t} \tag{2}$$

is the Hamiltonian. The constant tensor ω_{ij} can be conveniently diagonalized and normalized in such a way to coincide with a pseudo-Euclidean metric η_{ij} with signature (p, q). Usually the eigenvalues are all assumed being positive (i.e. $q = 0$), however examples can be given (see [7]), of physical systems whose supersymmetry algebra is characterized by an indefinite tensor. In the following I will discuss the simplest example of this kind.

Any given finite-dimensional representation multiplet of the above superalgebra can be represented in form of a chain of d bosons and d fermions

$$\Phi_{a_0}^0, \quad \Phi_{a_1}^1, \quad \cdots, \quad \Phi_{a_{M-1}}^{M-1}, \quad \Phi_{a_M}^M \tag{3}$$

whose components $\Phi_{a_I}^I$, $(a_I = 1, 2, \cdots, d_I)$ are real and alternatively bosonic and fermionic ($d = d_0 + d_2 + d_4 + ... = d_1 + d_3 + d_5 + ...$). For such a multiplet the short notation $\{d_0, d_1, \cdots, d_M\}$ will also be employed.

Due to dimensionality argument the $i - th$ supersymmetry transformation for the $\Phi_{a_I}^I$ components is given by

$$\delta_\varepsilon \Phi_{a_I}^I = \varepsilon^i (C_i^I)_{a_I}{}^{a_{I+1}} \Phi_{a_{I+1}}^{I+1} + \varepsilon^i (\tilde{C}_i^I)_{a_I}{}^{a_{I-1}} \frac{d}{d\tau} \Phi_{a_{I-1}}^{I-1}, \tag{4}$$

and it simplifies for the end-components (due to the absence of the $I = -1$ and $I = M + 1$ components).

In one dimension it is therefore possible to redefine the last components according to

$$\Phi_{a_M}^M = \frac{d}{d\tau} \Psi_{a_M}^{M-2} \tag{5}$$

in terms of some functions $\Psi_{a_M}^{M-2}$. The initial supermultiplet of length $M + 1$ is now re-expressed as the $\{d_0, d_1, \cdots, d_{M-2} + d_M, d_{M-1}, 0\}$ supermultiplet of length M. By repeating M times the same procedure the shortest supermultiplet $\{d, d\}$ of length 2 can be reached. The above argument outlines the proof of the statement that all supermultiplets are classified according to the irreducible representations of supermultiplets of length 2.

2. Extended supersymmetries and real valued Clifford algebras

The main result of the previous Section is that the problem of classifying all N-extended supersymmetric quantum mechanical systems is reduced to the problem of classifying the irreducible representations of length 2. Having this in mind let us simplify the notations. Let the indices $a, \alpha = 1, \cdots, d$ number the bosonic (and respectively fermionic) elements in the SUSY multiplet. All of them are assumed to depend on the time coordinate τ ($X_a \equiv X_a(\tau)$, $\theta_\alpha \equiv \theta_\alpha(\tau)$).

In order to be definite and without loss of generality let us take the bosonic elements to be the first ones in the chain $\{d, d\}$, which can be conveniently represented also as a column

$$\Psi = \begin{pmatrix} X_a \\ \theta_\alpha \end{pmatrix}, \tag{6}$$

the supersymmetry transformations are reduced to the following set of equations

$$\delta_\varepsilon X_a = \varepsilon^i (C_i)_a{}^\alpha \theta_\alpha \equiv i(\varepsilon^i Q_i \Psi)_a$$

$$\delta_\varepsilon \theta_\alpha = \varepsilon^i (\tilde{C}_i)_\alpha{}^b \frac{d}{d\tau} X_b \equiv i(\varepsilon^i Q_i \Psi)_\alpha \tag{7}$$

where, as a consequence of (1),

$$C_i \tilde{C}_j + C_j \tilde{C}_i = i\eta_{ij} \tag{8}$$

and

$$\tilde{C}_i C_j + \tilde{C}_j C_i = i\eta_{ij} \tag{9}$$

Since $\varepsilon_i, X_a, \theta_\alpha$ are real, the matrices C_i's, $\tilde{C}_i$'s have to be respectively imaginary and real. If we set (just for normalization)

$$C_i = \frac{i}{\sqrt{2}} \sigma_i$$

$$\tilde{C}_i = \frac{1}{\sqrt{2}} \tilde{\sigma}_i \tag{10}$$

and accommodate $\sigma_i, \tilde{\sigma}_i$ into a single matrix

$$\Gamma_i = \begin{pmatrix} 0 & \sigma_i \\ \tilde{\sigma}_i & 0 \end{pmatrix}, \tag{11}$$

they form a set of real-valued Clifford Γ-matrices of Weyl type (i.e. block antidiagonal), obeying the (pseudo-) Euclidean anticommutation relations

$$\{\Gamma_i, \Gamma_j\} = 2\eta_{ij}. \tag{12}$$

Therefore the classification of irreducible multiplets of representation of a (p, q) extended supersymmetry is in one-to-one correspondence with the classification of the real-valued Clifford algebras $C_{p,q}$ with the further property that the Γ matrices can be realized in Weyl (i.e. block antidiagonal) form.

Real-valued Clifford algebras have been classified in [8] for compact ($q = 0$) case, and in [9] for the non-compact one. I follow here the exposition in [10].

Three cases have to be distinguished for real representations, specified by the type of most general solution allowed for a real matrix S commuting with all the Clifford Γ_i matrices, i.e.

i) the normal case, realized when S is a multiple of the identity,

ii) the almost complex case, for S being given by a linear combination of the identity and of a real $J^2 = -1$ matrix,

iii) finally the quaternionic case, for S being a linear combination of real matrices satisfying the quaternionic algebra.

Real irreducible representations of normal type exist whenever the condition $p - q = 0, 1, 2 \mod 8$ is satisfied (their dimensionality being given by $2^{[\frac{N}{2}]}$, where $N = p+q$), while the almost complex and the quaternionic type representations are realized in the $p - q = 3, 7 \mod 8$ and in the $p - q = 4, 5, 6 \mod 8$ cases respectively. The dimensionality of these representations is given in both cases by $2^{[\frac{N}{2}]+1}$.

We further require the extra-condition that the real representations should admit a block antidiagonal realization for the Clifford Γ matrices. This condition is met for $p - q = 0 \mod 8$ in the normal case (it corresponds to the standard Majorana-Weyl requirement), $p - q = 7 \mod 8$ in the almost complex case and $p - q = 4, 6 \mod 8$ in the quaternionic case. In all these cases the real irreducible representation is unique.

It is therefore possible to furnish the dimensionality of the irreducible representations of the of the supersymmetry algebra or, conversely, the allowed (p, q) signatures associated to a given dimensionality of the bosonic and fermionic spaces. The latter result is conveniently expressed by introducing the notion of maximally extended supersymmetry. The $C_{p,q}$ ($p - q = 6 \mod 8$) real representation for the quaternionic case can be recovered from the $7 \mod 8$ almost complex $C_{p+1,q}$ representation by deleting one of the Γ matrices; in its turn the latter representation is recovered from the $C_{p+2,q}$ normal Majorana-Weyl representation by deleting another Γ matrix. The dimensionality of the three representations above being the same, the normal Majorana-Weyl representation realizes the maximal possible extension of supersymmetry compatible with the dimensionality of the representation. In search for the maximal extension of supersymmetry we can therefore limit ourselves to consider the normal Majorana-Weyl representations, as well as the quaternionic ones satisfying the $p - q = 4 \mod 8$ condition.

Let us therefore introduce a parameter ϵ, which assumes two values and is used to distinguish the Majorana-Weyl ($\epsilon = 0$) with respect to the quaternionic case ($\epsilon = 1$). A space of $d = 2^t$ bosonic and $d = 2^t$ fermionic states can carry the following set of maximally extended supersymmetries

$$(p = t - 4z + 5 - 3\epsilon, q = t + 4z + \epsilon - 3) \tag{13}$$

where the integer $z = k - l$ must take values in the interval

$$\frac{1}{4}(3 - t - \epsilon) \leq z \leq \frac{1}{4}(t + 5 - 3\epsilon) \tag{14}$$

in order to guarantee the $p \geq 0$ and $q \geq 0$ requirements.

3. An application and conclusions.

One of the most significant application of extended supersymmetric quantum mechanics concerns the 1-dimensional σ models evolving in a target spacetime manifold presenting both bosonic and fermionic coordinates. In general such models present a non-linear kinetic term and the extended supersymmetries put constraints on the metric of the target. In this section let us present here a very simplified model, which however is illustrative of how invariances under pseudo-Euclidean supersymmetry can arise. Let us in fact consider a model of d bosonic fields X_a and d spinors ψ_α freely moving in a flat d-dimensional target manifold, not necessarily Minkowskian or Euclidean, endorsed of a pseudo-euclidean η_{ab}. Let us furthermore introduce the free kinetic action being given by

$$S_K = \int dt \mathcal{L} = \frac{1}{2} \int dt \left(\dot{X}_a \dot{X}_b \eta^{ab} + i\delta \dot{\psi}_\alpha \psi_\beta \eta^{\alpha\beta} \right), \tag{15}$$

where the metric $\eta^{\alpha\beta}$ for the spinorial part is assumed to have the same signature as the metric η^{ab}, and δ is just a sign normalization ($\delta = \pm 1$).

A natural question to be asked is which supersymmetries are invariances of the above free kinetic action. The answer is furnished by accommodating the d bosonic and d fermionic coordinates into a (maximally extended) irreducible representation of the extended supersymmetries, and later counting how many such transformations survive as invariances of the action. The first non-trivial example concerns a 2-dimensional target($d = 2$), whose two bosonic and two fermionic degrees of freedom carry the $\{2, 2\}$ representation of $(2, 2)$ extended supersymmetry. However, only half of these supersymmetries are realized as invariances of the action. The action indeed is invariant under either the $(2, 0)$ or the $(1, 1)$ extended supersymmetries, whether the target space is respectively Euclidean or Minkowskian. Therefore already in the 2-dimensional Minkowskian case we observe the arising of a pseudo-Euclidean supersymmetry invariance. The next simplest example is realized by a 4-dimensional target. The four bosonic and four fermionic coordinates can be accommodated into three irreducible representations of maximally extended supersymmetry, according to formula (13), namely the $(4, 0)$, the $(0, 4)$ and the $(3, 3)$ extended supersymmetries. The action (15) turns out to be invariant, for Euclidean $(4 + 0)$, Minkowskian $(3 + 1)$ and $(2 + 2)$ signature for the metric η, according to the following table

		$(4, 0)$	$(0, 4)$	$(3, 3)$		
$(4 + 0)$	$(4, 0)$	$(0, 0)$	$(3, 0)$	$\delta = +1$
$(4 + 0)$	$(0, 0)$	$(0, 4)$	$(0, 3)$	$\delta = -1$
$(3 + 1)$	$(1, 0)$	$(0, 0)$	$(1, 0)$	$\delta = +1$
$(3 + 1)$	$(0, 0)$	$(0, 1)$	$(0, 1)$	$\delta = -1$
$(2 + 2)$	$(2, 0)$	$(0, 2)$	$(2, 1)$	$\delta = +1$
$(2 + 2)$	$(2, 0$	$(0, 2)$	$(1, 2)$	$\delta = -1$

which should be understood as follows. The central entries denote how many supersymmetries are realized as invariances of the (15) action for each one of the three irreducible representations of maximally extended supersymetry, in correspondence with the given signature of spacetime and sign for δ. In this particular case invariance under pseudo-Euclidean supersymmetry is guaranteed for the target of signature $(2 + 2)$.

In this talk I have presented some results concerning the representation theory for irreducible multiplets of the one-dimensional $N = (p, q)$ extended supersymmetry. A peculiar feature of the one-dimensional supersymmetric algebras consists in the fact that the supermultiplets formed by d bosonic and d fermionic degrees of freedom accommodated in a chain with $M + 1$ ($M \geq 2$) different spin states uniquely determines a 2-chain multiplet of the form $\{\mathbf{d}, \mathbf{d}\}$ which carries a representation of the N extended supersymmetry. Furthermore, it is shown that all such 2-chain irreducible multiplets of the (p, q) extended supersymmetry are fully classified; when e.g. the condition $p - q = 0 \bmod 8$ is satisfied, their classification is equivalent to that one of Majorana-Weyl spinors in any given space-time, the number $p + q$ of extended supersymmetries being associated to the dimensionality D of the spacetime, while the $2d$ supermultiplet dimensionality is the dimensionality of the corresponding Γ matrices. The more general case for arbitrary values of p and q has also been fully discussed.

These mathematical properties can find a lot of interesting applications in connection with the construction of Supersymmetric and Superconformal Quantum Mechanical Models. These theories are vastly studied due to their relevance in many different physical domains, to name just a few it can be mentioned the low-energy effective dynamics of black-hole models, the dimensional reduction of higher-dimensional superfield theories, which are a laboratory for the investigation of the spontaneous breaking of the supersymmetry, and so on.

Acknowledgments. It is a pleasure for me to acknowledge A. Pashnev. The results reported in this talk are fruit of our collaboration. I wish also acknowledge for useful discussions E.A. Ivanov, S. J. Gates Jr., S.O. Krivonos and V. Zima. Finally, let me express my gratitude to the organizers of the ARW conference for the invitation and the warm hospitality.

References

1. A. Pashnev and F. Toppan, *On the Classification of N-Extended Supersymmetric Quantum Mechanical Systems, CBPF, JINR preprint*, CBPF-NF-029/00, JINR E2-2000-193, hep-th/0010135, Dubna, Rio de Janeiro, 2000.
2. M. De Crombrugghe and V. Rittenberg, Ann. of Phys. **151** (1983), 99.
3. M.Claudson and M.B. Halpern, Nucl.Phys. **B250** (1985), 689.
4. B. de Wit, A.K. Tollsten and H. Nicolai, Nucl.Phys. **B392** (1993), 3.
5. S. James Gates, Jr. and Lubna Rana, Phys. Lett. **B352** (1995), 50; ibid. **B369** (1996), 262.
6. R.A. Coles and G. Papadopoulos, Class. Quant. Grav. **7** (1990), 427–438.
7. A. Pashnev, *Noncompact Extension of One-Dimensional Supersymmetry and Spinning Particle, JINR preprint*, E2-91-536, Dubna, 1991.
8. M. Atiyah, R. Bott and A. Shapiro, Topology 3, (Suppl. 1) (1964), 3.
9. I. Porteous, *Topological Geometry*, van Nostand Rheinhold, London, (1969).
10. S. Okubo, Jou. Math. Phys., **32** (1991), 1657; ibid. 1669.

PARA, PSEUDO, AND ORTHOSUPERSYMMETRIC
QUANTUM MECHANICS AND THEIR BOSONIZATION

CHRISTIANE QUESNE*

*PNTPM, Université Libre de Bruxelles, Campus de la Plaine
CP229, Boulevard du Triomphe, B-1050 Brussels, Belgium*

Abstract. We consider the problem of bosonizing supersymmetric quantum mechanics (SSQM) and some of its variants, i.e., of realizing them in terms of only boson-like operators without fermion-like ones. In the SSQM case, this is realized in terms of the generators of the Calogero-Vasiliev algebra (also termed deformed Heisenberg algebra with reflection). In that of the SSQM variants, this is done by considering generalizations of the latter algebra, namely the C_λ-extended oscillator algebras, where C_λ is the cyclic group of order λ.

1. Introduction

Supersymmetry has established an elegant symmetry between bosons and fermions and is one of the cornerstones of modern theoretical physics. Its application to quantum mechanics has provided a powerful method of generating solvable quantum mechanical models. On the other hand, exotic quantum statistics have received considerable attention due to their possible relevance to the fractional quantum Hall effect and anyon superconductivity.

By combining both concepts within the framework of quantum mechanics, one gets variants of SSQM: paraSSQM [1–3], pseudoSSQM [4, 5], and orthoSSQM [6]. They can be realized in terms of bosons and parafermions [7], pseudofermions [4, 5], or orthofermions [8], respectively.

By using the Calogero-Vasiliev algebra [9], Plyushchay showed [10] that SSQM can be described in terms of only boson-like operators without fermion-like ones (see also [11]).

In the present communication, we shall consider generalizations of the Calogero-Vasiliev algebra, namely the C_λ-extended oscillator algebras (where $C_\lambda = Z_\lambda$ is the cyclic group of order λ) [12–14]. We shall show that they have

---

* cquesne@ulb.ac.be

S. Duplij and J. Wess (eds.), Noncommutative Structures in Mathematics and Physics, 203–213.
© 2001 *Kluwer Academic Publishers. Printed in the Netherlands.*

some interesting applications to variants of SSQM [12, 14], as they provide a bosonization of the latter analogous to that obtained by Plyushchay for SSQM.

2. Generalized deformed and $\mathcal{G}$-extended oscillator algebras

The generalized deformed oscillator algebras (GDOAs) (see e.g. Refs. [15, 16] and references quoted therein) arose from successive generalizations of the Arik-Coon [17] and Biedenharn-Macfarlane [18, 19] q-oscillators. Such algebras, denoted by $\mathcal{A}_q(G(N))$, are generated by the unit, creation, annihilation, and number operators I, $a^\dagger$, a, N, satisfying the Hermiticity conditions $\left(a^\dagger\right)^\dagger = a$, $N^\dagger = N$, and the commutation relations

$$\left[N, a^\dagger\right] = a^\dagger, \qquad [N, a] = -a, \qquad \left[a, a^\dagger\right]_q \equiv aa^\dagger - qa^\dagger a = G(N), \qquad (1)$$

where q is some real number and $G(N)$ is some Hermitian, analytic function.

On the other hand, $\mathcal{G}$-extended oscillator algebras, where $\mathcal{G}$ is some finite group, appeared in connection with n-particle integrable models. For the Calogero model [20], for instance, $\mathcal{G}$ is the symmetric group S_n [21, 22].

For two particles, the S_2-extended oscillator algebra $\mathcal{A}_\kappa^{(2)}$, where $S_2 = \{I, K \mid K^2 = I\}$, is generated by the operators I, $a^\dagger$, a, N, K, subject to the Hermiticity conditions $\left(a^\dagger\right)^\dagger = a, N^\dagger = N, K^\dagger = K^{-1}$, and the relations

$$\left[N, a^\dagger\right] = a^\dagger, \qquad [N, K] = 0, \qquad K^2 = I,$$
$$\left[a, a^\dagger\right] = I + \kappa K \qquad (\kappa \in \mathbf{R}), \qquad a^\dagger K = -Ka^\dagger, \qquad (2)$$

together with their Hermitian conjugates.

When the S_2 generator K is realized in terms of the Klein operator $(-1)^N$, $\mathcal{A}_\kappa^{(2)}$ becomes a GDOA characterized by $q = 1$ and $G(N) = I + \kappa(-1)^N$, and known as the Calogero-Vasiliev oscillator algebra [9].

The operator K may be alternatively considered as the generator of the cyclic group C_2 of order two, since the latter is isomorphic to S_2. By replacing C_2 by the cyclic group of order λ, $C_\lambda = \{I, T, T^2, \ldots, T^{\lambda-1} \mid T^\lambda = I\}$, one then gets a new class of $\mathcal{G}$-extended oscillator algebras [12–14], generalizing that describing the two-particle Calogero model.

3. C_λ-extended oscillator algebras

Let us consider the algebras generated by the operators I, $a^\dagger$, a, N, T, satisfying the Hermiticity conditions $\left(a^\dagger\right)^\dagger = a$, $N^\dagger = N$, $T^\dagger = T^{-1}$, and the relations

$$\left[N, a^\dagger\right] = a^\dagger, \qquad [N, T] = 0, \qquad T^\lambda = I,$$

$$\left[a, a^\dagger\right] = I + \sum_{\mu=1}^{\lambda-1} \kappa_\mu T^\mu, \qquad a^\dagger T = e^{-i2\pi/\lambda} T a^\dagger, \qquad (3)$$

together with their Hermitian conjugates [12]. Here T is the generator of (a unitary representation of) the cyclic group C_λ (where $\lambda \in \{2, 3, 4, \dots\}$), and κ_μ, $\mu = 1$, $2, \dots, \lambda-1$, are some complex parameters restricted by the conditions $\kappa_\mu^* = \kappa_{\lambda-\mu}$ (so that there remain altogether $\lambda - 1$ independent real parameters).

C_λ has λ inequivalent, one-dimensional matrix unitary irreducible representations (unirreps) Γ^μ, $\mu = 0, 1, \dots, \lambda - 1$, which are such that $\Gamma^\mu(T^\nu) = \exp(i2\pi\mu\nu/\lambda)$ for any $\nu = 0, 1, \dots, \lambda - 1$. The projection operator on the carrier space of Γ^μ may be written as

$$P_\mu = \frac{1}{\lambda} \sum_{\nu=0}^{\lambda-1} e^{-i2\pi\mu\nu/\lambda} T^\nu, \qquad (4)$$

and conversely T^ν, $\nu = 0, 1, \dots, \lambda - 1$, may be expressed in terms of the P_μ's as

$$T^\nu = \sum_{\mu=0}^{\lambda-1} e^{i2\pi\mu\nu/\lambda} P_\mu. \qquad (5)$$

The algebra defining relations (3) may therefore be rewritten in terms of I, $a^\dagger$, a, N, and $P_\mu = P_\mu^\dagger$, $\mu = 0, 1, \dots, \lambda - 1$, as

$$\left[N, a^\dagger\right] = a^\dagger, \qquad [N, P_\mu] = 0, \qquad \sum_{\mu=0}^{\lambda-1} P_\mu = I,$$

$$\left[a, a^\dagger\right] = I + \sum_{\mu=0}^{\lambda-1} \alpha_\mu P_\mu, \qquad a^\dagger P_\mu = P_{\mu+1} a^\dagger, \qquad P_\mu P_\nu = \delta_{\mu,\nu} P_\mu, \qquad (6)$$

where we use the convention $P_{\mu'} = P_\mu$ if $\mu' - \mu = 0 \bmod \lambda$ (and similarly for other operators or parameters indexed by μ, μ'). Equation (6) depends upon λ real parameters $\alpha_\mu = \sum_{\nu=1}^{\lambda-1} \exp(i2\pi\mu\nu/\lambda)\kappa_\nu$, $\mu = 0, 1, \dots, \lambda - 1$, restricted by the condition $\sum_{\mu=0}^{\lambda-1} \alpha_\mu = 0$. Hence, we may eliminate one of them, for instance $\alpha_{\lambda-1}$, and denote C_λ-extended oscillator algebras by $\mathcal{A}_{\alpha_0\alpha_1\dots\alpha_{\lambda-2}}^{(\lambda)}$.

The cyclic group generator T and the projection operators P_μ can be realized in terms of N as

$$T = e^{i2\pi N/\lambda}, \qquad P_\mu = \frac{1}{\lambda} \sum_{\nu=0}^{\lambda-1} e^{i2\pi\nu(N-\mu)/\lambda}, \qquad \mu = 0, 1, \ldots, \lambda - 1, \qquad (7)$$

respectively. With such a choice, $A^{(\lambda)}_{\alpha_0\alpha_1\ldots\alpha_{\lambda-2}}$ becomes a GDOA, $A^{(\lambda)}(G(N))$, characterized by $q = 1$ and $G(N) = I + \sum_{\mu=0}^{\lambda-1} \alpha_\mu P_\mu$, where P_μ is given in Eq. (7).

For any GDOA $A_q(G(N))$, one may define a so-called structure function $F(N)$, which is the solution of the difference equation $F(N+1) - qF(N) = G(N)$, such that $F(0) = 0$ [15]. For $A^{(\lambda)}(G(N))$, we find

$$F(N) = N + \sum_{\mu=0}^{\lambda-1} \beta_\mu P_\mu, \quad \beta_0 \equiv 0, \quad \beta_\mu \equiv \sum_{\nu=0}^{\mu-1} \alpha_\nu \quad (\mu = 1, 2, \ldots, \lambda - 1).$$

$$(8)$$

At this point, it is worth noting that for $\lambda = 2$, we obtain $T = K$, $P_0 = (I + K)/2$, $P_1 = (I - K)/2$, and $\kappa_1 = \kappa_1^* = \alpha_0 = -\alpha_1 = \kappa$, so that $A^{(2)}_{\alpha_0}$ coincides with the S_2-extended oscillator algebra $A^{(2)}_\kappa$ and $A^{(2)}(G(N))$ with the Calogero-Vasiliev algebra.

In Ref. [14], it was shown that $A^{(\lambda)}(G(N))$ (and more generally $A^{(\lambda)}_{\alpha_0\alpha_1\ldots\alpha_{\lambda-2}}$) has only two different types of unirreps: infinite-dimensional bounded from below unirreps and finite-dimensional ones. Among the former, there is the so-called bosonic Fock space representation, wherein $a^\dagger a = F(N)$ and $aa^\dagger = F(N+1)$. Its carrier space $\mathcal{F}$ is spanned by the eigenvectors $|n\rangle$ of the number operator N, corresponding to the eigenvalues $n = 0, 1, 2, \ldots$, where $|0\rangle$ is a vacuum state, i.e., $a|0\rangle = N|0\rangle = 0$ and $P_\mu|0\rangle = \delta_{\mu,0}|0\rangle$. The eigenvectors can be written as

$$|n\rangle = \mathcal{N}_n^{-1/2} \left(a^\dagger\right)^n |0\rangle, \qquad n = 0, 1, 2, \ldots, \qquad (9)$$

where $\mathcal{N}_n = \prod_{i=1}^{n} F(i)$. The creation and annihilation operators act upon $|n\rangle$ in the usual way, i.e.,

$$a^\dagger|n\rangle = \sqrt{F(n+1)}\, |n+1\rangle, \qquad a|n\rangle = \sqrt{F(n)}\, |n-1\rangle, \qquad (10)$$

while P_μ projects on the μth component $\mathcal{F}_\mu \equiv \{ |k\lambda + \mu\rangle \mid k = 0, 1, 2, \ldots \}$ of the Z_λ-graded Fock space $\mathcal{F} = \sum_{\mu=0}^{\lambda-1} \oplus \mathcal{F}_\mu$. It is obvious that such a bosonic Fock space representation exists if and only if $F(\mu) > 0$ for $\mu = 1, 2, \ldots, \lambda - 1$. This gives the following restrictions on the algebra parameters α_μ,

$$\sum_{\nu=0}^{\mu-1} \alpha_\nu > -\mu, \qquad \mu = 1, 2, \ldots, \lambda - 1. \qquad (11)$$

In the bosonic Fock space representation, one may consider the bosonic oscillator Hamiltonian, defined as usual by

$$H_0 \equiv \tfrac{1}{2} \left\{ a, a^\dagger \right\}. \tag{12}$$

It can be rewritten as

$$H_0 = a^\dagger a + \frac{1}{2} \left(I + \sum_{\mu=0}^{\lambda-1} \alpha_\mu P_\mu \right) = N + \frac{1}{2} I + \sum_{\mu=0}^{\lambda-1} \gamma_\mu P_\mu, \tag{13}$$

where $\gamma_0 \equiv \tfrac{1}{2}\alpha_0$ and $\gamma_\mu \equiv \sum_{\nu=0}^{\mu-1} \alpha_\nu + \tfrac{1}{2}\alpha_\mu$ for $\mu = 1, 2, \ldots, \lambda - 1$.

The eigenvectors of H_0 are the states $|n\rangle = |k\lambda + \mu\rangle$, defined in Eq. (9), and their eigenvalues are given by

$$E_{k\lambda+\mu} = k\lambda + \mu + \gamma_\mu + \tfrac{1}{2}, \qquad k = 0, 1, 2, \ldots, \qquad \mu = 0, 1, \ldots, \lambda - 1. \tag{14}$$

In each $\mathcal{F}_\mu$ subspace of the Z_λ-graded Fock space $\mathcal{F}$, the spectrum of H_0 is therefore harmonic, but the λ infinite sets of equally spaced energy levels, corresponding to $\mu = 0, 1, \ldots, \lambda - 1$, may be shifted with respect to each other by some amounts depending upon the algebra parameters $\alpha_0, \alpha_1, \ldots, \alpha_{\lambda-2}$, through their linear combinations $\gamma_\mu, \mu = 0, 1, \ldots, \lambda - 1$.

For the Calogero-Vasiliev oscillator, i.e., for $\lambda = 2$, the relation $\gamma_0 = \gamma_1 = \kappa/2$ implies that the spectrum is very simple and coincides with that of a shifted harmonic oscillator. For $\lambda \geq 3$, however, it has a much richer structure. According to the parameter values, it may be nondegenerate, or may exhibit some $(\nu+1)$-fold degeneracies above some energy eigenvalue, where ν may take any value in the set $\{1, 2, \ldots, \lambda - 1\}$. In Ref. [13], the complete classification of nondegenerate, twofold and threefold degenerate spectra was obtained for $\lambda = 3$ in terms of α_0 and α_1.

In the remaining part of this communication, we will show that the bosonic Fock space representation of $\mathcal{A}^{(\lambda)}(G(N))$ and the corresponding bosonic oscillator Hamiltonian H_0 have some useful applications to variants of SSQM.

4. Application to parasupersymmetric quantum mechanics of order p

In SSQM with two supercharges, the supersymmetric Hamiltonian $\mathcal{H}$ and the supercharges $Q^\dagger, Q = \left(Q^\dagger \right)^\dagger$, satisfy the sqm(2) superalgebra, defined by the relations

$$Q^2 = 0, \qquad [\mathcal{H}, Q] = 0, \qquad \left\{ Q, Q^\dagger \right\} = \mathcal{H}, \tag{15}$$

together with their Hermitian conjugates. Such a superalgebra is most often realized in terms of mutually commuting boson and fermion operators.

Plyushchay [10], however, showed that it can alternatively be realized in terms of only boson-like operators, namely the generators of the Calogero-Vasiliev algebra $\mathcal{A}^{(2)}(G(N))$ (see also Ref. [11]). The SSQM bosonization can be performed in two different ways, by choosing either $Q = a^\dagger P_1$ (so that $\mathcal{H} = H_0 - \frac{1}{2}(K+\kappa)$) or $Q = a^\dagger P_0$ (so that $\mathcal{H} = H_0 + \frac{1}{2}(K+\kappa)$). The first choice corresponds to unbroken SSQM (all the excited states are twofold degenerate while the ground state is nondegenerate and at vanishing energy), and the second choice describes broken SSQM (all the states are twofold degenerate and at positive energy).

SSQM was generalized to parasupersymmetric quantum mechanics (PSSQM) of order two by Rubakov and Spiridonov [1], and later on to PSSQM of arbitrary order p by Khare [2]. In the latter case, Eq. (15) is replaced by

$$Q^{p+1} = 0 \qquad \text{(with } Q^p \neq 0\text{)},$$

$$[\mathcal{H}, Q] = 0,$$

$$Q^p Q^\dagger + Q^{p-1} Q^\dagger Q + \cdots + Q Q^\dagger Q^{p-1} + Q^\dagger Q^p = 2p Q^{p-1} \mathcal{H}, \qquad (16)$$

and is retrieved in the case where $p = 1$. The parasupercharges Q, $Q^\dagger$, and the parasupersymmetric Hamiltonian $\mathcal{H}$ are usually realized in terms of mutually commuting boson and parafermion operators.

A property of PSSQM of order p is that the spectrum of $\mathcal{H}$ is $(p+1)$-fold degenerate above the $(p-1)$th energy level. This fact and Plyushchay's results for $p = 1$ hint at a possibility of representing $\mathcal{H}$ as a linear combination of the bosonic oscillator Hamiltonian H_0 associated with $\mathcal{A}^{(p+1)}(G(N))$ and some projection operators.

In Ref. [14] (see also Ref. [12]), it was proved that PSSQM of order p can indeed be bosonized in terms of the generators of $\mathcal{A}^{(p+1)}(G(N))$ for any allowed (i.e., satisfying Eq. (11)) values of the algebra parameters $\alpha_0, \alpha_1, \ldots, \alpha_{p-1}$. For such a purpose, ansätze of the type

$$Q = \sum_{\nu=0}^{p} \sigma_\nu a^\dagger P_\nu, \qquad \mathcal{H} = H_0 + \frac{1}{2} \sum_{\nu=0}^{p} r_\nu P_\nu, \qquad (17)$$

were chosen. Here σ_ν and r_ν are some complex and real constants, respectively, to be determined in such a way that Eq. (16) is fulfilled. It was found that there are $p+1$ families of solutions, which may be distinguished by an index $\mu \in \{0, 1, \ldots, p\}$ and from which one may choose the following representative solutions

$$Q_\mu = \sqrt{2} \sum_{\nu=1}^{p} a^\dagger P_{\mu+\nu},$$

$$\mathcal{H}_\mu = N + \frac{1}{2}(2\gamma_{\mu+2} + r_{\mu+2} - 2p + 3)I + \sum_{\nu=1}^{p} (p+1-\nu)P_{\mu+\nu}, \qquad (18)$$

where

$$r_{\mu+2} = \frac{1}{p}\left[(p-2)\alpha_{\mu+2} + 2\sum_{\nu=3}^{p}(p-\nu+1)\alpha_{\mu+\nu} + p(p-2)\right]. \qquad (19)$$

The eigenvectors of $\mathcal{H}_\mu$ are the states (9) and the corresponding eigenvalues are easily found. All the energy levels are equally spaced. For $\mu = 0$, PSSQM is unbroken, otherwise it is broken with a $(\mu + 1)$-fold degenerate ground state. All the excited states are $(p+1)$-fold degenerate. For $\mu = 0, 1, \ldots, p-2$, the ground state energy may be positive, null, or negative depending on the parameters, whereas for $\mu = p-1$ or p, it is always positive.

Khare [2] showed that in PSSQM of order p, $\mathcal{H}$ has in fact $2p$ (and not only two) conserved parasupercharges, as well as p bosonic constants. In other words, there exist p independent operators Q_r, $r = 1, 2, \ldots, p$, satisfying with $\mathcal{H}$ the set of equations (16), and p other independent operators I_t, $t = 2, 3, \ldots, p+1$, commuting with $\mathcal{H}$, as well as among themselves. In Ref. [14], a realization of all such operators was obtained in terms of the $\mathcal{A}^{(p+1)}(G(N))$ generators.

As a final point, let us note that there exists an alternative approach to PSSQM of order p, which was proposed by Beckers and Debergh [3], and wherein the multilinear relation in Eq. (16) is replaced by the cubic equation

$$\left[Q, \left[Q^\dagger, Q\right]\right] = 2Q\mathcal{H}. \qquad (20)$$

In Ref. [12], it was proved that for $p = 2$, this PSSQM algebra can only be realized by those $\mathcal{A}^{(3)}(G(N))$ algebras that simultaneously bosonize Rubakov-Spiridonov-Khare PSSQM algebra.

5. Application to pseudosupersymmetric quantum mechanics

Pseudosupersymmetric quantum mechanics (pseudoSSQM) was introduced by Beckers, Debergh, and Nikitin [4, 5] in a study of relativistic vector mesons interacting with an external constant magnetic field. In the nonrelativistic limit, their theory leads to a pseudosupersymmetric oscillator Hamiltonian, which can be realized in terms of mutually commuting boson and pseudofermion operators, where the latter are intermediate between standard fermion and $p = 2$ parafermion operators.

It is then possible to formulate a pseudoSSQM [4, 5], characterized by a pseudosupersymmetric Hamiltonian $\mathcal{H}$ and pseudosupercharge operators Q, $Q^\dagger$, satisfying the relations

$$Q^2 = 0, \qquad [\mathcal{H}, Q] = 0, \qquad QQ^\dagger Q = 4c^2 Q\mathcal{H}, \qquad (21)$$

and their Hermitian conjugates, where c is some real constant. The first two relations in Eq. (21) are the same as those occurring in SSQM, whereas the third one is similar to the multilinear relation valid in PSSQM of order two. Actually, for $c = 1$ or $1/2$, it is compatible with Eq. (16) or (20), respectively.

In Ref. [14], it was proved that pseudoSSQM can be bosonized in two different ways in terms of the generators of $\mathcal{A}^{(3)}(G(N))$ for any allowed values of the parameters α_0, α_1. This time, the ansätze

$$Q = \sum_{\nu=0}^{2} \left(\xi_\nu a + \eta_\nu a^\dagger \right) P_\nu, \qquad \mathcal{H} = H_0 + \tfrac{1}{2} \sum_{\nu=0}^{2} r_\nu P_\nu, \qquad (22)$$

were chosen, and the complex constants ξ_ν, η_ν, and the real ones r_ν were determined in such a way that Eq. (21) is fulfilled.

The first type of bosonization corresponds to three families of two-parameter solutions, labeled by an index $\mu \in \{0, 1, 2\}$,

$$\begin{aligned}
Q_\mu(\eta_{\mu+2}, \varphi) &= \left(\eta_{\mu+2} a^\dagger + e^{i\varphi} \sqrt{4c^2 - \eta_{\mu+2}^2}\, a \right) P_{\mu+2}, \\
\mathcal{H}_\mu(\eta_{\mu+2}) &= N + \tfrac{1}{2}(2\gamma_{\mu+2} + r_{\mu+2} - 1)I + 2P_{\mu+1} + P_{\mu+2},
\end{aligned} \qquad (23)$$

where $0 < \eta_{\mu+2} < 2|c|$, $0 \leq \varphi < 2\pi$, and

$$r_{\mu+2} = \frac{1}{2c^2}(1 + \alpha_{\mu+2}) \left(|\eta_{\mu+2}|^2 - 2c^2 \right). \qquad (24)$$

Choosing for instance $\eta_{\mu+2} = \sqrt{2}|c|$, and $\varphi = 0$, hence $r_{\mu+2} = 0$ (producing an overall shift of the spectrum), leads to

$$\begin{aligned}
Q_\mu &= c\sqrt{2} \left(a^\dagger + a \right) P_{\mu+2}, \\
\mathcal{H}_\mu &= N + \tfrac{1}{2}(2\gamma_{\mu+2} - 1)I + 2P_{\mu+1} + P_{\mu+2}.
\end{aligned} \qquad (25)$$

A comparison between Eq. (23) or (25) and Eq. (18) shows that the pseudosupersymmetric and $p = 2$ parasupersymmetric Hamiltonians coincide, but that the corresponding charges are of course different. The conclusions relative to the spectrum and the ground state energy are therefore the same as in Sec. 4.

The second type of bosonization corresponds to three families of one-parameter solutions, again labeled by an index $\mu \in \{0, 1, 2\}$,

$$\begin{aligned}
Q_\mu &= 2|c|a P_{\mu+2}, \\
\mathcal{H}_\mu(r_\mu) &= N + \tfrac{1}{2}(2\gamma_{\mu+2} - \alpha_{\mu+2})I + \tfrac{1}{2}(1 - \alpha_{\mu+1} + \alpha_{\mu+2} + r_\mu)P_\mu \\
&\quad + P_{\mu+1},
\end{aligned} \qquad (26)$$

where $r_\mu \in R$ changes the Hamiltonian spectrum in a significant way. The levels are indeed equally spaced if and only if $r_\mu = (\alpha_{\mu+1} - \alpha_{\mu+2} + 3) \bmod 6$. If r_μ is small enough, the ground state is nondegenerate, and its energy is negative for $\mu = 1$, or may have any sign for $\mu = 0$ or 2. On the contrary, if r_μ is large enough, the ground state remains nondegenerate with a vanishing energy in the former case, while it becomes twofold degenerate with a positive energy in the latter. For some intermediate r_μ value, one gets a two or threefold degenerate ground state with a vanishing or positive energy, respectively.

6. Application to orthosupersymmetric quantum mechanics of order two

Mishra and Rajasekaran [8] introduced order-p orthofermion operators by replacing the Pauli exclusion principle by a more stringent one: an orbital state shall not contain more than one particle, whatever be the spin direction. The wave function is thus antisymmetric in spatial indices alone with the order of the spin indices frozen.

Khare, Mishra, and Rajasekaran [6] then developed orthosupersymmetric quantum mechanics (OSSQM) of arbitrary order p by combining boson operators with orthofermion ones, for which the spatial indices are ignored. OSSQM is formulated in terms of an orthosupersymmetric Hamiltonian $\mathcal{H}$, and $2p$ orthosupercharge operators Q_r, $Q_r^\dagger$, $r = 1, 2, \ldots, p$, satisfying the relations

$$Q_r Q_s = 0, \qquad [\mathcal{H}, Q_r] = 0, \qquad Q_r Q_s^\dagger + \delta_{r,s} \sum_{t=1}^{p} Q_t^\dagger Q_t = 2\delta_{r,s}\mathcal{H}, \qquad (27)$$

and their Hermitian conjugates, where r and s run over $1, 2, \ldots, p$.

In Ref. [14], it was proved that OSSQM of order two can be bosonized in terms of the generators of some well-chosen $\mathcal{A}^{(3)}(G(N))$ algebras. As ansätze, the expressions

$$Q_1 = \sum_{\nu=0}^{2} \left(\xi_\nu a + \eta_\nu a^\dagger \right) P_\nu, \qquad Q_2 = \sum_{\nu=0}^{2} \left(\zeta_\nu a + \rho_\nu a^\dagger \right) P_\nu,$$

$$\mathcal{H} = H_0 + \tfrac{1}{2} \sum_{\nu=0}^{2} r_\nu P_\nu, \qquad (28)$$

were used, and the complex constants ξ_ν, η_ν, ζ_ν, ρ_ν, and the real ones r_ν were determined in such a way that Eq. (27) is fulfilled. There exist two families of two-parameter solutions, labeled by $\mu \in \{0, 1\}$,

$$Q_{1,\mu}(\xi_{\mu+2}, \varphi) = \xi_{\mu+2} a P_{\mu+2} + e^{i\varphi} \sqrt{2 - \xi_{\mu+2}^2} \, a^\dagger P_\mu,$$

$$Q_{2,\mu}(\xi_{\mu+2}, \varphi) = -e^{-i\varphi} \sqrt{2 - \xi_{\mu+2}^2} \, a P_{\mu+2} + \xi_{\mu+2} a^\dagger P_\mu,$$

$$\mathcal{H}_\mu = N + \tfrac{1}{2}(2\gamma_{\mu+1} - 1)I + 2P_\mu + P_{\mu+1}, \qquad (29)$$

where $0 < \xi_{\mu+2} \leq \sqrt{2}$ and $0 \leq \varphi < 2\pi$, provided the algebra parameter $\alpha_{\mu+1}$ is taken as $\alpha_{\mu+1} = -1$. As a matter of fact, the absence of a third family of solutions corresponding to $\mu = 2$ comes from the incompatibility of this condition (i.e., $\alpha_0 = -1$) with conditions (11).

The orthosupersymmetric Hamiltonian $\mathcal{H}$ in Eq. (29) is independent of the parameters $\xi_{\mu+2}$, φ. All the levels of its spectrum are equally spaced. For $\mu = 0$, OSSQM is broken: the levels are threefold degenerate, and the ground state energy is positive. On the contrary, for $\mu = 1$, OSSQM is unbroken: only the excited states are threefold degenerate, while the nondegenerate ground state has a vanishing energy. Such results agree with the general conclusions of Ref. [6].

For p values greater than two, the OSSQM algebra (27) becomes rather complicated because the number of equations to be fulfilled increases considerably. A glance at the 18 independent conditions for $p = 3$ led to the conclusion that the $\mathcal{A}^{(4)}(G(N))$ algebra is not rich enough to contain operators satisfying Eq. (27). Contrary to what happens for PSSQM, for OSSQM the $p = 2$ case is therefore not representative of the general one.

7. Conclusion

In this communication, we showed that the S_2-extended oscillator algebra, which was introduced in connection with the two-particle Calogero model, can be extended to the whole class of C_λ-extended oscillator algebras $\mathcal{A}^{(\lambda)}_{\alpha_0 \alpha_1 \ldots \alpha_{\lambda-2}}$, where $\lambda \in \{2, 3, \ldots\}$, and $\alpha_0, \alpha_1, \ldots, \alpha_{\lambda-2}$ are some real parameters. In the same way, the GDOA realization of the former, known as the Calogero-Vasiliev algebra, is generalized to a class of GDOAs $\mathcal{A}^{(\lambda)}(G(N))$, where $\lambda \in \{2, 3, \ldots\}$, for which one can define a bosonic oscillator Hamiltonian H_0, acting in the bosonic Fock space representation.

For $\lambda \geq 3$, the spectrum of H_0 has a very rich structure in terms of the algebra parameters $\alpha_0, \alpha_1, \ldots, \alpha_{\lambda-2}$. This can be exploited to provide a bosonization of PSSQM of order $p = \lambda - 1$, and, for $\lambda = 3$, a bosonization of pseudoSSQM and OSSQM of order two.

References

1. V.A. Rubakov and V.P. Spiridonov, *Parasupersymmetric Quantum Mechanics*, Mod. Phys. Lett. **A3** (1988) 1337.
2. A. Khare, *Parasupersymmetry in Quantum Mechanics*, J. Math. Phys. **34** (1993) 1277.
3. J. Beckers and N. Debergh, *Parastatistics and Supersymmetry in Quantum Mechanics*, Nucl. Phys. **B340** (1990) 767.
4. J. Beckers, N. Debergh and A.G. Nikitin, *On Parasupersymmetries and Relativistic Descriptions for Spin one Particles: II. The Interacting Context with (Electro)Magnetic Fields*, Fortschr. Phys. **43** (1995) 81.
5. J. Beckers and N. Debergh, *From Relativistic Vector Mesons in Constant Magnetic Fields to Nonrelativistic (Pseudo)Supersymmetries*, Int. J. Mod. Phys. **A10** (1995) 2783.
6. A. Khare, A.K. Mishra, and G. Rajasekaran, *Orthosupersymmetric Quantum Mechanics*, Int. J. Mod. Phys. **A8** (1993) 1245.

7. Y. Ohnuki and S. Kamefuchi, *Quantum Field Theory and Parastatistics*, Springer-Verlag, Berlin, 1982.

8. A.K. Mishra and G. Rajasekaran, *Algebra for Fermions with a New Exclusion Principle*, Pramana - J. Phys. **36** (1991) 537.

9. M.A. Vasiliev, *Higher Spin Algebras and Quantization on the Sphere and Hyperboloid*, Int. J. Mod. Phys. **A6** (1991) 1115.

10. M.S. Plyushchay, *Deformed Heisenberg Algebra, Fractional Spin Fields, and Supersymmetry without Fermions*, Ann. Phys. (N.Y.) **245** (1996) 339.

11. J. Beckers, N. Debergh, and A.G. Nikitin, *Reducibility of Supersymmetric Quantum Mechanics*, Int. J. Theor. Phys. **36** (1997) 1991.

12. C. Quesne and N. Vansteenkiste, C_λ-*Extended Harmonic Oscillator and (Para)Supersymmetric Quantum Mechanics*, Phys. Lett. **A240** (1998) 21.

13. C. Quesne and N. Vansteenkiste, *Algebraic Realization of Supersymmetric Quantum Mechanics for Cyclic Shape Invariant Potentials*, Helv. Phys. Acta **72** (1999) 71.

14. C. Quesne and N. Vansteenkiste, C_λ-*Extended Oscillator Algebras and Some of Their Deformations and Applications to Quantum Mechanics*, Int. J. Theor. Phys. **39** (2000) 1175.

15. C. Quesne and N. Vansteenkiste, *Generalized q-Oscillators and Their Hopf Structures*, J. Phys. **A28** (1995) 7019.

16. C. Quesne and N. Vansteenkiste, *Representation Theory of Deformed Oscillator Algebras*, Helv. Phys. Acta **69** (1996) 141.

17. M. Arik and D.D. Coon, *Hilbert Spaces of Analytic Functions and Generalized Coherent States*, J. Math. Phys. **17** (1976) 524.

18. L.C. Biedenharn, *The Quantum Group $SU_q(2)$ and a q-Analogue of the Boson Operators*, J. Phys. **A22** (1989) L873.

19. A.J. Macfarlane, *On q-Analogues of the Quantum Harmonic Oscillator and the Quantum Group $SU(2)_q$*, J. Phys. **A22** (1989) 4581.

20. F. Calogero, *Solution of the One-Dimensional N-Body Problems with Quadratic and/or Inversely Quadratic Pair Potentials*, J. Math. Phys. **12** (1971) 419.

21. A.P. Polychronakos, *Exchange Operator Formalism for Integrable Systems of Particles*, Phys. Rev. Lett. **69** (1992) 703.

22. L. Brink, T.H. Hansson and M.A. Vasiliev, *Explicit Solution to the N-Body Calogero Problem*, Phys. Lett. **B286** (1992) 109.

SUPERSYMMETRIC ODD MECHANICAL SYSTEMS AND HILBERT Q-MODULE QUANTIZATION

ANDRZEJ FRYDRYSZAK *

Institute of Theoretical Physics, University of Wroclaw,
pl. Borna 9, 50-204 Wroclaw, Poland

1. Introduction

Supersymmetry can be implemented within a particle model in two ways. The first one is commonly exploited and assumes that we use conventional graded Lie algebra approach in the sense that on the classical level we have a $\mathbb{Z}_2$-graded Lie-Poisson algebra of observables which after quantization is replaced by a $\mathbb{Z}_2$-graded Lie algebra of operators. Both, graded Poisson bracket and $\mathbb{Z}_2$-graded commutator are even mappings. The second way of realization of supersymmetry in a particle model is related to the anti-bracket algebras. In this case Lagrangian as well as Hamiltonian of the supersymmetric system is an odd Grassmann algebra valued function and the Grassmannian parity of canonical momenta is opposite to the parity of related coordinates. The anti-bracket is an odd mapping. Realizations of the mentioned type we shall call the even supersymmetric mechanics and the odd supersymmetric mechanics, respectively [1, 2]. The odd mechanics allows particular deformation of geometry of the configuration superspace. The realization of the supersymmetry algebra after the passage to phase superspace in terms of the Dirac anti-bracket remains conventional [3]. The canonical quantization of both types of models can be done in parallel but in the case of the odd systems one can introduce a new $\mathbb{Z}_2$-graded algebra generalizing complex numbers in such a sense that we introduce additional imaginary unit of the odd Grassmannian parity [4, 5]. Such a structure we shall call oddons (referring to the name of quaternions, octonions etc.). The formalism, in both cases, allows to mimic the approach known from the harmonic analysis on the Heisenberg group [6]. In the

---

* amfry@ift.uni.wroc.pl

S. Duplij and J. Wess (eds.), Noncommutative Structures in Mathematics and Physics, 215–227.

even sector it is done for a pair - Heisenberg group and Fermionic Heisenberg group. In the odd sector it is done for so called Odd Heisenberg group.

In this presentation we shall briefly describe classical aspect of the even and odd models on the example of $\mathbb{Z}_2$-graded supersymmetric oscillators and moreover we shall display some issues of their quantization.

2. Supersymmetric (Superfield) Classical Mechanics.

To discuss even and odd mechanical systems on the same footing let us define the following notion of the superfield supersymmetric classical mechanics (SSCM) [7, 3]. Let (N_0, N_1) be a fixed pair of non-negative integers. (N_0, N_1)-dimensional SSCM is quadruple $(\Upsilon; \{Q_\alpha, D_\beta, T\}; (M, J, G); S)$ consisting of:

(a) Υ - a "supersymmetrized time" (t, θ_α), $\alpha = 1, 2$
(b) $\{Q_\alpha, D_\beta, T\}$ - super Lie algebra of supertranslations
and respective covariant derivatives on Υ

$$\{Q_\alpha, Q_\beta\} = 2i\delta_{\alpha\beta}T, \qquad \{D_\alpha, D_\beta\} = -2i\delta_{\alpha\beta}T \qquad (1)$$

$$\{Q_\alpha, T\} = 0 = \{D_\alpha, T\}, \qquad \{Q_\alpha, D_\beta\} = 0 \qquad (2)$$

(c) M - $\mathbb{Z}_2$-graded configuration space $dim M = (N_0, N_1)$ with gradation mapping

$$J : M \longrightarrow M, \quad M = M_0 + M_1 \quad J(\overset{s}{\phi}) = (-1)^s \overset{s}{\phi}, \quad \overset{s}{\phi} \in M_s, s = 1, 2$$

(d) G - $\mathbb{Z}_2$-graded metric in M

$$<\overset{s}{\phi}, \overset{s}{\phi}> = \sum_{i,j=1}^{N_s} \overset{s}{G}^{ij} \overset{s}{\phi_i} \overset{s}{\phi_j}, \qquad <\overset{s}{\phi}, \overset{s'}{\phi}> = 0 \quad for \quad s \neq s' \quad \overset{s}{G}^{ij} = (-1)^s \overset{s}{G}^{ji}$$

(e) S - an action. The action S is invariant under supertranslations
Trajectories in M are superfields with the following expansion

$$\overset{0}{\phi_j}(t, \vartheta) = x_j(t) + i\vartheta_\alpha x_j^\alpha(t) + \frac{1}{2}\vartheta^2 b_j(t) \qquad (3)$$

$$\overset{1}{\phi_j}(t, \vartheta) = y_j(t) + \vartheta_\alpha y_j^\alpha(t) + \frac{1}{2}\vartheta^2 f_j(t) \qquad (4)$$

The action S has the form

$$S = \int L(D_\alpha\phi, \phi)dt d\vartheta_1 d\vartheta_2 \qquad (5)$$

and yields the following equations of motion

$$\frac{\delta L}{\delta \overset{s}{\phi}} - (-1)^s D_\alpha \frac{\delta L}{\delta D_\alpha \overset{s}{\phi}} = 0. \tag{6}$$

It is worth mentioning that the mapping J in the above definition is the counter-part of the fermionic number operator $(-1)^F$ known in supersymmetric quantum mechanics.

Such SSCM has two natural realizations in $\mathbb{Z}_2$-graded configuration space. Namely,

Even realization: Graded Superfield Oscillator (GSO) [7].
Let $N_1 = 2k$.

$$S = \int \frac{1}{4} (\epsilon^{\alpha\beta} < D_\alpha \phi, D_\beta \phi > -2\omega < \phi, \phi >) dt d\vartheta_1 d\vartheta_2 \tag{7}$$

in components it gives

$$S = \frac{1}{2} \int dt \overset{0}{G}^{ij} [(\dot{x}_i \dot{x}_j - b_i b_j) - (\delta^\alpha_\beta x_{\alpha i} \dot{x}^\beta_j) - \omega(x_i b_j + b_i x_j + i\epsilon_{\alpha\beta} x^\alpha_i x^\beta_j)$$

$$\frac{1}{2} \int dt \overset{1}{G}^{ij} [-(\dot{y}_i \dot{y}_j + f_i f_j) + (\delta^\alpha_\beta y_{\alpha i} \dot{y}^\beta_j) - \omega(y_i f_j + f_i y_j + i\epsilon_{\alpha\beta} y^\alpha_i y^\beta_j). \tag{8}$$

We can summarize the component content of the model as follows:

- GSO consists of system of bosonic oscillators and rotators and system of fermionic oscillators and rotators
- full GSO has additional symmetry (mixing both sectors)
- momenta have the same grade as conjugate coordinates $\Rightarrow$ $\mathbb{Z}_2$-graded Poisson bracket in phase space is an even mapping. The phase space of this model we shall denote $(P_{(0,0)} \oplus P_{(1,1)}; \{.,.\}_0)$, where $(p,q) \in P_{(0,0)}$ and $(\Pi, \Theta) \in P_{(1,1)}$ Moreover the pair (p,q) describes even coordinates and their conjugated momenta and similarly (Π, Θ) denotes pairs of the odd coordinates and momenta.

Odd realization: Odd Graded Superfield Oscillator (OGSO) [3].
Let $N_1 = N_0$. Here we introduce the odd extension of covariant derivatives, in the sense that instead of considering $D_\alpha \otimes id_M$ we define

$$D_\alpha \otimes \Pi, \qquad \Pi^2 = id_M \Rightarrow \Pi_\alpha = \begin{pmatrix} 0 & q_\alpha^{-1} \\ q_\alpha & 0 \end{pmatrix}, \tag{9}$$

where $c_\alpha \in \mathbb{R}$. Now

$$S = \int (\frac{1}{2} < D_\alpha \phi, \Pi^{\alpha\beta} D_\beta \phi > -\omega < \phi, \Pi\phi >) dt d\vartheta_1 d\vartheta_2 \tag{10}$$

and

$$\Pi^{\alpha\beta} D_\beta \overset{0}{\phi}_i = \epsilon^{\alpha\beta} q_\beta^{-1} D_\beta \overset{1}{\phi}_i \tag{11}$$

$$\Pi^{\alpha\beta} D_\beta \overset{1}{\phi}_i = \epsilon^{\alpha\beta} q_\beta D_\beta \overset{0}{\phi}_i . \tag{12}$$

In components this action has the following form

$$S = \int dt \sum_{s=1,2} \overset{s}{G}{}^{ij} \{ \frac{1}{2} [Tr \overset{s}{\delta}_q (\dot{x}_i \dot{y}_j - b_i f_j) +$$

$$\frac{1}{2} (\overset{s}{\delta}_{q\beta} x_{\alpha i} \dot{y}_j^\beta - \overset{s}{\delta}_{q\alpha} \dot{x}_i^\alpha y_{\beta j})] + (-1)^s \omega (x_i f_j + i\epsilon_{\alpha\beta} x_i^\alpha y_j^\beta + b_i y_j) \}. \tag{13}$$

The component content of this model can be characterized as follows:

 - OGSO consists of system of bosonic oscillators and rotators and system of
   fermionic oscillators and rotators
 - momenta have opposite grade with respect to the grade of conjugate coordinates $\Rightarrow$ "Poisson" bracket in odd phase space i.e. anti-bracket is an odd
   mapping with shifted grade properties i.e.

$$\{A, B\}_1 = -(-1)^{(A+1)(B+1)} \{B, A\}_1 \tag{14}$$

$$\sum_{cycl} -(-1)^{(A+1)(C+1)} \{A, \{B, C\}_1\}_1 = 0 \tag{15}$$

The canonical relations for component fields, in this case, have the form

$$\{F^A, p_{F_B}\} = (-)^{|F^A|(|F^A|+1)} \delta_B^A , \tag{16}$$

where F is generic component field and p_F is its momentum. Here, in analogy
to the previous case we shall denote phase space for this system as $(P_{(0,1)} \oplus P_{(1,0)}; \{\cdot, \cdot\}_1)$, where $(p, \Theta) \in P_{(0,1)}$ and $(\Pi, q) \in P_{(1,0)}$. As before Greek letters
denote odd entities.

3. Generalization of the Heisenberg group

The Heisenberg group is connected to the structure of usual phase space. We shall
need two generalizations of this object. For the even mechanics: a generalization
related to the $(P_{(1,1)}, \{\cdot, \cdot\}_0)$ and for the odd mechanics: a generalization related
to the $(P_{(0,1)}, \{\cdot, \cdot\}_1)$.

Fermionic Heisenberg Group [8]. We shall consider here the odd part of the phase space i.e. $P_{(1,1)}$ extended by the time dimension. Hence, let P_n be the free Q-module $P_n = Q^{2n,1}$ with the fixed basis $\{e_i\}_{i=0}^{2n}$, $|e_0| = 0$, $|e_i| = 1$, $i = 1, 2, ..., 2n$ (Q is the Banach -Grassmann algebra and $|\cdot|$ denotes Grassmannian parity of an element [9]). Moreover, let $B(\cdot, \cdot)$ be the graded symplectic even form defined on $Q^{2n,0}$ with values in Q. In our basis

$$v = \sum_{i=1}^{n} \Pi_i e_i + \sum_{i=1}^{n} \Theta_i e_{n+i} \tag{17}$$

and we fix the form of $B(\cdot, \cdot)$ as follows

$$B(e_i, e_{j+n}) = \delta_{ij}. \tag{18}$$

Hence

$$B(v, v') = -\sum_{i=1}^{n}(\Pi^i \Theta_i' + \Theta_i \Pi^{i'}), \tag{19}$$

where $|\Pi^i| = |\Theta_i| = 1$. Now we consider the module P with coordinates

$$(v, t) \equiv (\Pi, \Theta, t) = \left(\Pi^1, ..., \Pi^n, \Theta_1, ..., \Theta_n, t\right) \tag{20}$$

(where $t \in Q_0$) and with the following multiplication law

$$(v, t) \circ (v', t') = \left(v + v', t + t' + \frac{1}{2}B(v, v')\right). \tag{21}$$

P equipped with this multiplication forms a group. It is called the Fermionic Heisenberg group and denoted by FH_n. As in the conventional case, this group has a matrix realization. For given $(\Pi, \Theta, t) \in Q^{2n,1}$ we define the matrix $\mu(\Pi, \Theta, t) \in M_{n+2}(Q)$ by

$$\mu(\Pi, \Theta, t) = \begin{pmatrix} 0 & \Pi^1 & ... & \Pi^n & t \\ 0 & 0 & ... & 0 & \Theta_1 \\ \vdots & \vdots & & \vdots & \vdots \\ 0 & 0 & ... & 0 & \Theta_n \\ 0 & 0 & ... & 0 & 0 \end{pmatrix} \tag{22}$$

We have

$$\exp\mu(\Pi, \Theta, t)\exp\mu(\Pi', \Theta', t')$$

$$= \exp\mu\left(\Pi + \Pi', \Theta + \Theta', t + t' - \frac{1}{2}(\Pi\Theta' + \Theta\Pi')\right), \tag{23}$$

what gives the multiplication law.

The elements $\mu\left(\Pi, \Theta, t\right)$ form a graded Lie algebra with one even generator T and 2n odd generators $e_i = \hat{\Pi}_i$ and $e_{i+n} = \hat{\Theta}_i$ and with the following structural relations

$$\left[\hat{\Pi}_i, \hat{\Pi}_j\right]_+ = \left[\hat{\Theta}_i, \hat{\Theta}_j\right]_+ = 0$$
$$\left[\hat{\Pi}_i, T\right]_- = \left[\hat{\Theta}_i, T\right]_- = 0 \qquad (24)$$
$$\left[\hat{\Pi}^i, \hat{\Theta}_j\right]_+ = \delta^i_j T.$$

Odd Heisenberg Group [5]. To describe the Odd Heisenberg group we shall use a new structure replacing the complex numbers i.e. the algebra of oddons. It provides the odd multiplication in the set of observables. The definition of oddons and some of their properties are collected in the Appendix. Let us consider as an extension of the phase space $P_{(0,1)}$ by the time dimension the free $Q^{\mathbb{RO}}$-module $T_n = Q_{\mathrm{RO}}^{n|n+1}$ with the basis $\{E_i, e_i, e_0\}_{i=1}^n$, where $|e_i| = |e_0| = 0$, $|E_i| = 1$; $i = 1, 2, \ldots, n$. Let $B(\cdot, \cdot)$ be the odd symplectic form defined on $Q_{\mathrm{RO}}^{n,n}$ with values in Q_{RO}. We shall consider vectors of the form

$$v = \sum_{i=1}^n p_i E_i + \sum_{i=1}^n \Theta_i e_i \qquad (25)$$

and we fix $B(\cdot, \cdot)$ as follows

$$B(E_i, e_i) = \delta_{ij} \qquad (26)$$

Therefore $B(v, v') = \sum_{i=1}^n (p^i \Theta'_i - \Theta_i p'^i)$. Now let OH_n be the set of vectors of the form

$$(v, \tau) = (p, \Theta, \tau) = (p^1, p^2, \ldots, p^n, \Theta_1, \Theta_2, \ldots, \Theta_n, \tau) \qquad (27)$$

where $\tau = t \cdot \hat{1}$, $t \in Q_0^R$, $\Theta_i \in Q_1^R$, $p^i \in Q_0^{\mathbb{RO}}$. In the set OH_n we define the action in the following form

$$(v, \tau) \star (v', \tau') = (v + v', \tau + \tau' + \frac{1}{2} B(v, v')) \qquad (28)$$

The $(OH_n, \star)$ is a group, we shall call it the Odd Heisenberg group. Its matrix realization can be written in the following form, for the (p, Θ, τ) we define matrix $\mu(p^i, \Theta_i, \tau) \in M_{n+2}(Q_{\mathrm{CO}})$

$$\mu\left(p, \Theta, \tau\right) = \begin{pmatrix} 0 & p^1 & \ldots & p^n & \tau \\ 0 & 0 & \ldots & 0 & \Theta_1 \\ \vdots & \vdots & & \vdots & \vdots \\ 0 & 0 & \ldots & 0 & \Theta_n \\ 0 & 0 & \ldots & 0 & 0 \end{pmatrix} \qquad (29)$$

The odd product of odd exponents gives the following relation

$$\exp_\star \mu\,(p, \Theta, \tau) \star \exp_\star \mu\,(p', \Theta', \tau')$$

$$= \exp_\star \mu\,\left(p + p', \Theta + \Theta', \tau + \tau' + \frac{1}{2}\,(p\Theta' - \Theta p')\right) \qquad (30)$$

Elements $\mu(p, \Theta, \tau)$ form a graded anti-bracket algebra with even generators $\hat{e}_i, \hat{e}_0$ and odd generators $\hat{E}_i$

$$\left[\hat{E}_i, \hat{e}_j\right]_1 = \delta_{ij}\hat{e}_0 \qquad (31)$$

4. Hilbert Q-module Quantization

To describe the quantization of the supersymmetric model we shall use the formalism of the Hilbert Q-modules.

Q-representations of the Fermionic Heisenberg Group. As in the case of the Heisenberg group, it is possible to consider the Schrödinger representation for FH_n. However, due to the nature of the Berezin integral [10] the essentially functional content of it is trivial and the analog of the representation in function space arising in this way is finite dimensional and of an algebraic kind. Let S^n be the set of $Q_{\mathbb{C}}$ valued functions of n = 2k real Grassmann variables $\eta_i \in Q^n_{\mathbb{R},1}$. Let "$\star$" denote conjugation in the $Q_{\mathbb{C}}$ - algebra extending complex conjugation. In S^n we introduce the $Q_{\mathbb{C}}$ - scalar product

$$\langle f, g \rangle_S = \int d\eta\, f^*(\eta)g(\eta) \qquad (32)$$

$$d\eta = d\eta_n \ldots d\eta_1$$

$(S^n, \langle \cdot, \cdot \rangle_S)$ is the Hilbert Q-module [6,10]. This is the counterpart of the conventional Hilbert space of square integrable functions. In this space the Hermitian conjugate operators to the differentiation and multiplication operators are

$$\frac{\partial}{\partial \eta_i} \equiv \partial_\eta, \qquad \partial_\eta{}^\dagger = i\partial_\eta \qquad (33)$$

$$\hat{\eta} \equiv \eta \cdot \qquad \hat{\eta}^\dagger = -i\eta \cdot \qquad (34)$$

This operators are not self-adjoint in S^n. However, for the construction of the Q-representation of FH_n we need the following Let $D = -i\partial_\eta$, $X^i = \eta^i$. The operator $\Pi^i D_i + \Theta_i X^i$ is self-adjoint in S^n and

$$\exp i\,(t + \Theta X + \Pi D)\,f(\eta) = \exp i\,\left(t + \Theta\eta + \frac{1}{2}\Theta\Pi\right)f\,(\eta + \Pi) \qquad (35)$$

or equivalently

$$e^{i(\Theta X + \Pi D)} = e^{\frac{i}{2}\Theta\Pi} e^{i\Theta X} e^{i\Pi D}. \tag{36}$$

The multiplication of $e^{iA'} e^{iA}$ yields the FH_n group multiplication (6). There exists a homomorphism

$$\pi_1 : FH_n \mapsto Op(S^n)$$

$$\pi_1(\Pi, \Theta, t) = \exp i(t + \Theta X + \Pi D) \tag{37}$$

which gives the Q - representation of FH_n on S^n, $n = 2k$. π_1 given by

$$\pi_1(\Pi, \Theta, t) f(\eta) = e^{i(t + \Theta\eta + \frac{1}{2}\Theta\Pi)} f(\eta + \Pi) \tag{38}$$

is a Q - irreducible Q - unitary representation of FH_n. The matrix coefficients of the representation π_1 for f, $g \in S^n$ are defined by

$$M(\Pi, \Theta) = \langle f, \pi_1(\Pi, \Theta) g \rangle \tag{39}$$

Analogously to the conventional theory we can introduce the function $V(f, g)$ on the S^n by

$$V(f, g)(\Pi, \Theta) = M(\Pi, \Theta) = \langle f, e^{i(\Theta X + \Pi D)} g \rangle =$$

$$\int d\eta\, f^*(\eta - \frac{1}{2}\Pi) e^{i\Theta\eta} g(\eta + \frac{1}{2}\Pi) \tag{40}$$

The mapping $V : S^n \times S^n \mapsto S^{2n}$ is the Grassmannian version of the Fourier- Wigner transform (the GFW-transform). In particular, GFW-transform for Grassmannian Gaussian $\omega_0 \in S^n$, $n = 2k$, can be written in the following form

$$V(\omega_0, f)(\Pi, \Theta) = (Pf)^{-\frac{1}{2}} e^{-\frac{i}{4} z^* G^{-1} z} \int d\eta\, e^{\frac{1}{2}\eta G\eta - \eta z - \frac{1}{4} z G^{-1} z} f(\eta), \tag{41}$$

with a new variable z defined as

$$z_k = G_{kj}\Pi^j + i\Theta_k, \tag{42}$$

and

$$\omega_0 = (Pf\, G)^{-\frac{1}{2}} e^{\frac{1}{2}\eta G\eta}, \tag{43}$$

where $G = (G_{ij})$ is an anti-symmetric matrix and $Pf\, G$ its Pfaffian and

$$\langle \omega_0, \omega_0 \rangle_S = 1. \tag{44}$$

This allows us to define the Grassmannian Bargmann transform (GB - transform) as

$$(\mathbf{B}f)(z) \equiv 2^{-\frac{n}{4}} \int d\eta \, e^{\frac{1}{2}\eta G \eta - \eta z - \frac{1}{4} z G^{-1} z} f(\eta). \tag{45}$$

For further convenience we shall denote

$$\|z\|^2 = \frac{i}{2} z^* G^{-1} z. \tag{46}$$

One can write the FH_n group multiplication for (z,t) in the form

$$(z,t) \circ (z',t') = \left(z + z', t + t' + \frac{1}{2} Im(-z^* \frac{i}{2} G^{-1} z) \right), \tag{47}$$

The transferred representation β can be defined as

$$\beta(z,t) \circ \mathbf{B} = \mathbf{B} \circ \pi_1(\Pi, \Theta, t) \tag{48}$$

where

$$V(\omega_0, f)(\Pi, \Theta) = e^{-\frac{1}{2}\|z\|^2} (\mathbf{B}f)(z) \tag{49}$$

Let us define the Grassmannian Bargmann-Fock space as

$$\mathcal{F}_n = \{ f \mid f \text{ is holomorphic on } Q_{C,1}^n \text{ and } \|f\|_{\mathcal{F}}^2 = \int |dz| \, e^{-\|z\|^2} f^*(z) f(z) \} \tag{50}$$

The basis in this space is formed by polynomials $\{z^{I_k}\}_{I_k, k}$, where I_k is a strongly ordered multi-index (with increasing entries). The Q - scalar product in $\mathcal{F}_n$ is defined as

$$\langle f, g \rangle_{\mathcal{F}} = \int |dz| \, e^{-\|z\|^2} f^*(z) g(z), \tag{51}$$

where

$$|dz| \equiv -(\frac{i}{2})^n dz dz^*.$$

$(\mathcal{F}_n, \langle \, , \, \rangle_{\mathcal{F}})$ is a Hilbert Q - module. The operator Hermitian conjugate to the differentiation ∂_z in $\mathcal{F}_n$ is

$$(\partial_z)^\dagger = -\frac{i}{2} G^{-1} z. \tag{52}$$

We can write representation β explicitly. Let $w = G\rho + i\sigma$, for $f \in S^n$ we have

$$(\beta(w)\mathbf{B}f)(z) = (\mathbf{B}\pi_1(\rho, \sigma)f)(z) = e^{\frac{1}{2}\|z\|^2} V(\omega_0, \pi_1(\rho, \sigma)f)(\Pi, \Theta) \tag{53}$$

what gives

$$(\beta(w)\mathbf{B}f)(z) = e^{-\frac{i}{2}\|w\|^2} e^{-\frac{i}{2}zG^{-1}w^*} \mathbf{B}f(z) . \tag{54}$$

For the Heisenberg group the Bargmann transform relates two distinguished bases in the Schrödinger representation space and in the Fock space. This property also holds for the FH_n. The Grassmannian Hermite polynomials are related to the z_{I_k} polynomials forming the basis of the Fock Q - module. The Grassmannian Hermite [11] polynomials can be taken in the form [12]

$$h_k^{i_1\ldots i_k} = K_k e^{-\frac{1}{2}\eta G\eta}\partial^{\alpha_k}\ldots\partial^{\alpha_1} e^{\eta G\eta} \tag{55}$$

where K_k are numerical factors. Then the GB transform of h_k, $0 \le k \le n$, yields

$$(\mathbf{B}h_k)_{i_1\ldots i_k}(z) = 2^{\frac{n}{4}} K_k z_{I_k}. \tag{56}$$

The GFW transform of the Grassmannian Hermite function gives a Grassmannian Laguerre polynomial [8]

$$\langle z_{I_k}, \beta(w)z_{I_k'}\rangle = e^{-\frac{1}{2}\|w\|^2} L_{I_k}^{(0)}, \quad L_{I_k}^{(0)} = \sum_{m,I_m\subset I_k} N_{I_m}^{I_m'} w_{I_{k-m}}^* w_{I_{k-m}'}. \tag{57}$$

Q-representations of the Odd Heisenberg Group [5]. The construction of the Schrödinger Q-representation known for the Fermionic Heisenberg group can be extended to the Q_{CO}-representation of the Odd Heisenberg group (we consider here only the sector (p,Θ)). Appropriate Grassmannian odd transforms and generalized Grassmannian odd polynomials fall to this scheme as well.

Let S_{OD}^n be the set of functions on Q_1^n with values in the Q_{CO}, let the Q_{CO} valued scalar product be given in the form

$$\langle f,g\rangle_S = \int d\eta\, f^*(\eta)g(\eta), \quad d\eta = d\eta_n\ldots d\eta_1,\, \eta_i \in Q_1 \tag{58}$$

$(S^n, \langle\cdot,\cdot\rangle_S)$ forms the Hilbert Q_{CO}- module.
Let $D_j = -\hat{\imath}\frac{\partial}{\partial\eta_j}$ and $X^i = \eta^i$ then the following relations give rise to the definition of representation π

$$\exp_\star \hat{\imath}\,(t + \Theta X + pD)\, f(\eta) = \tag{59}$$

$$\exp_\star \hat{\imath}\left(t + \Theta\eta + \frac{1}{2}\Theta \star p\right) f\left(\eta + \hat{\imath}p\right) = \pi(p,\Theta,t)$$

Let $p^i = \hat{\imath}\Pi^i$, $\Pi^i \in Q_1$. Therefore algebraic form of relations obtained in the first part of the report for the Fermionic Heisenberg group will be here preserved modulo odd exponents and odd units.

The Grassmannian Fourier-Wigner odd transform takes the following form

$$V(f,g)(p,\Theta) = \int d\eta \, f^\star(\eta - \frac{1}{2}\hat{1}p)e_\star^{i\Theta\eta}g(\eta + \frac{1}{2}\hat{1}p) \qquad (60)$$

Let $\hat{\omega}_O$ be a Grassmann odd Gaussian of the form

$$\hat{\omega}_0 = Ae_\star^{\frac{1}{2}\eta\hat{G}\eta}, \qquad (61)$$

where A is a normalization factor and $\hat{G} = \hat{1}(G_{ij})$ with $G = (G_{ij})$ being an anti-symmetric matrix in orthogonal form. Defining the new variable z as

$$z_k = G_{kl}p^l + i\Theta_k \qquad (62)$$

we can introduce the Grassmannian Bargmann odd transform as follows

$$(\hat{B}f)(z) \equiv 2^{-\frac{n}{4}} \int d\eta \, e_\star^{\frac{1}{2}\eta\hat{G}\eta - \eta z - \frac{1}{4}z\hat{G}^{-1}z} f(\eta). \qquad (63)$$

Analogously as for the FH_n the group product of the OH_n can be expressed as

$$(z,t) \star (z',t') = \left(z + z', \tau + \tau' + \frac{1}{2}Im_{OD}(z^\star\hat{G}^{-1}z) \right), \qquad (64)$$

where Im_{OD} denotes the oddonic imaginary part.

Modification of Grassmannian Hermite polynomials to the odd case is given by the formula

$$\hat{h}_k^{i_1\ldots i_k} = H_k e_\star^{-\frac{1}{2}\eta\hat{G}\eta} \partial^{i_k} \ldots \partial^{i_1} e_\star^{\eta\hat{G}\eta} \qquad (65)$$

where in comparison to the fermionic case, here the odd exponents enter the definition. H_k are normalization factors.

Grassmannian Bargmann odd transform relates Grassmannian Hermite odd polynomials to the z_{I_k} basis of the Fock Q_{CO}-module. Grassmannian Laguerre odd polynomials take values in complex Oddons as well and have the form

$$\hat{L}_{I_k} = \sum_{m, I_m \subset I_k} W_{I_m}^{I'_m} z_{I_{k-m}}^\star z_{I'_{k-m}}, \quad z_{I_{k-m}}, z_{I'_{k-m}} \in Q_{CO} \qquad (66)$$

where $W_{I_m}^{I'_m} \in Q_{CO}$ are normalization factors.

5. Final Remarks

We have discussed some issues of the Q-module quantization of the $\mathbb{Z}_2$-graded mechanical systems, taking as the example, realizations of the same supersymmetry in the even (GSO) and odd (OGSO) superfield model yielding the phase superspace with even superPoisson-bracket and anti-bracket, respectively. The formalism for the odd system can be developed analogously to the one known for the even systems, provided that in the odd case we introduce an odd multiplication of observables. This has been done here by means of the algebra of oddons.

Appendix

Real Oddons. Let $\hat{1}$ be an element such that, for homogeneous $q_s \in Q_s^{\mathbb{R}}$

$$1\hat{1} = \hat{1} \quad \hat{1}^2 = 1 \quad q_s \hat{1} = (-1)^s \hat{1} q_s \tag{67}$$

The expressions of the form

$$r = q + \hat{1} q', \qquad q, q' \in Q_{\mathbb{R}} \tag{68}$$

we shall call the real oddons. They form a graded algebra Q_{RO}. This algebra in not graded commutative. Despite the extension of the usual product we can define a new odd product

$$r \star r' \equiv r \cdot \hat{1} \cdot r' \tag{69}$$

The $\hat{1}$ is a unit with respect to the $\star$-multiplication, having the same parity as the multiplication.

Complex Oddons. Similarly we can consider the complexification of above structure, in the sense that $Q_{\mathrm{CO}} \equiv Q_{\mathrm{C}} \oplus \hat{i} Q_{\mathrm{C}}$ and

$$\hat{i}^2 = -1, \quad \hat{i} \cdot 1 = \hat{i}, \quad \hat{i} \cdot \hat{1} = i \quad \hat{i} \cdot q_s = (-1)^s q_s \hat{i}, \tag{70}$$

where $q_s \in Q_s^{\mathrm{C}}$. Obviously $\hat{i} \cdot i = -\hat{1}$. The product of two homogeneous complex oddons takes the form

$$z_s \cdot z'_r = a_s a'_r - (-1)^{s+1} b_s b'_r + \hat{i}((-1)^s a_s b'_r + b_s a'_r) \neq (-1)^{rs} z'_r \cdot z_s, \tag{71}$$

where $z_r = a_r + \hat{i} b_r \in Q_r^{\mathrm{CO}}$. The component a we shall call the oddonic real part and the b - the oddonic imaginary part. The Q_{CO} can be considered as an algebra with the odd $\star$ product. The even mapping

$$* : Q_{\mathrm{CO}} \longrightarrow Q_{\mathrm{CO}} \tag{72}$$

such that

$$z = a + \hat{i} b \longrightarrow z^* = a^* - b^* \hat{i} \tag{73}$$

we shall consider as oddonic conjugation. Note that we use the same symbol for the extension of complex conjugation in the Q_C.

References

1. D. A. Leites, Dokl. Akad. Nauk. SSSR **236** (1977), 804
2. D. A. Leites, Supplement 3 in F. A. Berezin, M. A. Shubin *"The Schrödinger Equation"* (Kluwer Academic Publisher, Dordrecht, 1992)
3. A. Frydryszak, J. Phys. A: Math. Gen. **26** (1993), 7227
4. D. V. Volkov, V. A. Soroka, Sov. J. Nucl. Phys **46** (1988), 110
5. A. Frydryszak, Lett. Math. Phys. **44** (1998), 89
6. M. E. Taylor, *"Noncommutative Harmonic Analysis"* (AMS Providence, Rhode Island, 1986)
7. A. Frydryszak, Lett. Math. Phys. **18** (1989), 87
8. A. Frydryszak, Lett. Math. Phys. **26** (1992), 105
9. A. Frydryszak, L. Jakóbczyk, Lett. Math. Phys. **16** (1988), 101
10. F. A. Berezin, *"The Method of Second Quantization"* (Academic Press, New York, 1966)
11. R. Finkelstein, M. Villasante, Phys. Rev. **D 6** (1986), 1666
12. A. Frydryszak, Lett. Math. Phys. **20** (1990), 159

LOCALLY ANISOTROPIC SUPERGRAVITY AND GAUGE GRAVITY ON NONCOMMUTATIVE SPACES

SERGIU VACARU *
*Institute of Applied Physics, Academy of Sciences, Academy str. 5,
Chisinău MD2028, Republic of Moldova*

IURIE CHIOSA, NADEJDA VICOL
Faculty of Mathematics and Informatics, gr. 33 MI, State University of Moldova, Mateevici str. 60, Chisinău MD2009, Republic of Moldova

Abstract. We outline the the geometry of locally anisotropic (la) superspaces and la–supergravity. The approach is backgrounded on the method of anholonomic superframes with associated nonlinear connection structure. Following the formalism of enveloping algebras and star product calculus we propose a model of gauge la–gravity on noncommutative spaces. The corresponding Seiberg–Witten maps are established which allow the definition of dynamics for a finite number of gravitational gauge field components on noncommutative spaces.

1. Introduction

Locally anisotropic supergravity was developed as a model of supergravity with anholonomic superframes and associated nonlinear connection (N–connection) structure [13]. This model contain as particular cases supersymmetric Kaluza–Klein and generalized Lagrange and/or Finsler gravities and for nontrivial curvatures the N–connection describes splittings from higher to lower dimensions of (super) spaces and generic anholonomic local anisotropies.

In order to avoid the problem of formulation of gauge theories on noncommutative spaces [3, 10, 5, 7] with Lie algebra valued infinitesimal transformations and with Lie algebra valued gauge fields the authors of [6] suggested to use enveloping algebras of the Lie algebras for setting this type of gauge theories and showed that in spite of the fact that such enveloping algebras are infinite–dimensional one can restrict them in a way that it would be a dependence on the Lie algebra valued

---

* sergiu.vacaru@phys.asm.md, sergiu_vacaru@yahoo.com

S. Duplij and J. Wess (eds.), *Noncommutative Structures in Mathematics and Physics,* 229–243.

parameters and the Lie algebra valued gauge fields and their spacetime derivatives only.

A still presented drawback of noncommutative geometry and physics is that there is not yet formulated a generally accepted approach to interactions of elementary particles coupled to gravity. There are improved Connes–Lott and Chamsedine–Connes models of nocommutative geometry [2] which yielded action functionals typing together the gravitational and Yang–Mills interactions and gauge bosons the Higgs sector (see also the approaches [4] and [8]).

In this paper we outline the geometry of locally anisotropoc supergravity and follow the method of restricted enveloping algebras [5, 6] and construct gauge gravitational theories by stating corresponding structures with semisimple or nonsemisimple Lie algebras and their extensions. We consider power series of generators for the affine and non linear realized de Sitter gauge groups and compute the coefficient functions of all the higher powers of the generators of the gauge group which are functions of the coefficients of the first power. Such constructions are based on the Seiberg–Witten map [10] and on the formalism of $*$–product formulation of the algebra [18] when for functional objects, being functions of commuting variables, there are associated some algebraic noncommutative properties encoded in the $*$–product. The concept of gauge gravity theory on noncommutative spaces is introduced in a geometric manner [7] by defining the covariant coordinates without speaking about derivatives and this formalism was developed for quantum planes [17]. We prove the existence for noncommutative spaces of gauge models of gravity which agrees with usual gauge gravity theories [14] being equivalent, or extending, the general relativity theory (see works [9, 11] for locally isotropic spaces and corresponding reformulations and generalizations respectively for anholonomic frames [15] and locally anisotropic (super) spaces [16]) in the limit of commuting spaces.

2. Locally Anisotropic Supergravity

Let us consider a vector superbundle (vs–bundle) $\widetilde{E}$ over a supermanifold (s–manifold)$\widetilde{M}$ with surjective projection $\pi_E : \widetilde{E} \to \widetilde{M}$ (for simplicity, all constructions are locally trivial). The local supersymmetric coordinates (s–coordinates) on $\widetilde{E}$ and $\widetilde{M}$ are denoted respectively $u = (x, y) = \{u^\alpha = \left(x^I, y^A\right)$, where $x = \{x^I = \left(x^i, x^{\widehat{i}}\right)\}$ are (even,odd) coordinates on $\widetilde{M}$ and $y = \{y^A = \left(y^a, y^{\widehat{a}}\right)\}$ are (even,odd) coordinates in fibers of π_E (indices run values defined by even and odd dimensions of corresponding submanifolds). Latin s–indices $I, J, K, L, M, ...$ and $A, B, C, D, ...$ will be used respectively for base and fiber components.

A nonlinear connection (N–connection) structure which defines a global decomposition of $T\widetilde{E}$ into horizontal, $H\widetilde{E}$, and vertical parts, $V\widetilde{E}$,

$$N : T\widetilde{E} = H\widetilde{E} \oplus V\widetilde{E}. \tag{1}$$

The coefficients of a N–connection $N_I^A(u)$ determine the locally adapted s–frame (basis, in brief la–frame)

$$\delta_\alpha = \delta/\delta u^\alpha = \left(\delta_I = \delta/\delta x^I = \partial_I - N_I^B(u)\,\partial_B, \partial_A\right), \qquad (2)$$

where $\partial_I = \partial/\partial x^I, \partial_A = \partial/\partial y^A$ are partial s–derivatives, and the dual s–frame

$$\delta^\alpha = \delta u^\alpha = \left(d^I = \delta x^I = dx^I, \delta^A = \delta y^A = dy^A + N_I^A(u)\,dx^I\right). \qquad (3)$$

The s–frame (2) is anholonomic

$$[\delta_J, \delta_K\} = \delta_J \delta_K - (-)^{|JK|}\delta_K \delta_J = \Omega^A_{\ JK}\partial_A,$$

where $|JK| = |J| \cdot |K|$ is defined by the parity of indices and we write $(-)^{|JK|}$ instead $(-1)^{|JK|}$, with anholonomy coefficients coinciding with the N-connection curvature

$$\Omega^A_{\ JK} = \delta_K N_J^A - (-)^{|JK|}\delta_J N_K^A.$$

The geometrical objects on $\widetilde{E}$ are given with respect to la–basis (2) and (3) or their tensor products and called ds–tensors, ds–connections (for some additional linear connections), d–spinors and so on. For instance, a metric ds–tensor is written

$$\widetilde{g} = g_{\alpha\beta}\delta^\alpha \otimes \delta^\beta = g_{IJ}d^I \otimes d^J + g_{AB}\delta^A \otimes \delta^B. \qquad (4)$$

The Lagrange and Finsler ds–metrics can be modeled on a locally anisotropic superspace if vs–bundle $\widetilde{E}$ over a s–manifold $\widetilde{M}$ is substituted by the tangent s–bundle $T\widetilde{M}$ and the coefficients of ds–metric (4) are taken respectively

$$g_{IJ}(u) = \frac{1}{2}\frac{\partial^2 \mathcal{L}(u)}{\partial y^I \partial y^L} \text{ and } g_{IJ}(u) = \frac{1}{2}\frac{\partial^2 F^2(u)}{\partial y^I \partial y^L}$$

where the s–Lagrangian $\mathcal{L}:T\widetilde{M} \to \Lambda$ is a s–differentiable function on $T\widetilde{M}$, and F is a Finsler s–metric function on $T\widetilde{M}$.

A linear distinguished connection D, d–connection, in sv–bundle $\widetilde{E}$ is a linear connection which preserves by parallelism the horizontal (h) and vertical (v) distribution (1).

A d–connection $D\Gamma = \{\Gamma^\alpha_{\ \beta\gamma} = \left(L, \widetilde{L}, \widetilde{C}, C\right)\}$, is determined by its invariant hh-, hv-, vh- and vv–components, where

$$D_{(\delta_K)}\delta_J = L^I_{\ JK}(u)\,\delta_I, \ D_{(\delta_K)}\partial_B = L^A_{\ BK}(u)\,\partial_A, \qquad (5)$$

$$D_{(\partial_C)}\delta_J = C^I_{\ JC}(u)\,\delta_I, \ D_{(\partial_C)}\partial_B = C^A_{\ BC}(u)\,\partial_A.$$

There is a canonical d–connection $^{(c)}\Gamma$ defined by the coefficients of d–metric (4) and of N–connection and satisfying the metricity condition $D\tilde{g} = 0$,

$$^{(c)}L^I_{JK} = \frac{1}{2}g^{IH}\left(\delta_K g_{HJ} + \delta_J g_{HK} - \delta_H g_{JK}\right),$$

$$^{(c)}L^A_{BK} = \partial_B N^A_K + \frac{1}{2}h^{AC}\left(\delta_K H_{BC} - (\partial_B N^D_K)h_{DC} - (\partial_C N^D_K)h_{DB}\right),$$

$$^{(c)}C^I_{JC} = \frac{1}{2}g^{IK}\partial_C g_{JK}, \quad ^{(c)}C^A_{BC} = \frac{1}{2}h^{AD}\left(\partial_C h_{DB} + \partial_B h_{DC} - \partial_D h_{BC}\right).$$

The torsion $T^\alpha_{\beta\gamma}$ of a d–connection, $T(X,Y) = [X, DY\} - [X, Y\}$, where X and Y are ds–vectors and by $[...\}$ we denote the s–anticommutator, is decomposed into hv–invariant ds–torsions

$$hT(\delta_K, \delta_J) = T^I_{JK}\delta_I, \; vT(\delta_K, \delta_J) = \tilde{T}^A_{JK}\delta_I, \; hT(\partial_A, \delta_J) = \tilde{P}^I_{JA}\delta_I,$$

$$vT(\partial_B, \delta_J) = P^A_{JB}\partial_A, \quad vT(\partial_C, \partial_B) = S^A_{BC}\partial_A,$$

with coefficients

$$T^I_{JK} = L^I_{JK} - (-)^{|JK|}L^I_{KJ}, \; \tilde{T}^A_{JK} = \delta_K N^A_J - (-)^{|KJ|}\delta_J N^A_K, \qquad (6)$$

$$\tilde{P}^I_{JA} = C^I_{JA}, \; P^A_{JB} = \partial_B N^A_J - L^A_{BJ}, \; S^A_{BC} = C^A_{BC} - (-)^{|BC|}C^A_{CB}.$$

The even and odd components of ds–torsions (6) can be specified in explicit form by using decompositions of indices into even and odd parts, $I = (i, \hat{i})$, $A = (a, \hat{a})$ and so on.

The curvature $R^\alpha_{\beta\gamma\tau}$ of a d–connection, $R(X,Y)Z = D_{[X}D_{Y\}}Z - D_{[X,Y\}}Z$, where X, Y, Z are ds–vectors, splits into hv–invariant ds–torsions

$$R(\delta_K, \delta_J)\delta_H = R^I_{HJK}\delta_I, \; R(\delta_K, \delta_J)\partial_B = R^A_{BJK}\partial_A, \qquad (7)$$

$$R(\partial_C, \delta_K)\delta_J = \tilde{P}^I_{JKC}\delta_I, \; R(\partial_C, \delta_K)\partial_B = P^A_{BKC},$$

$$R(\partial_C, \partial_B)\delta_J = \tilde{S}^I_{JBC}\delta_I, \; R(\partial_D, \partial_C)\partial_B = S^A_{BCD}\partial_A$$

where the coefficients are computed

$$R^I_{MJK} = \delta_{[K}L^I_{|M|J\}} + L^W_{MJ}L^I_{WK} - (-)^{|KJ|}L^W_{MK}L^I_{WJ} + C^I_{KA}W^A_{JK}$$

$$\hat{R}^A_{BJK} = \delta_{[K}L^A_{|B|J\}} + L^C_{BJ}L^A_{CK} - (-)^{|KJ|}L^C_{BK}L^A_{CJ} + C^A_{BC}W^C_{JK}$$

$$\tilde{S}^I_{JBC} = \partial_C C^I_{JB} - (-)^{|BC|}\partial_B C^I_{JC} + C^H_{JB}C^I_{HC} - (-)^{|BC|}C^H_{JC}C^I_{HB}$$

$$S^A_{BCD} = \partial_D C^A_{BC} - (-)^{|CD|}\partial_C C^A_{BD} + C^E_{BC}C^A_{ED} - (-)^{|CD|}C^E_{BD}C^A_{EC}$$

$$\tilde{P}^I_{JKA} = \partial_A L^I_{JK} - C^I_{JA|K} + C^I_{JB}P^B_{KA}$$

$$P^A_{BKC} = \partial_C L^A_{BK} - C^A_{BC|K} + C^A_{BD}P^D_{KC}$$

where, for instance,

$$\delta_{[K}L^I_{|M|J\}} = \delta_K L^I_{MJ} - (-)^{|KJ|}\delta_J L^I_{MK},$$

$$C^I_{JA|K} = \delta_K C^I_{JA} + L^I_{MK}C^M_{IA} - L^M_{JK}C^I_{MA} - L^B_{AK}C^I_{JB}.$$

The even and odd components of ds–curvatures are computed by splitting indices into even and odd parts.

The torsion and curvature of a d–connection D on a sv–bundle satisfy the identities

$$\sum_{SC}[(D_X T)(Y,Z) - R(X,Y)Z + T(T(X,Y),Z)] = 0,$$

$$\sum_{SC}[(D_X R)(U,Y,Z) - R(T(X,Y),Z)U] = 0,$$

where $\sum_{SC}$ means supersymmetric cyclic sums over ds–vectors X, Y, Z and U, from which the generalized Bianchi and Ricci identities follow [1-3].

The Ricci ds–tensor $R_{\beta\gamma} = R^{\alpha}_{\beta\gamma\alpha}$ has hv–invariant components

$$R_{IJ} = R^K_{IJK}, \quad R_{IA} = -\,^{(2)}P_{IA} = -(-)^{|KA|}\tilde{P}^K_{IKA}, \qquad (8)$$
$$R_{AI} = \,^{(1)}P_{AI} = P^B_{AIB}, \quad R_{AB} = S^C_{ABC} = S_{AB}.$$

If a ds–metric (4) is defined on $\tilde{E}$, we can introduce the supersymmetric scalar curvature

$$\hat{R} = g^{\alpha\beta}R_{\alpha\beta} = R + S,$$

where $R = g^{IJ}R_{IJ}$ and $S = h^{AB}S_{AB}$.

The simplest model of locally anisotropic supergravity (la–supergravity) was constructed by postulating a variant of supersymmetric Einstein--Cartan theory on locally anisotropic superspace $\tilde{E}$, which in invariant hv–components has the fundamental s–field equations

$$R_{IJ} - \frac{1}{2}(R + S - \lambda)g_{IJ} = k_1 \Upsilon_{IJ}, \quad {}^{(1)}P_{AI} = k_1 \Upsilon_{AI}, \qquad (9)$$

$$S_{AB} - \frac{1}{2}(R + S - \lambda)h_{AB} = k_1 \Upsilon_{AB}, \quad {}^{(2)}P_{IA} = -k_1 \Upsilon_{IA},$$

and

$$T^{\alpha}_{\beta\gamma} + \delta^{\alpha}_{\beta}T^{\tau}_{\gamma\tau} - (-)^{|\beta\gamma|}\delta^{\alpha}_{\gamma}T^{\tau}_{\beta\tau} = k_2 Q^{\alpha}_{\beta\gamma},$$

where λ is the cosmological constant, $k_{1,2}$ are respective interaction constants $\Upsilon_{\alpha\beta}$ is the energy–momentum ds–tensor and $Q^{\alpha}_{\beta\gamma}$ is defined by the supersymmetric spin–density.

The bulk of theories of locally isotropic s–gravity are formulated as gauge supersymmetric models based on supervielbein formalism. Similar approaches to la–supergravity on vs–bundles can be developed by considering arbitrary s–frames $B_{\underline{\alpha}}(u) = (B_{\underline{I}}(u), B_{\underline{C}}(u))$ adapted to the N–connection structure on a vs–bundle $\tilde{E} = \tilde{E}^{m,l}$ over s–manifold $\tilde{M} = \tilde{M}^{n,k}$ where (m,l) and (n,k) are respective (even, odd) dimensions of s–manifolds. A s–frame $B_{\underline{\alpha}}(u)$ is related with a standard la–frame (2) via transforms $\delta_\alpha = A_{\bar{\alpha}}^{\underline{\alpha}}(u) B_{\underline{\alpha}}(u)$, where s–matrices $A_{\bar{\alpha}}^{\underline{\alpha}}(u) = \begin{pmatrix} A_{\bar{I}}^{\underline{I}} & 0 \\ 0 & A_{\bar{C}}^{\underline{C}} \end{pmatrix}$ take values into a super Lie group $GL_{n,k}^{m,l}(\Lambda) = GL(n,k,\Lambda) \oplus GL(m,l,\Lambda)$ (on superspaces the graded Grassmann algebra with Euclidean topology, denoted by Λ, substitutes the real and complex number fields).

We denote by $LN(\tilde{E})$ the set of all adapted to N–connection s–frames in all points of vs–bundle $\tilde{E}$ and consider the s–bundle of linear adapted s–frames on $\tilde{E}$ defined as the principal s–bundle

$$\mathcal{L}N(\tilde{E}) = \left(LN(\tilde{E}), \pi_L : LN(\tilde{E}) \to \tilde{E}, GL_{n,k}^{m,l}(\Lambda) \right),$$

for a surjective s–map π_L. The canonical basis of standard distinguished s–generators $I_{\hat{\alpha}} \to I_{\hat{\beta}}^{\hat{\alpha}} = \begin{pmatrix} I_{\bar{J}}^{\bar{I}} & 0 \\ 0 & I_{\bar{B}}^{\bar{A}} \end{pmatrix}$ for the super Lie algebra $\mathcal{G}L_{n,k}^{m,l}(\Lambda)$ of the structural s–group $GL_{n,k}^{m,l}(\Lambda)$ satisfy s–commutation rules $[I_{\hat{\alpha}}, I_{\hat{\beta}}\} = f_{\hat{\alpha}\hat{\beta}}{}^{\hat{\gamma}} I_{\hat{\gamma}}$.
On $\mathcal{L}N(\tilde{E})$ we consider the d–connection 1–form

$$\mathcal{F} = \Gamma_{\underline{\beta}\gamma}^{\underline{\alpha}}(u) I_{\underline{\alpha}}^{\underline{\beta}} \delta u^\gamma,$$

where

$$\Gamma_{\underline{\beta}\gamma}^{\underline{\alpha}}(u) = A_{\alpha}^{\underline{\alpha}} A_{\underline{\beta}}^{\beta} \Gamma_{\beta\gamma}^{\alpha} + A_{\underline{\beta}}^{\alpha} \delta_\gamma A_{\underline{\beta}}^{\beta}, \tag{10}$$

$\Gamma_{\beta\gamma}^{\alpha}$ are the components of canonical variant of d–connection (5) and the s–matrix $A_{\underline{\beta}}^{\beta}$ is inverse to $A_{\underline{\alpha}}^{\underline{\alpha}}$.

The curvature $\mathcal{B}$ of the d–connection (10)

$$\mathcal{B} = \delta\mathcal{F} + \mathcal{F} \wedge \mathcal{F} = R_{\underline{\alpha}\gamma\tau}^{\underline{\beta}} I_{\underline{\beta}}^{\underline{\alpha}} \delta u^\gamma \wedge \delta u^\tau \tag{11}$$

has the coefficients $R_{\underline{\alpha}\gamma\tau}^{\underline{\beta}} = A_{\underline{\alpha}}^{\alpha}(u) A_{\beta}^{\underline{\beta}}(u) R_{\alpha\gamma\tau}^{\beta}$, where $R_{\alpha\gamma\tau}^{\beta}$ are defined by ds–curvatures (7).

Aside from $\mathcal{L}N(\tilde{E})$ with vs–bundle $\tilde{E}$ is naturally related another s–bundle, the bundle of adapted to N–connection affine s–frames

$$\mathcal{A}N(\tilde{E}) = \left(AN(\tilde{E}), \pi_A : AN(\tilde{E}) \to \tilde{E}, AF_{n,k}^{m,l}(\Lambda) \right),$$

with the affine strucural s–group $AF_{n,k}^{m,l}(\Lambda) = GL_{n,k}^{m,l}(\Lambda) \odot \Lambda^{n,k} \oplus \Lambda^{m,l}$.

The d–connection $\mathcal{F}$ (10) in $\mathcal{L}N(\tilde{E})$ induces in a linear Cartan d–connection $\overline{\mathcal{F}} = (\mathcal{F}, \chi)$, in $\mathcal{A}N(\tilde{E})$, where $\chi = e_{\underline{\alpha}} \otimes A_{\alpha}^{\underline{\alpha}}(u) \delta u^\alpha$, $e_{\underline{\alpha}}$ is the standard basis in $\Lambda^{n,k} \oplus \Lambda^{m,l}$, and, in consequence, the curvature $\mathcal{B}$ (11) in $\mathcal{LN}\left(\widetilde{E}\right)$ induces the pair (curvature, torsion) $\overline{\mathcal{B}} = (\mathcal{B}, \mathcal{T})$ in $\mathcal{AN}\left(\widetilde{E}\right)$, where

$$\mathcal{T} = \delta\chi + [\mathcal{F} \wedge \gamma\} = \mathcal{T}^\alpha_{\beta\gamma} e_{\underline{\alpha}} \delta u^\beta \wedge \delta u^\gamma,$$

when $\mathcal{T}^\alpha_{\beta\gamma} = A^\alpha_{\underline{\alpha}} T^\alpha_{\beta\gamma}$ is defined by the coefficients of d–torsions (6).

By using the ds–metric (4) in $\widetilde{E}$ one defines the (dual for s–forms) Hodge operator $*_{\widetilde{g}}$. Let the operator $*_{\widetilde{g}}^{-1}$ be inverse to $*_{\widetilde{g}}$ and $\widehat{\delta}_{\widetilde{g}}$ be the adjoint to the absolute derivation $\widehat{\delta}$ (associated to the scalar product of ds–forms) specified for (r,s)–forms $\widehat{\delta}_{\widetilde{g}} = (-1)^{r+s} *_{\widetilde{g}}^{-1} \circ \widehat{\delta} \circ *_{\widetilde{g}}$.

The supersymmetric variant of the Killing form of the s–group $AF^{m,l}_{n,k}(\Lambda)$ is degenerate. In order to generate a metric structure $\widetilde{g}_A$ in the total spaces of the s–bundle $\mathcal{AN}\left(\widetilde{E}\right)$ we use and auxiliary nondegenerate bilinear s–form which gives rise to the possibility to define the Hodge operator $*_{\widetilde{g}_A}$ and $\widehat{\delta}_{\widetilde{g}_A}$. Applying the operator of horizontal projection $\widehat{H}$ one defines the operator $\triangle \doteq \widehat{H} \circ \widehat{\delta}_{\widetilde{g}_A}$ which does not depend on components of auxiliary biliniar s–form in the fiber.

Following an abstract geometric calculus, by using operators $*_{\widetilde{g}}, *_{\widetilde{g}_A}, \widehat{\delta}_{\widetilde{g}}, \widehat{\delta}_{\widetilde{g}_A}$ and $\triangle$ one computers

$$\triangle\overline{\mathcal{B}} = (\triangle\mathcal{B}, \mathcal{R}t + \mathcal{R}i), \tag{12}$$

where the one s–forms

$$\mathcal{R}t = \widehat{\delta}_{\widetilde{g}}\mathcal{T} + *_{\widetilde{g}}^{-1}[\mathcal{F}, *_{\widetilde{g}}\mathcal{T}\},$$

$$\mathcal{R}i = *_{\widetilde{g}}^{-1}[\chi, *_{\widetilde{g}}\mathcal{B}\} = (-1)^{n+k+l+m} R_{\alpha\beta} g^{\alpha\widehat{\beta}} e_{\widehat{\beta}} \delta u^\beta$$

are constructed respectively by using the ds–torsions (6) and Ricci ds–tensors (8).

Let us introduce the locally anisotropic supersymmetric matter source $\mathcal{J}$ constructed by using the same formulas from (12) when instead of $R_{\alpha\beta}$ is taken $k_1(\Upsilon_{\alpha\beta} - \frac{1}{2} g_{\alpha\beta}\Upsilon) - \lambda\left(g_{\alpha\beta} - \frac{1}{2} g_{\alpha\beta}\delta^\tau_\tau\right)$. By straightforward calculations we can proof [3,4] that the Yang–Mills equations

$$\triangle\overline{\mathcal{B}} = \overline{\mathcal{J}} \tag{13}$$

for d–connection $\overline{\mathcal{F}} = (\mathcal{F}, \chi)$ in s–bundle $\mathcal{AN}\left(\widetilde{E}\right)$, projected on the base s–manifold, are equivalent to the Einstein equations (9) on $\widetilde{E}$. We emphasize that the equations (13) were introduced in a "pure" geometric manner by using operators $*, \widehat{\delta}$ and the horizontal projection $\widehat{H}$ but such gauge s–field equations are not variational because of degeneration of the Killing s–form. To construct a variational gauge like supersymmetric la–supergravitational model is possible, for instance, by considering a minimal extension of the gauge s–group $AF_{n,k}^{m,l}(\Lambda)$ to the de Sitter s–group $S_{n,k}^{m,l}(\Lambda) = SO_{n,k}^{m,l}(\Lambda)$, acting on the s–space $\Lambda_{n,k}^{m,l} \oplus \Lambda$ and formulating a nonlinear version of de Sitter gauge s–gravity.

There are analyzed models of supergravity with generic local anisotropy [13] when instead of s–field equations and constraints (9) there are considered an anholonomic generalization of the Wess–Zumino supergravity and some variants induced in low energy limit from superstring theory. The N–connection s–field allows us to model generic la–interactions with dynamics and constraints induced by nontrivial (not only via toroidal compactifications) from higher dimensions and this results in a geometrical unification of the so–called generalized Finsler–Kaluza–Klein theories.

3. *–Products and Enveloping Algebras in Noncommutative Spaces

For a noncommutative space the coordinates $\hat{u}^i$, $(i = 1, ..., N)$ satisfy some noncommutative relations of type

$$[\hat{u}^i, \hat{u}^j] = \begin{cases} i\theta^{ij}, & \theta^{ij} \in \mathbb{C}, \text{ canonical structure;} \\ if_k^{ij}\hat{u}^k, & f_k^{ij} \in \mathbb{C}, \text{ Lie structure;} \\ iC_{kl}^{ij}\hat{u}^k\hat{u}^l, & C_{kl}^{ij} \in \mathbb{C}, \text{ quantum plane structure} \end{cases} \quad (14)$$

where $\mathbb{C}$ denotes the complex number field.

The noncommutative space is modeled as the associative algebra of $\mathbb{C}$; this algebra is freely generated by the coordinates modulo ideal $\mathcal{R}$ generated by the relations (one accepts formal power series) $\mathcal{A}_u = \mathbb{C}[[\hat{u}^1, ..., \hat{u}^N]]/\mathcal{R}$. One restricts attention [6] to algebras having the (so–called, Poincare–Birkhoff–Witt) property that any element of $\mathcal{A}_u$ is defined by its coefficient function and vice versa,

$$\hat{f} = \sum_{L=0}^{\infty} f_{i_1,...,i_L} : \hat{u}^{i_1} ... \hat{u}^{i_L} : \quad \text{when } \hat{f} \sim \{f_i\},$$

where $: \hat{u}^{i_1} ... \hat{u}^{i_L} :$ denotes that the basis elements satisfy some prescribed order (for instance, the normal order $i_1 \leq i_2 \leq ... \leq i_L$, or, another example, are totally symmetric). The algebraic properties are all encoded in the so–called diamond ($\Diamond$) product which is defined by

$$\hat{f}\hat{g} = \hat{h} \sim \{f_i\} \Diamond \{g_i\} = \{h_i\}.$$

In the mentioned approach to every function $f(u) = f(u^1, ..., u^N)$ of commuting variables $u^1, ..., u^N$ one associates an element of algebra $\hat{f}$ when the commuting variables are substituted by anticommuting ones,

$$f(u) = \sum f_{i_1 \dots i_L} u^1 \cdots u^N \to \hat{f} = \sum_{L=0}^{\infty} f_{i_1, \dots, i_L} : \hat{u}^{i_1} \cdots \hat{u}^{i_L} :$$

when the $\lozenge$–product leads to a bilinear $*$–product of functions (see details in [7])

$$\{f_i\} \lozenge \{g_i\} = \{h_i\} \sim (f * g)(u) = h(u).$$

The $*$–product is defined respectively for the cases (14)

$$f * g = \begin{cases} \exp[\frac{i}{2} \frac{\partial}{\partial u^i} \theta^{ij} \frac{\partial}{\partial u'^j}] f(u) g(u')|_{u' \to u}, & \text{canonical str.;} \\ \exp[\frac{i}{2} u^k g_k(i \frac{\partial}{\partial u'}, i \frac{\partial}{\partial u''})] f(u') g(u'')|_{u'' \to u}^{u' \to u}, & \text{Lie str.;} \\ q^{\frac{1}{2}(-u' \frac{\partial}{\partial u'} v \frac{\partial}{\partial v} + u \frac{\partial}{\partial u} v' \frac{\partial}{\partial v'})} f(u, v) g(u', v')|_{v' \to v}^{u' \to u}, & \text{quantum plane,} \end{cases}$$

where there are considered values of type

$$e^{ik_n \hat{u}^n} e^{ip_{nl} \hat{u}^n} = e^{i\{k_n + p_n + \frac{1}{2} g_n(k,p)\} \hat{u}^n},$$ (15)

$$g_n(k,p) = -k_i p_j f_n^{ij} + \frac{1}{6} k_i p_j (p_k - k_k) f_m^{ij} f_n^{mk} + \cdots,$$

$$e^A e^B = e^{A + B + \frac{1}{2}[A,B] + \frac{1}{12}([A,[A,B]] + [B,[B,A]])} + \cdots$$

and for the coordinates on quantum (Manin) planes one holds the relation $uv = qvu$.

A non–abelian gauge theory on a noncommutative space is given by two algebraic structures, the algebra $\mathcal{A}_u$ and a non–abelian Lie algebra $\mathcal{A}_I$ of the gauge group with generators $I^1, \dots, I^S$ and the relations

$$[I^{\underline{s}}, I^{\underline{p}}] = i f_{\underline{t}}^{sp} I^t.$$ (16)

In this case both algebras are treated on the same footing and one denotes the generating elements of the big algebra by $\hat{u}^i$,

$$\hat{z}^i = \{\hat{u}^1, \dots, \hat{u}^N, I^1, \dots, I^S\},$$
$$\mathcal{A}_z = \mathbb{C}[[\hat{u}^1, \dots, \hat{u}^{N+S}]]/\mathcal{R},$$

and the $*$–product formalism is to be applied for the whole algebra $\mathcal{A}_z$ when there are considered functions of the commuting variables u^i ($i, j, k, \dots = 1, \dots, N$) and $I^{\underline{s}}$ ($s, p, \dots = 1, \dots, S$).

For instance, in the case of a canonical structure for the space variables u^i we have

$$(F * G)(u) = e^{\frac{i}{2} \left(\theta^{ij} \frac{\partial}{\partial u'^i} \frac{\partial}{\partial u''^j} + t^s g_s(i \frac{\partial}{\partial t'}, i \frac{\partial}{\partial t''}) \right)} F(u', t') G(u'', t'')|_{t' \to t, t'' \to t}^{u' \to u, u'' \to u}.$$ (17)

This formalism was developed in [6] for general Lie algebras. In this paper we shall consider those cases when in the commuting limit one obtains the gauge gravity and general relativity theories.

4. Enveloping Algebras for Gravitational Gauge Connections

To define gauge gravity theories on noncommutative space we first introduce gauge fields as elements the algebra $\mathcal{A}_u$ that form representation of the generator I–algebra for the de Sitter gauge group. For commutative spaces it is known [9, 11, 16] that an equivalent reexpression of the Einstein theory as a gauge like theory implies, for both locally isotropic and anisotropic spacetimes, the nonsemisimplicity of the gauge group, which leads to a nonvariational theory in the total space of the bundle of locally adapted affine frames (to this class one belong the gauge Poincare theories; on metric–affine and gauge gravity models see original results and reviews in [12]). By using auxililiary biliniear forms, instead of degenerated Killing form for the affine structural group, on fiber spaces, the gauge models of gravity can be formulated to be variational. After projection on the base spacetime, for the so–called Cartan connection form, the Yang–Mills equations transforms equivalently into the Einstein equations for general relativity [9]. A variational gauge gravitational theory can be also formulated by using a minimal extension of the affine structural group $\mathcal{A}f_{3+1}(\mathcal{R})$ to the de Sitter gauge group $S_{10} = SO(4+1)$ acting on $\mathcal{R}^{4+1}$ space. For simplicity, in this paper we restrict our consideration only with the even components of frames, connections and curvatures of gauge la–supergavity outlined in previous section.

Let now consider a noncommutative space. In this case the gauge fields are elements of the algebra $\widehat{\psi} \in \mathcal{A}_I^{(dS)}$ that form the nonlinear representation of the de Sitter Lie algebra $so_{(\eta)}(5)$ when the whole algebra is denoted $\mathcal{A}_z^{(dS)}$. Under a nonlinear de Sitter transformation the elements transform as follows

$$\delta\widehat{\psi} = i\widehat{\gamma}\widehat{\psi}, \widehat{\psi} \in \mathcal{A}_u, \widehat{\gamma} \in \mathcal{A}_z^{(dS)}.$$

So, the action of the generators on $\widehat{\psi}$ is defined as this element is supposed to form a nonlinear representation of $\mathcal{A}_I^{(dS)}$ and, in consequence, $\delta\widehat{\psi} \in \mathcal{A}_u$ despite $\widehat{\gamma} \in \mathcal{A}_z^{(dS)}$. It should be emphasized that independent of a representation the object $\widehat{\gamma}$ takes values in enveloping de Sitter algebra and not in a Lie algebra as would be for commuting spaces. The same holds for the connections that we introduce (similarly to [7]) in order to define covariant coordinates

$$\widehat{U}^\nu = \widehat{u}^\nu + \widehat{\Gamma}^\nu, \widehat{\Gamma}^\nu \in \mathcal{A}_z^{(dS)}.$$

The values $\widehat{U}^\nu\widehat{\psi}$ transforms covariantly, $\delta\widehat{U}^\nu\widehat{\psi} = i\widehat{\gamma}\widehat{U}^\nu\widehat{\psi}$, if and only if the connection $\widehat{\Gamma}^\nu$ satisfies the transformation law of the enveloping nonlinear realized de Sitter algebra,

$$\delta\widehat{\Gamma}^\nu\widehat{\psi} = -i[\widehat{u}^\nu, \widehat{\gamma}] + i[\widehat{\gamma}, \widehat{\Gamma}^\nu],$$

where $\delta\widehat{\Gamma}^\nu \in \mathcal{A}_z^{(dS)}$. The enveloping algebra–valued connection has infinitely many component fields. Nevertheless, it was shown that all the component fields can be induced from a Lie algebra–valued connection by a Seiberg–Witten map ([10, 5, 6] and [1] for $SO(n)$ and $Sp(n)$). In this subsection we show that similar constructions could be proposed for nonlinear realizations of de Sitter algebra when the transformation of the connection is considered

$$\delta\widehat{\Gamma}^\nu = -i[u^\nu, {}^* \widehat{\gamma}] + i[\widehat{\gamma}, {}^* \widehat{\Gamma}^\nu].$$

For simplicity, we treat in more detail the canonical case with the star product (17). The first term in the variation $\delta\widehat{\Gamma}^\nu$ gives

$$-i[u^\nu, {}^* \widehat{\gamma}] = \theta^{\nu\mu}\frac{\partial}{\partial u^\mu}\gamma.$$

Assuming that the variation of $\widehat{\Gamma}^\nu = \theta^{\nu\mu}Q_\mu$ starts with a linear term in θ we have

$$\delta\widehat{\Gamma}^\nu = \theta^{\nu\mu}\delta Q_\mu, \quad \delta Q_\mu = \frac{\partial}{\partial u^\mu}\gamma + i[\widehat{\gamma}, {}^* Q_\mu].$$

We follow the method of calculation from the papers [7, 6] and expand the star product (17) in θ but not in g_a and find to first order in θ,

$$\gamma = \gamma_{\underline{a}}^1 I^{\underline{a}} + \gamma_{\underline{ab}}^1 I^{\underline{a}}I^{\underline{b}} + ..., \text{ and } Q_\mu = q_{\mu,\underline{a}}^1 I^{\underline{a}} + q_{\mu,\underline{ab}}^2 I^{\underline{a}}I^{\underline{b}} + ... \qquad (18)$$

where $\gamma_{\underline{a}}^1$ and $q_{\mu,\underline{a}}^1$ are of order zero in θ and $\gamma_{\underline{ab}}^1$ and $q_{\mu,\underline{ab}}^2$ are of second order in θ. The expansion in $I^{\underline{b}}$ leads to an expansion in g_a of the $*$–product because the higher order $I^{\underline{b}}$–derivatives vanish. For de Sitter case as $I^{\underline{b}}$ we take the generators, see commutators (16), with the corresponding de Sitter structure constants $f_{\underline{d}}^{\underline{bc}} \simeq f_{\underline{\beta}}^{\alpha\beta}$ (in our further identifications with spacetime objects like frames and connections we shall use Greek indices).

The result of calculation of variations of (18), by using g_a to the order given in (15), is

$$\delta q_{\mu,\underline{a}}^1 = \frac{\partial\gamma_{\underline{a}}^1}{\partial u^\mu} - f_{\underline{a}}^{bc}\gamma_{\underline{b}}^1 q_{\mu,\underline{c}}^1,$$

$$\delta Q_\tau = \theta^{\mu\nu}\partial_\mu\gamma_{\underline{a}}^1\partial_\nu q_{\tau,\underline{b}}^1 I^{\underline{a}}I^{\underline{b}} + ...,$$

$$\delta q_{\mu,\underline{ab}}^2 = \partial_\mu\gamma_{\underline{ab}}^2 - \theta^{\nu\tau}\partial_\nu\gamma_{\underline{a}}^1\partial_\tau q_{\mu,\underline{b}}^1 - 2f_{\underline{a}}^{bc}\{\gamma_{\underline{b}}^1 q_{\mu,\underline{cd}}^2 + \gamma_{\underline{bd}}^2 q_{\mu,\underline{c}}^1\}.$$

Next we introduce the objects ε, taking the values in de Sitter Lie algebra and W_μ, being enveloping de Sitter algebra valued,

$$\varepsilon = \gamma_{\underline{a}}^1 I^{\underline{a}} \text{ and } W_\mu = q_{\mu,\underline{ab}}^2 I^{\underline{a}}I^{\underline{b}}$$

with the variation δW_μ satisfying the equation [7, 6]

$$\delta W_\mu = \partial_\mu(\gamma^2_{\underline{ab}}I^{\underline{a}}I^{\underline{b}}) - \frac{1}{2}\theta^{\tau\lambda}\{\partial_\tau\varepsilon, \partial_\lambda q_\mu\} + i[\varepsilon, W_\mu] + i[(\gamma^2_{\underline{ab}}I^{\underline{a}}I^{\underline{b}}), q_\nu]. \quad (19)$$

The equation (19) has the solution (found in [7, 10])

$$\gamma^2_{\underline{ab}} = \frac{1}{2}\theta^{\nu\mu}(\partial_\nu\gamma^1_{\underline{a}})q^1_{\mu,\underline{b}}, \text{ and } q^2_{\mu,\underline{ab}} = -\frac{1}{2}\theta^{\nu\tau}q^1_{\nu,\underline{a}}\left(\partial_\tau q^1_{\mu,\underline{b}} + R^1_{\tau\mu,\underline{b}}\right)$$

where $R^1_{\tau\mu,\underline{b}} = \partial_\tau q^1_{\mu,\underline{b}} - \partial_\mu q^1_{\tau,\underline{b}} + f^{\underline{ec}}_{\underline{d}}q^1_{\tau,\underline{e}}q^1_{\mu,\underline{e}}$ can be identified with the coefficients $\mathcal{R}^{\underline{\alpha}}_{\ \underline{\beta}\mu\nu}$ of de Sitter nonlinear gauge gravity curvature if in the commutative limit

$$q^1_{\mu,\underline{b}} \simeq \begin{pmatrix} \Gamma^{\underline{\alpha}}_{\ \underline{\beta}} & l_0^{-1}\chi^{\underline{\alpha}} \\ l_0^{-1}\chi_{\underline{\beta}} & 0 \end{pmatrix}.$$

The presented procedure can be generalized to all higher powers of θ [6].

5. Noncommutative Gauge Gravity Covariant Dynamics

The constructions from the previous section are summarized by the conclusion that the de Sitter algebra valued object $\varepsilon = \gamma^1_{\underline{a}}(u)I^{\underline{a}}$ determines all the terms in the enveloping algebra

$$\gamma = \gamma^1_{\underline{a}}I^{\underline{a}} + \frac{1}{4}\theta^{\nu\mu}\partial_\nu\gamma^1_{\underline{a}}\,q^1_{\mu,\underline{b}}\left(I^{\underline{a}}I^{\underline{b}} + I^{\underline{b}}I^{\underline{a}}\right) + \dots$$

and the gauge transformations are defined by $\gamma^1_{\underline{a}}(u)$ and $q^1_{\mu,\underline{b}}(u)$, when

$$\delta_{\gamma^1}\psi = i\gamma\left(\gamma^1, q^1_\mu\right) * \psi.$$

For de Sitter enveloping algebras one holds the general formula for compositions of two transformations

$$\delta_\gamma\delta_\varsigma - \delta_\varsigma\delta_\gamma = \delta_{i(\varsigma*\gamma-\gamma*\varsigma)}$$

which holds also for the restricted transformations defined by γ^1,

$$\delta_{\gamma^1}\delta_{\varsigma^1} - \delta_{\varsigma^1}\delta_{\gamma^1} = \delta_{i(\varsigma^1*\gamma^1-\gamma^1*\varsigma^1)}.$$

Applying the formula (17) we computer

$$[\gamma, * \varsigma] = i\gamma^1_{\underline{a}}\varsigma^1_{\underline{b}}f^{\underline{ab}}_{\underline{c}}I^{\underline{c}} + \frac{i}{2}\theta^{\nu\mu}\{\partial_\nu\left(\gamma^1_{\underline{a}}\varsigma^1_{\underline{b}}f^{\underline{ab}}_{\underline{c}}\right)q_{\mu,\underline{c}} \\ + \left(\gamma^1_{\underline{a}}\partial_\nu\varsigma^1_{\underline{b}} - \varsigma^1_{\underline{a}}\partial_\nu\gamma^1_{\underline{b}}\right)q_{\mu,\underline{b}}f^{\underline{ab}}_{\underline{c}} + 2\partial_\nu\gamma^1_{\underline{a}}\partial_\mu\varsigma^1_{\underline{b}}\}I^{\underline{d}}I^{\underline{c}}.$$

Such commutators could be used for definition of tensors [7]

$$\widehat{S}^{\mu\nu} = [\widehat{U}^\mu, \widehat{U}^\nu] - i\widehat{\theta}^{\mu\nu}, \quad (20)$$

where $\hat{\theta}^{\mu\nu}$ is respectively stated for the canonical, Lie and quantum plane structures. Under the general enveloping algebra one holds the transform

$$\delta \hat{S}^{\mu\nu} = i[\hat{\gamma}, \hat{S}^{\mu\nu}].$$

For instance, the canonical case is characterized by

$$
\begin{aligned}
S^{\mu\nu} &= i\theta^{\mu\tau}\partial_\tau\Gamma^\nu - i\theta^{\nu\tau}\partial_\tau\Gamma^\mu + \Gamma^\mu * \Gamma^\nu - \Gamma^\nu * \Gamma^\mu \\
&= \theta^{\mu\tau}\theta^{\nu\lambda}\{\partial_\tau Q_\lambda - \partial_\lambda Q_\tau + Q_\tau * Q_\lambda - Q_\lambda * Q_\tau\}.
\end{aligned}
$$

By introducing the gravitational gauge strength (curvature)

$$R_{\tau\lambda} = \partial_\tau Q_\lambda - \partial_\lambda Q_\tau + Q_\tau * Q_\lambda - Q_\lambda * Q_\tau, \tag{21}$$

which could be treated as a noncommutative extension of de Sitter nonlinear gauge gravitational curvature one computers

$$R_{\tau\lambda,\underline{a}} = R^1_{\tau\lambda,\underline{a}} + \theta^{\mu\nu}\{R^1_{\tau\mu,\underline{a}}R^1_{\lambda\nu,\underline{b}} - \frac{1}{2}q^1_{\mu,\underline{a}}\left[(D_\nu R^1_{\tau\lambda,\underline{b}}) + \partial_\nu R^1_{\tau\lambda,\underline{b}}\right]\}I^{\underline{b}},$$

where the gauge gravitation covariant derivative is introduced,

$$(D_\nu R^1_{\tau\lambda,\underline{b}}) = \partial_\nu R^1_{\tau\lambda,\underline{b}} + q_{\nu,\underline{c}}R^1_{\tau\lambda,\underline{d}}f^{\underline{cd}}_{\underline{b}}.$$

Following the gauge transformation laws for γ and q^1 we find

$$\delta_{\gamma^1} R^1_{\tau\lambda} = i\left[\gamma, {}^* R^1_{\tau\lambda}\right]$$

with the restricted form of γ.

Such formulas were proved in references [6, 10] for usual gauge (nongravitational) fields. Here we reconsidered them for gravitational gauge fields.

Following the nonlinear realization of de Sitter algebra and the *–formalism we can formulate a dynamics of noncommutative spaces. Derivatives can be introduced in such a way that one does not obtain new relations for the coordinates. In this case a Leibniz rule can be defined [6] that

$$\hat{\partial}_\mu \hat{u}^\nu = \delta^\nu_\mu + d^{\nu\tau}_{\mu\sigma}\, \hat{u}^\sigma\, \hat{\partial}_\tau,$$

where the coefficients $d^{\nu\tau}_{\mu\sigma} = \delta^\nu_\sigma \delta^\tau_\mu$ are chosen to have not new relations when $\hat{\partial}_\mu$ acts again to the right hand side. In consequence one holds the *–derivative formulas

$$\partial_\tau * f = \frac{\partial}{\partial u^\tau}f + f * \partial_\tau,$$

$$[\partial_\iota, {}^* (f * g)] = ([\partial_\iota, {}^* f]) * g + f * ([\partial_\iota, {}^* g])$$

and the Stokes theorem $\int [\partial_l, f] = \int d^N u [\partial_l, {}^* f] = \int d^N u \frac{\partial}{\partial u^l} f = 0$, where, for the canonical structure, the integral is defined,

$$\int \widehat{f} = \int d^N u f \left(u^1, ..., u^N \right).$$

An action can be introduced by using such integrals. For instance, for a tensor of type (20), when $\delta \widehat{L} = i \left[\widehat{\gamma}, \widehat{L} \right]$, we can define a gauge invariant action

$$W = \int d^N u \, Tr \widehat{L}, \ \delta W = 0,$$

were the trace has to be taken for the group generators.

For the nonlinear de Sitter gauge gravity a proper action is

$$L = \frac{1}{4} R_{\tau \lambda} R^{\tau \lambda},$$

where $R_{\tau \lambda}$ is defined by the even part of (11). In this case the dynamic of noncommutative space is entirely formulated in the framework of quantum field theory of gauge fields. The method works for matter fields as well to restrictions to the general relativity theory (see references [11, 9]).

6. Acknowledgment

The first author (S. V.) is grateful to the organizers of NATO ARW in Kiev, where the results of this work were communicated, for kind hospitality and support.

References

1. L. Bonora, M. Schnabl, M. M. Sheikh–Jabbari and A. Tomasiello, *Noncommutative SO(n) and Sp(n) gauge theories*, hep–th/0006091.
2. A. Connes, J. Math. Phys. 36 (1995) 6194; A. H. Chamseddine, A. Connes, Phys. Rev. Letters 77 (1996) 4868; L. Carminati, B. Iochum, T. Schöcer, *Noncommutative Yang–Mills and noncommutative relativity: a bridge over troubled water*, hep–th/9706105.
3. A. Connes, M. R. Douglas and A. Schwarz, JHEP **9802** (1998) 003, hep–th/0001203.
4. E. Hawkins, Comm. Math. Phys. 187 (1997) 471; gr–qc / 9605068; G. Landi, N. A. Viet, K. C. Wali, Phys. Lett. B326 (1994) 45; hep–th / 9402046; A, Sitarz, Class. Quant. Grav. 11 (1994) 2127; hep–th / 9401145; J. Madore and J. Mourad, Int. J. Mod. Phys. D3 (1994) 221.
5. B. Jurčo and P. Schupp, Eur. Phys. J. C **14**, (2000) 367, hep–th/0001032.
6. B. Jurčo, S. Schraml, P. Shupp and J. Wess, *Enveloping algebra valued gauge transformations for non–abelian gauge groups on non–commutative spaces*, hep–th/0006246
7. J. Madore, S. Schraml, P. Schupp and J. Wess, *Gauge theory on noncommutative spaces*, Eur. Phys. J. C, in press, hep–th/0001203.
8. S. Majid, *Conceptual Issues for Noncommutative gravity and algebras and finite sets*, math.QA / 0006152; A. H. Chamseddine, *Complexified gravity and noncommutative spaces*, hep–th / 0005222; R. Kerner, *Noncommutative extensions of classical theories in physics*, hep–th / 0004033.

9. D. A. Popov, Theor. Math. Phys. **24** (1975) 347; D. A. Popov and L. I. Dikhin, Doklady Akademii Nauk SSSR **245** (1975) 347 [in Russian].

10. N. Seiberg and E. Witten, JHEP **9909** (1999) 032, hep–th/9908142.

11. A. A. Tseytlin, Phys. Rev. D **26** (1982) 3327.

12. R. Utiyama, Phys. Rev. **101** (1956) 1597; V. N. Ponomariov, A. Barvinsky and Yu. N. Obukhov, *Geometrodynamical Methods and the Gauge Approach to the Gravitational Interactions* (Energoatomizdat, Moscow, 1985); E. W. Mielke, *Geometrodynamics of Gauge Fields – on the Geometry of Yang–Mills and Gravitational Gauge Theories* (Academic–Verlag, Berlin, 1987); F. Hehl, J. D. McGrea, E. W. Mielke and Y. Ne'eman, Phys. Rep. **258** (1995) 1; H. Dehnen and E. Hitzer, *Int. J. Theor. Phys.* **34** (1995) 1981.

13. S. Vacaru, Nucl. Phys. B **434** (1997) 590; gr–qc/9604016; hep–th / 9604194–9604196; Interactions, Strings and Isotopies in Higher Order Anisotropic Superspaces (Hadronic Press, 1998).

14. S. Vacaru, *Gauge and Eistein gravity from non–Abelian gauge models on noncommutative spaces,* hep-th/0009163

15. S. Vacaru and H. Dehnen, *Locally Anisotropic Structures and Nonlinear Connections in Einstein and Gauge Gravity,* gr–qc / 0009039.

16. S. Vacaru and Yu. Goncharenko, Int. J. Theor. Phys. **34** (1995) 1955.

17. J. Wess and B. Zumino, Nucl. Phys. Proc. Suppl. **18B** (1991) 302; J. Wess, in *Proceeding of the 38 International Universitätswochen für Kern– und Teilchenphysik*, no. 543 in Lect. Notes in Phys., Springer–Verlag, 2000, Schladming, January 1999, eds. H. Gusterer, H. Grosse and L. Pitner; math–ph / 9910013.

18. H. Weyl, Z. Physik **46** (1927) 1; *The theory of groups and quantum mechanics* (Dover, New–York, 1931), translated from *Gruppentheorie and Quantenmechanik* (Hirzel Verlag, Leipzig, 1928); E. P. Wigner, Phys. Rev. **40** (1932) 749; J. E. Moyal, Proc. Cambridge Phil. Soc. **45** (1949) 99; F. Bayen, M. Flato, C. Fronsdal, A. Lichnerowicz, D. Sternheimer, Ann. Physics **111** (1978) 61; M. Kontsevitch, *Deformation quantization of Poisson manifolds, I.* q–alg/9709040; D. Sternheimer, *Deformation Quantization: Twenty Years After,* math / 9809056.

FINITENESS IN CONVENTIONAL $N = 1$ GUTS

TATSUO KOBAYASHI *
Dept. of Phys, Kyoto Univ., Kyoto 606-8502, Japan

JISUKE KUBO †
Dept. of Physics, Kanazawa Univ., Kanazawa 920-1192, Japan

MYRIAM MONDRAGÓN ‡
Inst. de Física, UNAM, Apdo. Postal 20-364, México 01000 D.F., México

GEORGE ZOUPANOS §
Physics Dept., Nat. Technical Univ., GR-157 80 Zografou, Athens, Greece

Abstract. Finite Unified Theories (FUTs) are $N = 1$ supersymmetric GUT's which have the remarkable feature of being all-loop finite beyond the unification point. They also have impressive predictive power. We present here a review of the recent developments of the softly broken sector of $N = 1$ FUTs. The new characteristic predictions of FUTs are: 1) The lightest Higgs boson mass is predicted to be in the window 120-130 GeV, in case the LSP is neutralino, while in case the LSP is the $\tilde{\tau}$ (which can be consistently accommodated in presence of bilinear R-parity violating terms) it can be as light as 111 GeV. 2) The s-spectrum starts above several hundreds of GeV.

1. Introduction

In recent years new frameworks have been developed aiming to provide a unified description of all interactions including gravity. Theories based on superstrings, non-commutative geometry and quantum groups, although at a different stage of development in each area, have common unification targets and share similar hopes for exhibiting improved renormalization properties in the ultraviolet as compared to ordinary field theories. Moreover, recent progress shows that all above theoretical endeavors could be related and thus they might be understood

* kobayash@gauge.scphys.kyoto-u.ac.jp
† jik@hep.s.kanazawa-u.ac.jp
‡ myriam@ft.ifisicacu.unam.mx
§ george.zoupanos@cern.ch

S. Duplij and J. Wess (eds.), Noncommutative Structures in Mathematics and Physics, 245–257.
© 2001 *Kluwer Academic Publishers. Printed in the Netherlands.*

in a unified manner too. However in spite the importance of having frameworks to discuss quantum gravity in a self consistent way, the main goal expected from a unified description of interactions by the particle physics community is to understand the present day free parameters of the Standard Model (SM) in terms of a few fundamental ones, or in other words to achieve *reduction of couplings* at a more fundamental level. Unfortunately all the above theoretical frameworks did not offer anything in the understanding of the free parameters of the SM, and in the best case they have managed to accommodate earlier tools such as supersymmetry and ideas like Grand Unified Theories (GUTs) but without providing any further predictive power in these constructions.

In our recent studies [1]-[6], [7] we have developed a complementary strategy in searching for a more fundamental theory possibly at the Planck scale, whose basic ingredients are GUTs and supersymmetry, but its consequences certainly go beyond the known ones. Our method consists of hunting for renormalization group invariant (RGI) relations holding below the Planck scale, which in turn are preserved down to the GUT scale. This programme, called Gauge–Yukawa unification scheme, applied in the dimensionless couplings of supersymmetric GUTs, such as gauge and Yukawa couplings, had already noticable successes by predicting correctly, among others, the top quark mass in the finite and in the minimal $N = 1$ supersymmetric SU(5) GUTs. An impressive aspect of the RGI relations is that one can guarantee their validity to all-orders in perturbation theory by studying the uniqueness of the resulting relations at one-loop, as was proven in the early days of the programme of *reduction of couplings* [8]. Even more remarkable is the fact that it is possible to find RGI relations among couplings that guarantee finiteness to all-orders in perturbation theory [9, 10].

Although supersymmetry seems to be an essential feature for a successful realization of the above programme, its breaking has to be understood too, since it has the ambition to supply the SM with predictions for several of its free parameters. Indeed, the search for RGI relations has been extended to the soft supersymmetry breaking sector (SSB) of these theories [4, 11], which involves parameters of dimension one and two. More recently a very interesting progress has been made [12]- [17] concerning the renormalization properties of the SSB parameters based conceptually and technically on the work of ref. [18]. In ref. [18] the powerful supergraph method [19] for studying supersymmetric theories has been applied to the softly broken ones by using the "spurion" external space-time independent superfields [20]. In the latter method a softly broken supersymmetric gauge theory is considered as a supersymmetric one in which the various parameters such as couplings and masses have been promoted to external superfields that acquire "vacuum expectation values". Based on this method the relations among the soft term renormalization and that of an unbroken supersymmetric theory have been derived. In particular the β-functions of the parameters of the softly broken theory are expressed in terms of partial differential operators involving the dimensionless parameters of the unbroken theory. The key point in the strategy of refs. [15]-[17] in solving the set of coupled differential equations so as to be able to express all parameters in a RGI way, was to transform the partial differential operators involved to total derivative operators. This is indeed possible to be done on the RGI surface which is defined by the solution of the reduction equations.

On the phenomenological side there exist some serious developments too. Previously an appealing "universal" set of soft scalar masses was asummed in the SSB sector of supersymmetric theories, given that apart from economy and simplicity (1) they are part of the constraints that preserve finiteness up to two-loops [21, 22], (2) they are RGI up to two-loops in more general supersymmetric gauge theories, subject to the condition known as $P = 1/3\, Q$ [11] and (3) they appear in the attractive dilaton dominated supersymmetry breaking superstring scenarios [23]. However, further studies have exhibited a number of problems all due to the restrictive nature of the "universality" assumption for the soft scalar masses. For instance (a) in finite unified theories the universality predicts that the lightest supersymmetric particle is a charged particle, namely the superpartner of the τ lepton $\tilde{\tau}$ (b) the MSSM with universal soft scalar masses is inconsistent with the attractive radiative electroweak symmetry breaking [24] and (c) which is the worst of all, the universal soft scalar masses lead to charge and/or colour breaking minima deeper than the standard vacuum [25]. Therefore, there have been attempts to relax this constraint without loosing its attractive features. First an interesting observation was made that in $N = 1$ Gauge–Yukawa unified theories there exists a RGI sum rule for the soft scalar masses at lower orders; at one-loop for the non-finite case [5] and at two-loops for the finite case [6]. The sum rule manages to overcome the above unpleasant phenomenological consequences. Moreover it was proven [17] that the sum rule for the soft scalar masses is RGI to all-orders for both the general as well as for the finite case. Finally the exact β-function for the soft scalar masses in the Novikov-Shifman-Vainstein-Zakharov (NSVZ) scheme [26] for the softly broken supersymmetric QCD has been obtained [17]. Armed with the above tools and results we are in a position to study the spectrum of the full finite and minimal supersymmetric SU(5) models in terms of few free parameters with emphasis on the predictions for the masses of the lightest Higgs and LSP and on the constraints imposed by having a large $\tan\beta$.

2. Reduction of Couplings and Finiteness in $N = 1$ SUSY Gauge Theories

A RGI relation among couplings, $\Phi(g_1, \cdots, g_N) = 0$, has to satisfy the partial differential equation (PDE) $\mu\, d\Phi/d\mu = \sum_{i=1}^{N} \beta_i\, \partial\Phi/\partial g_i = 0$, where β_i is the β-function of g_i. There exist $(N - 1)$ independent Φ's, and finding the complete set of these solutions is equivalent to solve the so-called reduction equations (REs), $\beta_g\, (dg_i/dg) = \beta_i$, $i = 1, \cdots, N$, where g and β_g are the primary coupling and its β-function. Using all the $(N - 1)$ Φ's to impose RGI relations, one can in principle express all the couplings in terms of a single coupling g. The complete reduction, which formally preserves perturbative renormalizability, can be achieved by demanding a power series solution, whose uniqueness can be investigated at the one-loop level. The completely reduced theory contains only one independent coupling with the corresponding β-function. This possibility of coupling unification is attractive, but it can be too restrictive and hence unrealistic. In practice one may use fewer Φ's as RGI constraints.

It is clear by examining specific examples, that the various couplings in supersymmetric theories have easily the same asymptotic behaviour. Therefore searching for a power series solution to the REs is justified. This is not the case in non-supersymmetric theories.

Let us then consider a chiral, anomaly free, $N = 1$ globally supersymmetric gauge theory based on a group G with gauge coupling constant g. The superpotential of the theory is given by

$$W = \frac{1}{2} m^{ij} \Phi_i \Phi_j + \frac{1}{6} C^{ijk} \Phi_i \Phi_j \Phi_k , \tag{1}$$

where m^{ij} and C^{ijk} are gauge invariant tensors and the matter field Φ_i transforms according to the irreducible representation R_i of the gauge group G.

The one-loop β-function of the gauge coupling g is given by

$$\beta_g^{(1)} = \frac{dg}{dt} = \frac{g^3}{16\pi^2} \left[\sum_i l(R_i) - 3\, C_2(G) \right] , \tag{2}$$

where $l(R_i)$ is the Dynkin index of R_i and $C_2(G)$ is the quadratic Casimir of the adjoint representation of the gauge group G. The β-functions of C^{ijk}, by virtue of the non-renormalization theorem, are related to the anomalous dimension matrix γ_i^j of the matter fields Φ_i as

$$\beta_C^{ijk} = \frac{d}{dt} C^{ijk} = C^{ijp} \sum_{n=1} \frac{1}{(16\pi^2)^n} \gamma_p^{k(n)} + (k \leftrightarrow i) + (k \leftrightarrow j) . \tag{3}$$

At one-loop level the γ_i^j are given by

$$\gamma_i^{j(1)} = \frac{1}{2} C_{ipq} C^{jpq} - 2\, g^2\, C_2(R_i)\delta_i^j , \tag{4}$$

where $C_2(R_i)$ is the quadratic Casimir of the representation R_i, and $C^{ijk} = C_{ijk}^*$.

As one can see from Eqs. (2) and (4) all the one-loop β-functions of the theory vanish if $\beta_g^{(1)}$ and $\gamma_i^{j(1)}$ vanish, i.e.

$$\sum_i \ell(R_i) = 3C_2(G), \qquad \frac{1}{2} C_{ipq} C^{jpq} = 2\delta_i^j g^2 C_2(R_i). \tag{5}$$

A very interesting result is that the conditions (5) are necessary and sufficient for finiteness at the two-loop level.

The one- and two-loop finiteness conditions (5) restrict considerably the possible choices of the irreps. R_i for a given group G as well as the Yukawa couplings in the superpotential (1). Note in particular that the finiteness conditions cannot be applied to the supersymmetric standard model (SSM), since the presence of a $U(1)$ gauge group is incompatible with the condition (5), due to $C_2[U(1)] = 0$. This naturally leads to the expectation that finiteness should be attained at the grand unified level only, the SSM being just the corresponding, low-energy, effective theory.

A natural question to ask is what happens at higher loop orders. There exists a very interesting theorem [9] which guarantees the vanishing of the β-functions to all orders in perturbation theory, if we demand reduction of couplings, and that all the one-loop anomalous dimensions of the matter field in the completely and uniquely reduced theory vanish identically.

3. Soft Supersymmetry Breaking-Sum Rule of soft scalar masses

The above described method of reducing the dimensionless couplings has been extended [4] to the soft supersymmetry breaking (SSB) dimensionful parameters of $N = 1$ supersymmetric theories. In addition it was found [5] that RGI SSB scalar masses in Gauge-Yukawa unified models satisfy a universal sum rule. Here we will describe first how the use of the available two-loop RG functions and the requirement of finiteness of the SSB parameters up to this order leads to the soft scalar-mass sum rule [6].

Consider the superpotential given by (1) along with the Lagrangian for SSB terms

$$-\mathcal{L}_{\text{SB}} = \frac{1}{6}h^{ijk}\phi_i\phi_j\phi_k + \frac{1}{2}b^{ij}\phi_i\phi_j + \frac{1}{2}(m^2)^j_i\phi^{*i}\phi_j + \frac{1}{2}M\lambda\lambda + \text{h.c.}, \quad (6)$$

where the ϕ_i are the scalar parts of the chiral superfields Φ_i, λ are the gauginos and M their unified mass. Since we would like to consider only finite theories here, we assume that the gauge group is a simple group and the one-loop β-function of the gauge coupling g vanishes. We also assume that the reduction equations admit power series solutions of the form

$$C^{ijk} = g \sum_{n=0} \rho^{ijk}_{(n)} g^{2n}. \quad (7)$$

According to the finiteness theorem of ref. [9], the theory is then finite to all orders in perturbation theory, if, among others, the one-loop anomalous dimensions $\gamma^{j(1)}_i$ vanish. The one- and two-loop finiteness for h^{ijk} can be achieved by

$$h^{ijk} = -MC^{ijk} + \cdots = -M\rho^{ijk}_{(0)} g + O(g^5). \quad (8)$$

With the above assumptions (and a couple of minor ones [6]) we find the following soft scalar-mass sum rule

$$(m_i^2 + m_j^2 + m_k^2)/MM^\dagger = 1 + \frac{g^2}{16\pi^2} \Delta^{(1)} + O(g^4) \tag{9}$$

for i, j, k with $\rho_{(0)}^{ijk} \neq 0$, where $\Delta^{(1)}$ is the two-loop correction

$$\Delta^{(1)} = -2 \sum_l [(m_l^2/MM^\dagger) - (1/3)] \, T(R_l), \tag{10}$$

which vanishes for the universal choice in accordance with the previous findings of ref. [22].

If we know higher loop β-functions explicitly, we can follow the same procedure and find higher loop RGI relations among SSB terms. However, the β-functions of the soft scalar masses are explicitly known only up to two loops. In order to obtain higher loop results, we need something else instead of knowledge of explicit β-functions, e.g. some relations among β-functions.

The recent progress made using the spurion technique [19, 20] leads to the following all-loop relations among SSB β-functions [12]-[16],

$$\beta_M = 2O \left(\frac{\beta_g}{g} \right), \tag{11}$$

$$\begin{aligned} \beta_h^{ijk} = {}& \gamma^i{}_l h^{ljk} + \gamma^j{}_l h^{ilk} + \gamma^k{}_l h^{ijl} \\ & - 2\gamma_1{}^i{}_l C^{ljk} - 2\gamma_1{}^j{}_l C^{ilk} - 2\gamma_1{}^k{}_l C^{ijl}, \end{aligned} \tag{12}$$

$$(\beta_{m^2})^i{}_j = \left[\Delta + X \frac{\partial}{\partial g} \right] \gamma^i{}_j, \tag{13}$$

$$O = \left(Mg^2 \frac{\partial}{\partial g^2} - h^{lmn} \frac{\partial}{\partial C^{lmn}} \right), \tag{14}$$

$$\Delta = 2OO^* + 2|M|^2 g^2 \frac{\partial}{\partial g^2} + \tilde{C}_{lmn} \frac{\partial}{\partial C_{lmn}} + \tilde{C}^{lmn} \frac{\partial}{\partial C^{lmn}}, \tag{15}$$

where $(\gamma_1)^i{}_j = O\gamma^i{}_j, C_{lmn} = (C^{lmn})^*$, and

$$\tilde{C}^{ijk} = (m^2)^i{}_l C^{ljk} + (m^2)^j{}_l C^{ilk} + (m^2)^k{}_l C^{ijl}. \tag{16}$$

It was also found [16] that the relation

$$h^{ijk} = -M(C^{ijk})' \equiv -M \frac{dC^{ijk}(g)}{d\ln g}, \tag{17}$$

among couplings is all-loop RGI. Furthermore, using the all-loop gauge β-function of Novikov $et\ al.$ [26] given by

$$\beta_g^{\text{NSVZ}} = \frac{g^3}{16\pi^2} \left[\frac{\sum_l T(R_l)(1 - \gamma_l/2) - 3C(G)}{1 - g^2 C(G)/8\pi^2} \right], \tag{18}$$

it was found the all-loop RGI sum rule [17],

$$m_i^2 + m_j^2 + m_k^2 = |M|^2 \{ \frac{1}{1 - g^2 C(G)/(8\pi^2)} \frac{d \ln C^{ijk}}{d \ln g} + \frac{1}{2} \frac{d^2 \ln C^{ijk}}{d(\ln g)^2} \}$$
$$+ \sum_l \frac{m_l^2 T(R_l)}{C(G) - 8\pi^2/g^2} \frac{d \ln C^{ijk}}{d \ln g} . \tag{19}$$

In addition the exact β-function for m^2 in the NSVZ scheme has been obtained [17] for the first time and is given by

$$\beta_{m_i^2}^{\text{NSVZ}} = \left[|M|^2 \{ \frac{1}{1 - g^2 C(G)/(8\pi^2)} \frac{d}{d \ln g} + \frac{1}{2} \frac{d^2}{d(\ln g)^2} \} \right.$$
$$\left. + \sum_l \frac{m_l^2 T(R_l)}{C(G) - 8\pi^2/g^2} \frac{d}{d \ln g} \right] \gamma_i^{\text{NSVZ}} . \tag{20}$$

4. Finite Unified Theories

In this section we examine two concrete $SU(5)$ finite models, where the reduction of couplings in the dimensionless and dimensionful sector has been achieved. A predictive Gauge-Yukawa unified $SU(5)$ model which is finite to all orders, in addition to the requirements mentioned already, should also have the following properties:

1. One-loop anomalous dimensions are diagonal, i.e., $\gamma_i^{(1) j} \propto \delta_i^j$.
2. Three fermion generations, $\bar{5}_i$ ($i = 1, 2, 3$), obviously should not couple to **24**. This can be achieved for instance by imposing $B - L$ conservation.
3. The two Higgs doublets of the MSSM should mostly be made out of a pair of Higgs quintet and anti-quintet, which couple to the third generation.

In the following we discuss two versions of the all-order finite model.
A: The model of ref. [1].
B: A slight variation of the model **A**, whose differences from **A** will become clear in the following.
The superpotential which describes the two models takes the form [1, 6]

$$W = \sum_{i=1}^{3} [\frac{1}{2} g_i^u \, 10_i 10_i H_i + g_i^d \, 10_i \bar{5}_i \, \overline{H}_i] + g_{23}^u \, 10_2 10_3 H_4 \tag{21}$$

$$+ g_{23}^d \, 10_2 \bar{5}_3 \, \overline{H}_4 + g_{32}^d \, 10_3 \bar{5}_2 \, \overline{H}_4 + \sum_{a=1}^{4} g_a^f \, H_a \, 24 \, \overline{H}_a + \frac{g^\lambda}{3} \, (24)^3 ,$$

where H_a and $\overline{H}_a$ ($a = 1, \ldots, 4$) stand for the Higgs quintets and anti-quintets.

The non-degenerate and isolated solutions to $\gamma_i^{(1)} = 0$ for the models {A , B} are:

$$(g_1^u)^2 = \{\frac{8}{5}, \frac{8}{5}\}g^2 \ , \ (g_1^d)^2 = \{\frac{6}{5}, \frac{6}{5}\}g^2 \ , \ (g_2^u)^2 = (g_3^u)^2 = \{\frac{8}{5}, \frac{4}{5}\}g^2 \ , \tag{22}$$

$$(g_2^d)^2 = (g_3^d)^2 = \{\frac{6}{5}, \frac{3}{5}\}g^2 \ , \ (g_{23}^u)^2 = \{0, \frac{4}{5}\}g^2 \ , \ (g_{23}^d)^2 = (g_{32}^d)^2 = \{0, \frac{3}{5}\}g^2 \ ,$$

$$(g^\lambda)^2 = \frac{15}{7}g^2 \ , \ (g_2^f)^2 = (g_3^f)^2 = \{0, \frac{1}{2}\}g^2 \ , \ (g_1^f)^2 = 0 \ , \ (g_4^f)^2 = \{1, 0\}g^2 \ .$$

According to the theorem of ref. [9] these models are finite to all orders. After the reduction of couplings the symmetry of W is enhanced [1, 6].

The main difference of the models A and B is that three pairs of Higgs quintets and anti-quintets couple to the 24 for B so that it is not necessary to mix them with H_4 and $\overline{H}_4$ in order to achieve the triplet-doublet splitting after the symmetry breaking of $SU(5)$.

In the dimensionful sector, the sum rule gives us the following boundary conditions at the GUT scale [6]:

$$m_{H_u}^2 + 2m_{10}^2 = m_{H_d}^2 + m_{\overline{5}}^2 + m_{10}^2 = M^2 \ \text{for A} \ , \tag{23}$$

$$m_{H_u}^2 + 2m_{10}^2 = M^2 \ , \ m_{H_d}^2 - 2m_{10}^2 = -\frac{M^2}{3} \ ,$$

$$m_{\overline{5}}^2 + 3m_{10}^2 = \frac{4M^2}{3} \ \text{for B,} \tag{24}$$

where we use as free parameters $m_{\overline{5}} \equiv m_{\overline{5}_3}$ and $m_{10} \equiv m_{10_3}$ for the model A, and m_{10} for B, in addition to M.

5. Predictions of Low Energy Parameters

Since the gauge symmetry is spontaneously broken below M_{GUT}, the finiteness and Gauge-Yukawa unification conditions do not restrict the renormalization property at low energies, and all it remains are boundary conditions on the gauge and Yukawa couplings (22), the $h = -MC$ relation (8) and the soft scalar-mass sum rule (9) at M_{GUT}, as applied in the various models. So we examine the evolution of these parameters according to their renormalization group equations at two-loop for dimensionless parameters and at one-loop for dimensionful ones with the relevant boundary conditions. Below M_{GUT} their evolution is assumed to be governed by the MSSM. We further assume a unique supersymmetry breaking scale M_s so that below M_s the SM is the correct effective theory.

The predictions for the top quark mass M_t are ~ 183 and ~ 174 GeV in models A and B respectively. Comparing these predictions with the most recent experimental value $M_t = (173.8 \pm 5.2)$ GeV, and recalling that the theoretical values for M_t may suffer from a correction of less than $\sim 4\%$ [7], we see that

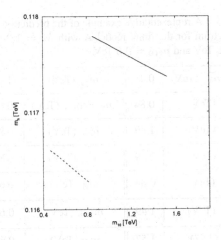

Figure 8. m_h as function of m_{10} for $M = 0.8$ (dashed) 1.0 (solid) TeV for the finite model **B**.

they are consistent with the experimental data. In addition the value of $\tan \beta$ is obtained as $\tan \beta = 54$ and 48 for models **A** and **B** respectively.

In the SSB sector, besides the constraints imposed by reduction of couplings and finiteness, we also look for solutions which are compatible with radiative electroweak symmetry breaking.

Concerning the SSB sector of the finite theories **A** and **B**, besides the gaugino mass we have two and one more free parameters respectively, as previously mentioned. Thus, we look for the parameter space in which the lighter $\tilde{\tau}$ mass squared $m_{\tilde{\tau}}^2$ is larger than the lightest neutralino mass squared m_χ^2 (which is the LSP). In the case where all the soft scalar masses are universal at the unification scale, there is no region of $M_s = M$ below O(few) TeV in which $m_{\tilde{\tau}}^2 > m_\chi^2$ is satisfied. But once the universality condition is relaxed this problem can be solved naturally (provided the sum rule). More specifically, using the sum rule (9) and imposing the conditions a) successful radiative electroweak symmetry breaking b) $m_{\tilde{\tau}^2} > 0$ and c) $m_{\tilde{\tau}^2} > m_{\chi^2}$, we find a comfortable parameter space for both models (although model **B** requires large $M \sim 1$ TeV).

In Tables 1 and 2 we present representative examples of the values obtained for the sparticle spectra in each of the models. The value of the lightest Higgs physical mass m_h has already the one-loop radiative corrections included, evaluated at the appropriate scale [27].

Finally, we calculate $BR(b \rightarrow s\gamma)$ [28], whose experimental value is $1 \times 10^{-4} < BR(b \rightarrow s\gamma) < 4 \times 10^{-4}$. The SM predicts $BR(b \rightarrow s\gamma) = 3.1 \times 10^{-4}$. This imposes a further restriction in our parameter space, namely $M \sim 1$ TeV if $\mu < 0$ for all three models. This restriction is less strong in the case that $\mu > 0$. For

TABLE II. A representative example of the predictions for the s-spectrum for the finite model **A** with $M = 1.0$ TeV, $m_{\bar{5}} = 0.8$ TeV and $m_{10} = 0.6$ TeV.

$m_X = m_{\chi_1}$ (TeV)	0.45	$m_{\tilde{b}_2}$ (TeV)	1.76
m_{χ_2} (TeV)	0.84	$m_{\tilde{\tau}} = m_{\tilde{\tau}_1}$ (TeV)	0.63
m_{χ_3} (TeV)	1.49	$m_{\tilde{\tau}_2}$ (TeV)	0.85
m_{χ_4} (TeV)	1.49	$m_{\tilde{\nu}_1}$ (TeV)	0.88
$m_{\chi_1^\pm}$ (TeV)	0.84	m_A (TeV)	0.64
$m_{\chi_2^\pm}$ (TeV)	1.49	$m_{H^\pm}$ (TeV)	0.65
$m_{\tilde{t}_1}$ (TeV)	1.57	m_H (TeV)	0.65
$m_{\tilde{t}_2}$ (TeV)	1.77	m_h (TeV)	0.122
$m_{\tilde{b}_1}$ (TeV)	1.54		

TABLE III. A representative example of the predictions of the s-spectrum for the finite model **B** with $M = 1$ TeV and $m_{10} = 0.65$ TeV.

$m_X = m_{\chi_1}$ (TeV)	0.45	$m_{\tilde{b}_2}$ (TeV)	1.70
m_{χ_2} (TeV)	0.84	$m_{\tilde{\tau}} = m_{\tilde{\tau}_1}$ (TeV)	0.47
m_{χ_3} (TeV)	1.30	$m_{\tilde{\tau}_2}$ (TeV)	0.67
m_{χ_4} (TeV)	1.31	$m_{\tilde{\nu}_1}$ (TeV)	0.88
$m_{\chi_1^\pm}$ (TeV)	0.84	m_A (TeV)	0.73
$m_{\chi_2^\pm}$ (TeV)	1.31	$m_{H^\pm}$ (TeV)	0.73
$m_{\tilde{t}_1}$ (TeV)	1.51	m_H (TeV)	0.73
$m_{\tilde{t}_2}$ (TeV)	1.73	m_h (TeV)	0.118
$m_{\tilde{b}_1}$ (TeV)	1.56		

example, the minimal model with $M = 1$ TeV leads to $BR(b \to s\gamma) = 3.8 \times 10^{-4}$ for $\mu < 0$.

6. Conclusions

The programme of searching for exact RGI relations among dimensionless couplings in supersymmetric GUTs, started few years ago, has now supplemented with the derivation of similar relations involving dimensionful parameters in the SSB sector of these theories. In the earlier attempts it was possible to derive RGI relations among gauge and Yukawa couplings of supersymmetric GUTs, which could lead even to all-loop finiteness under certain conditions. These theoretically attractive theories have been shown not only to be realistic but also to lead to a successful prediction of the top quark mass. The new theoretical developments include the existence of a RGI sum rule for the soft scalar masses in the SSB sector of $N = 1$ supersymmetric gauge theories exhibiting gauge-Yukawa unification. The all-loop sum rule substitutes now the universal soft scalar masses and overcomes its phenomenological problems. Of particular theoretical interest is the fact that the finite unified theories, which could be made all-loop finite in the supersymmetric sector can now be made completely *finite*. In addition it is interesting to note that the sum rule coincides with that of a certain class of string models in which the massive string modes are organized into $N = 4$ super-multiplets. Last but not least in ref. [17], the exact β-function for the soft scalar masses in the NSVZ scheme was obtained for the first time. On the other hand the above theories have a remarkable predictive power leading to testable predictions of their spectrum in terms of very few parameters. In addition to the prediction of the top quark mass, which holds unchanged, the characteristic features that will judge the viability of these models in the future are 1) the lightest Higgs mass is found to be around 120 GeV and the s-spectrum starts beyond several hundreds of GeV. Therefore the next important test of Gauge-Yukawa and Finite Unified theories will be given with the measurement of the Higgs mass, for which these models show an appreciable stability, which is alarmingly close to the IR quasi fixed point prediction of the MSSM for large $\tan \beta$ [29]. Our preliminary search in the available parameter space of the above models shows that in case we relax the requirement that the mass of the s-tau should be smaller than the neutralinos masses, we obtain a wider window in the prediction of the lightest Higgs mass starting from 111 GeV. This possibility has no obvious problem in case we introduce bilinear R-parity violating terms that preserve finiteness. Actually, the introduction of such terms might be unavoidable given that it is a necessary ingredient of the only known mechanism to introduce neutrino masses in these models [30].

Acknowledgements

It is a pleasure to thank the Organizing Committee for the very warm hospitality offered to one of us (G.Z.). Supported by the projects PAPIIT-125298 and ERBFMRXCT960090.

References

1. Kapetanakis D., Mondragón, M. and Zoupanos, G., (1993) *Zeit. f. Phys.* **C60** 181; Mondragón, M. and Zoupanos, G. (1995) *Nucl. Phys. B* (Proc. Suppl.) **37C**) 98.
2. Kubo, J., Mondragón, M. and Zoupanos, G. (1994) *Nucl. Phys.* **B424** 291.
3. Kubo, J., Mondragón, M., Tracas, N.D. and Zoupanos, G. (1995) *Phys. Lett.* **B342** 155; Kubo, J., Mondragón, M., Shoda, S. and Zoupanos, G., (1996) *Nucl. Phys.* **B469** 3; Kubo, J., Mondragón, M., Olechowski, M. and Zoupanos, G. (1996) *Nucl. Phys.* **B479** 25.
4. Kubo, J., Mondragón, M. and Zoupanos, G. (1996) *Phys. Lett.* **B389** 523.
5. Kawamura, T., Kobayashi, T. and Kubo, J. (1997) *Phys. Lett.* **B405** 64.
6. Kobayashi, T., Kubo, J., Mondragón, M. and Zoupanos, G. (1998) *Nucl. Phys.* **B511** 45.
7. For an extended discussion and a complete list of references see: Kubo, J., Mondragón, M. and Zoupanos, G. (1997) *Acta Phys. Polon.* **B27** 3911.
8. Zimmermann, W., (1985) *Com. Math. Phys.* **97** 211; Oehme, R. and Zimmermann, W. (1985) *Com. Math. Phys.* **97** 569.
9. Lucchesi, C., Piguet, O. and Sibold, K. (1988) *Helv. Phys. Acta* **61** 321; Piguet, O. and Sibold, K. (1986) *Intr. J. Mod. Phys.* **A1** 913; (1986) *Phys. Lett.* **B177** 373; see also Lucchesi, C. and Zoupanos, G. (1997) *Fortsch. Phys.* **45** 129.
10. Ermushev, A.Z., Kazakov, D.I. and Tarasov, O.V. (1987) *Nucl. Phys.* **281** 72; Kazakov, D.I. (1987) *Mod. Phys. Lett.* **A9** 663.
11. Jack, I. and Jones, D.R.T. (1995) *Phys. Lett.* **B349** 294.
12. Hisano, J. and Shifman, M. (1997) *Phys. Rev.* **D56** 5475.
13. Jack, I. and Jones, D.R.T. (1997) *Phys. Lett.* **B415** 383.
14. Avdeev, L.V., Kazakov, D.I. and Kondrashuk, I.N. (1998) *Nucl. Phys.* **B510** 289; Kazakov, D.I. (1999) *Phys.Lett.* **B449** 201.
15. Kazakov, D.I. (1998) *Phys. Lett.* **B412** 21.
16. Jack, I., Jones, D.R.T. and Pickering, A. (1998) *Phys. Lett.* **B426** 73.
17. Kobayashi, T., Kubo, J. and Zoupanos, G. (1998) *Phys. Lett.* **B427** 291.
18. Yamada, Y. (1994) *Phys.Rev.* **D50** 3537.
19. Delbourgo, R. (1975) *Nuovo Cim* **25A** 646; Salam, A. and Strathdee, J. (1975) *Nucl. Phys.* **B86** 142; Fujikawa, K. and Lang, W. (1975) *Nucl. Phys.* **B88** 61; Grisaru, M.T., Rocek, M. and Siegel, W. (1979) *Nucl. Phys.* **B59** 429.
20. Girardello, L. and Grisaru, M.T. (1982) *Nucl. Phys.* **B194** 65; Helayel-Neto, J.A. (1984) *Phys. Lett.* **B135** 78; Feruglio, F., Helayel-Neto, J.A. and Legovini, F. (1985) *Nucl. Phys.* **B249** 533; Scholl, M. (1985) *Zeit. f. Phys.* **C28** 545.
21. Jones, D.R.T., Mezincescu, L. and Yao, Y.-P. (1984) *Phys. Lett.* **B148** 317.
22. Jack, I. and Jones, D.R.T. (1994) *Phys. Lett.* **B333** 372.
23. Ibáñez, L.E. and Lüst, D. (1992) *Nucl. Phys.* **B382** 305; Kaplunovsky, V.S. and Louis, J. (1993) *Phys. Lett.* **B306** 269; Brignole, A., Ibañez, L.E. and Muñoz, C. (1994) *Nucl. Phys.* **B422** 125 [Erratum: (1995) **B436** 747].
24. Brignole, A., Ibáñez, L.E. and Muñoz, C. (1996) *Phys. Lett.* **B387** 305.
25. Casas, J.A., Lleyda, A. and Muñoz, C. (1996) Phys. Lett. **B380** 59.

26. Novikov, V., Shifman, M., Vainstein, A., and Zakharov, V. (1983) *Nucl. Phys.* **B229** 381; (1986) *Phys. Lett.* **B166** 329; Shifman, M. (1996) *Int.J. Mod. Phys.* **A11** 5761 and references therein.

27. Gladyshev, A.V., Kazakov, D.I., de Boer, W., Burkart, G. and Ehret, R. (1997) *Nucl. Phys.* **B498** 3; Carena, M. et. al., (1995) *Phys. Lett.* **B355** 209.

28. Bertolini, S., Borzumati, F., Masiero, A. and Ridolfi, G. (1991) *Nucl. Phys.* **B353** 591.

29. Jurčišin, M and Kazakov, D.I. (1999) *Mod. Phys. Lett.* **A14** 671.

30. Hirsch,M. et.al., (2000) *Phys. Rev.* **D62** 113008.

WORLD VOLUME REALIZATION OF AUTOMORPHISMS

JOAN SIMON *

Departament of Particle Physics, The Weizmann Institute of Science,
2 Herzl Street, 76100 Rehovot, Israel

Abstract. The relation among spacetime supersymmetry algebras and the world volume approach to string theory is reviewed. The realization of some of the automorphism transformations of these superalgebras on the world volume theory is discussed. We distinguish among linear realizations and non-local ones. The consistency of the latter with duality in M/string theory is checked.

1. Introduction

Our contribution to the NATO Advanced Research Workshop on 'NonCommutative Structures in Mathematics and Physics' is devoted to the relation among supersymmetry algebras and reparametrization invariant field theories describing the low energy dynamics of branes. In particular, we shall concentrate on branes propagating in SuperPoincaré, and consequently, on maximally extended SuperPoincaré algebras.

The study of M/String theory spectrums can be done along purely algebraic methods or field theory ones. The *algebraic approach* is based on the assumption that the $\mathcal{N} = 1$ supersymmetry in eleven dimensions (or the corresponding $\mathcal{N} = 2$ supersymmetries in ten dimensions) is valid at any energy, so that the M-theory (string theory) spectrum must be organized into representations of the SuperPoincaré algebra. This approach entirely characterizes *BPS states*, those preserving some amount of supersymmetry, thus filling in short irreducible representations of the forementioned algebra. Given a maximally extended supersymmetry algebra [1], [2]

$$\{Q_\alpha, Q_\beta\} = -\mathcal{M} I_{\alpha\beta} + \Gamma(\mathcal{Z})_{\alpha\beta}, \qquad (1)$$

where $\Gamma(\mathcal{Z})_{\alpha\beta}$ stands for the traceless part of the supersymmetry anticommutation relations, and given any state $|\alpha >$, the positivity of the matrix $<$

* jsimon@weizmann.ac.il

S. Duplij and J. Wess (eds.), Noncommutative Structures in Mathematics and Physics, 259–270.

$\alpha|\{Q_\alpha, Q_\beta\}|\alpha >$ [1] implies a bound on the rest mass $\mathcal{M}$. When the latter is saturated, there is a linear combination of the supersymmetry generators annihilating the state. This means that the symmetric matrix $\{Q_\alpha, Q_\beta\}$ has at least one zero eigenvalue (det $\{Q_\alpha, Q_\beta\} = 0$). Thus, generically, the search for such BPS states is equivalent to the resolution of the eigenvalue problem [4]

$$\Gamma(\mathcal{Z})|\alpha >= \mathcal{M}|\alpha > . \tag{2}$$

Any solution to equation (2) describes a Clifford valued BPS state $|\alpha >$ by its mass $\mathcal{M}$ and the amount of supersymmetry preserved (ν), which will generically be determined by some set of mutually commuting constant operators $\{\mathcal{P}_i\}$ such that $\mathcal{P}_i|\alpha >= |\alpha > \forall i$. Both depend on the charges $\mathcal{Z}$ carried by $|\alpha >$. A partial analysis of equation (2) was done in [5], where a whole family of BPS states, called factorizable states, were classified. We refer the reader to [5] for a discussion on equation (2) and some of their solutions.

The *world volume approach* is based on brane effective actions, which are supposed to describe the low energy dynamics of string theory when the string scale vanishes $(\alpha' \to 0)$ and gravity decouples. The dynamics of branes propagating in SuperPoincaré are described by reparametrization susy-kappa invariant field theories providing us with a field theory realization of the previous superalgebras. The algebraic saturation of the BPS bound has its field theory counterpart in the saturation of the Bogomolny' type bound derived from the energy density computed on the brane [6]. Only certain field theory configurations do saturate such bounds, these are the so called *BPS configurations*. One way of systematically looking for such configurations is the resolution of the *kappa symmetry preserving condition*. This method is based on the search for the subset of supersymmetry transformations that leave bosonic configurations $(\theta = 0)$ invariant. Since fermions do transform inhomogeneously in brane effective actions,

$$\delta\theta = \epsilon + (1 + \Gamma_\kappa)\kappa + O(\theta) \tag{3}$$

where ϵ is the global supersymmetry parameter (the Killing spinor of the background geometry) and κ is the local kappa symmetry one, the above invariance requirement is satisfied whenever [7]

$$\Gamma_\kappa \epsilon = \epsilon . \tag{4}$$

Γ_κ is a spinor valued matrix being field and background dependent. It satisfies $\Gamma_\kappa^2 = I$ and tr $\Gamma_\kappa = 0$, conditions that allow kappa symmetry to remove half of the fermionic degrees of freedom on the brane, a necessary condition to get a supersymmetric field theory on the brane, but not a sufficient one.

---

[1] It is assumed that Q_α satisfies the necessary requirements for this positivity to hold. In M-theory, the Majorana charges do certainly satisfy them. See [3], for a discussion on this point in arbitrary spacetime signatures.

In the case of SuperPoincaré backgrounds, ϵ is a 32 constant spinor. In less symmetric background superspaces, it will generically depend on the point. Solving equation (4) gives rise to

1. some constraints on the configuration space $f_i[\phi^j] = 0$
2. some supersymmetry preserving conditions $\mathcal{P}_i\epsilon = \epsilon \;\forall i$

where $f_i[\phi^j]$ stands for some functional relation involving the dynamical fields on the brane $\{\phi^i\}$ and their derivatives $\{\partial\phi^i, \partial\partial\phi^i, \dots\}$. On the other hand, $\mathcal{P}_i$ is a constant spinor valued matrix satisfying $\mathcal{P}_i^2 = 1$ and $\mathrm{tr}\,\mathcal{P}_i = 0$. If $\mathcal{P}_i = \Gamma_{[a_1\dots a_i]}$ equals the antisymmetrized product of gamma matrices, we shall call it single projector.

Constraints 1. become *BPS equations*. This can be checked by computing the energy density functional of the field theory which can always be written as [2]

$$\mathcal{E}^2 = (\mathcal{E}_0 + \mathcal{Z})^2 + \sum_i \left(t^i f_i[\phi^j]\right)^2 \tag{5}$$

if we are describing a BPS state at threshold (intersection of branes) or as

$$\mathcal{E}^2 = \mathcal{E}_0^2 + \mathcal{Z}^2 + \sum_i \left(t^i f_i[\phi^j]\right)^2 \tag{6}$$

for a non-threshold BPS state. Both expressions show the BPS equation character of the constraints $f_i[\phi^j] = 0$.

Conditions 2. determine the amount of supersymmetry preserved (ν) and the kind of branes involved in the state due to the one to two correspondence among single branes and single projectors [3].

Thus, all in all, one gets a field theory realization of the previous algebraic BPS states ($|\alpha>$). They are indeed the same because they are characterized by the same supersymmetry projection conditions ($\mathcal{P}_i$) and they do have the same energy ($\mathcal{M} = \mathcal{E}$).

Once the connection among brane effective actions and supersymmetry algebras has been established, it is natural to ask about the extent of such a connection regarding the maximal automorphism groups of SuperPoincaré algebras. In particular, the $\mathcal{N} = 1\; D = 11$ superalgebra admits a $GL(32, R)$ automorphism group [5][8][9][10][11]. One of the first consequences of such automorphism structure is the existence of $SO(32)$ transformations relating $\nu = \frac{1}{2}$ non-threshold bound states with $\nu = \frac{1}{2}$ bound states at threshold, having the same mass [4]. Without loss

---

[2] We have assumed the existence of a single $\mathcal{Z}$ charge in the above derivation, but the extension to more general configurations is straightforward. $\mathcal{E}_0$ stands for the vacuum energy of the configuration.

[3] This is because given any single projector $\mathcal{P}_i$, there always exists $\tilde{\mathcal{P}}_i$ such that $\mathcal{P}_i\tilde{\mathcal{P}}_i = I$. So, if $\mathcal{P}_i\epsilon = \epsilon \Rightarrow \tilde{\mathcal{P}}_i\epsilon = \epsilon$.

[4] there exist similar phenomena for less supersymmetric BPS states, see [5].

of generality, consider a non-threshold bound state described by

$$(\cos \beta \, \Gamma_1 + \sin \beta \, \Gamma_2) |\alpha> \, = \, |\alpha> \quad , \quad \{\Gamma_1, \Gamma_2\} = 0 \qquad (7)$$

$$\mathcal{M} = \sqrt{\mathcal{Z}_1^2 + \mathcal{Z}_2^2} \qquad (8)$$

where $\Gamma_i \; i = 1, 2$ satisfies analogous properties to those of $\mathcal{P}_j$ and β is a constant parameter. There always exists $U_\beta = e^{\beta \Gamma_2 \Gamma_1 / 2} \in SO(32)$, such that (7) becomes

$$U_\beta \, \Gamma_1 \, U_\beta^t |\alpha> \, = \, |\alpha> \quad \Leftrightarrow \quad \Gamma_1 |\alpha'> \, = \, |\alpha'>, \qquad (9)$$

which allows us to reinterpret it in terms of an SO(32) related BPS state $|\alpha'>$ at threshold having the same mass (8).

Motivated by the previous discussion, it seems rather natural to look for world volume realizations of such automorphisms. Since the Lorentz group in eleven dimensions can be seen as a subgroup of $GL(32, R)$, it is obvious that such subgroup will be linearly realized on the brane (before any gauge fixing). This is because any brane effective action propagating in SuperPoincaré is manifestly (quasi-)invariant under the superisometries of the background [12]. In section 2, we will discuss a particular example of such linear realizations and the way they act on BPS configurations, showing explicitly the connection among non-threshold and threshold bound states illustrated in the algebraic approach. Besides this linear realizations, the analysis done in [5] shows that central charges $\mathcal{Z}$'s are generically 'rotated' among themselves under automorphism transformations. Since for bosonic configurations, such topological charges are given by world space integrals involving derivatives of the brane dynamical fields, one should also expect, if any, the existence of non-local transformations leaving certain brane theories invariant. We review the results of [13] concerning that point in section 3. Starting from the non-local transformations leaving the D3-brane action invariant [14], which are the world volume realization of the S-duality automorphism for the $\mathcal{N} = 2 \; D = 10$ type IIB SuperPoincaré algebra, we perform a T-duality along a world volume direction to get some new non-local transformations of the D2-brane in type IIA. The latter have a natural M-theory interpretation as rotations involving the world volume scalar (y) which becomes a one form $(V_{(1)})$ after the world volume dualization relating both effective theories in three dimensions [15, 16]. This dualization explains the origin of such non-local transformations in type IIA theory.

These results illustrate that part of the automorphism group is realized on the world volume field theory, either as linear realizations or as non-local ones. It would be interesting to clarify which is the symmetry structure that is being realized on brane effective actions. Along the same lines, it would also be interesting to understand the existing relation among the automorphism group and U-duality groups. As it was pointed out in [13], the $\mathcal{N} = 2 \; D = 10$ type IIB SuperPoincaré algebra admits $SL(2, R)$ in its maximal automorphism group, the latter being the U-duality group for type IIB superstring theory. When compactifying several dimensions and using T-duality adequately, one may suspect of deriving some relation among the corresponding U-duality group and the automorphism group of the dimensionally reduced superalgebra.

2. Linear realizations

Given any brane effective action $S[\phi^i]$, the set of dynamical fields can always be splitted into $\{\phi^i\} = \{x^m, \theta, V_{(p)}\}$, x^m and θ being superspace coordinates and $V_{(p)}$ some p-form degrees of freedom on the brane. These actions are invariant $(\delta S[\phi^i] = 0)$ under some set of global and local transformations. We shall concentrate on the global ones. These include the superisometries of the background geometry, so since we are considering SuperPoincaré backgrounds, it certainly includes the $SO(1, D - 1)$ Lorentz transformations

$$\delta\theta = \frac{1}{4}\omega^{mn}\Gamma_{mn}\theta \quad , \quad \delta x^m = \omega^{mn}\eta_{np}x^p \quad , \quad \delta V_{(p)} = 0. \tag{10}$$

Let us concentrate on M2-brane effective actions in M-theory. We are thus considering three dimensional field theories probing eleven dimensional Super-Poincaré space [17]. To illustrate previous ideas, we shall look for a world volume soliton on an M2-brane corresponding to the non-threshold bound state

$$
\begin{array}{c}
M2: 1\ 2\ \_\ \_\ \_\ \_\ \_\ \_\ \_\ \_ \\
M2: \_\ 2\ 3\ \_\ \_\ \_\ \_\ \_\ \_\ \_ \\
M2: 1\ \_\ 3\ \_\ \_\ \_\ \_\ \_\ \_\ \_ \ .
\end{array}
$$

By setting the static gauge ($x^\mu = \sigma^\mu$ $\mu = 0, 1, 2$) and exciting one transverse scalar ($x^3 = x$), one can check that the kappa symmetry preserving condition (4) is solved by

$$x = \tan\alpha\left(\cos\beta\sigma^1 + \sin\beta\sigma^2\right), \tag{11}$$

where α and β are arbitrary constants, whenever ϵ satisfies

$$\{\cos\alpha\Gamma_{012} + \sin\alpha\left(\cos\beta\Gamma_{023} + \sin\beta\Gamma_{013}\right)\}\epsilon = \epsilon, \tag{12}$$

which indeed corresponds to the forementioned $\nu = \frac{1}{2}$ non-threshold bound state.

According to our discussion in the introduction, there must exist an $SO(32)$ transformation relating such a configuration with a $\nu = \frac{1}{2}$ bound state at threshold, corresponding in this particular case, to a single membrane lying in the 12-plane. We will explicitly check that this is indeed the case by considering the following $SO(32)$ group element

$$U = U_\alpha U_\beta = e^{-\alpha\Gamma_{13}/2}e^{-\beta\Gamma_{12}/2}. \tag{13}$$

By computing its finite transformation on the scalar coordinates, we derive

$$\tilde{x}^2 = \cos\beta\,\sigma^2 + \sin\beta\,\sigma^1 \ , \quad \tilde{x}^1 = \frac{\cos\beta}{\cos\alpha}\sigma^1 - \frac{\sin\beta}{\cos\alpha}\sigma^2$$

$$\tilde{x} = 0 \tag{14}$$

which shows there is no transverse scalar excited in the rotated configuration ($\tilde{x} = 0$). This is understood as having no more membranes in the configuration than just the defining one. This interpretation is further confirmed by rewriting the supersymmetry projection condition in terms of the transformed Killing spinor

$$\Gamma_{012}\epsilon' = \epsilon' \quad , \quad \epsilon' = U^t\epsilon. \tag{15}$$

Equation (15) describes a single membrane in the 12-plane, as expected.

3. Non-local realizations

In this section we shall review the results reported in [13]. We shall start our analysis by studying D3-brane effective actions. These provide a field theory realization of some truncation of $\mathcal{N} = 2\ D = 10$ type IIB SuperPoincaré algebra [3]

$$\{Q^i, Q^j\} = \mathcal{P}^+\Gamma^M Y_M^{ij} + \mathcal{P}^+\frac{1}{3!}\Gamma^{MNP}\epsilon^{ij}Y_{MNP}$$

$$+\mathcal{P}^+\frac{1}{5!}\Gamma^{M_1...M_5}Y_{M_1...M_5}^{+ij}\,, \tag{16}$$

where the central charges are given by $Y_M^{ij} = \delta^{ij}Y_M^{(0)} + \tau_1^{ij}Y_M^{(1)} + \tau_3^{ij}Y_M^{(3)}$ and $Y_{M_1...M_5}^{+ij} = \delta^{ij}Y_{M_1...M_5}^{+(0)} + \tau_1^{ij}Y_{M_1...M_5}^{+(1)} + \tau_3^{ij}Y_{M_1...M_5}^{+(3)}$.

If we consider an $SL(2,R)$ transformation $\tilde{Q}^i = (U\,Q)^i$, $U_\lambda = e^{\lambda i\tau_2/2} \in SL(2,R)$, the latter belongs to the type IIB automorphism group if the charges transform as

$$\tilde{\mathcal{Z}}^{ij} = \left(U\,\mathcal{Z}\,U^t\right)^{ij}\,. \tag{17}$$

Notice that $U_\lambda \in SO(2)$ subgroup of $SL(2,R)$ which rotates $\left(Y_M^{(1)}, Y_M^{(3)}\right)$ and $\left(Y_{M_1...M_5}^{+(1)}, Y_{M_1...M_5}^{+(3)}\right)$ as doublets, whereas Y_{mnp} and $Y_{m_1...m_5}^{+(0)}$ remain invariant. This is consistent with the S-duality interpretation of $U_{\pi/2}$, which interchanges D-strings and fundamental strings, D5-branes and NS5-branes, while leaving D3 and KK5B monopoles self-dual.

This $SO(2)$ transformation is reminiscent of the electro-magnetic duality in four dimensions, and it was indeed proved in [14] that the off-shell transformations giving rise to such a rotation are given by

$$\delta x^m = 0 \quad , \quad \delta\theta = \frac{\lambda}{2}i\tau_2\theta \tag{18}$$

$$\delta F_{\mu\nu} = \lambda K_{\mu\nu} \quad , \quad \delta K_{\mu\nu} = -\lambda F_{\mu\nu} \tag{19}$$

where $K_{\mu\nu} = -\frac{1}{2}\varepsilon_{\mu\nu\rho\sigma}\tilde{K}^{\rho\sigma}$ [5] and $\tilde{K}^{\rho\sigma} = \frac{1}{\sqrt{-\det \mathcal{G}}}\frac{\partial \mathcal{L}_{D3}}{\partial F_{\rho\sigma}}$, $\mathcal{L}_{D3}$ being the Lagrangian density for an abelian D3-brane propagating in SuperPoincaré [18–21]. It is remarkable that the infinitesimal transformation for the fermionic field agrees with the infinitesimal transformation of the supersymmetry generator. Notice that it is $F = dV$ the one entering in previous linear transformations (19). So, when rewritten in terms of the gauge potential V, they become non-local transformations [22].

To get a more physical understanding of these transformations, we shall evaluate them on-shell; in particular, on Bion configurations [23, 24]. These are $\nu = 1/4$ solitons representing fundamental strings ending on the brane. As all BPS configurations, they are characterized by some BPS equations $F_{0a} = \partial_a y$ $a = 1, 2, 3$ and some supersymmetry conditions

$$\Gamma_{0123}\, i\tau_2\, \epsilon = \epsilon \tag{20}$$

$$\Gamma_{0y}\tau_3\, \epsilon = \epsilon \tag{21}$$

corresponding to the array

$$D3 : 1\ 2\ 3\ \_\ \_\ \_\ \_\ \_\ \_$$
$$F1 : \_\ \_\ \_\ 4\ \_\ \_\ \_\ \_\ \_\ .$$

If we compute $K_{\mu\nu}$ when we are on-shell, we get $K_{0a} = 0$, $K_{ab} = \epsilon_{abc}F_{0c}$, which give rise to $\delta E^a = 0$ and $\delta B^a = \lambda E^a$, whose finite form generates an $SO(2)$ rotation $\tilde{E}^a = \cos\lambda\, E^a$, $\tilde{B}^a = \sin\lambda\, E^a$, where E^a and B^a correspond to the electric and magnetic fields, respectively. Thus the rotated configuration is both electrically and magnetically charged: it is a dyon. This interpretation is further confirmed by rewriting the supersymmetry condition (21) in terms of the transformed Killing spinor, $\tilde{\epsilon} = U^t\epsilon$

$$\Gamma_{0y}\left(\cos\alpha\,\tau_3 + \sin\alpha\,\tau_1\right)\tilde{\epsilon} = \tilde{\epsilon}, \tag{22}$$

which indeed describes a non-threshold bound state of fundamental strings (τ_3 factor) and D-strings (τ_1 factor).

We could have also analyzed the energy of such configurations. The starting BIon verifies $E_{BIon} = E_{D3} + Y_4^{(3)}$, where $Y_4^{(3)} = \int_{D3} \vec{E}\cdot\vec{\nabla}y$ is the charge carried by the fundamental string along the y (x^4) direction, whereas E_{D3} stands for the energy of an infinite planar D3-brane. After the $SO(2)$ transformation, $E_{dyon} =$

$$E_{BIon} = E_{D3} + \sqrt{\left(\tilde{Y}_4^{(3)}\right)^2 + \left(\tilde{Y}_4^{(1)}\right)^2}, \text{ where } \tilde{Y}_4^{(3)} = \int_{D3}\cos\lambda\,\vec{E}\cdot\vec{\nabla}y \quad \text{and}$$

$\tilde{Y}_4^{(1)} = \int_{D3}\sin\lambda\,\vec{B}\cdot\vec{\nabla}y$. In this way, we check that the field theory $SO(2)$

---

[5] $\varepsilon_{\mu\nu\rho\sigma}$ denotes the covariantly constant antisymmetric tensor with indices raised and lowered by $\mathcal{G}_{\mu\nu}$.

transformations (18-19) indeed rotate the charges of the spacetime supersymmetry algebra.

In the following we shall check the consistency of the previous set of transformations with the known web of dualities in M/string theory. The first step will be to perform a longitudinal T-duality transformation, that is, along one of the D3-brane world volume directions, to study the corresponding symmetry structure in type IIA. Finally, the M-theory origin for such type IIA symmetry will be explained. As before, these checks can be studied either from an algebraic perspective or from a field theory one.

The realization of T-duality at the level of superalgebras is known to be a mapping relating the supersymmetry charges as follows

$$Q^+ = Q^2 \quad , \quad Q^- = \Gamma_s Q^1 , \tag{23}$$

where $Q^\pm$ are the type IIA supercharges and s stands for the spacelike direction along we perform the transformation. Such a mapping, does change the chirality of one of the generators and induces some transformation on the charges $\mathcal{Z}$'s [3] which agrees with the known T-duality rules among BPS single branes. In this way, the previous U_λ automorphism can be rewritten as $U_s = e^{\lambda/2\,\Gamma_s\Gamma_{11}}$, which indeed belongs to SO(32), the subgroup of type IIA automorphisms preserving energy. The latter statement can be straightforwardly derived from the M-algebra analysis done in [5]. Notice that Γ_{11} is the ten dimensional chirality operator, so that U_s can not be interpreted as an spacetime rotation. This transformation will "rotate" several doublets of charges appearing in type IIA, while keeping some others invariant. In particular, charges $\mathcal{Z}_{sm}$ and $\mathcal{Z}_m$ corresponding to D2-branes and fundamental strings will form an SO(2) doublet under U_s transformations.

Moving back to the world volume approach, the analysis done in [25, 26] will be used to derive the symmetry structure inherited on the D2-brane after performing the longitudinal T-duality. Since $\delta x^m = 0$ in (18), there will be no compensating diffeomorphism transformation coming from the partial gauge fixing locally identifying ($x^s = \rho$) one world volume direction (ρ) with one target space direction (x^s). It is then straightforward to derive a set of non-local transformations leaving the D2-brane invariant, just by double dimensional reduction of (18-19)

$$\delta\theta = \tfrac{\lambda}{2}\Gamma_m\Gamma_{11}\theta , \tag{24}$$

$$\delta K^m_{\hat\mu\hat\nu} = -\lambda^m F_{\hat\mu\hat\nu} \quad , \quad \delta F_{\hat\mu\hat\nu} = \lambda^m K^m_{\hat\mu\hat\nu} \tag{25}$$

$$\delta K_{\hat\mu\rho} = -\lambda^m \partial_{\hat\mu}\tilde x^m \quad , \quad \delta\partial_{\hat\mu}\tilde x^m = \lambda^m K_{\hat\mu\rho} \tag{26}$$

where $K^m_{\hat\mu\hat\nu}$ and $K_{\hat\mu\rho}$ where computed explicitly in [13].

Notice that whereas in type IIB there was a single transformation (λ), in type IIA we have a set of them (λ^m). This enhancement of symmetry is typical of T-duality on symmetric backgrounds. The performance of T-duality is manifestly non-covariant, but in the limit $R \to \infty$, the isometries of the background allow us to recover target space covariance. A much more algebraic way to reach the same conclusion is to compute the commutator of a rotation (ω) with our previous non-local transformation (λ^s)

$$[\delta_\omega, \delta_{\lambda^s}] = \delta_{\lambda^{\tilde{s}}} \quad \tilde{s} \neq s, \tag{27}$$

which generates all the forementioned transformations. Another difference between this set of transformations and type IIB ones, is that bosonic matter fields do transform $(\delta x^m \neq 0)$, its origin being the component of the original gauge field (V_ρ) along which we perform the T-duality.

Just as for the D3-brane case, we shall analyze the behaviour of some particular BPS configuration under these new transformations. We shall consider the T-dual configuration of a type IIB dyon. This is given by the array

$$
\begin{array}{llllllllll}
D2: & 1 & 2 & \_ & \_ & \_ & \_ & \_ & \_ & \_ \\
F1: & \_ & \_ & \_ & 4 & \_ & \_ & \_ & \_ & \_ \\
D2: & \_ & \_ & 3 & 4 & \_ & \_ & \_ & \_ & \_ \ .
\end{array}
$$

This supersymmetric configuration is described by the BPS equations

$$E^{\hat{a}} = \cos\alpha\, \partial_{\hat{a}} y \tag{28}$$

$$\epsilon^{\hat{a}\hat{b}} \partial_{\hat{b}} \tilde{x}^3 = \sin\alpha\, \delta^{\hat{a}\hat{b}} \partial_{\hat{b}} y \quad \hat{a}, \hat{b} = 1, 2 \tag{29}$$

and supersymmetry projection conditions

$$\Gamma_{012}\epsilon = \epsilon \tag{30}$$

$$(\cos\alpha\, \Gamma_{0y}\Gamma_{11} + \sin\alpha\, \Gamma_{03y})\epsilon = \epsilon. \tag{31}$$

The further condition $F_{12} = 0$ states that there are no D0-branes being described by our configuration as can be seen from inspection of equations (30-31). Notice that when $\alpha = 0$, we recover the usual BIon describing a fundamental string ending on the D2-brane, whereas for $\alpha = \frac{\pi}{2}$, we recover the Cauchy-Riemann equations describing the intersection of two D2-branes at a point, $D2 \perp D2(0)$. Both configurations are related to each other by application of transformations (25) and (26). Computing them when (28)-(29) are satisfied we get

$$\delta \vec{E} = -\lambda \star \nabla \tilde{x}^3 \quad, \quad \delta\left(\star\nabla \tilde{x}^3\right) = \lambda \vec{E}, \tag{32}$$

where we are using the standard two dimensional calculus notation, that is, $\vec{\nabla} = (\partial_1, \partial_2)$ and $\star\vec{\nabla} = (\partial_2, -\partial_1)$. Its finite transformation is

$$\vec{E}' = \cos(\alpha + \lambda)\, \vec{\nabla} y \quad, \quad \star\vec{\nabla}\tilde{x}'^3 = \sin(\alpha + \lambda)\, \vec{\nabla} y \tag{33}$$

Thus, as expected, by fine tuning the global parameter λ, we interpolate between BIon configurations and $D2 \perp D2(0)$ intersections.

The SO(2) rotation described by (32) fits with the supersymmetry algebra picture. In this case, the charge carried by the fundamental string is given by the worldspace integral $\mathcal{Z}_y = \int_{D2} \vec{E} \cdot \vec{\nabla} y$, whereas the charge carried by the second D2-brane admits the field theory realization $\mathcal{Z}_{3y} = \int_{D2} \star \vec{\nabla} x^3 \cdot \vec{\nabla} y$. Thus we see that $\mathcal{Z}_y$, $\mathcal{Z}_{3y}$ are indeed rotated under (32) transformations, as the pure algebraic digression was suggesting to us.

We shall conclude with the M-theory interpretation of the latter set of transformations. Since the eleven dimensional supersymmetry generator decomposes as $Q = Q_+ + Q_-$, it is pretty clear that the previous type IIA automorphism transformations become rotations in eleven dimensions, and as such, they should be linearly realized on the membrane effective action as in (10). It is actually quite simple to understand the relation among these linear transformations and the non-local ones found in the D2-brane action. As it is known [15, 16], the world volume dualization of a scalar in three dimensions gives rise to a one form. When doing such a dualization on the membrane action, the relation among their field strengths is given by $\partial_{\hat{\mu}} y = K_{\hat{\mu}\rho}$. Thus, linear transformations among eleven dimensional scalar fields generate linear transformations among $K_{\hat{\mu}\rho}$ and $\partial_{\hat{\mu}} x^m$. The above relation explains the origin of the non-local symmetries in type IIA. Furthermore, it matches with the enhancement of symmetry derived previously from T-duality.

We shall conclude by analyzing the uplifted configuration corresponding to the type IIA one discussed above. This is described by the array

$$
\begin{array}{llllllllllll}
M2: & 1 & 2 & \_ & \_ & \_ & \_ & \_ & \_ & \_ & \_ & \_ \\
M2: & \_ & \_ & \_ & 4 & 5 & \_ & \_ & \_ & \_ & \_ & \_ \\
M2: & \_ & \_ & \_ & 3 & 4 & \_ & \_ & \_ & \_ & \_ & \_ .
\end{array}
$$

Setting the static gauge $x^\mu = \sigma^\mu$ $\mu = 0, 1, 2$ and exciting three transverse scalars x^i $i = 3, 4, 5$ one can check that a solution to the kappa symmetry preserving condition is found when the following BPS equations are satisfied

$$
\cos\alpha \, \vec{\nabla} x^4 = \star\vec{\nabla} x^5 \quad , \quad \sin\alpha \, \vec{\nabla} x^4 = \star\vec{\nabla} x^3 , \tag{34}
$$

whenever ϵ satisfies

$$
\Gamma_{012}\epsilon = \epsilon \tag{35}
$$

$$
(\cos\alpha \, \Gamma_{045} + \sin\alpha \, \Gamma_{034}) \epsilon = \epsilon . \tag{36}
$$

Notice that (34) interpolate among $M2 \perp M2(0)$ configurations in definite directions for $\alpha = 0, \frac{\pi}{2}$.

It is straightforward to check that the rotation in the 35-plane generated by $U = e^{\alpha\Gamma_{35}/2}$ relates the previous configuration with one in which $\tilde{x}^3$ has a constant value, and is no longer excited. Such a configuration corresponds to two membranes intersecting at a point. This interpretation can also be checked by rewriting equation (36) in terms of the transformed Killing spinor $\epsilon' = U^t \epsilon$

$$\Gamma_{045}\epsilon' = \epsilon', \qquad (37)$$

which indeed corresponds to a membrane along 45-plane, while equation (35) is not modified ($\Gamma_{012}\epsilon' = \epsilon'$). Furthermore, all previous results on the D2-brane can be easily recovered from M-theory, this being the last check of consistency between the presented non-local transformations and dualities in M/string theory.

Acknowledgements

JS was supported by a fellowship from Comissionat per a Universitats i Recerca de la Generalitat de Catalunya and is presently being supported by a fellowship from the Feinberg Graduate School. This work was supported in part by AEN98-0431 (CICYT), GC 1998SGR (CIRIT). J.S. thanks NATO for partial support.

References

1. P. K. Townsend, *M-theory from its superalgebra*, in 'Strings, branes and dualities', Cargèse 1997, ed. L. Baulieu et al., Kluwer Academic Publ. 1999, p.141, hep-th/9712004.
2. P. K. Townsend, *M(embrane) theory on T^9*, Nucl. Phys. Proc. Suppl. **67** (1998) 88-92, hep-th/9708034.
3. E. Bergshoeff and A. van Proeyen, *The many faces of OSp(1—32)* Class. Quant. Grav. **17** (2000) 3277-3304, hep-th/0003261.
4. N. A. Obers and B. Pioline, *U-duality and M-theory* Phys. Rep. 318 (1999) 113-225, hep-th/9809039.
5. J. Molins and J. Simón, *BPS states and Automorphisms*, to appear in Phys. Rev. D. hep-th/0007253.
6. J. Gauntlett, J. Gomis and P. K. Townsend, *BPS bounds for Worldvolume Branes* JHEP9801 (1998) 003, hep-th/9711205.
7. E. Bergshoeff, R. Kallosh, T. Ortín and G. Papadopoulos, *Kappa-Symmetry, SuperSymmetry and Intersecting Branes* Nucl. Phys. **B502** (1997) 149, hep-th/9705040.
8. O. Baerwald and P. West, *Brane Rotating Symmetries and the Fivebrane Equations of Motion* Phys. Lett. **B476** (2000) 157, hep-th/9912226.
9. P. West, *Automorphisms, Non-Linear Realizations and Branes*, hep-th/0001216.
10. P. West, *Hidden Superconformal Symmetry in M-theory* JHEP 0008 (2000) 007, hep-th/0005270.
11. J. P. Gauntlett, G. W. Gibbons, C. M. Hull and P. K. Townsend, *BPS states of $D = 4$ $N = 1$ supersymmetry*, hep-th/0001024.
12. E. Bergshoeff and P. K. Townsend, *Super D-brane revisited* Nucl. Phys. **B531** (1998) 226, hep-th/9804011.
13. J. Simón, *Automorphisms as brane non-local transformations*, hep-th/0010242.
14. Y. Igarashi, K. Itoh and K. Kamimura, *Self-Duality in Super D3-brane Action* Nucl. Phys. **B536** (1998) 469, hep-th/9806161.
15. P. K. Townsend, *D-branes from M-branes* Phys. Lett. **B373** (1996) 68-75, hep-th/9512062.

16. P. K. Townsend, *Four lectures in M-theory*, Proceedings of the ICTP Summer School on High Energy Physics and Cosmology, Trieste, June 1996, hep-th/9612121.

17. E. Bergshoeff, E. Sezgin and P. K. Townsend, *Properties of the eleven dimensional Super Membrane theory* Annals of Physics **185** (1988) 330.

18. M. Cederwall, A. von Gussich, B. E. W. Nilsson and A.Westerberg, *The Dirichlet Super-Three-Brane in Ten-Dimensional Type IIB Supergravity* Nucl. Phys. **B490** (1997) 163-178, hep-th/9610148.

19. M. Cederwall, A. von Gussich, B. E. W. Nilsson, P. Sundell and A. Westerberg, *The Dirichlet Super-p-Branes in Ten-Dimensional Type IIA and type IIB Supergravity* Nucl. Phys. **B490** (1997) 179-201, hep-th/9611159.

20. M. Aganagic, C. Popescu and J. H. Schwarz, *D-Brane Actions with Local Kappa Symmetry* Phys. Lett. **B393** (1997) 311-315, hep-th/9610249.

21. M. Aganagic, C. Popescu and J. H. Schwarz, *Gauge Invariant and Gauge-Fixed D-Brane Actions* Nucl. Phys. **B495** (1997) 99-126, hep-th/9612080.

22. Y. Igarashi, K. Itoh and K. Kamimura, *Electric-Magnetic Duality Rotations and Invariance of Actions* Nucl. Phys. **B536** (1998) 454-468, hep-th/9806160.

23. C. Callan and J. M. Maldacena, *Brane Dynamics From Born-Infeld Action* Nucl. Phys. **B513** (1998) 198-212, hep-th/9708147.

24. G. W. Gibbons, *Born-Infeld particles and Dirichlet p-branes* Nucl. Phys. **B514** (1998) 603-659, hep-th/9709027.

25. J. Simón, *T-duality and Effective D-Brane Actions* Phys. Rev. **D61** 047702 (2000), hep-th/9812095.

26. K. Kamimura and J. Simón, *T-duality Covariance of SuperD-branes* Nucl. Phys. **B585** (2000) 219-252, hep-th/0003211.

SOME METRICS ON THE MANIN PLANE

GAETANO FIORE *
Dip. di Matematica e Applicazioni, Fac. di Ingegneria
Università di Napoli, V. Claudio 21, 80125 Napoli

MARCO MACEDA †
Laboratoire de Physique Théorique et Hautes Energies
Université de Paris-Sud, Bâtiment 211, F-91405 Orsay

JOHN MADORE ‡
Laboratoire de Physique Théorique et Hautes Energies
Université de Paris-Sud, Bâtiment 211, F-91405 Orsay

1. Introduction and notation

Let $\mathcal{A}$ be a *-algebra with differential calculus $\Omega^1(\mathcal{A})$ [1] and suppose that it has a frame [2], a set of 1-forms θ^i dual to a set of inner derivations $e_i = \mathrm{ad}\,\lambda_i$ and which therefore commutes with the elements of the algebra:

$$\theta^i f = f \theta^i. \tag{1}$$

The differential calculus will be real [4] if the λ_i are anti-hermitian. Using the frame we can set

$$df = e_i f \theta^i \tag{2}$$

from which it follows that the module structure of $\Omega^1(\mathcal{A})$ is given by

$$f\,dg = (f e_i g)\theta^i, \qquad dg f = (e_i g) f \theta^i.$$

* gaetano.fiore@na.infn.it
† marco.maceda@th.u-psud.fr
‡ john.madore@th.u-psud.fr

S. Duplij and J. Wess (eds.), *Noncommutative Structures in Mathematics and Physics*, 271–282.
© 2001 *Kluwer Academic Publishers. Printed in the Netherlands.*

If a frame exists the module $\Omega^1(\mathcal{A})$ is free of rank n as a left or right module. It can therefore be identified with the direct sum

$$\Omega^1(\mathcal{A}) = \bigoplus_1^n \mathcal{A} \tag{3}$$

of n copies of $\mathcal{A}$. In this representation θ^i is given by the element of the direct sum with the unit in the i-th position and zero elsewhere. We shall refer to the integer n as the dimension of the geometry. Using the frame formalism we consider some possible metrics on the Manin plane. We require that the metric be real and symmetric. In practice this means that we use the freedom of noncommutative geometry to impose a different 'σ-symmetry', which is chosen so that a complex metric is hermitian and an un-symmetric metric is σ-symmetric. The notion of reality and symmetry are changed so that the definition of hermitian does not change. We refer to a longer article [3] for more details as well as for a comparison with other definitions of metrics.

Let π be the product in $\Omega^*(\mathcal{A})$ and set

$$\pi(\theta^i \otimes \theta^j) = P^{ij}{}_{kl}\theta^k \otimes \theta^l, \qquad P^{ij}{}_{kl} \in \mathcal{Z}(\mathcal{A}).$$

Since π is a projection we have

$$P^{ij}{}_{mn}P^{mn}{}_{kl} = P^{ij}{}_{kl} \tag{4}$$

and the product $\theta^i\theta^j$ satisfies

$$\theta^i\theta^j = P^{ij}{}_{kl}\theta^k\theta^l. \tag{5}$$

If the θ^i anti-commute then

$$P^{ij}{}_{kl} = \frac{1}{2}(\delta^i_k\delta^j_l - \delta^j_k\delta^i_l). \tag{6}$$

Since the exterior derivative of θ^i is a 2-form it can necessarily be written as

$$d\theta^i = -\frac{1}{2}C^i{}_{jk}\theta^j\theta^k.$$

where, because of (5), the structure elements can be chosen to satisfy the constraints

$$C^i{}_{jk}P^{jk}{}_{lm} = C^i{}_{lm}.$$

From the generators θ^i we can construct a 1-form

$$\theta = -\lambda_i\theta^i \tag{7}$$

in $\Omega^1(\mathcal{A})$ which plays the role [1] of a Dirac operator:

$$df = -[\theta, f].$$

From the identity $d^2 = 0$ one finds that

$$d(\theta f - f\theta) = [d\theta, f] + [\theta, [\theta, f]] = [d\theta + \theta^2, f] = 0.$$

It follows that if we write

$$d\theta + \theta^2 = -\frac{1}{2}K_{ij}\theta^i\theta^j \tag{8}$$

the coefficients K_{ij} must lie in $\mathcal{Z}(\mathcal{A})$. Again from (5) they can be chosen to satisfy the constraints

$$K_{jk}P^{jk}{}_{lm} = K_{lm}.$$

It will also be convenient to introduce the quantities

$$C^{ij}{}_{kl} = \delta^i_k\delta^j_l - 2P^{ij}{}_{kl}. \tag{9}$$

Then from (4) we find that

$$C^{ij}{}_{kl}C^{kl}{}_{mn} = \delta^i_m\delta^j_n. \tag{10}$$

From the condition $d^2 = 0$ it can be shown that

$$2P^{ij}{}_{kl}\lambda_K\lambda_l - F^i{}_{kl}\lambda_i - K_{ij} = 0$$

for some array of numbers $F^i{}_{jk}$.

We introduce a flip σ:

$$\Omega^1(\mathcal{A}) \otimes_{\mathcal{A}} \Omega^1(\mathcal{A}) \xrightarrow{\ \sigma\ } \Omega^1(\mathcal{A}) \otimes_{\mathcal{A}} \Omega^1(\mathcal{A}). \tag{11}$$

In terms of the frame it is given by $S^{ij}{}_{kl} \in \mathcal{Z}(\mathcal{A})$ defined by

$$\sigma(\theta^i \otimes \theta^j) = S^{ij}{}_{kl}\theta^k \otimes \theta^l$$

and which must satisfy the constraint

$$(S^{ji}{}_{kl})^* S^{lk}{}_{mn} = \delta^i_m\delta^j_n. \tag{12}$$

We use σ to impose the reality condition

$$S^{ij}{}_{kl}g^{kl} = (g^{ji})^* \tag{13}$$

on the metric. This is a combination of a 'twisted' symmetry condition and the ordinary condition of hermiticity on a complex matrix. A covariant derivative on the module $\Omega^1(\mathcal{A})$ must satisfy both a left and a right Leibniz rule. We use the ordinary left Leibniz rule and define the right Leibniz rule as

$$D(\xi f) = \sigma(\xi \otimes df) + (D\xi)f \tag{14}$$

for arbitrary $f \in \mathcal{A}$ and $\xi \in \Omega^1(\mathcal{A})$. Using σ one can also impose [5] a reality condition on the curvature.

For every differential calculus and flip one can construct the linear connection

$$\omega^i{}_{jk} = \lambda_l(S^{il}{}_{jk} - \delta^l_j \delta^i_k). \tag{15}$$

The connection 1-form is given by

$$\omega^i{}_k = \lambda_l S^{il}{}_{jk}\theta^j + \delta^i_k \theta. \tag{16}$$

When $F^i{}_{jk} = 0$ the curvature of the covariant derivative D defined in (15) can be readily calculated. One finds the expression

$$\frac{1}{2}R^i{}_{jkl} = S^{im}{}_{rn}S^{np}{}_{sj}P^{rs}{}_{kl}\lambda_m\lambda_p - \frac{1}{2}\delta^i_j K_{kl}.$$

This can also be written in the form

$$\frac{1}{2}R^i{}_{jkl} = -S^{im}{}_{rn}S^{np}{}_{sj}S^{rs}{}_{uv}P^{uv}{}_{kl}\lambda_m\lambda_p - \frac{1}{2}\delta^i_j K_{kl}.$$

The relation (18) suggests that we define a Ricci map by the action

$$\text{Ric}(\theta^i) = \frac{1}{2}R^i{}_k\theta^k, \qquad R^i{}_k = R^i{}_{jkl}g^{lj}$$

on the frame.

In complete analogy with the commutative case a metric g can be defined as an $\mathcal{A}$-bilinear, nondegenerate map [6]

$$\Omega^1(\mathcal{A}) \otimes_\mathcal{A} \Omega^1(\mathcal{A}) \xrightarrow{g} \mathcal{A} \tag{17}$$

and as such it can [7] be used to define a 'distance' between 'points'. It is important to notice here that the bilinearity is an alternative way of expressing locality. In ordinary differential geometry if ξ and η are 1-forms then the value of $g(\xi \otimes \eta)$ at a given point depends only on the values of ξ and η at that point. Bilinearity is an exact expression of this fact. In general the algebra introduces a certain amount of non-locality via the commutation relations and it is important to assure that all geometric quantities be just that nonlocal and not more. Without the bilinearity condition it is not possible to distinguish for example in ordinary space-time a metric which assigns a function to a vector field in such a way that the value at a given point depends only on the vector at that point from one which is some sort of convolution over the entire manifold.

We define frame components of the metric by

$$g^{ij} = g(\theta^i \otimes \theta^j).$$

They lie necessarily in the center $\mathcal{Z}(\mathcal{A})$ of the algebra. The condition that (15) be metric-compatible can be written as

$$S^{im}{}_{ln}g^{np}S^{jk}{}_{mp} = g^{ij}\delta_l^k. \tag{18}$$

One can understand this odd condition by introducing a 'covariant derivative' $D_i X^j$ of a constant 'vector' by the formula

$$D_i X^j = \omega^j{}_{ik}X^k.$$

The covariant derivative $D_i(X^j Y^k)$ of the product of two such 'vectors' must be defined as

$$D_i(X^j Y^k) = D_i X^j Y^k + S^{jl}{}_{im}X^m D_l Y^k$$

since there is a 'flip' as the index on the derivation crosses the index on the first 'vector'. The condition (18) becomes then simply

$$D_i g^{jk} = 0.$$

We shall require that the metric be symmetric in the sense

$$g \circ \pi = 0 \tag{19}$$

that it annihilates the 2-forms. We shall impose also the condition

$$\pi \circ (\sigma + 1) = 0 \tag{20}$$

that the antisymmetric part of a symmetric tensor vanish. This can be considered as a condition on the product or on the flip. In ordinary geometry it is the definition of π; a 2-form can be considered as an antisymmetric tensor. Because of this condition the torsion is a bilinear map [6]. The most general solution can be written in the form

$$1 + \sigma = (1 - \pi) \circ \tau \tag{21}$$

where τ is arbitrary. Suppose that τ is invertible. Then because of the identity

$$1 = \pi + (1 + \sigma) \circ \tau^{-1}$$

one can identify the second term on the right-hand side as the projection onto the symmetric part of the tensor product. The choice $\tau = 2$ yields the value $\sigma = 1 - 2\pi$. If τ is not invertible then there arises the possibility that part of the tensor product is neither symmetric nor antisymmetric.

It is sometimes convenient to write the metric as a sum

$$g^{ij} = g_S^{ij} + g_A^{ij}$$

of a symmetric and an antisymmetric part (in the usual sense of the word) The inverse matrix we write as a sum

$$g_{ij} = \eta_{ij} + B_{ij}$$

of a symmetric and an antisymmetric term. We shall choose as normalization when possible the condition that η_{ij} be the standard Minkowski or euclidean form.

2. The Wess-Zumino calculus

The extended quantum plane is the $*$-algebra $\mathcal{A}$ generated by the hermitian elements u and v with their inverses u^{-1} and v^{-1} and the relation

$$uv = qvu, \qquad q = e^{i\alpha} \tag{22}$$

as well as the usual relations between inverses. We define, for $q^4 \neq 1$,

$$\lambda_1 = \frac{q^4}{q^4 - 1} u^{-2}v^2, \qquad \lambda_2 = -\frac{q^2}{q^4 - 1} u^{-2}.$$

The important fact is that the λ_a are singular in the limit $q \to 1$ and that they are anti-hermitian if q is of unit modulus. We find for $q^2 \neq -1$

$$e_1 u = -\frac{q^2}{(q^2 + 1)} u^{-1}v^2, \quad e_1 v = -\frac{q^4}{q^2 + 1} u^{-2}v^3,$$

$$e_2 u = 0, \qquad\qquad e_2 v = \frac{q^2}{q^2 + 1} u^{-2}v. \tag{23}$$

These derivations are again extended to arbitrary polynomials in the generators by the Leibniz rule. Using them we find

$$du = -\frac{q^2}{(q^2 + 1)} u^{-1}v^2\theta^1, \quad dv = -\frac{q^2}{q^2 + 1} u^{-2}v(q^2v^2\theta^1 - \theta^2) \tag{24}$$

and solving for the θ^i we obtain

$$\theta^1 = -q^2(q^2 + 1)uv^{-2}du, \qquad \theta^2 = -(q^2 + 1)u(uv^{-1}dv - du).$$

The module structure which follows from the condition (1) that the θ^i commute with the elements of the algebra is given by [8]

$$udu = q^2 duu, \quad udv = qdvu + (q^2 - 1)duv,$$

$$vdu = qduv, \quad vdv = q^2 dvv. \tag{25}$$

One can show that they are invariant under the coaction of the quantum group $SL_q(2, \mathbb{C})$. This invariance was encoded in the choice of λ_a.

Consider the change of generators defined by

$$u = \tilde{u}^{-2}, \qquad v = \tilde{q}^2 \tilde{u}^{-2} \tilde{v}^2.$$

If one sets also $q = \tilde{q}^{-4}$ then one finds that the Wess-Zumino relations (25) written using the generators $\tilde{u}$ and $\tilde{v}$ become

$$udu = qduu, \qquad udv = qdvu,$$
$$vdu = q^{-1}duv, \quad vdv = q^{-1}dvv. \qquad (26)$$

What we have done in fact is use the λ_a as generators of the algebra and the differential calculus; otherwise nothing has been changed. Properly renormalized then we have

$$\lambda_1 = \frac{q^{1/2}}{q-1}v, \qquad \lambda_2 = -\frac{q^{1/2}}{q-1}u.$$

and solving for the θ^i one obtains

$$\theta^1 = -q^{-1/2}(u^{-1}v)^{-1}d(u^{-1}), \qquad \theta^2 = q^{1/2}(u^{-1}v)d(v^{-1}).$$

It follows that the volume element is an exact form:

$$\theta^1\theta^2 = -d(u^{-1})d(v^{-1}).$$

This formula has been obtained by a straight-forward change of generators and, independent of the perhaps not-too-convincing arguments of the following sections, suggests that u^{-1} and v^{-1} are light-cone coordinates in the commutative limit. The frame is singular along the light cone through the origin. If in a representation one forces the original $\tilde{u}$ and $\tilde{v}$ to be hermitian then the u and v must be positive operators. One concludes then that $|t| > |x|$ and x must therefore be a bounded operator.

The structure of the differential algebra is given by the relations

$$(\theta^1)^2 = 0, \qquad (\theta^2)^2 = 0, \qquad \theta^1\theta^2 + q\theta^2\theta^1 = 0.$$

This can be written in the form (5) with $C^{12}{}_{21} = q$ and $C^{21}{}_{12} = q^{-1}$. The reality of the differential implies that the structure elements must satisfy the conditions

$$((C^i{}_{jk})^* + C^i{}_{jk})P^{jk}{}_{lm} = 0$$

from which follows that

$$(C^i{}_{21})^* = -C^i{}_{12} = q^{-1}C^i{}_{21}, \qquad (C^i{}_{12})^* = -C^i{}_{21} = qC^i{}_{12}.$$

are given by

$$C^1{}_{12} = (q^{-1} - 1)\lambda_2, \qquad C^2{}_{12} = (q^{-1} - 1)\lambda_1.$$

With the change of generators

$$t = \frac{1}{\sqrt{2}}(u^{-1} - v^{-1}), \qquad x = \frac{1}{\sqrt{2}}(u^{-1} + v^{-1}). \tag{27}$$

the commutation relation can be written as

$$[t, x] = -i\tan(\alpha/2)(t^2 - x^2).$$

3. The metrics and their connections

With our index conventions the metric is written as $g^{ij} = (g^1, g^2, g^3, g^4)$ and so the condition (18) can be written in the matrix form

$$\begin{pmatrix} S^1{}_1 & S^1{}_2 & S^1{}_3 & S^1{}_4 \\ S^2{}_1 & S^2{}_2 & S^2{}_3 & S^2{}_4 \\ S^3{}_1 & S^3{}_2 & S^3{}_3 & S^3{}_4 \\ S^4{}_1 & S^4{}_2 & S^4{}_3 & S^4{}_4 \end{pmatrix} \left(S_{(g)} \right) = \begin{pmatrix} g^1 & 0 & g^3 & 0 \\ 0 & g^1 & 0 & g^3 \\ g^2 & 0 & g^4 & 0 \\ 0 & g^2 & 0 & g^4 \end{pmatrix} \tag{28}$$

where we have introduced the matrix $S_{(g)}$ defined by

$$S_{(g)} = \begin{pmatrix} S^1{}_1 g^1 + S^1{}_2 g^3 & \cdots & \cdots & S^3{}_3 g^1 + S^3{}_4 g^3 \\ S^1{}_1 g^2 + S^1{}_2 g^4 & \cdots & \cdots & S^3{}_3 g^2 + S^3{}_4 g^4 \\ S^2{}_1 g^1 + S^2{}_2 g^3 & \cdots & \cdots & S^4{}_3 g^1 + S^4{}_4 g^3 \\ S^2{}_1 g^2 + S^2{}_2 g^4 & \cdots & \cdots & S^4{}_3 g^2 + S^4{}_4 g^4 \end{pmatrix}. \tag{29}$$

If we introduce the matrix

$$P = \frac{1}{2} \begin{pmatrix} 0 & 0 & 0 & 0 \\ 0 & 1 & -q & 0 \\ 0 & -q^{-1} & 1 & 0 \\ 0 & 0 & 0 & 0 \end{pmatrix} \tag{30}$$

of frame components for π then the condition (19) is equivalent to the relation

$$g^2 = qg^3. \tag{31}$$

The consistency condition (20) is equivalent to the conditions

$$S^1{}_3 = qS^1{}_2, \qquad S^2{}_3 = q(S^2{}_2 + 1), \qquad S^3{}_3 = qS^3{}_2 - 1, \qquad S^4{}_3 = qS^4{}_2. \tag{32}$$

The equations to be solved then are Equations (28), (31) and (32). We are especially interested in real solutions, which satisfy therefore also (13). We have found that there are several types of solutions [3], four of which we shall describe in the following subsections. One can show that there are no solutions with $\tau = 2$.

A complete classification has been given [9] of the solutions to the braid equation as well [10, 11] as of those which satisfy a weaker modified equation. In any case to within four arbitrary constants we can write the coefficients of the metric with respect to the basis $d\tilde{u}$ and $d\tilde{v}$. If we introduce the components $\tilde{g}^{ij} = g(d\tilde{u}^i \otimes d\tilde{u}^j)$ then we find from (24) that in the limit $q \to 1$

$$\tilde{g}^{ij} = \frac{1}{4}\tilde{u}^{-4}\tilde{v}^4 \begin{pmatrix} g^1\tilde{u}^2 & \tilde{u}(g^2\tilde{v} + g^3\tilde{v}^{-1}) \\ \tilde{u}(g^2\tilde{v} + g^3\tilde{v}^{-1}) & g^2\tilde{v}^2 - 2g^3 + g^4\tilde{v}^{-2} \end{pmatrix}.$$

The line element is determined by the inverse of this matrix. A metric g' defined by setting

$$\tilde{g}'^{ij} = \begin{pmatrix} 1 & 0 \\ 0 & 1 \end{pmatrix}$$

necessarily then cannot be bilinear.

3.1. SOLUTION I

A family of solutions can be found with a Minkowski-signature metric. These are the most interesting solutions. With the convenient normalization of the metric so that $g^3 = q^{-1/2}$ the flip is given by the matrix

$$S = \begin{pmatrix} q & -q^{-1/2}(q-1)g^1 & -q^{1/2}(q-1)g^1 & q^{-1}(q+1)^{-1}(q-1)(q^2+1) \\ 0 & 0 & q & -q^{-1/2}(q-1)g^1 \\ 0 & q^{-1} & 0 & q^{-3/2}(q-1)g^1 \\ 0 & 0 & 0 & q^{-1} \end{pmatrix}.$$

It tends to the ordinary flip as $q \to 1$ and for $g^1 = 0$ is a solution to the braid equation. The corresponding metric given by

$$g^{ij} = \begin{pmatrix} g^1 & q^{1/2} \\ q^{-1/2} & 0 \end{pmatrix}. \tag{33}$$

From (31) one sees that it is σ-symmetric for all g^1 and hermitian if $g^1 = 0$. In this case σ is given by

$$S = \begin{pmatrix} q & 0 & 0 & 0 \\ 0 & 0 & q & 0 \\ 0 & q^{-1} & 0 & 0 \\ 0 & 0 & 0 & q^{-1} \end{pmatrix}. \tag{34}$$

The σ and π are related as in (21) with (using the same conventions)

$$T = \begin{pmatrix} 1+q & 0 & 0 & 0 \\ 0 & 2 & 0 & 0 \\ 0 & 0 & 2 & 0 \\ 0 & 0 & 0 & 1+q^{-1} \end{pmatrix}. \tag{35}$$

The fact that T is not proportional to the identity is due to the fact that the map $(1 + \sigma)/2$ is not a projector and that we would like it to act as such and be the complementary to π. The metric is of indefinite signature and in 'light-cone' coordinates. If we use the expression $q = e^{i\alpha}$ we find that

$$g_S^{ij} = \cos(\frac{\alpha}{2}) \begin{pmatrix} 0 & 1 \\ 1 & 0 \end{pmatrix}, \qquad g_A^{ij} = i\sin(\frac{\alpha}{2}) \begin{pmatrix} 0 & 1 \\ -1 & 0 \end{pmatrix}. \tag{36}$$

The inverse metric components are defined by the equation

$$g_{ij}g^{jk} = \delta_i^k.$$

This matrix also can be split. If we rescale so that the symmetric part is of the standard form we find

$$\eta_{ij} = \begin{pmatrix} 0 & 1 \\ 1 & 0 \end{pmatrix}, \qquad B_{ij} = i\tan(\frac{\alpha}{2}) \begin{pmatrix} 0 & 1 \\ -1 & 0 \end{pmatrix}.$$

The metric connection has vanishing curvature. The linear connection (15) is given by

$$\omega^i{}_j = (1 - q) \begin{pmatrix} 1 & 0 \\ 0 & -q^{-1} \end{pmatrix} \theta.$$

Because of the identities

$$d\theta = 0, \qquad \theta^2 = 0$$

the curvature vanishes; with the choice (34) of flip the quantum plane is flat. In the commutative limit the line element is given by

$$ds^2 = g_{ij}\theta^i \otimes_S \theta^b = 2\theta^1 \otimes_S \theta^2 = d(u^{-1}) \otimes_S d(v^{-1}) = dt^2 - dx^2.$$

The subscript S indicates a symmetrized tensor product.

3.2. SOLUTION II

A family of solutions defined by flips which are not solutions to the braid equation is given by

$$S = \begin{pmatrix} -q^2 & 0 & 0 & 0 \\ 0 & 0 & q & 0 \\ 0 & -q^{-2} & -1-q^{-1} & 0 \\ 0 & 0 & 0 & q^{-1} \end{pmatrix} \tag{37}$$

The metric is given again by (33). The curvature Curv is defined by

$$\Omega^i{}_j = -(q^2 - 1)q^{-3}(1 + q + q^2) \begin{pmatrix} 0 & 0 \\ 1 & 0 \end{pmatrix} (\lambda_1)^2 \theta^1 \theta^2.$$

It diverges as $(q-1)^{-1}$ when $q \to 1$. This is then the case of a regular metric which has a singular metric connection.

3.3. SOLUTION III

A third family satisfies no reality conditions

$$S = \frac{1}{q^2 + 1} \begin{pmatrix} 2q & 0 & 0 & 1 - q^2 \\ 0 & 1 - q^2 & 2q & 0 \\ 0 & 2q & q^2 - 1 & 0 \\ q^2 - 1 & 0 & 0 & 2q \end{pmatrix}. \tag{38}$$

A σ-symmetric metric is given by

$$S^{12}{}_{21} = S^{21}{}_{12} = \frac{2q}{q^2 + 1}.$$

In the limit $q \to 1$ this becomes

$$\Omega^i{}_j = \begin{pmatrix} 0 & -1 \\ 1 & 0 \end{pmatrix} (u^2 + v^2)\theta^1\theta^2.$$

3.4. NON-SOLUTIONS

There are a certain number of partial solutions which are unsatisfactory for some reason or other. As an example, to underline the possibility of exotic metrics which are neither symmetric nor anti-symmetric according to our definitions, we consider σ defined by the matrix

$$S = \begin{pmatrix} 0 & 0 & 0 & \gamma \\ 0 & -1 & 0 & 0 \\ 0 & 0 & -1 & 0 \\ \gamma^{-1} & 0 & 0 & 0 \end{pmatrix}$$

where $\gamma \in \mathbb{R}$ is a parameter. This value of S is a solution to the braid equation. The σ and π are related as in (21) with (using the same conventions)

$$T = \begin{pmatrix} 1 & 0 & 0 & \gamma \\ 0 & 0 & 0 & 0 \\ 0 & 0 & 0 & 0 \\ \gamma^{-1} & 0 & 0 & 1 \end{pmatrix}. \tag{39}$$

This means that τ is not invertible and the case is degenerate. The problem here is that $(1 + \sigma)/2$ cannot even be twisted to a projector. The metric is given by

$$g^{ij} = i \begin{pmatrix} 1 & 0 \\ 0 & -\gamma^{-1} \end{pmatrix}. \tag{40}$$

One has $\tau = 1 + \sigma$ and the flip is degenerate. Instead of interchanging g^2 and g^3 as does the ordinary flip, it interchanges g^1 and g^4. It also changes the sign, which accounts for the i in the metric components. Also $g \circ (1 + \sigma) = 0$ so in a certain sense the metric has vanishing symmetric as well as antisymmetric parts. We refer to σ nonetheless as a 'flip' because it satisfies (20). The curvature is given by

$$\Omega^i{}_j = q^{-1}(q^2 - 1)\delta^i_j \lambda_1 \lambda_2 \theta^1 \theta^2$$

It is singular in the commutative limit.

Finally we notice that here is no solution using the $\hat{R}$-matrix to construct σ. A similar problem was found by Cotta-Ramusino & Rinaldi in trying to construct holonomy groups [12].

Acknowledgment

The authors would like to thank A. Chakrabarti for enlightening conversations. One of them (JM) would like to thank Dieter Lüst for his hospitality at the Institut für Physik, Berlin, were part of this research was carried out.

References

1. A. Connes, *Noncommutative Geometry*, Academic Press, 1994.
2. A. Dimakis and J. Madore, *Differential calculi and linear connections*, J. Math. Phys. **37** (1996), 4647–4661.
3. G. Fiore, M. Maceda and J. Madore, *Metrics on the Manin Plane*, Preprint (to appear).
4. A. Connes, *Noncommutative geometry and reality*, J. Math. Phys. **36** (1995), 6194.
5. G. Fiore and J. Madore, *Leibniz rules and reality conditions*, Euro. Phys. Jour. C math/980607 (to appear).
6. M. Dubois-Violette, J. Madore, T. Masson, and J. Mourad, *On curvature in noncommutative geometry*, J. Math. Phys. **37** (1996), 4089–4102.
7. B. L. Cerchiai, R. Hinterding, J. Madore, and J. Wess, *The geometry of a q-deformed phase space*, Euro. Phys. Jour. C **8** (1999), 533–546.
8. J. Wess and B. Zumino, *Covariant differential calculus on the quantum hyperplane*, Nucl. Phys. (Proc. Suppl.) **18B** (1990), 302.
9. J. Hietarinta, *Solving the two-dimensional constant quantum Yang-Baxter equation*, J. Math. Phys. **34** (1993), 1725.
10. M. Gerstenhaber and A. Giaquinto, *Boundary solutions of the quantum Yang-Baxter equation and solutions in three dimentions*, q-alg/9710033 (to appear).
11. B. Aneva, D. Arnaudon, A. Chakrabarti, V. Dobrev, and S. Mihov, *On combined standard-nonstandard or hybrid (q, h)-deformations*, math.QA/0006206 (to appear).
12. P. Cotta-Ramusino and M. Rinaldi, *Link-diagrams, Yang-Baxter equation and quantum holonomy*, in Quantum Groups with applications to Physic, (M. Gerstenhaber and J. Stasheff, eds.), Vol. 134 Amer. Math. Soc., Providence, Rhode Island, 1992, pp. 19–44.

COHERENCE ISOMORPHISMS FOR A HOPF CATEGORY

VOLODYMYR LYUBASHENKO *
Institute of Mathematics, Kyiv, Ukraine

Crane and Frenkel proposed a notion of a Hopf category in [1]. It was motivated by Lusztig's approach to quantum groups – his theory of canonical bases. In particular, Lusztig obtains braided deformations $U_q \mathfrak{n}_+$ of universal enveloping algebras $U \mathfrak{n}_+$ for some nilpotent Lie algebras $\mathfrak{n}_+$ together with canonical bases of these braided Hopf algebras [2–4]. The elements of the canonical basis are identified with certain objects of equivariant derived categories, contained in semisimple abelian subcategories of semisimple complexes. Conjectural properties of these categories were collected into a system of axioms of a Hopf category, equipped with functors of multiplication and comultiplication, isomorphisms of associativity, coassociativity and coherence which satisfy four equations [1]. Crane and Frenkel gave an example of a Hopf category resembling the semisimple category encountered in Lusztig's theory corresponding to one-dimensional Lie algebra $\mathfrak{n}_+$ – nilpotent subalgebra of $\mathfrak{sl}(2)$. The mathematical framework and some further examples of Hopf categories were provided by Neuchl [5].

We discuss an example of a related notion – triangulated Hopf category – the whole equivariant derived category equipped with operations-functors and structure isomorphisms. The additive relations between operations proposed in [1] are replaced with distinguished triangles. The preliminary study of the subject can be found in [6, 7]. In the present paper we construct the coherence isomorphisms in full required generality. The essential ingredient – the equation for coherence isomorphisms is still not proven.

1. Operations in a graded Hopf algebra

Let Q_+ be a commutative monoid additively generated by elements of a finite set I. Denote $R = \mathbb{Z}[q, q^{-1}]$. Let $\mathcal{H}$ be a Q_+-graded braided Hopf R-algebra, for instance, the algebra $U_q \mathfrak{n}_+$ of Lusztig [4]. The comultiplication in $\mathcal{H} = \oplus_{v \in Q_+} \mathcal{H}_v$

* lub@imath.kiev.ua

S. Duplij and J. Wess (eds.), Noncommutative Structures in Mathematics and Physics, 283–294.

can be written as

$$\Delta = \sum_{u,v \in Q_+} \Delta_{u,v}, \qquad \Delta_{u,v} : \mathcal{H}_{u+v} \to \mathcal{H}_u \otimes_R \mathcal{H}_v.$$

Similarly for iterated comultiplication $\Delta^{(b)} = (\Delta^{(b-1)} \otimes 1) \circ \Delta : \mathcal{H} \to \mathcal{H}^{\otimes b}$,

$$\Delta^{(b)} = \sum_{v_j \in Q_+} \Delta^{(b)}_{v_1,\ldots,v_b}, \qquad \Delta^{(b)}_{v_1,\ldots,v_b} : \mathcal{H}_{v_1+\cdots+v_b} \to \mathcal{H}_{v_1} \otimes_R \cdots \otimes_R \mathcal{H}_{v_b}.$$

The associativity, the coassociativity and the bialgebra axiom imply the equation

$$\Delta^{(b)}(x^1) \cdot \ldots \cdot \Delta^{(b)}(x^a) = \Delta^{(b)}(x^1 \cdot \ldots \cdot x^a) \tag{1}$$

for arbitrary elements $x^i \in \mathcal{H}$. Note that the multiplication in the left hand side uses the braiding. Apply equation (1) to homogeneous elements x^i of degree v^i and write down its homogeneous component of multidegree $(v_1, \ldots, v_b) \in Q_+^b$:

$$\sum_{\substack{\sum_j v_j^i = v^i \\ \sum_i v_j^i = v_j}} \Delta^{(b)}_{v_1^1,\ldots,v_b^1}(x^1) \cdot \ldots \cdot \Delta^{(b)}_{v_1^a,\ldots,v_b^a}(x^a) = \Delta^{(b)}_{v_1,\ldots,v_b}(x^1 \cdot \ldots \cdot x^a). \tag{2}$$

Each summand in the left hand side can be viewed as an operation with a inputs and b outputs. These operations are not distinguished in algebra setup. However, in graded Hopf categories their explicit use seems advantageous.

2. The main ingredients

Categories will be equivariant derived categories $\frac{X}{G} := D_G^{b,c}(X)$, where X is a complex algebraic variety, equipped with the action of a complex algebraic group G, as defined by Bernstein and Lunts [8].

The functors will be compositions of functors of the three types (see [8]). Let $\phi : G \to H$ be a group homomorphism, let X be a G-space, let Y be an H-space, and let $f : X \to Y$ be a ϕ-equivariant map. Then there are
— the inverse image functor $\frac{f^*}{\phi} : \frac{Y}{H} \to \frac{X}{G}$,
— if $\phi : G \to H$ is surjective, $K = \text{Ker}(\phi)$, X is K-free, and $Y = K\backslash X$, the direct image functor (in this case it is an equivalence) $\frac{f}{\phi_*} : \frac{X}{G} \to \frac{Y}{H}$.
— if $\phi = 1 : G = H$ is the identity, the direct image functor with proper supports $\frac{f}{1!} : \frac{X}{G} \to \frac{Y}{G}$.

Quiver. Let (H, I) be a finite oriented graph with the set of vertices I, the set of edges H, the structure map $H \to I \times I$, $h \mapsto (h', h'')$, where $h' \in I$ is the source of $h \in H$, and $h'' \in I$ is the target of $h \in H$, such that $h' \neq h''$.

Let V be a finite dimensional I-graded $\mathbb{C}$-vector space, (a function $V : I \to$ Ob $\mathbb{C}$-vect, $i \mapsto V(i)$). Its automorphism group is

$$G_V = \text{Aut}_{I\text{-grad-vect}} V = \prod_{i \in I} GL(V(i)).$$

Define a linear space

$$E_V = \oplus_{h \in H} \text{Hom}_{\mathbb{C}}(V(h'), V(h'')).$$

The union of all E_V is the class of representations of the quiver. The group G_V acts on E_V by $(g.x)_h = g_{h''} x_h g_{h'}^{-1}$. The union of all $G_V \backslash E_V$ is the set of isomorphism classes of representations of the quiver. We consider the collection of equivariant derived categories $\frac{E_V}{G_V}$ as our Hopf category.

Filtrations. To introduce operations we need to consider decompositions of V

$$\mathcal{V}: \qquad V^1 \oplus V^2 \oplus \cdots \oplus V^k = V$$

into I-graded subspaces. Associate with it a filtration of V

$$0 = V^{(0)} \subset V^{(1)} \subset \cdots \subset V^{(k)} = V, \qquad V^{(m)} = V^1 \oplus \cdots \oplus V^m.$$

Associate with it the parabolic group $P_{\mathcal{V}}$

$$P_{\mathcal{V}} = \{g \in G_V \mid \forall m \ g(V^{(m)}) \subset V^{(m)}\}.$$

The unipotent radical of $P_{\mathcal{V}}$ is denoted $U_{\mathcal{V}}$. The group

$$L_{\mathcal{V}} = \{g \in G_V \mid \forall m \ g(V^m) \subset V^m\} = \prod_{m=1}^{k} G_{V^m} \simeq P_{\mathcal{V}}/U_{\mathcal{V}}$$

is a Levi subgroup of $P_{\mathcal{V}}$.

Let $F_{\mathcal{V}}$ be the linear subspace of E_V respecting the filtration:

$$F_{\mathcal{V}} = \{x \in E_V \mid \forall m, h \ x_h(V^{(m)}(h')) \subset V^{(m)}(h'')\}.$$

The group $P_{\mathcal{V}}$ acts in $F_{\mathcal{V}}$.

Operations. Let two decompositions of V into a direct sum be given:

$$\mathcal{V}: \qquad V^1 \oplus V^2 \oplus \cdots \oplus V^k \xrightarrow{\sim} V,$$
$$\mathcal{W}: \qquad W_1 \oplus W_2 \oplus \cdots \oplus W_l \xrightarrow{\sim} V.$$

Let $\mathcal{O} \subset G_V$ be a left P_W-invariant and right P_V-invariant subset. We associate with it an operation

$$
\mathbf{X}_W^{\mathcal{O};V} = \quad\cdots\quad = \quad\cdots\quad = \pitchfork_W^V \circ \Psi_W^{\mathcal{O};V}.
$$

The components of it are the generalized multiplication and comultiplication functors.

Multiplication half. The multiplication half operation is

$$
\cdots = \Psi_W^{\mathcal{O};V}
$$

$$
= \left(\frac{\prod_{i=1}^{k} E_{V^i}}{L_V} \xrightarrow{\phi^*} \frac{\mathcal{O} \times F_V}{P_W \times P_V} \xrightarrow{\pi_*} \frac{\mathcal{O} \times_{P_V} F_V}{P_W} \xrightarrow{\alpha_!} \frac{E_V}{P_W} \right).
$$

The scheme of multiplication is similar to that of Lusztig [2–4]:

$$
\prod_{i=1}^{k} E_{V^i} \xleftarrow{\phi} \mathcal{O} \times F_V \xrightarrow{\pi} \mathcal{O} \times_{P_V} F_V \xrightarrow{\alpha} E_V,
$$

where $\phi(o,f) = \kappa(f)$ is the forgetful map, $\kappa : F_V \to \prod_{i=1}^{k} E_{V^i}$ is the natural projection, π is the canonical projection, $\alpha(o,f) = o.\iota(f)$ is induced from the action map, and $\iota : F_V \to E_V$ is the natural embedding.

Comultiplication half. The comultiplication half operation functor is

$$
\cdots = \pitchfork_W^V
$$

$$
= \left(\frac{E_V}{P_W} \xrightarrow{\iota^*} \frac{F_W}{L_W} \xrightarrow{\kappa_!} \frac{\prod_{j=1}^{l} E_{W^j}}{L_W} \xrightarrow{\tau} \frac{\prod_{j=1}^{l} E_{W^j}}{L_W} \right),
$$

where τ is the shift

$$
\tau L = L[2 \sum_{r>s;h} \dim W^r(h') \cdot \dim W^s(h'')].
$$

The scheme of comultiplication is made of the natural embedding ι and the natural projection κ (as in Lusztig [2–4]):

$$
E_V \xleftarrow{\iota} F_W \xrightarrow{\kappa} \prod_{j=1}^{l} E_{W^j}.
$$

Braiding.   For a module M over $G_{A_1} \times \cdots \times G_{A_k}$ and a module N over $G_{B_1} \times \cdots \times G_{B_l}$ where A_m, B_n are some I-graded vector spaces, we define the braiding as the functor

$$\frac{M \times N}{\prod G_{A_m} \times \prod G_{B_n}} \xrightarrow{\tau} \frac{M \times N}{\prod G_{A_m} \times \prod G_{B_n}} \xrightarrow{\sigma^*} \frac{N \times M}{\prod G_{B_n} \times \prod G_{A_m}},$$

where σ is the permutation isomorphism of groups and modules and the functor τ is the shift

$$\tau(L) = L\left[-2 \sum_{\substack{m,n \\ i \in I}} \dim A_m(i) \dim B_n(i) + 2 \sum_{\substack{m,n \\ h \in H}} \dim A_m(h') \dim B_n(h'')\right].$$

Distinguished triangles.   To clarify the meaning of operations $\mathbf{X}^{\mathcal{O}}$, notice that the orbits of the action of $P_W \times P_V$ in G_V are in natural bijection with the orbits of the action of G_V in the space of pairs of filtrations $P_W \backslash G_V \times G_V / P_V$. By [9] these orbits are in bijection with $a \times b$-matrices (v_j^i) with elements in $Q_+ = \mathbb{Z}_+^I$, such that $\sum_j v_j^i = v^i$ is the dimension of V^i and $\sum_i v_j^i = v_j$ is the dimension of W_j. Thus, the orbits are in bijection with the summands in the left hand side of equation (2). The $P_W \times P_V$-invariant subsets are unions of orbits, thereby, they are represented by sums of several summands in (2).

The additive relation (2) in algebra is replaced for our Hopf category by a system of functorial distinguished triangles

$$\mathbf{X}^{\mathcal{O}_U} \to \mathbf{X}^{\mathcal{O}_X} \to \mathbf{X}^{\mathcal{O}_F} \to$$

given for any bi-invariant subset $\mathcal{O}_X \subset G_V$ and a bi-invariant closed subset $\mathcal{O}_F \subset \mathcal{O}_X$ with $\mathcal{O}_U = \mathcal{O}_X - \mathcal{O}_F$. The following diagram made with given distinguished triangles is an octahedron

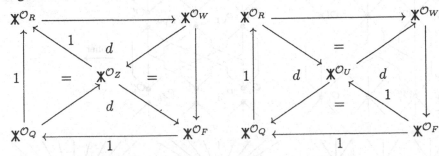

for any pair of closed embeddings $\mathcal{O}_F \subset \mathcal{O}_Z \subset \mathcal{O}_W$, where $\mathcal{O}_U = \mathcal{O}_W - \mathcal{O}_F$, $\mathcal{O}_Q = \mathcal{O}_Z - \mathcal{O}_F$, $\mathcal{O}_R = \mathcal{O}_W - \mathcal{O}_Z$. This means commutativity of two squares formed by diagonal maps and of the four triangles marked "=".

Coherence isomorphism.   Both associativity isomorphism and coassociativity isomorphism of [7] are particular cases of the general coherence isomorphism.

For any collection of indices and for any collection of bi-invariant subsets $(\mathcal{O}'_1, \ldots, \mathcal{O}'_a, \mathcal{O}''_1, \ldots, \mathcal{O}''_b)$, which may occur in the following diagram, there exists a bi-invariant subset $\mathcal{O}$ and a coherence isomorphism

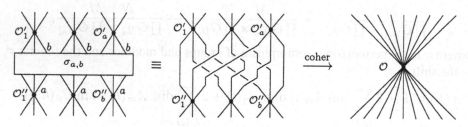

Here $\sigma_{a,b} = (s_{a,b})^{\sim}_{+}$ is the braid, corresponding to the permutation $s_{a,b}$ of the set $\{1, 2, \ldots, ab\}$,

$$s_{a,b}(1 + r + kb) = 1 + k + ra \quad \text{for} \quad 0 \leq r < b, 0 \leq k < a,$$

under the standard splitting $S_{ab} \to B_{ab}$, which maps the elementary transpositions to the generators of the braid group. The subset $\mathcal{O}$ is computed as follows

$$\overline{\mathcal{O}} = U_{\mathcal{V}} \cdot \prod_m \mathcal{O}'_m = \prod_m \mathcal{O}'_m \cdot U_{\mathcal{V}} \subset P_{\mathcal{V}},$$

$$\underline{\mathcal{O}} = U_{\mathcal{W}} \cdot \prod_r \mathcal{O}''_r = \prod_r \mathcal{O}''_r \cdot U_{\mathcal{W}} \subset P_{\mathcal{W}},$$

$$\mathcal{O} = \underline{\mathcal{O}} P_{\mathcal{U}} \times_{P_{\mathcal{U}}} \overline{\mathcal{O}} = \underline{\mathcal{O}} \times_{P_{\mathcal{W}} \cap P_{\mathcal{U}}} \overline{\mathcal{O}} = \underline{\mathcal{O}} \cdot \overline{\mathcal{O}} \subset G_{\mathcal{V}}.$$

The general coherence isomorphism is built as the composition

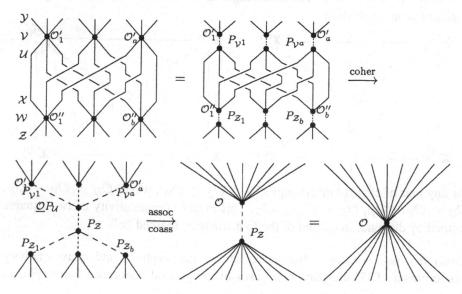

The three components of the coherence isomorphism are defined next.

$$\prod \frac{E_{Y^m_s}}{L_{ym}} \xrightarrow{\phi^*} \prod \frac{O'_m \times F_{ym}}{P_{ym} \times P_{ym}} \xrightarrow{\pi_*} \prod \frac{O'_m \times_{P_{ym}} F_{ym}}{P_{ym}} \xrightarrow{\alpha_!} \prod \frac{E_{vm}}{P_{ym}}$$

The isomorphism coher is presented in Figure 9, where the numbers

$$A = \sum_{m<n;r>s} \sum_{i \in I} \dim V^m_r(i) \cdot \dim V^n_s(i),$$

$$B = \sum_{m>n;r>s} \sum_{h \in H} \dim V^m_r(h') \cdot \dim V^n_s(h'')$$

are, actually, dimensions of the spaces

$$A = U_W/(U_W \cap P_V) = \oplus_{m<n;r>s} \operatorname{Hom}(V^m_r, V^n_s),$$

$$B = \oplus_{h \in H; m>n; r>s} \operatorname{Hom}_{\mathbb{C}}(V^m_r(h'), V^n_s(h'')) \subset F_W \cap F_V,$$

and we use the notation $F = F_W \cap F_V/B$.

The whole coherence isomorphism is presented in Figure 10.

3. Elementary isomorphisms.

The coherence isomorphisms are pastings of isomorphisms and their inverses of the following 10 types:

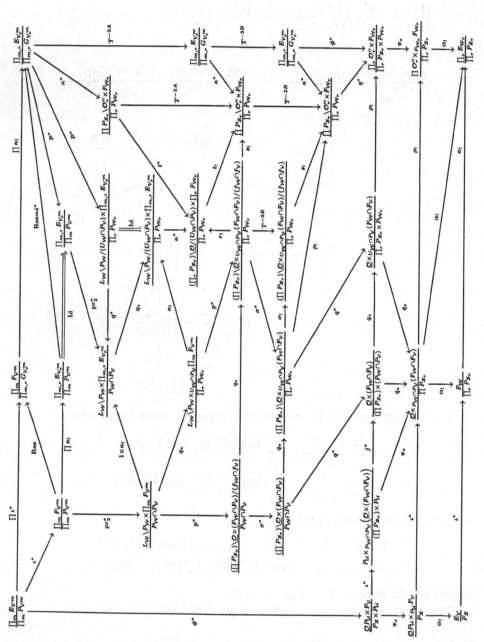

Figure 9. The isomorphism coher.

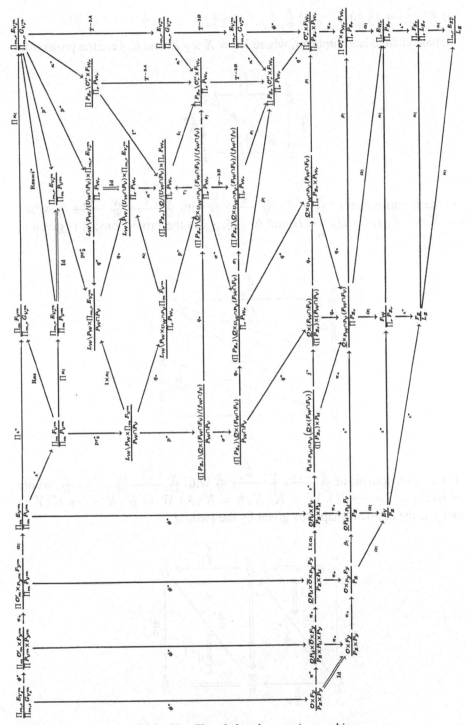

Figure 10. The whole coherence isomorphism.

a) $\frac{g}{\psi}^* \frac{f}{\phi}^* \xrightarrow{\sim} (\frac{f}{\phi}\frac{g}{\psi})^*$;　　b) $\frac{f}{\phi}_* \frac{g}{\psi}_* \xrightarrow{\sim} (\frac{f}{\phi}\frac{g}{\psi})_*$;　　c) $\frac{f}{1!}\frac{g}{1!} \xrightarrow{\sim} (\frac{f}{1}\frac{g}{1})_!$;

d) base change isomorphism, where $W = X \times_Y Z$, and h, j are the projections

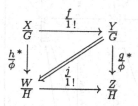

e) the isomorphism of $\frac{X}{G} \xrightarrow{f} \frac{Y}{H} \xrightarrow{g^*} \frac{Z}{B}$ with $\frac{X}{G} \xrightarrow{h^*} \frac{W}{K} \xrightarrow{j} \frac{Z}{B}$, where $W = X \times_Y Z$, $K = G_\phi \times_H {}_\psi B$ and h, j, ξ, χ are the projections; it is given by the pasting

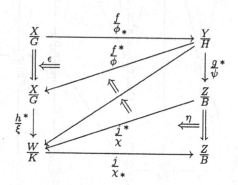

f) the isomorphism of $\frac{X}{G} \xrightarrow{f} \frac{Y}{G} \xrightarrow{g} \frac{Z}{H}$ with $\frac{X}{G} \xrightarrow{j} \frac{W}{H} \xrightarrow{h} \frac{Z}{H}$, where $K = \mathrm{Ker}(\phi : G \longrightarrow H), W = K\backslash X, h = K\backslash f : W = K\backslash X \longrightarrow K\backslash Y = Z$, and j is the quotient map; it is given by the pasting

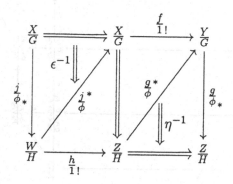

And 4 more types of elementary isomorphisms:

i) whenever $j|\phi$ is an induction map and π, $q = \pi \circ (j|\phi)$ are quotient maps, there is an isomorphism

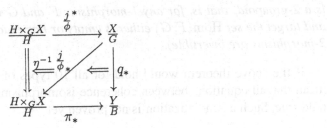

q) whenever $\pi : \frac{X}{G} \to \frac{Y}{H}$ is a quotient map, there is an isomorphism

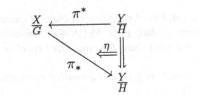

s) whenever $\tilde{P}$ is a split extension of $\tilde{L}$, $U = \mathrm{Ker}(P \xrightarrow{p} L)$ is contractible and $\tilde{E}$ is a $\tilde{P}$-space, on which U acts trivially, then $\frac{1}{p}^* : \frac{\tilde{E}}{\tilde{L}} \to \frac{\tilde{E}}{\tilde{P}}$ is an equivalence and there is an isomorphism

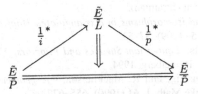

v) whenever G-map $h : E \to B$ is a vector bundle, there is an isomorphism

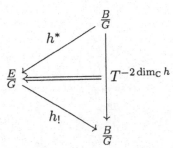

Theorem 17. *The 2-category formed by*

— *objects: equivariant derived categories;*
— *1-morphisms: compositions of functors of 3 types: inverse image functors, direct image functors for quotient maps, direct image functors with proper supports;*

— *2-morphisms: compositions of isomorphisms of 6 types a)–f) or their inverses*

is a 2-groupoid, that is, for any 1-morphisms F and G with the common source and target the set $\mathrm{Hom}(F, G)$ *either is empty or has exactly one element (and all 2-morphisms are invertible).*

If the above theorem would hold for all 10 types of isomorphisms, it would mean that all equations between coherence isomorphisms, which can be written, hold true. Such a generalization is not proven yet.

References

1. L. Crane and I. B. Frenkel, *Four dimensional topological quantum field theory, Hopf categories, and the canonical bases*, J. Math. Phys. **35** (1994), 5136–5154.
2. G. Lusztig, *Canonical bases arising from quantized enveloping algebras*, J. American Math. Soc. **3** (1990), 447–498.
3. ———, *Quivers, perverse sheaves, and quantized enveloping algebras*, J. American Math. Soc. **4** (1991), 365–421.
4. ———, *Introduction to Quantum groups*, Birkhäuser, Boston, 1993.
5. M. Neuchl, *Representation Theory of Hopf Categories*, PhD thesis. To appear in Adv. in Math. under the title Higher-dimensional algebra VI: Hopf categories. Available at http://www.mathematik.uni-muenchen.de/~neuchl.
6. V. V. Lyubashenko, *Example of a triangulated Hopf category*, Vīsnik Kiïv. Unīv. Ser. Fīz.-Mat. Nauki **2** (1999), 50–58 (in Ukrainian).
7. ———, *Operations and isomorphisms in a triangulated Hopf category*, Methods of Func. Analysis and Topology **5** (1999), 37–53.
8. J. Bernstein and V. Lunts, *Equivariant Sheaves and Functors*, Vol. 1578 of *Lecture Notes in Math.*, Springer, Berlin, Heidelberg, 1994.
9. A. A. Beilinson, G. Lusztig, and R. MacPherson, *A geometric setting for the quantum deformations of gl_n*, Duke Math. J. **61** (1990), 655–677.

FUSION RINGS AND TENSOR CATEGORIES

ALEXANDER GANCHEV *
INRNE, Tsarigradsko chausse 72, BG 1784 Sofia

The definition of a fusion ring F [1], [2], [3] is an abstraction of the properties of the Grothendieck ring $K_0(C)$ of a rigid braided semisimple monoidal category C. For certain issues it is convenient to pass to an algebra (over the complex numbers) thus a fusion algebra F is a unital associative and commutative algebra with a chosen basis I such that the fusion rules N_{ab}^c, $a, b, c \in I$, i.e., the structure constants in this basis, $a \cdot b = \sum_c N_{ab}^c c$, are in $\mathbf{Z}_+$ and their is an involutive automorphism $a \to \bar{a}$ such that $N_{ab}^1 = \delta_{\bar{a},b}$. The set I corresponds to the sectors, i.e., the equivalence classes of simple objects or irreps, the monoidal structure in C is responsible for the structure of unital associative ring, the braiding for the commutativity, while the rigidity translates in the involutative automorphism.

Fusion rings/algebras appear in many occasions (we consider only finite dimensional ones) : the category C in $K_0(C)$ could be **Rep**(finite (quantum) group); **Rep**$(U_q(g))/Z$ with $q^p = 1$, g a simple Lie algebra, and Z the ideal of zero quantum dimensional modules ; Also C could be the Moore-Siberg category of 2-dimensional rational conformal field theory (2D-RCFT) or the Doplicher-Roberts category of localizable automorphisms of the algebra of observables of a 2D-QFT (quantum field theory) with I labeling the superselection sectors (the generalized charges). The last three are typically non Tannakian categories and in particular the statistical dimensions(=ranks) of the sectors are in general only algebraic integers. Most generally C is the rep category of a quasitriangular weak Hopf algebra (or quantum gropoid). On many occasions (2D-RCFT, 2D-QFT) one has more structure with C being ribbon(=tortile) and in fact a Turaev modular category with I comprising a representation of the modular group $SL_2(\mathbf{Z})$ with modular S and T matrices. The S plays the role of characters and diagonalizes the fusion rules (Verlinde's famous formula) while T is diagonal with the balancing phases on the diagonal.

Now I briefly mention several in my view important problems: structure theory of fusion rings and tensor categories, classification of particular cases of fusion

---

* ganchev@inrne.bas.bg

S. Duplij and J. Wess (eds.), Noncommutative Structures in Mathematics and Physics, 295–298.

rings and tensor categories, categorification, i.e., reconstructing a tensor category from its fusion ring and finally explicit formulas for certain fusion rings.

Fusion rules with a generator of dimension < 2 are classified in [1]. For modular fusion algebras (i.e., reps of $SL_2(\mathbf{Z})$ with Verlinde giving fusion rules) there are initial steps towards a classification [4]. Most anything else is open.

Fusion algebras are particular cases of table algebras [5]. For a table algebra the requirements that the structure constants N_{ab}^c are positive integers and $N_{ab}^1 = \delta_{\bar{a},b}$ are relaxed to $N_{ab}^c \in \mathbf{R}_+$ and $N_{ab}^1 \neq 0$ iff $\bar{a} = b$. Table algebras have been extensively studied by Arad, Blau and coworkers. Particular cases of table algebras with generators of dimension 2 or 3 have been classified. Though they are not directly relevant to fusion rule algebra classification one again encounters for the fusion graphs a 1-dimensional structure (affine Dynkin diagrams) for the case of a dimension 2 generator and a 2-dimensional structure (the fusion graph of the fundamental irrep of $sl(3)$, a tringular tesselation of the corresponding Weyl chamber, or foldings of it) [6].

For finite groups it is clear that simple groups have fusion rules algebras which have no notrivial subfuison rule algebras, hence such fusion rule algebras is natural to call simple. More generally if a group G has a normal subgroup H then $K_0(G/H)$ is a subfusion rule algebra of $K_0(G)$. This extends to Hopf algebras [7] and [8]. For table algebras there is a more developed structure theory [9] — in particular one has composition series for table algebras. What is the theory of extensions for fusion rule algebras is an open subject. Since K_0 is only half exact one will probably have to use the higher K functors and the long exact sequence in K theory to relate information about the structure of tensor categories and their fusion rule algebras.

Categorification, i.e., reversing the K_0 functor, is a very challenging problem. Some very initial "experimental" work of solving the pentagon equations to obtain categories from given fusion rules was done in [10]. For the fusion rules of truncated $sl(n)$ with the relevant Hecke algebra the corresponding braided tensor categories were reconstructed in [11]. The pentagon is a (in general a nonabelian) 3-cocycle condition — a preliminary sketch of how to attack the relevant non-abelian cohomology problem is given in [12]. For the case of abelian fusion rules (F is the group algebra of an abelian group) it is an ordinary group cohomology problem solved in [1]. The categorification of the fusion rules of the quaternionic or the rank 8 dihedral group and their generalizations (where all but one of the sectors are abelian) was done in [13]. In general, for a tensor category with a nontrivial abelian subfusion algebra one can characterize all $6j$ symbols involving an abelian lable in terms of abelian group cohomology and moreover there is also an action on all $6j$ symbols by the abelian group (work in preparation). One would like to characterize the image of K_0 in the category of all fusion rule algebras and find "moduli" distinguishing categories with the same fusion rules. In the case of modular categories one is tempted to conjecture that the balancing phases (the T matrix) separates categories with the same fusion rules (="character table"=modular S matrix) and that K_0 is a bijection from (equivalent classes of) modular categories to modular fusion algebras [14]. Since it seems to be the case that two different simple finite groups cannot have the same fusion rules one can try to explore a conjecture that if a simple fusion rule algebra has a categorification than it is unique.

For certain classes of fusion rules, e.g. fusion rules of WZW models based on affine Kac-Moody algebras $\hat{g}_k$ at integer levels k (same as truncated $U_q(g)$ for $q^{k+h^\vee} = 1$) one has nice formulas for N_{ab}^c generalizing a classical formula of Weyl [15], [16], [17]. For the much harder and less studied case of fractiona level WZW one knows the fusion rules only for $g = sl(2)$ and $sl(3)$ ([18] and [19] respectively). Very little is known for the fusion rules of more general models of 2D-RCFT. In particular one would like to know the fusion rules of fractional level affine $sl(n)$ and on the other hand to relate them to the fusion rules of W-algebras obtained from these models by quantum hamiltonian reduction or cosetting. Even for the case of the Polyakov-Bershadski $W_3^{(2)}$ which is obtained as the nonprincipal reduction of $sl(3)$ at fractional levels the fusion rules are not known in general. A better understanding of the structure theory of fusion rules hopefully could help in such problems. On the other hand the fusion rules of fractional $sl(3)$ do not look like anything coming from a known algebraic object (finite group, Lie (super) algebra) hence it is interesting to try to categorify these fusion rules.

References

1. J. Frölich and T. Kerler, *Quantum groups, quantum categories and quantum field theory*, Lecture Notes in Math. **1524**, Springer-Verlag, Berlin , 1993.
2. J. Fuchs, *Fusion rules in conformal field theory*, Fortschr. Phys. **42** (1994), 1.
3. P. Di Francesco, P. Mathieu, and D. Sénéchal, *Conformal Field Theory*, Springer-Verlag, Berlin, 1997.
4. W. Eholzer, *On the classification of modular fusion algebras*, Commun. Math. Phys. **172** (1995), 623.
5. Z. Arad and H. Blau, *Table algebras and applications to finite group theory*, J. Algebra, **138** (1991), 137-185.
6. H. Blau, B. Xu, Z. Arad, E. Fisman, V. Miloslavsky, and M. Muzychuk, *Homogeneous Integral Table Algebras of Degree Three: A Trilogy* Memoairs of the AMS, vol. 144, no. 684 (2000).
7. W. Nichols, M. B. Richmond, *The Grothendieck Group of a Hopf Algebra*, J. Pure and Appl. Algebra **106** (1996), 297-306.
8. D. Nikshych, *K_0-Rings and Twisting of Finite Dimensional Semisimple Hopf Algebras* Commun. Algebra **26** (1998), 321-342.
9. H. Blau, *Quotient Structures in C-Algebras*, J. Algebra **175** (1995), 24-64.
10. J. Fuchs, A. Ganchev, and P. Vecsernyés, *Rational Hopf algebras: polynomial equations, gauge fixing, and low dimensional examples*, Int. J. Mod. Phys. **A10** (1995), 3431;

11. D. Kazhdan and H. Wenzl *Reconstructing Monoidal Categories* Adv. Sov. Math. **16** (1993), 111-136.

12. A. Davydov,   *On some Hochschild cohomology classes of fusion algebras*,   q-alg/9711025.

13. D. Tambara and S. Yamagami, *Tensor Categories with Fusion Rules of Self-Duality for Finite Abelian Groups*, J. Algebra, **209** (1998), 692-707.

14. A. Ganchev, *Fusion rules, modular categories and conformal models*, in: New trends in QFT, (A. Ganchev, R. Kerler, and I. Todorov, eds.), Heron Press, Sofia, 1996, pp.142-145.

15. M. Walton, *Fusion rules for WZW models*, Nucl. Phys. **340** (1990), 777.

16. P. Furlan, A. Ganchev, and V. Petkova, *Quantum groups and fusion rule multiplicities*, Nucl. Phys. **343** (1990), 205.

17. V. Kac, *Infinite dimensional Lie algebras (3rd edition)*, Cambridge Univ. Press, Cambridge, 1990.

18. H. Awata and Y. Yamada, *Fusion rules for fractional level $sl(2)$ algebra*, Mod. Phys. Lett. **A7** (1992), 1185.

19. P. Furlan, A. Ganchev, and V. Petkova, *An extension of the character ring of $\widehat{sl}(3)$ and its quantization*, Commun. Math. Phys. **202** (1999), 701.

ON CATEGORIES OF GELFAND-ZETLIN MODULES

VOLODYMYR MAZORCHUK *
Göteborg University, Sweden

1. The origins

Although the theory of Gelfand-Zetlin modules can be developed for all serial complex simple finite-dimensional Lie algebras and their (non-standard) quantum analogs, in this paper we will discuss the most classical case of the Lie algebra $\mathfrak{g} = \mathfrak{gl}(n, \mathbb{C})$ and will give a short overview of known results in other cases in the end of the paper. We will denote by $e_{i,j}$, $1 \leq i, j \leq n$, the matrix units and will always abbreviate Gelfand-Zetlin by GZ.

This theory starts from the famous original paper [9] by Gelfand and Zetlin, in which, using a step by step reduction to the smaller subalgebras, the authors constructed a very special and nice basis in each simple finite-dimensional $\mathfrak{g}$-module. It is well-known that simple finite-dimensional $\mathfrak{g}$-modules are parametrized by the vectors $\mathbf{m} = (m_1, m_2, \ldots, m_n)$ with complex coefficients, satisfying $m_i - m_{i+1} \in \mathbb{N}$. These vectors represent the (shifted) highest weight of the corresponding simple module with respect to the standard Cartan subalgebra $\mathfrak{h}$ of $\mathfrak{g}$ consisting of diagonal matrices. We will denote the simple module, which corresponds to $\mathbf{m}$, by $V(\mathbf{m})$. To formulate the result of Gelfand and Zetlin we have to introduce the notion of tableau. By a *tableau*, $[l]$, we will mean a doubly-indexed complex vector $(l_{i,j})$, where $1 \leq i \leq n$ and $1 \leq j \leq i$.

Theorem 18. $V(\mathbf{m})$ *possesses a basis, indexed by all tableaux* $[l]$, *satisfying the following conditions:* $l_{n,j} = m_j$, $1 \leq j \leq n$, *and* $l_{i,j} \geq l_{i-1,j} > l_{i,j+1}$, $1 < i \leq n$, $1 \leq j < i$. *Moreover, the action of the generators of* $\mathfrak{g}$ *in this basis is given by the*

---

* volody@math.chalmers.se

S. Duplij and J. Wess (eds.), Noncommutative Structures in Mathematics and Physics, 299–307.

following Gelfand-Zetlin *formulae:*

$$e_{i,i+1}[l] = -\sum_{j=1}^{i} \frac{\prod\limits_{k=1}^{i+1}(l_{i,j} - l_{i+1,k})}{\prod\limits_{\substack{k=1 \\ k \neq i}}(l_{i,j} - l_{i,k})}[l + \delta^{i,j}],$$

$$e_{i+1,i}[l] = \sum_{j=1}^{i} \frac{\prod\limits_{k=1}^{i-1}(l_{i,j} - l_{i-1,k})}{\prod\limits_{\substack{k=1 \\ k \neq i}}(l_{i,j} - l_{i,k})}[l - \delta^{i,j}],$$

$$e_{i,i}[l] = \left(\sum_{j=1}^{i} l_{i,j} - \sum_{j=1}^{i-1} l_{i,j}\right)[l].$$

2. Generic Gelfand-Zetlin modules

The idea to use Theorem 18 to construct new $\mathfrak{g}$-modules goes back to Drozd, Ovsienko and Futorny ([4, 5]). This was based on the observation that GZ-formulae contain only rational functions in parameters, so, if one takes a set of tableaux, closed under the shifts, coming from the action of generators, such that all functions in GZ-formulae will be well-defined, the resulting space should be a $\mathfrak{g}$-module. This can be formally presented in the following statement.

Theorem 19. *Let* $[t]$ *be a tableau satisfying* $t_{i,j} - t_{i,k} \notin \mathbb{Z}$ *for all* $1 \leq i < n$ *and* $1 \leq j \neq k \leq i$. *Denote by* $P([t])$ *the set of all tableaux* $[l]$ *satisfying* $l_{n,j} = t_{n,j}$, $1 \leq j \leq n$ *and* $l_{i,j} - t_{i,j} \in \mathbb{Z}$ *for all possible* i, j. *Let* $V([t])$ *denote a vectorspace, where* $P([t])$ *is a basis. Then GZ-formulae define on* $V([t])$ *the structure of a* $\mathfrak{g}$-*module of finite length.*

Idea of the proof of the first statement. To prove the first part of the theorem (that $V([t])$ is a $\mathfrak{g}$-module) it is sufficient to check that any relation in $U(\mathfrak{g})$ is satisfied on $V([l])$. In our fixed basis $P([t])$ this relation can be rewritten as a collection of rational functions in entries of tableaux, which have to be shown to be zero. The last is easy cause finite-dimensional modules give sufficiently many points, in which these functions take zero values. The last argument uses crucially Theorem 18.   □

To prove the second part we need to recall one more property of the GZ-basis of $V(\mathbf{m})$, which will lead us to the notion of Gelfand-Zetlin subalgebra.

3. Gelfand-Zetlin subalgebra

As we have already mentioned, Theorem 18 was obtained using step by step reduction to the smaller subalgebras. Now we make this statement more precise. We consider a chain of subalgebras

$$\mathfrak{gl}(1, \mathbb{C}) \subset \mathfrak{gl}(2, \mathbb{C}) \subset \cdots \subset \mathfrak{gl}(n, \mathbb{C})$$

embedded with respect to the left upper corner. This chain induces the chain of the corresponding universal enveloping algebras

$$U(\mathfrak{gl}(1, \mathbb{C})) \subset U(\mathfrak{gl}(2, \mathbb{C})) \subset \cdots \subset U(\mathfrak{gl}(n, \mathbb{C})).$$

Denote by Z_k the center $Z(\mathfrak{gl}(k, \mathbb{C}))$ of the algebra $U(\mathfrak{gl}(k, \mathbb{C}))$, $1 \le k \le n$.

The idea to get the GZ-basis of $V(\mathbf{m})$ was the following: we take $V(\mathbf{m})$ and consider it as $\mathfrak{gl}(n - 1, \mathbb{C})$-module. The last is completely reducible and we can consider all components as $\mathfrak{gl}(n - 2, \mathbb{C})$-module, decompose them and proceed till $\mathfrak{gl}(1, \mathbb{C})$. Now we recall that simple finite-dimensional $\mathfrak{gl}(k, \mathbb{C})$-modules are completely determined by their central character. It is also important that, if we decompose a simple finite-dimensional $\mathfrak{gl}(k, \mathbb{C})$-module into a direct sum of simple $\mathfrak{gl}(k - 1, \mathbb{C})$ submodules, all latter will occur with multiplicity 1. Altogether this mean that the resulting GZ basis will be an eigenbasis for all algebras Z_k, or, in other words, for the commutative subalgebra $\Gamma \subset U = U(\mathfrak{gl}(n, \mathbb{C}))$, generated by all Z_k. Moreover, the remark about the multiplicities implies that Γ in fact separates the elements of the GZ-basis of $V(\mathbf{m})$.

Drozd, Ovsienko and Futorny called Γ the Gelfand-Zetlin subalgebra of U. It is well-known that Γ is a polynomial algebra in $n(n + 1)/2$ variables. It was observed by Zhelobenko ([22]), that there is a set of generators, $\gamma_{i,j}$, $1 \le i \le n$, $1 \le j \le i$, of Γ such that the eigenvalue of the action of $\gamma_{i,j}$ on a tableaux, $[l]$, occurring in $V(\mathbf{m})$, should be computed as the j-th symmetric polynomial in variables $(l_{i,1}, l_{i,2}, \ldots, l_{i,i})$. Using the arguments analogous to that, presented in Section 2, one gets that the same is true in all $V([t])$.

Idea of the proof of the second statement of Theorem 19. As we saw, the basis $P([t])$ of $V([t])$ is an eigenbasis for Γ. Moreover, it is easy to get that Γ in fact separates the elements of $P([t])$. Hence, any submodule of $V([t])$ has a basis, which is a subset of $P([t])$. Now if one draws a graph with elements of $P([t])$ as vertices and joins thous pairs, who mutually appear with non-zero coefficients in GZ-formulae, one gets a graph with a finite number of connected components (this number can be easily computed). This finishes the proof.   □

Remark that a complete proof of Theorem 19 can be found in [16].

4. Category of Gelfand-Zetlin modules

The introduction of GZ-subalgebra caused a natural definition of an abstract notion of Gelfand-Zetlin modules, analogous to the notion of the weight module. This was also done by Drozd, Ovsienko and Futorny. They proposed to call a *Gelfand-Zetlin module* any g-module, V, which decomposes into a direct sum of finite-dimensional modules, when viewed as Γ-module. Then by the category, $\mathcal{G}Z$, of Gelfand-Zetlin modules it is natural to understand the full subcategory of the category of all g-modules, consisting of all GZ-modules. As examples of Gelfand-Zetlin modules one can take finite-dimensional modules, $\mathfrak{h}$-weight modules with finite-dimensional weight spaces (in particular, all highest weight modules) or generic Gelfand-Zetlin modules.

Now we recall that tableaux naturally parameterize (not bijectively!) simple finite-dimensional Γ-modules, moreover, non-isomorphic Γ-simples do not have non-trivial extensions. Hence, any GZ-module, V, comes together with its *Gelfand-Zetlin support*, gzsupp(V), i.e. the set of all tableaux parameterizing all simple Γ-modules, occurring in V. We have to note that the product $G = S_1 \times S_2 \times \cdots \times S_n$ of symmetric groups naturally acts on the space of all tableaux permuting the components in the rows. Any fundamental domain of this action bijectively parameterizes Γ-simples and, by definition, gzsupp(V) is invariant under this action. Hence the orbits of G acting on gzsupp(V) bijectively parameterize Γ-simples appearing in V.

Call two tableaux, $[l]$ and $[t]$, equivalent provided $l_{n,j} = t_{n,j}$ and $l_{i,j} - t_{i,j} \in \mathbb{Z}$ for all i, j. Let $\mathfrak{D}$ denote the set of equivalence classes of tableaux. First basic result about the category of Gelfand-Zetlin modules was the following statement, due to Drozd, Ovsienko and Futorny ([6]).

Theorem 20. *The category $\mathcal{G}Z$ decomposes into a direct sum,*

$$\mathcal{G}Z = \oplus_{P \in \mathfrak{D}} \mathcal{G}Z_P,$$

of full subcategories, where the category $\mathcal{G}Z_P$ consists of all Gelfand-Zetlin modules V such that gzsupp(V) $\subset G \circ P$.

Proof. Is not difficult if one reminds that GZ-formulae preserve the equivalence classes of tableaux. $\square$

In fact, Drozd, Ovsienko and Futorny embedded this special case of $U - \Gamma$ relative situation in a wide framework of *Harish-Chandra subalgebras*, which is very convenient (and very general) for study of the whole category of Gelfand-Zetlin modules. It is not our aim to discuss this approach and we refer the reader to the original paper [5].

5. A few theorems of Ovsienko

As soon as one has formulated the notion of a GZ-module, there is a natural and basic question arising: Is it true that each character of Γ can be continued to a $\mathfrak{g}$-module. Equivalently: is it true that each GZ_P is not empty. It is easy to answer "yes" for $n = 1, 2$. For $n = 3$ the same was prooved in [4]. The general case was recently completed by Ovsienko ([21]), but the paper has not appeared yet.

Theorem 21. *Each GZ_P is not empty.*

Idea of the proof. The proof is hard and technical. In fact, the result appears as a biproduct to a special geometrical statement. One should look at the image of Γ in $gr(U)$. This image of $\{\gamma_{i,j}\}$ defines a certain algebraic variety, which is the variety of the so-called *strongly nilpotent matrices* (i.e. matrices, all main minors of which are nilpotent). The statement will follow from abstract nonsense if one proves that the sequence $\{\gamma_{i,j}\}$ is regular. The last can be derived if one proves that the variety of strongly nilpotent matrices is a complete intersection, i.e. that all the irreducible components of it have the same dimension. The last is the most difficult and technical part of the proof and is the main result of the mentioned paper of Ovsienko. □

From Theorem 21 it follows that for any tableau $[l]$ there exists a simple GZ-module, V, such that $[l] \in \text{gzsupp}(V)$. Using the convenient technique of Harish-Chandra subalgebras, mentioned above, Ovsienko managed to give much more useful information about simple GZ-modules.

Theorem 22. *1. For each $[l]$ there exists only finitely many (up to isomorphism) simple GZ-modules V with $[l] \in \text{gzsupp}(V)$.*

2. Let V be a simple finite-dimensional $\mathfrak{g}$-module and F be a simple finite-dimensional Γ-module. Then the multiplicity of F in V (the last is viewed as Γ-module) is finite.

I have also to note that [21] contains a complete proof of the statement that Γ is a maximal commutative subalgebra of $U(\mathfrak{g})$. This statement can be found (without proof!) in all classical monographs (e.g. [22]). The proof in [21] is the first complete I have seen.

6. Generalized Verma modules and Gelfand-Zetlin modules

It seems that the first time, when it was understood that generic Gelfand-Zetlin modules are very convenient for computations was the paper [18], where the authors investigated the question about the structure of the so-called generalized Verma modules. Consider the inclusion $\mathfrak{gl}(k, \mathbb{C}) \subset \mathfrak{gl}(n, \mathbb{C}) = \mathfrak{g}$ with respect to the left upper corner. Let $\mathfrak{P}$ denote the parabolic subalgebra of $\mathfrak{g}$, generated by

$\mathfrak{gl}(k, \mathbb{C})$ and the standard Borel subalgebra op upper-triangular matrices. Take a simple $\mathfrak{gl}(k, \mathbb{C})$-module, V, set that the rest of the Cartan subalgebra acts on it via some character, say λ, and the rest of the Borel subalgebra annihilates it. Thus V becomes a $\mathfrak{P}$-module. The induced module $M(V, \lambda) = U \otimes_{U(\mathfrak{P})} V$ is called a *generalized Verma module*. It turned out that taking V to be a simple generic GZ-module, $V([t])$, the structure of $M(V([t]), \lambda)$ can be described in terms of the Weyl group acting on the space of parameters, as it was done for the classical Verma modules by Bernstein, I.Gelfand and S.Gelfand ([2]).

It is trivial that $M(V([t]), \lambda)$ is a GZ-module over $\mathfrak{g}$. One can also see that it is generated by the elements (annihilated by the nilpotent radical of $\mathfrak{P}$), corresponding to the tableaux $[l]$, satisfying the following condition: $l_{i,j} = l_{i-1,j}$, $k < i \leq n$. The Weyl group S_n acts naturally on the set of such tableaux, permuting the elements of the upper row (which also causes the corresponding changes in all rows with $i > k$). For a transposition, $(i, j) \in S_n$, $i < j$, write $(i, j)[l] \leq [l]$ provided $l_{n,i} - l_{n,j} \in \mathbb{Z}_+$ and close the relation $\leq$ transitively. The next statement is the main result of [18].

Theorem 23. *Let $[l]$ (resp. $[l']$) be the tableau of a canonical generator of $M(V([t]), \lambda)$ (resp. $M(V([t']), \lambda'))$. Assume that $l_{i,j} = l'_{i,j}$ for all $i < k$ and all j. Then the following statements are equivalent:*

1. $M(V([t]), \lambda) \subset M(V([t']), \lambda')$.
2. *The unique irreducible quotient of $M(V([t]), \lambda)$ is a composition subquotient of $M(V([t']), \lambda')$.*
3. $[l] \leq [l']$.

The proof of this theorem, presented in [18] goes the general line of the original proof in [2], but uses some calculations with generic GZ-modules. In particular, one of the mains things one needs here is a more or less precise description of $M(V([t]), \lambda)$ as a $\mathfrak{gl}(k, \mathbb{C})$-module. This question easily reduces to the calculation of $F \otimes V([t])$, where F is a simple finite-dimensional $\mathfrak{gl}(k, \mathbb{C})$-module. If one recalls that simple generic GZ-modules correspond to certain characters of Γ and the last one is generated by a sequence of centers, one can use the famous Theorem of Kostant ([12]), which tells how one can compute the action of the center on $F \otimes V([t])$. In this way one easily derives all potential subquotients of $F \otimes V([t])$. This (and existence of some of them, which is easy) was enough for the goals of Theorem 23.

7. Categories of $\mathfrak{gl}(n, \mathbb{C})$-modules generated by a simple generic Gelfand-Zetlin module

The necessity to study $F \otimes V([t])$ deeper was understood in [8], where some categories of Lie algebra modules where constructed, which are based on the categories of modules behaving well under tensoring with finite dimensional modules.

As the main example of the latter, a category, generated by a simple generic GZ-module, was presented. Let $V([t])$ be a simple generic GZ-module. Denote by $C([t])$ the full subcategory, consisting of all subquotients of modules $F \otimes V([t])$, where F is simple finite-dimensional. It turned out that this category has relatively easy structure.

Theorem 24. $C([t])$ *decomposes into a direct sum of full subcategories, each of which is equivalent to the module category of a finite-dimensional associative and local algebra. In particular,* $C([t])$ *has enough projective objects.*

Idea of the proof. One of the main ingredients of the proof is the following lemma:

Lemma 25. *The module* $F \otimes V([t])$ *has length* $\dim(F)$, *all simple subquotients of it are simple generic GZ-modules and the multiplicity of* $V([s])$ *in* $F \otimes V([t])$, *where* $s_{i,j} = t_{i,j}$, $i < n$, *equals* $\sum \dim(F_\mu)$, *where the sum is taken over all* μ *such that the vector* $(t_{n,j})_{j=1,\dots,n} + \mu$ *coincides with a permutation of* $(s_{n,j})_{j=1,\dots,n}$.

Lemma 25 is proved by a direct calculation, using GZ-formulae and the Littelwood-Richardson rule. It also represents a "generic behaviour" of simple generic GZ-modules in contrast with finite-dimensional modules.

After Lemma 25 one can first describe all simple modules in $C([t])$. These will be $V([s])$, with $s_{i,j} - t_{i,j} \in \mathbb{Z}$. Then it is easy to find among them a projective module and prove the existence of projectives using the exactness os $F \otimes \_$. Decomposition with respect to central characters completes the proof. $\qquad\square$

In two subsequent papers ([13, 14]) it was noticed that the category $C([t])$ closely connected to various categories of $\mathfrak{g}$-modules, independently appeared in different contexts. The results of these two papers can be collected in the following statement.

Theorem 26. *Assume that* $t_{n,j} \in \mathbb{Z}$ *for all* j. *Then the following categories of* $\mathfrak{g}$-*modules are equivalent:*

1. *The category* $C([t])$.
2. *The category of complete (in the sense of Enright, [7]) weight extensions of highest weight modules with integral support.*
3. *A certain category of algebraic Harish-Chandra bimodules in the sense of Bernstein and S.Gelfand ([1]).*

Idea of the proof. The equivalence of the first and the second categories is the content of [13]. It is based on a precise construction of the equivalence functor, which is a generalization of the Mathieu's twist functor ([15]). The equivalence of the second and the third categories is proved in [14], using an intermediate equivalence of the second category with a category of injectively copresented modules in the Bernstein-Gelfand-Gelfand category $\mathcal{O}$ ([3]). $\qquad\square$

8. Case of classical and quantum algebras and open problems

An analogue of Theorem 18 for orthogonal algebras (simple finite dimensional complex Lie algebras of type B_n and D_n) was obtained also by Gelfand and Zetlin in [10]. The corresponding generic modules were constructed in [17]. For symplectic Lie algebras (type C_n) an analogue of Theorem 18 is a recent result of Molev, [20]. For $U_q(\mathfrak{gl}_n)$ the classical result was obtained by Jimbo ([11]) and generic modules were constructed by Turowska and the author ([19]). For non-standard quantum deformations of orthogonal algebras the classical construction of Gelfand-Zetlin basis in finite-dimensional modules can be found in a series of recent papers by Klimyk and Jorgov, available at "xxx.lanl.gov", where one can also find information about corresponding results for root of unity case.

Finally, we want to give a list of some questions and open problems related to Gelfand-Zetlin modules:

1. Classify and give a precise construction of all simple GZ-modules.
2. Find a criterion, when a given character of Γ has only one extension to a simple $\mathfrak{g}$-module.
3. Let F be a simple finite dimensional $\mathfrak{gl}(n, \mathbb{C})$-module. Consider two Gelfand-Zetlin basis of it, with respect to the inclusions of subalgebras into left upper and into right lower corners. What will be the transformation matrix?
4. Let V be a simple Gelfand-Zetlin module and F be a finite-dimensional module. Does $V \otimes F$ have a finite length? Is it possible to compute composition subquotients and multiplicities of $V \otimes F$?
5. Are there any analogues of Gelfand-Zetlin construction for exceptional Lie algebras?
6. Extend all already known for $\mathfrak{gl}(n, \mathbb{C})$ results to the case of orthogonal and symplectic algebras. Also find in those cases solutions to the above problems.

References

1. I.Bernstein and S.Gelfand, *Tensor product of finite and infinite-dimensional representations of semisimple Lie algebras*, Compositio Math., **41** (1980), 245–285.
2. I.Bernstein, I.Gelfand and S.Gelfand, *Structure of representations that are generated by vectors of highest weight*, Funktsional. Anal. i Prilozhen., **5** (1971), 1–9.
3. I.Bernstein, I.Gelfand and S.Gelfand, *A certain category of* $\mathfrak{g}$*-modules*, Funktsional. Anal. i Prilozhen., **10** (1976), 1–8.
4. Yu.A.Drozd, S.A.Ovsienko and V.M.Futorny, *Irreducible weighted sl(3)-modules*, Funktsional. Anal. i Prilozhen., **23** (1989), 57–58.
5. ———, *Harish-Chandra subalgebras and Gelfand-Zetlin modules*, Math. and Phys. Sci., **424** (1994), 72–89.
6. ———, *On Gelfand-Zetlin modules*, Rend. Circ. Mat. Palermo (2) Suppl., **26** (1991), 143–147.
7. T.Enright, *On the fundamental series of a real semisimple Lie algebra: their irreducibility, resolutions and multiplicity formulae*, Annals Math., **110** (1979), 1–82.

8. V.Futorny, S.König and V.Mazorchuk, *Categories of induced modules and projectively stratified algebras*, Univ. *Bielefeld preprint*, 99-024, Bielefeld, 1999, to appear in Algebr. Represent. Theory.

9. I.M.Gelfand, M.L.Zetlin, *Finite-dimensional representations of the group of unimodular matrices*, Doklady Akad. Nauk SSSR (N.S.), **71** (1950), 825–828.

10. ———, *Finite-dimensional representations of the group of orthogonal matrices*, Doklady Akad. Nauk SSSR (N.S.), **71** (1950), 1017–1020.

11. M.Jimbo, *Quantum R-matrix for the generalized Toda system: an algebraic approach*, in Field Theory, quantum gravity and strings, Lecture notes in Physics, 246, Springer-Verlag, Berlin-New York, 1986, pp. 335–361.

12. B.Kostant, *On the tensor product of a finite and infinite dimensional representations*, Journal of Func. analisis, **20** (1975), 257–285.

13. S.König and V.Mazorchuk, *An equivalence of two categories of* $\mathfrak{sl}(n, \mathbb{C})$*-modules*, Univ. *Bielefeld preprint*, 99-114, Bielefeld, 1999, to appear in Algebr. Represent. Theory.

14. ———, *Enright's completions and injectively copresented modules*, Univ. *Bielefeld preprint*, 99-130, Bielefeld, 1999.

15. O.Mathieu, *Classification of simple weight modules*, Annales Inst. Fourier, **50** (2000), 537–592.

16. V.Mazorchuk, *Generalized Verma modules*, Univ. *Bielefeld preprint*, E-99-006, Bielefeld, 1999, to be published as a monograph by Lviv Scientific Publisher.

17. ———, *On Gelfand-Zetlin modules over orthogonal Lie algebras*, Univ. *Bielefeld preprint*, 98-106, Bielefeld, 1998, to appear in Algebra Colloq.

18. V.Mazorchuk, S.Ovsienko, *Submodule structure of generalized Verma modules induced from generic Gelfand-Zetlin modules*, Algebr. Represent. Theory, **1** (1998), 3–26.

19. V.Mazorchuk, L.Turowska, *On Gelfand-Zetlin modules over* $U_q(gl(n))$, Czech. J. Phys., **50** (2000), 139–144.

20. A.Molev, *A basis for representations of symplectic Lie algebras*, preprint, math.QA/9804127.

21. S.Ovsienklo, *Some finiteness statements for Gelfand-Zetlin modules*, to appear.

22. D.P.Zhelobenko, *Compact Lie groups and their representation*, Translation of Mathematical Monographs, Vol. 40, American MAthematical Society, Providence, R.I., 1973.

HIDDEN SYMMETRY OF SOME ALGEBRAS OF

q-DIFFERENTIAL OPERATORS

DMITRY SHKLYAROV, SERGEY SINEL'SHCHIKOV*   and
LEONID VAKSMAN†
*Institute for Low Temperature Physics & Engineering, 47 Lenin ave,
61164 Kharkov, Ukraine*

1. Introduction

Let us explain the meaning of the words "q-differential operators" and "hidden symmetry". Let $\mathbb{C}[z]_q$ be the algebra of polynomials in z over the field of rational functions $\mathbb{C}(q^{1/2})$ (we assume this field to be the ground field throughout the paper). We denote by $\Lambda^1(\mathbb{C})_q$ the $\mathbb{C}[z]_q$-bimodule with the generator dz such that

$$z \cdot dz = q^{-2}dz \cdot z.$$

Let d be the linear map $\mathbb{C}[z]_q \to \Lambda^1(\mathbb{C})_q$ given by the two conditions:

$$d : z \mapsto dz,$$

$$d(f_1(z)f_2(z)) = d(f_1(z))f_2(z) + f_1(z)d(f_2(z)).$$

(The later condition is just the Leibniz rule). The bimodule $\Lambda^1(\mathbb{C})_q$ (together with the map d) is a well known first order differential calculus over the algebra $\mathbb{C}[z]_q$. The differential d allows one to introduce an operator of "partial derivative" $\frac{d}{dz}$ in $\mathbb{C}[z]_q$:

$$d(f(z)) = dz \cdot \frac{df}{dz}(z).$$

Let us introduce also the notation $\hat{z}$ for the operator in $\mathbb{C}[z]_q$ of multiplication by z:

$$\hat{z} : f(z) \mapsto zf(z).$$

* sinelshchikov@ilt.kharkov.ua
† vaksman@ilt.kharkov.ua

S. Duplij and J. Wess (eds.), Noncommutative Structures in Mathematics and Physics, 309–320.
© 2001 *Kluwer Academic Publishers. Printed in the Netherlands.*

Let $D(\mathbb{C})_q$ be the subalgebra in the algebra $\mathrm{End}_{\mathbb{C}(q^{1/2})}(\mathbb{C}[z]_q)$ (of all endomorphisms of the linear space $\mathbb{C}[z]_q$) containing 1 and generated by $\frac{d}{dz}$, $\hat{z}$. It is easy to check that

$$\frac{d}{dz} \cdot \hat{z} = q^{-2}\hat{z} \cdot \frac{d}{dz} + 1.$$

Thus the algebra $D(\mathbb{C})_q$ is an analogue of the Weyl algebra $A_1(\mathbb{C})$.

Let $\lambda \in \mathbb{C}(q^{1/2})$. One checks that the map

$$\hat{z} \mapsto \lambda \cdot \hat{z}, \qquad \frac{d}{dz} \mapsto \lambda^{-1} \cdot \frac{d}{dz}$$

is extendable up to an automorphism of the algebra $D(\mathbb{C})_q$. Such automorphisms are "evident" symmetries of $D(\mathbb{C})_q$. It turn out that they belong to a wider set of symmetries of $D(\mathbb{C})_q$. This set does not consists of automorphisms only. Let us turn to precise formulations.

To start with, we recall one the definition of the quantum universal enveloping algebra $U_q\mathfrak{sl}_2$ [5]. It is

i)the algebra given by the generators E, F, K, K^{-1}, and the relations

$$KK^{-1} = K^{-1}K = 1, \quad KE = q^2EK, \quad KF = q^{-2}FK,$$

$$EF - FE = \frac{K - K^{-1}}{q - q^{-1}};$$

ii) the Hopf algebra : the comultiplication Δ, the antipode S, and the counit ε are determined by

$$\Delta(E) = E \otimes 1 + K \otimes E, \quad \Delta(F) = F \otimes K^{-1} + 1 \otimes F, \quad \Delta(K) = K \otimes K,$$

$$S(E) = -K^{-1}E, \qquad S(F) = -FK, \qquad S(K) = K^{-1},$$

$$\varepsilon(E) = \varepsilon(F) = 0, \qquad \varepsilon(K) = 1.$$

There is a well known structure of $U_q\mathfrak{sl}_2$-module in the space $\mathbb{C}[z]_q$. Let us describe it explicitly:

$$E : f(z) \mapsto -q^{1/2}z^2\frac{f(z) - f(q^2z)}{z - q^2z},$$

$$F : f(z) \mapsto q^{1/2}\frac{f(z) - f(q^{-2}z)}{z - q^{-2}z},$$

$$K^{\pm 1} : f(z) \mapsto f(q^{\pm 2}z).$$

It can be checked that $\mathbb{C}[z]_q$ is a $U_q\mathfrak{sl}_2$-module algebra, i.e. for any $\xi \in U_q\mathfrak{sl}_2$, $f_1, f_2 \in \mathbb{C}[z]_q$

$$\xi(1) = \varepsilon(\xi) \cdot 1, \tag{1}$$

$$\xi(f_1 f_2) = \sum_j \xi_j'(f_1) \xi_j''(f_2), \qquad (2)$$

with $\Delta(\xi) = \sum_j \xi_j' \otimes \xi_j''$.

REMARK. This observation is an analogue of the following one. The group $SL_2(\mathbb{C})$ acts on $\mathbb{CP}^1$ via the fractional-linear transformations. Thus the universal enveloping algebra $U\mathfrak{sl}_2$ acts via differential operators in the space of holomorphic functions on the open cell $\mathbb{C} \subset \mathbb{CP}^1$.

Let V be a $U_q\mathfrak{sl}_2$-module. Then the algebra $\mathrm{End}(V)$ admits a "canonical" structure of $U_q\mathfrak{sl}_2$-module: for $\xi \in U_q\mathfrak{sl}_2, T \in \mathrm{End}(V)$

$$\xi(T) = \sum_j \xi_j' \cdot T \cdot S(\xi_j''), \qquad (3)$$

where $\Delta(\xi) = \sum_j \xi_j' \otimes \xi_j''$, S is the antipode, and the elements in the right-hand side are multiplied within the algebra $\mathrm{End}(V)$. It is well known that this action of $U_q\mathfrak{sl}_2$ in $\mathrm{End}(V)$ makes $\mathrm{End}(V)$ into a $U_q\mathfrak{sl}_2$-module algebra (i.e. for $\xi \in U_q\mathfrak{sl}_2$, $T_1, T_2 \in \mathrm{End}(V)$ (1), (2) hold with f_1, f_2 being replaced by T_1, T_2, respectively).

The objects considered above are the simplest among ones we deal with in the present paper. In this simplest case our main result can be formulated as follows: the algebra $D(\mathbb{C})_q$ is a $U_q\mathfrak{sl}_2$-module subalgebra in the $U_q\mathfrak{sl}_2$-module algebra $\mathrm{End}_{\mathbb{C}(q^{1/2})}(\mathbb{C}[z]_q)$ (where the $U_q\mathfrak{sl}_2$-action is given by (3)). This $U_q\mathfrak{sl}_2$-module structure in the algebra $D(\mathbb{C})_q$ is what we call "hidden symmetry" of $D(\mathbb{C})_q$.

REMARK. In the setting of the previous Remark the analogous fact is evident: for $\xi \in \mathfrak{sl}_2$ the action (3) is just the commutator of the differential operators ξ and T in the space of holomorphic functions on $\mathbb{C}$. The commutator is again a differential operator.

We can describe the $U_q\mathfrak{sl}_2$-action in $D(\mathbb{C})_q$ explicitly:

$$E(\hat{z}) = -q^{1/2}\hat{z}^2, \quad F(\hat{z}) = q^{1/2}, \quad K^{\pm 1}(\hat{z}) = q^{\pm 2}\hat{z},$$

$$E\left(\frac{d}{dz}\right) = q^{-3/2}(q^{-2} + 1)\hat{z}\frac{d}{dz}, \quad F\left(\frac{d}{dz}\right) = 0, \quad K^{\pm 1}\left(\frac{d}{dz}\right) = q^{\mp 2}\frac{d}{dz}.$$

(The action of $U_q\mathfrak{sl}_2$ on an arbitrary element of $D(\mathbb{C})_q$ can be produced via the rule (2).)

The paper is organized as follows.

In Section 2 we recall one definitions of the quantum universal enveloping algebra $U_q\mathfrak{sl}_N$, a $U_q\mathfrak{sl}_N$-module algebra $\mathbb{C}[\mathrm{Mat}_{m,n}]_q$ of holomorphic polynomials on a quantum matrix space $\mathrm{Mat}_{m,n}$, and a well known first order differential calculus $\Lambda^1(\mathrm{Mat}_{m,n})_q$ over $\mathbb{C}[\mathrm{Mat}_{m,n}]_q$ (in this Introduction the case $m = n = 1$ was considered). Then we introduce an algebra $D(\mathrm{Mat}_{m,n})_q$ of q-differential operators in $\mathbb{C}[\mathrm{Mat}_{m,n}]_q$ and formulate a main theorem concerning a hidden symmetry of this algebra.

Section 3 contains a sketch of the proof of the main theorem.

In Section 4 we discuss briefly possible generalizations of our results. Specifically, the space $\mathrm{Mat}_{m,n}$ is an example of a prehomogeneous vector space of commutative parabolic type [6]. In [9] q-analogs of all such vector spaces were introduced. Our results admit a generalization on the case of an arbitrary quantum prehomogeneous vector space of commutative parabolic type.

We take this opportunity to thank Prof. H. P. Jakobsen and Prof. T. Tanisaki who attracted our attention to other approaches to the notion of quantum differential operators.

This research was partially supported by Award No.UM1-2091 of the U.S. Civilian Research and Development Foundation.

2. The main theorem

In this Section we deal with a well known q-analogue of the polynomial algebra on the space $\mathrm{Mat}_{m,n}$ of $m \times n$ matrices (in the Introduction we considered the case $m = n = 1$). Let the ground field be the field of rational functions $\mathbb{C}(q^{1/2})$. The algebra $\mathbb{C}[\mathrm{Mat}_{m,n}]_q$ is the unital algebra given by its generators z_a^α, $a = 1, \ldots n$, $\alpha = 1, \ldots m$, and the following relations

$$z_a^\alpha z_b^\beta =$$

$$= \begin{cases} q z_b^\beta z_a^\alpha & , \; a = b \,\&\, \alpha < \beta \quad \text{or} \quad a < b \,\&\, \alpha = \beta \\ z_b^\beta z_a^\alpha & , \; a < b \,\&\, \alpha > \beta \\ z_b^\beta z_a^\alpha + (q - q^{-1}) z_a^\beta z_b^\alpha & , \; a < b \,\&\, \alpha < \beta \end{cases} , \qquad (4)$$

The Hopf algebra $U_q \mathfrak{sl}_N$ is determined by the generators E_i, F_i, K_i, K_i^{-1}, $i = 1, \ldots, N - 1$, and the relations

$$K_i K_j = K_j K_i, \qquad K_i K_i^{-1} = K_i^{-1} K_i = 1, \qquad K_i E_j = q^{a_{ij}} E_j K_i,$$

$$K_i F_j = q^{-a_{ij}} F_j K_i, \qquad E_i F_j - F_j E_i = \delta_{ij}(K_i - K_i^{-1})/(q - q^{-1})$$

$$E_i^2 E_j - (q + q^{-1}) E_i E_j E_i + E_j E_i^2 = 0, \qquad |i - j| = 1 \qquad (5)$$

$$F_i^2 F_j - (q + q^{-1}) F_i F_j F_i + F_j F_i^2 = 0, \qquad |i - j| = 1$$

$$[E_i, E_j] = [F_i, F_j] = 0, \qquad |i - j| \neq 1.$$

The comultiplication Δ, the antipode S, and the counit ε are determined by

$$\Delta(E_i) = E_i \otimes 1 + K_i \otimes E_i, \; \Delta(F_i) = F_i \otimes K_i^{-1} + 1 \otimes F_i, \; \Delta(K_i) = K_i \otimes K_i,$$

$$\qquad (6)$$

$$S(E_i) = -K_i^{-1} E_i, \qquad S(F_i) = -F_i K_i, \qquad S(K_i) = K_i^{-1}, \qquad (7)$$

$$\varepsilon(E_i) = \varepsilon(F_i) = 0, \qquad \varepsilon(K_i) = 1.$$

The algebra $\mathbb{C}[\mathrm{Mat}_{m,n}]_q$ possesses a structure of $U_q\mathfrak{sl}_N$-module algebra with $N = m + n$. Explicit formulae for the action of $U_q\mathfrak{sl}_N$ in $\mathbb{C}[\mathrm{Mat}_{m,n}]_q$ are as follows (see [7]):

$$K_n z_a^\alpha = \begin{cases} q^2 z_a^\alpha & , \ a = n \ \& \ \alpha = m \\ q z_a^\alpha & , \ a = n \ \& \ \alpha \neq m \quad \text{or} \quad a \neq n \ \& \ \alpha = m \\ z_a^\alpha & , \ \text{otherwise} \end{cases}, \qquad (8)$$

$$F_n z_a^\alpha = q^{1/2} \cdot \begin{cases} 1 & , \ a = n \ \& \ \alpha = m \\ 0 & , \ \text{otherwise} \end{cases}, \qquad (9)$$

$$E_n z_a^\alpha = -q^{1/2} \cdot \begin{cases} q^{-1} z_a^m z_n^\alpha & , \ a \neq n \ \& \ \alpha \neq m \\ (z_n^m)^2 & , \ a = n \ \& \ \alpha = m \\ z_n^m z_a^\alpha & , \ \text{otherwise} \end{cases}, \qquad (10)$$

and with $k \neq n$
$$K_k z_a^\alpha =$$

$$= \begin{cases} q z_a^\alpha & , \ k < n \ \& \ a = k \quad \text{or} \quad k > n \ \& \ \alpha = N - k \\ q^{-1} z_a^\alpha & , \ k < n \ \& \ a = k+1 \quad \text{or} \quad k > n \ \& \ \alpha = N - k + 1 \\ z_a^\alpha & , \ \text{otherwise} \end{cases}, \qquad (11)$$

$$F_k z_a^\alpha = q^{1/2} \cdot \begin{cases} z_{a+1}^\alpha & , \ k < n \ \& \ a = k \\ z_a^{\alpha+1} & , \ k > n \ \& \ \alpha = N - k \\ 0 & , \ \text{otherwise} \end{cases}, \qquad (12)$$

$$E_k z_a^\alpha = q^{-1/2} \cdot \begin{cases} z_{a-1}^\alpha & , \ k < n \ \& \ a = k+1 \\ z_a^{\alpha-1} & , \ k > n \ \& \ \alpha = N - k + 1 \\ 0 & , \ \text{otherwise} \end{cases}. \qquad (13)$$

REMARKS.i) In the classical case the corresponding action of $U\mathfrak{sl}_N$ in the space of holomorphic functions on $\mathrm{Mat}_{m,n}$ can be produced via an embedding $\mathrm{Mat}_{m,n}$ into the Grassmanian $Gr_{m,N}$ as an open cell (we describe a q-analogue of the embedding in [7]).

ii) Using the structure of $U_q\mathfrak{sl}_N$-module in $\mathbb{C}[\mathrm{Mat}_{m,n}]_q$ we can define the structure of $U_q\mathfrak{sl}_N$-module algebra in $\mathrm{End}_{\mathbb{C}(q^{1/2})}(\mathbb{C}[\mathrm{Mat}_{m,n}]_q)$ via (3) with $\xi \in U_q\mathfrak{sl}_N$, $T \in \mathrm{End}_{\mathbb{C}(q^{1/2})}(\mathbb{C}[\mathrm{Mat}_{m,n}]_q)$.

Now let us recall a definition of a well known first order differential calculus over $\mathbb{C}[\mathrm{Mat}_{m,n}]_q$. Let $\Lambda^1(\mathrm{Mat}_{m,n})_q$ be the $\mathbb{C}[\mathrm{Mat}_{m,n}]_q$-bimodule given by its generators dz_a^α, $a = 1, \ldots n$, $\alpha = 1, \ldots m$, and the relations

$$z_b^\beta \, dz_a^\alpha = \sum_{\alpha',\beta'=1}^m \sum_{a',b'=1}^n R_{\beta\alpha}^{\beta'\alpha'} R_{ba}^{b'a'} dz_{a'}^{\alpha'} \cdot z_{b'}^{\beta'}, \tag{14}$$

with

$$R_{ba}^{b'a'} = \begin{cases} q^{-1} & , \ a = b = a' = b' \\ 1 & , \ a \neq b \ \ \& \ \ a = a' \ \ \& \ \ b = b' \\ q^{-1} - q & , \ a < b \ \ \& \ \ a = b' \ \ \& \ \ b = a' \\ 0 & , \ \text{otherwise} \end{cases} . \tag{15}$$

The map $d : z_a^\alpha \mapsto dz_a^\alpha$ can be extended up to a linear operator $d : \mathbb{C}[\mathrm{Mat}_{m,n}]_q \to \Lambda^1(\mathrm{Mat}_{m,n})_q$ satisfying the Leibniz rule. It was noted for the first time in [8], that there exists a unique structure of a $U_q\mathfrak{sl}_N$-module in $\Lambda^1(\mathrm{Mat}_{m,n})_q$ such that the map d is a morphism of $U_q\mathfrak{sl}_N$-modules. The pair $(\Lambda^1(\mathrm{Mat}_{m,n})_q, d)$ is the first order differential calculus over $\mathbb{C}[\mathrm{Mat}_{m,n}]_q$.

Let us introduce an algebra $D(\mathrm{Mat}_{m,n})_q$ of q-differential operators on $\mathrm{Mat}_{m,n}$. For this purpose, we define the linear operators $\frac{\partial}{\partial z_a^\alpha}$ in $\mathbb{C}[\mathrm{Mat}_{m,n}]_q$ via the differential d:

$$df = \sum_{a=1}^n \sum_{\alpha=1}^m dz_a^\alpha \cdot \frac{\partial f}{\partial z_a^\alpha}, \qquad f \in \mathbb{C}[\mathrm{Mat}_{m,n}]_q,$$

and the operators $\widehat{z_a^\alpha}$ by

$$\widehat{z_a^\alpha} f = z_a^\alpha \cdot f, \qquad f \in \mathbb{C}[\mathrm{Mat}_{m,n}]_q.$$

Then $D(\mathrm{Mat}_{m,n})_q$ is the unital subalgebra in $\mathrm{End}_{\mathbb{C}(q^{1/2})}(\mathbb{C}[\mathrm{Mat}_{m,n}]_q)$ generated by the operators $\frac{\partial}{\partial z_a^\alpha}, \widehat{z_a^\alpha}, a = 1, \dots n, \alpha = 1, \dots m$.

To start with, we describe $D(\mathrm{Mat}_{m,n})_q$ in terms of generators and relations.

Proposition 2.1. *The complete list of relations between the generators* $\widehat{z_a^\alpha}$, $\frac{\partial}{\partial z_a^\alpha}$, $a = 1, \dots n, \alpha = 1, \dots m$, *of* $D(\mathrm{Mat}_{m,n})_q$ *is as follows*
$$\widehat{z_a^\alpha}\widehat{z_b^\beta} =$$

$$= \begin{cases} q\widehat{z_b^\beta}\widehat{z_a^\alpha} & , \ a = b \ \& \ \alpha < \beta \ \ \text{or} \ \ a < b \ \& \ \alpha = \beta \\ \widehat{z_b^\beta}\widehat{z_a^\alpha} & , \ a < b \ \& \ \alpha > \beta \\ \widehat{z_b^\beta}\widehat{z_a^\alpha} + (q - q^{-1})\widehat{z_a^\beta}\widehat{z_b^\alpha} & , \ a < b \ \& \ \alpha < \beta \end{cases} , \tag{16}$$

$$\frac{\partial}{\partial z_b^\beta}\frac{\partial}{\partial z_a^\alpha} =$$

$$= \begin{cases} q\frac{\partial}{\partial z_a^\alpha}\frac{\partial}{\partial z_b^\beta} & , \ a=b \ \& \ \alpha < \beta \quad \text{or} \quad a < b \ \& \ \alpha = \beta \\ \frac{\partial}{\partial z_a^\alpha}\frac{\partial}{\partial z_b^\beta} & , \ a < b \ \& \ \alpha > \beta \\ \frac{\partial}{\partial z_a^\alpha}\frac{\partial}{\partial z_b^\beta} + (q-q^{-1})\frac{\partial}{\partial z_a^\beta}\frac{\partial}{\partial z_b^\alpha} & , \ a < b \ \& \ \alpha < \beta \end{cases} , \tag{17}$$

$$\frac{\partial}{\partial z_a^\alpha}\hat{z}_b^\beta = \sum_{a',b'=1}^{n} \sum_{\alpha',\beta'=1}^{m} R_{ba'}^{b'a} R_{\beta\alpha'}^{\beta'\alpha} \hat{z}_{b'}^{\beta'} \frac{\partial}{\partial z_{a'}^{\alpha'}} + \delta_{ab}\delta^{\alpha\beta}, \tag{18}$$

with $\delta_{ab}, \delta^{\alpha\beta}$ being the Kronecker symbols, and $R_{ba'}^{b'a}$ given by (15).

Now we present the main result of the paper

Theorem 2.2. *i) The algebra* $D(\mathrm{Mat}_{m,n})_q$ *is a* $U_q\mathfrak{sl}_N$*-module subalgebra in the* $U_q\mathfrak{sl}_N$*-module algebra* $\mathrm{End}_{\mathbb{C}(q^{1/2})}(\mathbb{C}[\mathrm{Mat}_{m,n}]_q)$.

ii) The $U_q\mathfrak{sl}_N$*-module structure in* $D(\mathrm{Mat}_{m,n})_q$ *is described explicitly as follows:*

$U_q\mathfrak{sl}_N$ *acts on the generators* $\hat{z}_a^\alpha$ *via formulae (8)-(13) (where* z_a^α *should be replaced by* $\hat{z}_a^\alpha$*); for the generators* $\frac{\partial}{\partial z_a^\alpha}$ *the formulae are*

$$K_n \frac{\partial}{\partial z_a^\alpha} = \begin{cases} q^{-2}\frac{\partial}{\partial z_a^\alpha} & , \ a=n \ \& \ \alpha = m \\ q^{-1}\frac{\partial}{\partial z_a^\alpha} & , \ a=n \ \& \ \alpha \neq m \quad \text{or} \quad a \neq n \ \& \ \alpha = m \\ \frac{\partial}{\partial z_a^\alpha} & , \ \text{otherwise} \end{cases} , \tag{19}$$

$$F_n \frac{\partial}{\partial z_a^\alpha} = 0 \qquad a = 1, \ldots n, \qquad \alpha = 1, \ldots m, \tag{20}$$

$$E_n \frac{\partial}{\partial z_a^\alpha} = q^{-3/2}.$$

$$\begin{cases} \sum_{b=1}^{n} \hat{z}_b^m \frac{\partial}{\partial z_b^m} + \sum_{\beta=1}^{m} \hat{z}_n^\beta \frac{\partial}{\partial z_n^\beta} + (q^{-2}-1)\sum_{b=1}^{n}\sum_{\beta=1}^{m} \hat{z}_b^\beta \frac{\partial}{\partial z_b^\beta} & , \ a=n \ \& \ \alpha = m \\ \sum_{\beta=1}^{m} \hat{z}_n^\beta \frac{\partial}{\partial z_a^\beta} & , \ a \neq n \ \& \ \alpha = m \\ \sum_{b=1}^{n} \hat{z}_b^m \frac{\partial}{\partial z_b^\alpha} & , \ a=n \ \& \ \alpha \neq m \\ 0 & , \ \text{otherwise} \end{cases} \tag{21}$$

and with $k \neq n$

$$K_k \frac{\partial}{\partial z_a^\alpha} =$$

$$= \begin{cases} q^{-1} \frac{\partial}{\partial z_a^\alpha} & , \quad k < n \ \& \ a = k \quad \text{or} \quad k > n \ \& \ \alpha = N - k \\ q \frac{\partial}{\partial z_a^\alpha} & , \quad k < n \ \& \ a = k+1 \quad \text{or} \quad k > n \ \& \ \alpha = N - k + 1 \\ \frac{\partial}{\partial z_a^\alpha} & , \quad \text{otherwise} \end{cases}, \qquad (22)$$

$$F_k \frac{\partial}{\partial z_a^\alpha} = -q^{3/2} \cdot \begin{cases} \frac{\partial}{\partial z_{a-1}^\alpha} & , \quad k < n \ \& \ a = k+1 \\ \frac{\partial}{\partial z_a^{\alpha-1}} & , \quad k > n \ \& \ \alpha = N - k + 1 \\ 0 & , \quad \text{otherwise} \end{cases}, \qquad (23)$$

$$E_k \frac{\partial}{\partial z_a^\alpha} = -q^{-3/2} \cdot \begin{cases} \frac{\partial}{\partial z_{a+1}^\alpha} & , \quad k < n \ \& \ a = k \\ \frac{\partial}{\partial z_a^{\alpha+1}} & , \quad k > n \ \& \ \alpha = N - k \\ 0 & , \quad \text{otherwise} \end{cases}. \qquad (24)$$

3. Sketch of the proof

Let us outline an idea of the proof of the main theorem. To prove the statement i) of the theorem we have to explain why for arbitrary $\xi \in U_q \mathfrak{sl}_N$, $T \in D(\mathrm{Mat}_{m,n})_q$

$$\xi(T) \in D(\mathrm{Mat}_{m,n})_q. \qquad (25)$$

The map $z_a^\alpha \mapsto \widehat{z}_a^\alpha$, $a = 1, \ldots n$, $\alpha = 1, \ldots m$, is extendable up to an embedding of algebras $J : \mathbb{C}[\mathrm{Mat}_{m,n}]_q \hookrightarrow \mathrm{End}_{\mathbb{C}(q^{1/2})}(\mathbb{C}[\mathrm{Mat}_{m,n}]_q)$. Evidently, J intertwines the actions of $U_q \mathfrak{sl}_N$ in $\mathbb{C}[\mathrm{Mat}_{m,n}]_q$ and $\mathrm{End}_{\mathbb{C}(q^{1/2})}(\mathbb{C}[\mathrm{Mat}_{m,n}]_q)$ (this is a corollary of the fact that $\mathbb{C}[\mathrm{Mat}_{m,n}]_q$ is a $U_q \mathfrak{sl}_N$-module algebra). This observation proves (25) for T of the form $J(f)$, $f \in \mathbb{C}[\mathrm{Mat}_{m,n}]_q$, as well as the first part of the statement ii) of the theorem. What remains is to prove (25) for $T = \frac{\partial}{\partial z_a^\alpha}$, $a = 1, \ldots n$, $\alpha = 1, \ldots m$.

The space $\mathrm{End}_{\mathbb{C}(q^{1/2})}(\mathbb{C}[\mathrm{Mat}_{m,n}]_q)$ can be made into a left $\mathbb{C}[\mathrm{Mat}_{m,n}]_q$-module as follows:

$$z_a^\alpha(T) = \widehat{z}_a^\alpha \cdot T,$$

with $a = 1, \ldots n$, $\alpha = 1, \ldots m$, $T \in \mathrm{End}_{\mathbb{C}(q^{1/2})}(\mathbb{C}[\mathrm{Mat}_{m,n}]_q)$. This structure is compatible with the action of $U_q \mathfrak{sl}_N$. Define the $U_q \mathfrak{sl}_N$-module

$$\Lambda^1(\mathrm{Mat}_{m,n})_q \bigotimes_{\mathbb{C}[\mathrm{Mat}_{m,n}]_q} \mathrm{End}_{\mathbb{C}(q^{1/2})}(\mathbb{C}[\mathrm{Mat}_{m,n}]_q).$$

The differential $d : \mathbb{C}[\mathrm{Mat}_{m,n}]_q \to \Lambda^1(\mathrm{Mat}_{m,n})_q$ is a morphism of the $U_q\mathfrak{sl}_N$-modules. This implies $U_q\mathfrak{sl}_N$-invariance of the element

$$\sum_{a=1}^{n}\sum_{\alpha=1}^{m} dz_a^\alpha \otimes \frac{\partial}{\partial z_a^\alpha} \in \Lambda^1(\mathrm{Mat}_{m,n})_q \underset{\mathbb{C}[\mathrm{Mat}_{m,n}]_q}{\bigotimes} \mathrm{End}_{\mathbb{C}(q^{1/2})}(\mathbb{C}[\mathrm{Mat}_{m,n}]_q),$$

i.e. for all $\xi \in U_q\mathfrak{sl}_N$

$$\sum_{a=1}^{n}\sum_{\alpha=1}^{m}\sum_{j} \xi_j' dz_a^\alpha \otimes \xi_j'' \frac{\partial}{\partial z_a^\alpha} = \varepsilon(\xi) \sum_{a=1}^{n}\sum_{\alpha=1}^{m} dz_a^\alpha \otimes \frac{\partial}{\partial z_a^\alpha} \qquad (26)$$

with ε being the counit of $U_q\mathfrak{sl}_N$, $\Delta(\xi) = \sum_j \xi_j' \otimes \xi_j''$ (Δ is the coproduct in $U_q\mathfrak{sl}_N$). As was proved in [7], $\Lambda^1(\mathrm{Mat}_{m,n})_q$ is the free right $\mathbb{C}[\mathrm{Mat}_{m,n}]_q$-module with the generators dz_a^α, $a = 1,\dots n$, $\alpha = 1,\dots m$. Thus, for $\xi \in U_q\mathfrak{sl}_N$ there exists a unique set $f_{\beta,a}^{b,\alpha}(\xi) \in \mathbb{C}[\mathrm{Mat}_{m,n}]_q$, $a = 1,\dots n, \alpha = 1,\dots m, b = 1,\dots n$, $\beta = 1,\dots m$, such that

$$\xi dz_a^\alpha = \sum_{b=1}^{n}\sum_{\beta=1}^{m} dz_b^\beta f_{\beta,a}^{b,\alpha}(\xi).$$

Using the later equality, we can rewrite (26) as follows:

$$\sum_{a,b=1}^{n}\sum_{\alpha,\beta=1}^{m}\sum_{j} dz_b^\beta \otimes f_{\beta,a}^{b,\alpha}(\xi_j')\xi_j'' \frac{\partial}{\partial z_a^\alpha} = \varepsilon(\xi) \sum_{a=1}^{n}\sum_{\alpha=1}^{m} dz_a^\alpha \otimes \frac{\partial}{\partial z_a^\alpha}. \qquad (27)$$

Now one can obtain formulae (19) - (24) (and thus prove (25) for $T = \frac{\partial}{\partial z_a^\alpha}$, $a = 1,\dots n, \alpha = 1,\dots m$) via applying (27) to the generators E_i, F_i, K_i, K_i^{-1} of $U_q\mathfrak{sl}_N$.

4. Concluding notes

The space $\mathrm{Mat}_{m,n}$ of $m \times n$ matrices considered in the present paper is the simplest example of a prehomogeneous vector space of commutative parabolic type [6]. Such vector spaces are closely related to non-compact Hermitian symmetric spaces. Specifically, any non-compact Hermitian symmetric space can be realized (via the so-called Harish-Chandra embedding) as a bounded symmetric domain in some prehomogeneous vector space of commutative parabolic type.

In [9] a q-analogue of an arbitrary prehomogeneous vector space of commutative parabolic type was constructed. More precisely, let U be a bounded symmetric domain, $\mathfrak{g}_{-1}$ the corresponding prehomogeneous vector space, and $\mathfrak{g}$ the Lie algebra of the automorphism group of U. In the paper [9] a $U_q\mathfrak{g}$-module algebra $\mathbb{C}[\mathfrak{g}_{-1}]_q$ and a covariant first order differential calculus $(\Lambda^1(\mathfrak{g}_{-1}), d)$ over $\mathbb{C}[\mathfrak{g}_{-1}]_q$

were introduced. Using the first order differential calculus, one can produce a definition of q-differential operators in $\mathbb{C}[\mathfrak{g}_{-1}]_q$ just as it was done in Section 2 in the case $\mathfrak{g}_{-1} = \mathrm{Mat}_{m,n}$.

Let $D(\mathfrak{g}_{-1})_q$ be the algebra of q-differential operators in $\mathbb{C}[\mathfrak{g}_{-1}]_q$. In this general setting it can also be proved that $D(\mathfrak{g}_{-1})_q$ is a $U_q\mathfrak{g}$-module subalgebra in the $U_q\mathfrak{g}$-module algebra $\mathrm{End}(\mathbb{C}[\mathfrak{g}_{-1}]_q)$. Indeed, it easy to see that the proof of our main theorem (Section 3) does not use a specific nature of the case $\mathfrak{g}_{-1} = \mathrm{Mat}_{m,n}$.

5. Appendix: q-Differential operators in holomorphic q-bundles.

In this Appendix '$\mathbb{C}[\mathrm{Mat}_{m,n}]_q$-module' means right $\mathbb{C}[\mathrm{Mat}_{m,n}]_q$-module.

Let Γ be a finitely generated free $\mathbb{C}[\mathrm{Mat}_{m,n}]_q$-module, i.e. there exists an isomorphism of the $\mathbb{C}[\mathrm{Mat}_{m,n}]_q$-modules

$$\pi : \Gamma \to V \bigotimes \mathbb{C}[\mathrm{Mat}_{m,n}]_q,$$

with V being a finite dimensional vector space. The isomorphism π will be called a trivialization of Γ. Elements of Γ are q-analogs of sections of a holomorphic bundle on $\mathrm{Mat}_{m,n}$. Let us consider two such $\mathbb{C}[\mathrm{Mat}_{m,n}]_q$-modules Γ_1, Γ_2 together with their trivializations $\pi_1 : \Gamma_1 \to V_1 \bigotimes \mathbb{C}[\mathrm{Mat}_{m,n}]_q$, $\pi_2 : \Gamma_2 \to V_2 \bigotimes \mathbb{C}[\mathrm{Mat}_{m,n}]_q$. Set
$D(\Gamma_1, \Gamma_2)_q =$

$$\{D \in \mathrm{Hom}(\Gamma_1, \Gamma_2) | \pi_2 \cdot D \cdot \pi_1^{-1} \in \mathrm{Hom}(V_1, V_2) \bigotimes D(\mathrm{Mat}_{m,n})_q\}.$$

Elements of $D(\Gamma_1, \Gamma_2)_q$ can be treated as q-analogues of differential operators in sections of holomorphic bundles.

To see that $D(\Gamma_1, \Gamma_2)_q$ is well defined, we need to verify its independence of the choice of trivializations. Let $\pi_1' : \Gamma_1 \to V_1' \bigotimes \mathbb{C}[\mathrm{Mat}_{m,n}]_q$, $\pi_2' : \Gamma_2 \to V_2' \bigotimes \mathbb{C}[\mathrm{Mat}_{m,n}]_q$ be other trivializations of Γ_1 and Γ_2, respectively. Evidently, it is sufficient to prove, that for an arbitrary $D' \in \mathrm{Hom}(V_1, V_2) \bigotimes D(\mathrm{Mat}_{m,n})_q$ the map $\pi_2' \cdot \pi_2^{-1} \cdot D' \cdot \pi_1 \cdot (\pi_1')^{-1}$ belongs to $\mathrm{Hom}(V_1', V_2') \bigotimes D(\mathrm{Mat}_{m,n})_q$. But this follows from the fact that π_1, π_2, π_1', π_2' are morphisms of the $\mathbb{C}[\mathrm{Mat}_{m,n}]_q$-modules, and, thus, $\pi_1 \cdot (\pi_1')^{-1} \in \mathrm{Hom}(V_1', V_1) \bigotimes J(\mathbb{C}[\mathrm{Mat}_{m,n}]_q)$ and $\pi_2' \cdot \pi_2^{-1} \in \mathrm{Hom}(V_2, V_2') \bigotimes J(\mathbb{C}[\mathrm{Mat}_{m,n}]_q)$ (with $J(\mathbb{C}[\mathrm{Mat}_{m,n}]_q)$ being the unital subalgebra in $D(\mathrm{Mat}_{m,n})_q$ generated by $\widehat{z}_a^\alpha$, $a = 1, \dots n$, $\alpha = 1, \dots m$).

In applications finitely generated free $\mathbb{C}[\mathrm{Mat}_{m,n}]_q$-modules with some additional properties arise. We will discuss two special types of such $\mathbb{C}[\mathrm{Mat}_{m,n}]_q$-modules.

The first type consists of those finitely generated free $\mathbb{C}[\mathrm{Mat}_{m,n}]_q$-modules Γ which, in addition, are $U_q\mathfrak{sl}_N$-module. It means that Γ is a $U_q\mathfrak{sl}_N$-module and the multiplication map $\Gamma \bigotimes \mathbb{C}[\mathrm{Mat}_{m,n}]_q \to \Gamma$ is a morphism of the $U_q\mathfrak{sl}_N$-modules.

For $\mathbb{C}[\mathrm{Mat}_{m,n}]_q$-modules of this type a result analogous to the main theorem (Section 2) can be obtained. Let us turn to precise formulations.

If V_1, V_2 are modules over a Hopf algebra A then the space $\mathrm{Hom}(V_1, V_2)$ admits the following "canonical" structure of an A-module: for $\xi \in A$, $T \in \mathrm{Hom}(V_1, V_2)$

$$\xi(T) = \sum_j \xi'_j \cdot T \cdot S(\xi''_j), \tag{28}$$

where $\Delta(\xi) = \sum_j \xi'_j \otimes \xi''_j$ (Δ is the coproduct), S is the antipode, and the product in the right-hand side means the composition of the maps $S(\xi''_j) \in \mathrm{End}(V_1)$, $T \in \mathrm{Hom}(V_1, V_2)$, $\xi'_j \in \mathrm{End}(V_2)$. It is well known that this action makes $\mathrm{Hom}(V_1, V_2)$ into an A-module left $\mathrm{End}(V_2)$-module and an A-module right $\mathrm{End}(V_1)$-module, i.e. the composition map

$$\mathrm{End}(V_2) \bigotimes \mathrm{Hom}(V_1, V_2) \bigotimes \mathrm{End}(V_1) \to \mathrm{Hom}(V_1, V_2)$$

is a morphism of the A-modules.

We can use the above construction to equip $\mathrm{Hom}(\Gamma_1, \Gamma_2)$ (where Γ_1, Γ_2 are $U_q \mathfrak{sl}_N$-module finitely generated free $\mathbb{C}[\mathrm{Mat}_{m,n}]_q$-modules) with the structure of a $U_q \mathfrak{sl}_N$-module. Using our main theorem, one can prove that

the subspace $D(\Gamma_1, \Gamma_2)_q \subset \mathrm{Hom}(\Gamma_1, \Gamma_2)$ is $U_q \mathfrak{sl}_N$-invariant; thus, the composition map

$$D(\Gamma_2)_q \bigotimes D(\Gamma_1, \Gamma_2)_q \bigotimes D(\Gamma_1)_q \to D(\Gamma_1, \Gamma_2)_q$$

(here $D(\Gamma)_q$ denotes $D(\Gamma, \Gamma)_q$) makes $D(\Gamma_1, \Gamma_2)_q$ into a $U_q \mathfrak{sl}_N$-module left $D(\Gamma_2)_q$-module and a $U_q \mathfrak{sl}_N$-module right $D(\Gamma_1)_q$-module.

The second type of $\mathbb{C}[\mathrm{Mat}_{m,n}]_q$-modules consists of those $U_q \mathfrak{sl}_N$-module $\mathbb{C}[\mathrm{Mat}_{m,n}]_q$-modules which admit good trivializations. Let $U_q(\mathfrak{f} + \mathfrak{p}_-)$ be the Hopf subalgebra in $U_q \mathfrak{sl}_N$ generated by F_i, $K_i^{\pm 1}$, $i = 1, \ldots N - 1$, and E_j, $j = 1, \ldots n - 1, n + 1, \ldots N - 1$. Suppose that a finitely generated free $\mathbb{C}[\mathrm{Mat}_{m,n}]_q$-module Γ is $U_q \mathfrak{sl}_N$-module (in particular, Γ is a $U_q(\mathfrak{f} + \mathfrak{p}_-)$-module $\mathbb{C}[\mathrm{Mat}_{m,n}]_q$-module). A trivialization $\pi : \Gamma \to V \otimes \mathbb{C}[\mathrm{Mat}_{m,n}]_q$ is called good trivialization if it satisfies the following conditions: i) V is a finite dimensional $U_q(\mathfrak{f} + \mathfrak{p}_-)$-module with the property $F_n v = 0$ for any $v \in V$; ii) π is a morphism of the $U_q(\mathfrak{f} + \mathfrak{p}_-)$-modules (here $V \otimes \mathbb{C}[\mathrm{Mat}_{m,n}]_q$ is endowed with $U_q(\mathfrak{f} + \mathfrak{p}_-)$-module structure via the coproduct $\Delta : U_q(\mathfrak{f} + \mathfrak{p}_-) \to U_q(\mathfrak{f} + \mathfrak{p}_-) \otimes U_q(\mathfrak{f} + \mathfrak{p}_-)$).

It turn out that the set of good trivializations of a $\mathbb{C}[\mathrm{Mat}_{m,n}]_q$-module Γ is not too wide: if $\pi_1 : \Gamma \to V_1 \otimes \mathbb{C}[\mathrm{Mat}_{m,n}]_q$, $\pi_2 : \Gamma \to V_2 \otimes \mathbb{C}[\mathrm{Mat}_{m,n}]_q$ are two good trivializations, then

$$\pi_2 \cdot \pi_1^{-1} = T \otimes 1 \tag{29}$$

with $T \in \mathrm{Hom}_{U_q(\mathfrak{f} + \mathfrak{p}_-)}(V_1, V_2)$.

We distinguish this type of $\mathbb{C}[\mathrm{Mat}_{m,n}]_q$-modules because for them the notion of a q-differential operator with constant coefficients is well-defined. Specifically, let $D(\mathrm{Mat}_{m,n})_q^0$ be the unital subalgebra in $D(\mathrm{Mat}_{m,n})_q$ generated by $\frac{\partial}{\partial z_a^\alpha}$, $a = 1, \ldots n$, $\alpha = 1, \ldots m$. Suppose that Γ_1, Γ_2 are $U_q(\mathfrak{k} + \mathfrak{p}_-)$-module $\mathbb{C}[\mathrm{Mat}_{m,n}]_q$-modules with good trivializations $\pi_1 : \Gamma_1 \to V_1 \otimes \mathbb{C}[\mathrm{Mat}_{m,n}]_q$, $\pi_2 : \Gamma_2 \to V_2 \otimes \mathbb{C}[\mathrm{Mat}_{m,n}]_q$. We set

$$D(\Gamma_1, \Gamma_2)_q^0 = \{D \in D(\Gamma_1, \Gamma_2)_q | \pi_2 \cdot D \cdot \pi_1^{-1} \in \mathrm{Hom}(V_1, V_2) \bigotimes D(\mathrm{Mat}_{m,n})_q^0\}.$$

Elements of $D(\Gamma_1, \Gamma_2)_q^0$ can be treated as q-analogues of the differential operators with constant coefficients in sections of holomorphic bundles. Independence $D(\Gamma_1, \Gamma_2)_q^0$ of trivializations directly follows from the relationship (29) between two arbitrary good trivializations of a $\mathbb{C}[\mathrm{Mat}_{m,n}]_q$-module.

References

1. E. E. Demidov, *Modules over quantum Weyl algebras*, Vestnik MGU, Mathematics and Mechanics, **1** (1993), 53–56.

2. H. P. Jakobsen, *Quantized Hermitian Symmetric Spaces*, In "Lie theory and its applications in physics" (Clausthal, 1995), 105 – 116.

3. ———, *Q-Differential Operators*, E-print: math.QA/9907009, 1999.

4. A. Kamita, Y. Morita, and T. Tanisaki, *Quantum deformations of certain prehomogeneous spaces I*, Hiroshima Math. J., **28** (1998), 527 – 540.

5. M. Rosso, *Representations des groups quantiques*, Seminaire BOURBAKI, **744** (1991), 443 – 483.

6. H. Rubenthaler, *Les paires duales dans les algèbres de Lie réductives*, Asterisque, **219** (1994).

7. D. Shklyarov, S. Sinel'shchikov, and L. Vaksman, *Quantum matrix ball: differential and integral calculi*, E-print: math.QA/9905035, 1999.

8. S. Sinel'shchikov and L. Vaksman, *Hidden symmetry of the differential calculus on the quantum matrix space*, J. Phys. A. **30** (1997), 23 – 26.

9. ———, *On q-analogues of bounded symmetric domains and Dolbeault complexes*, Math. Phys., Anal., and Geom., **1** (1998), 75 – 100; E-print: q-alg/9703005, 1997.

A FAMILY OF ∗-ALGEBRAS ALLOWING WICK ORDERING: FOCK

REPRESENTATIONS AND UNIVERSAL ENVELOPING C^*-ALGEBRAS

PALLE JORGENSEN *

Department of Mathematics, The University of Iowa, Iowa City, Iowa 52242-1419 U.S.A.

DANIIL PROSKURIN †

Kyiv Taras Shevchenko University, Cybernetics Department, Volodymyrska, 64, Kyiv, 01033, Ukraine

YURII SAMOILENKO ‡

Institute of Mathematics, National Academy of Sciences, Tereschenkivska, 3, Kyiv, 01601, Ukraine

Abstract. We consider an abstract Wick ordering as a family of relations on elements a_i and define ∗-algebras by these relations. The relations are given by a fixed operator $T: \mathfrak{h} \otimes \mathfrak{h} \to \mathfrak{h} \otimes \mathfrak{h}$, where $\mathfrak{h}$ is one-particle space, and they naturally define both a ∗-algebra and an inner-product space $\mathcal{H}_T$, $\langle \cdot, \cdot \rangle_T$. If a_i^* denotes the adjoint, i.e., $\langle a_i \varphi, \psi \rangle_T = \langle \varphi, a_i^* \psi \rangle_T$, then we identify when $\langle \cdot, \cdot \rangle_T$ is positive semidefinite (the positivity question!). In the case of deformations of the CCR-relations (the q_{ij}-CCR and the twisted CCR's), we work out the universal C^*-algebras $\mathfrak{A}$, and we prove that, in these cases, the Fock representations of the $\mathfrak{A}$'s are faithful.

1. Introduction

In recent papers [1–6], the applications of Lie superalgebras, quantum groups, q-algebras in mathematical physics have stimulated interest in the ∗-algebras defined by generators and relations and their representations by Hilbert space operators. For example, the representations of various deformations of canonical commutation relations (CCR), in particular Fock representaion, were used to construct non-classical models of theoretical physics and probability, such as the free quon gas (see [7]), q-Gaussian processes (see [8]) etc.

* jorgen@math.uiowa.edu
† prosk@imath.kiev.ua
‡ yurii_sam@imath.kiev.ua

321

S. Duplij and J. Wess (eds.), Noncommutative Structures in Mathematics and Physics, 321–329.
© 2001 *Kluwer Academic Publishers. Printed in the Netherlands.*

The constructions are interesting from both physical and mathematical points of view. They give a canonical realisation of a given deformed relation like the Fock representation, or a realisation by differential operators. When the relations can be realised by bounded operators, it is useful to study the universal enveloping C^*-algebras for them and the stability of isomorphism classes of these C^*-algebras on parameters (see for example [9, 10]). The stability question [10] refers to how the C^*-isomorphism classes depend on variations in the deformation variables; in some cases there are open regions in parameter space where the C^*-isomorphism class is constant.

In the present paper we give a review of some results concerning a wide class of deformed relations of the following form

$$a_i^* a_j = \delta_{ij} 1 + \sum_{k,l=1}^{d} T_{ij}^{kl} a_l a_k^*, \ i,j = 1,\dots,d, \tag{1}$$

where $T_{ij}^{kl} \in \mathbb{C}$, such that $T_{ij}^{kl} = \bar{T}_{ji}^{lk}$. These relations generate a $*$-algebra allowing Wick ordering or **Wick algebra** (see [4, 11–13]). The $*$-algebra $\mathfrak{A}_T$ has a naturally defined Fock vacuum "state" or functional and there is a corresponding inner-product space $\mathcal{H}_T, \langle \cdot, \cdot \rangle_T$, such that, in the associated GNS-representation, the identity $\langle a_i \varphi, \psi \rangle_T = \langle \varphi, a_i^* \psi \rangle_T$ holds. But the vacuum functional is generally not positive, and the operators in the representation not bounded, and therefore the Hermitian inner product $\langle \cdot, \cdot \rangle_T$ is then generally not positive semidefinite. The positivity question, and the faithfulness of the Fock representation, are the foci of this paper.

Note that (1) generalizes some well-known types of deformed commutation relations, quantum groups, etc. (see [1, 3, 5, 6, 8, 12, 14, 15]). The basic examples for us will be the q_{ij}-CCR introduced and studied by M. Bożejko and R. Speicher (see [8, 12]), and the twisted canonical commutation relations (TCCR) constructed by W. Pusz and S.L. Woronowicz (see [6]). They were further studied in [16] where the traditional Cuntz algebra of [17] was considered as a base-point, corresponding to $q_{ij} = 0$, and the variation of the C^*-isomorphism class was considered as a function of q_{ij}.

EXAMPLE 1. q_{ij}-CCR, $2d$ generators:

$$\mathbb{C}\langle a_i, a_i^* \mid a_i^* a_j = \delta_{ij} 1 + q_{ij} a_j a_i^*, \ i,j = 1,\dots,d,$$
$$q_{ji} = \bar{q}_{ij} \in \mathbb{C}, |q_{ij}| \le 1 \rangle$$

EXAMPLE 2. The Wick algebra for TCCR:

$$a_i^* a_i = 1 + \mu^2 a_i a_i^* - (1 - \mu^2) \sum_{k<i} a_k a_k^*, \ i = 1,\dots,d$$
$$a_i^* a_j = \mu a_j a_i^*, \ i \neq j, \quad 0 < \mu < 1$$

We present some sufficient conditions on the coefficients $\{T_{ij}^{kl}\}$ for the existence of the Fock representation, and we describe the structure of the Fock space. We also give conditions for the faithfulness of Fock representation and describe its kernel in the degenerated case (see Sec. 3).

Further we consider the universal C^*-algebras for the examples above. Specifically we show that the universal C^*-algebras for q_{ij}-CCR (TCCR) can be generated by isometries (partial isometries) satisfying a certain algebraic relation. The description of the C^*-isomorphism classes for different values of parameters is presented.

We also show that the Fock representations of q_{ij}-CCR for some values of parameters, and TCCR for any value of parameter, are faithful on the C^*-level, i.e., the Fock representations of the corresponding C^*-algebras are faithful (see Sec. 4).

The complete proofs of all results presented here can be found in [4, 10, 11, 18, 19]. For detailed information about *-representations of finitely generated *-algebras see [20].

2. Basic definitions

Firstly let us construct a canonical realization of Wick algebra, i.e., the *-algebra on the relations (1), with coefficients $\{T_{ij}^{kl}\}$: we denote it by $W(T)$. To do it consider a finite-dimensional Hilbert space $\mathcal{H} = \langle e_1, \ldots, e_d \rangle$. Construct the full tensor algebra over $\mathcal{H}$, $\mathcal{H}^*$, denoted by $\mathcal{T}(\mathcal{H}, \mathcal{H}^*)$. Then

$$W(T) \cong \mathcal{T}(\mathcal{H}, \mathcal{H}^*)/\langle e_i^* \otimes e_j - \delta_{ij}1 - \sum T_{ij}^{kl} e_i \otimes e_j^* \rangle, \qquad (2)$$

dividing out by the two-sided ideal on the relations (1). Note that in this realization the subalgebra of $W(T)$ generated by $\{a_i\}$ is identified with the $\mathcal{T}(\mathcal{H})$.

The following operators were presented in [11] as a useful tool for computation with Wick algebras and their Fock representations.

$$T: \mathcal{H} \otimes \mathcal{H} \mapsto \mathcal{H} \otimes \mathcal{H}, \quad T e_k \otimes e_l = \sum_{i,j} T_{ik}^{lj} e_i \otimes e_j, \quad T = T^*$$

$$T_i: \mathcal{H}^{\otimes n} \mapsto \mathcal{H}^{\otimes n}, \quad T_i = \underbrace{1 \otimes \cdots \otimes 1}_{i-1} \otimes T \otimes \underbrace{1 \otimes \cdots \otimes 1}_{n-i-1},$$

$$R_n: \mathcal{H}^{\otimes n} \mapsto \mathcal{H}^{\otimes n}, \quad R_n = 1 + T_1 + T_1 T_2 + \cdots + T_1 T_2 \cdots T_{n-1},$$

$$P_n: \mathcal{H}^{\otimes n} \mapsto \mathcal{H}^{\otimes n}, \quad P_2 = R_2, \quad P_{n+1} = (1 \otimes P_n) R_{n+1}. \qquad (3)$$

The sequences of operators $P_0 = 1_{vac}$, $R_1 = 1 + T$, $P_1 = (1 \otimes 1)(1 + T) \cong 1 + T$, $R_2, \ldots, P_n$ are defined recursively. It is the sequence P_n which enters into the positivity question. The other one is only intermediate. The Hermitian inner product $\langle \cdot, \cdot \rangle_T$ on $\mathcal{T}_n(\mathcal{H})$ is then

$$\langle \phi, \psi \rangle_{\mathcal{T}_n(\mathcal{H})} := \langle \varphi, P_n \psi \rangle_{tensor} \qquad (4)$$

where $\langle \cdot, \cdot \rangle_{\text{tensor}}$ is just the usual inner product on $\mathcal{T}_n(\mathcal{H})$ induced by $\langle \cdot, \cdot \rangle$ on $\mathcal{H}$. Hence, we need conditions on $T: \mathcal{H} \otimes \mathcal{H} \to \mathcal{H} \otimes \mathcal{H}$ which make the operators P_n positive for all n. For example, in terms of these operators we can describe the procedure of Wick ordering, i.e., the commutation formula for fixed generator a_i^* and any homogeneous polynomial in a_k, $k = 1, \ldots, d$ (see [21]).

Proposition 27. *Let* $X \in \mathcal{H}^{\otimes n}$. *Then*

$$e_i^* \otimes X = \mu(e_i^*) R_n X + \mu(e_i^*) \sum_{k=1}^{d} T_1 T_2 \cdots T_n (X \otimes e_k) e_k^*, \qquad (5)$$

where $\mu(e_i^*): \mathcal{T}(\mathcal{H}) \mapsto \mathcal{T}(\mathcal{H})$ *is defined as follows*

$$\mu(e_i^*)1 = 0, \quad \mu(e_i^*)e_{i_1} \otimes \cdots \otimes e_{i_n} = \delta_{i i_1} e_{i_2} \otimes \cdots \otimes e_{i_n}.$$

For our examples the operator T have the following form:

EXAMPLE 3.

$$T e_i \otimes e_j = q_{ij} e_j \otimes e_i, i, j = 1, \ldots, d.$$

EXAMPLE 4.

$$T e_i \otimes e_i = \mu^2 e_i \otimes e_i$$
$$T e_i \otimes e_j = \mu e_j \otimes e_i, \ i < j$$
$$T e_i \otimes e_j = -(1 - \mu^2) e_i \otimes e_j + \mu e_j \otimes e_i, \ i > j.$$

Note that for both examples, the operator T satisfies a **braid condition**, i.e., on the $\mathcal{H}^{\otimes 3}$ we have

$$T_1 T_2 T_1 = T_2 T_1 T_2. \qquad (6)$$

The operators presented above appear naturally in construction of Fock representation of $W(T)$. This notion is induced in the obvious way from the classical one for CCR, however, in general, the Fock space is not always symmetric (see [11]).

Definition 28. The representation λ_0 acting on the space $\mathcal{T}(\mathcal{H})$ by formulas

$$\lambda_0(a_i)e_{i_1} \otimes \cdots \otimes e_{i_n} = e_i \otimes e_{i_1} \otimes \cdots \otimes e_{i_n}, \quad n \in \mathbb{N} \cup \{0\}$$
$$\lambda_0(a_i^*)1_{\text{vac}} = 0$$

where the action of $\lambda_0(a_i^*)$ on the monomials of degree $n \geq 1$ is determined inductively using the basic relations, is called the Fock representation.

It is easy to see that $\lambda_0(a_i)$ are the classical creation operators and $\lambda_0(a_i^*)$ are twisted annihilation ones. Evidently in this way we have constructed a representation of $W(T)$, but not yet a $*$-representation. To do it one has to supply the $\mathcal{T}(\mathcal{H})$ by the appropriate inner product (see [11]). This is where formula (4) comes in.

Definition 29. The Fock inner product (see [11]) is the unique semilinear Hermitian form $\langle \, , \, \rangle_T$ on $T(\mathcal{H})$ such that

$$\langle \lambda_0(a_i)X, Y \rangle_T = \langle X, \lambda_0(a_i^*)Y \rangle_T, \quad X, Y \in T(\mathcal{H}).$$

Similarly to the definition of Fock representation, the Fock inner product on $T(\mathcal{H})$ can be computed inductively. It is easy to see that for $X \in \mathcal{H}^{\otimes m}, Y \in \mathcal{H}^{\otimes n}$, $n \neq m$, we have $\langle X, Y \rangle_T = 0$. On the components of powers $0, 1$, the Fock inner product concides with the standard one. For any $X, Y \in \mathcal{H}^{\otimes n}, n \geq 2$, we have

$$\langle X, Y \rangle_T = \langle X, P_n Y \rangle,$$

which agrees with (4) above. The operator $P_n = P_n(T)$ are given in (3).

Evidently, if we want to extend the Fock representation of $W(T)$ to the ∗-representation by Hilbert-space operators, we should require that all the operators $P_n, n = 2, \ldots$, be positive semidefinite, and that the subspace

$$\mathcal{I} = \bigoplus_{n \geq 2} \text{Ker } P_n$$

determines the kernel of the Fock inner product. Consequently the Hilbert-space structure of the Fock space emerges.

3. The structure of the Fock representation

In this section we present some sufficient conditions posed on the operator T for the positive-definite property of the Fock inner product, and we show that the kernel of the Fock representation is generated as a ∗-*ideal* by the kernel of the Fock inner product. In particular, when the Fock inner product is strictly positive definite (i.e., when it has zero kernel), the Fock representation π_F is faithful, i.e., $\text{Ker}(\pi_F) = 0$.

There are several sufficient conditions on the operator T for the Fock inner product to be positive. It was shown in [10] that for sufficiently small coefficients we have strict positivity of the Fock inner product. This result is a corollary of the stability of the universal enveloping C^*-algebra for the Wick algebra around the zero base point (see Sec. 4).

Theorem 30. *If the operator T satisfies the norm bound $\|T\| < \sqrt{2} - 1$, then $P_n > 0, n \geq 2$, where $>$ refers to strict positivity.*

Another kind of sufficient condition is positivity of operator T (see [11]).

Theorem 31. *If $T \geq 0$ then $P_n > 0, n \geq 2$.*

In the present paper we will suppose that the operator T satisfies the **braid** condition (6). It was shown by M. Bożejko and R. Speicher (see [12]) that, in this case, the operators P_n, $n \geq 2$, have a natural description in terms of quasi-multiplicative operator-valued mappings on the Coxeter group S_n. The following is a corollary of a much more general result proved in [12] for mappings on the general Coxeter group.

Theorem 32. *Let T satisfy the braid condition (6) and suppose $-1 \leq T \leq 1$. Then $P_n \geq 0$. Moreover, if $\|T\| \leq 1$, then $P_n > 0$, and the operators of the Fock representation are bounded, i.e., the Fock representation is by bounded operators. (Recall, the Fock representation of the undeformed CCR-algebra is unbounded.)*

We present a more precise version of this theorem. Namely, we give the description of kernel of P_n in the degenerate case. As an immediate corollary of this result we have the strict positivity of P_n, $n \geq 2$, for braided T satisfying the inequality $-1 < T \leq 1$ (see [4]).

Theorem 33. *Let $W(T)$ be a Wick algebra with braided operator T satisfying the norm bound $\|T\| \leq 1$. Then for any $n \geq 1$,*

$$Ker\, P_{n+1} = \sum_{k+l=n-1} \mathcal{H}^{\otimes k} \otimes Ker\,(1+T) \otimes \mathcal{H}^{\otimes l} = \sum_{k=1}^{n} Ker\,(1+T_k).$$

Let us illustrate this result on the examples.

EXAMPLE 5. For q_{ij}-CCR we have the alternatives:

  – $|q_{ij}| < 1$ for any $i, j = 1, \ldots, d$.
   In this case $-1 < T < 1$ and the Fock inner product is strictly positive.
  – $|q_{ij}| = 1$, $i \neq j$.
   For these values of parameters we have $-1 \leq T \leq 1$ and

$$Ker\,(1+T) = \langle a_j a_i - q_{ij} a_i a_j,\ i < j \rangle.$$

EXAMPLE 6. For the TCCR Wick algebra, we have $-1 \leq T \leq 1$, and

$$Ker\,(1+T) = \langle a_j a_i - \mu a_i a_j,\ i < j \rangle.$$

The following proposition shows that, for algebras with braided operator T, the kernel of the Fock representation is generated as a $*$-ideal by the kernel of the Fock inner product, i.e.,

$$\mathcal{I} = \bigoplus_{n \geq 2} Ker\, P_n.$$

Proposition 34. *Let $W(T)$ be a Wick algebra with braided operator T and let the Fock representation λ_0 be positive (i.e., the Fock inner product is positive definite). Then*

$$Ker\, \lambda_0 = \mathcal{I} \otimes T(\mathcal{H}^*) + T(\mathcal{H}) \otimes \mathcal{I}^*.$$

Combining this proposition with Theorem 33, we get:

Theorem 35. *Let $W(T)$ be a Wick algebra with the braided operator T, $-1 \leq T \leq 1$. Then the kernel of the Fock representation is generated as a *-ideal by $\text{Ker}(1 + T)$.*

This theorem implies that, for q_{ij}-CCR, $|q_{ij}| < 1$, the Fock representation is faithful. For the TCCR Wick algebra, and for q_{ij}-CCR, the kernels of the Fock representations are generated by the families $a_j a_i - \mu a_i a_j$, $i < j$, and $a_j a_i - q_{ij} a_i a_j$, $i < j$, respectively; and hence the Fock representations of quotients of these algebras by the *-ideals generated by these families are faithful.

4. Universal bounded representation

In this section we discuss universal enveloping C^*-algebras for q_{ij}-CCR and Wick TCCR.

Let us recall that the universal C^*-algebra for a certain *-algebra $\mathcal{A}$ is also called the universal bounded representation. It is the C^*-algebra $\mathbf{A}$ with natural homomorphism $\psi: \mathcal{A} \to \mathbf{A}$ such that, for any homomorphism $\varphi: \mathcal{A} \to B$, where B is a C^*-algebra, there exists a unique homomorphism $\theta: \mathbf{A} \to B$ satisfying $\theta \circ \psi = \varphi$. It can be obtained by the completion of $\mathcal{A}/J$ with the following C^*-seminorm on $\mathcal{A}$:

$$\|a\| = \sup_{\pi} \|\pi(a)\|,$$

where sup is taken over all bounded representations of $\mathcal{A}$, and J is the kernel of this seminorm. Obviously this process requires that $\sup_{\pi} \|\pi(a)\| < \infty$ for any $a \in \mathcal{A}$. Note that for our examples this condition is satisfied.

The universal bounded representation for q_{ij}-CCR was studied in [9, 10]. The following proposition follows from the main result of paper [10].

Proposition 36. *Let $\mathbf{A}_{\{q_{ij}\}}$ be the universal enveloping C^*-algebra for q_{ij}-CCR, $|q_{ij}| < \sqrt{2} - 1$. Then there exists the natural isomorphism*

$$\mathbf{A}_{\{q_{ij}\}} \cong \mathbf{A}_0,$$

where $\mathbf{A}_0$ is a C^*-algebra generated by the isometries s_i, $i = 1, \ldots, d$, satisfying

$$s_i^* s_j = 0, \; i \neq j$$

i.e., isomorphism with the Cuntz-Toeplitz algebra.

This implies that the Fock representation of $\mathbf{A}_{\{q_{ij}\}}$ is faithful.

Let us consider the $\mathbf{A}_{\{q_{ij}\}}$, $|q_{ij}| = 1$, for any $i \neq j$ and $q_{ii} := q_i$, $|q_i| < 1$ (i.e., unimodular off-diagonal terms). In this case, we do not have stability on the whole set of parameters (see [18]).

Proposition 37. *If for any $i \neq j$ we have $|q_{ij}| = 1$, then $\mathbf{A}_{\{q_{ij}\}}$ is isomorphic to the C^*-algebra $\mathbf{A}_{0,\{q_i\}}$ generated by isometries $\{s_i, \ i = 1, \ldots, d\}$ satisfying*

$$s_i^* s_j = q_{ij} s_j s_i^*, \quad s_j s_i = q_{ij} s_i s_j, \quad i \neq j,$$

and the Fock representation of $\mathbf{A}_{\{q_{ij}\}}$ is faithful.

Finally for the universal C^*-algebra $\mathbf{A}_\mu$ for the Wick TCCR, we have the isomorphism $\mathbf{A}_\mu \cong \mathbf{A}_0$ for any $-1 < \mu < 1$, where the C^*-algebra $\mathbf{A}_0$ is generated by the partial isometries $\{s_i, \ i = 1, \ldots, d\}$ satisfying the relations

$$s_i^* s_j = \delta_{ij} \left(1 - \sum_{k<i} s_k s_k^* \right), \quad i, j = 1, \ldots, d.$$

The Fock representation of $\mathbf{A}_\mu$ is faithful also (see [19]).

ACKNOWLEDGEMENTS. P. J. was partially supported by the NSF under grants DMS-9700130 and INT-9722779.

References

1. L.C. Biedenharn, *The quantum group* $\mathrm{SU}_q(2)$ *and a q-analogue of the boson operators*, J. Phys. A **22** (1989), L873–L878.

2. I.M. Burban and A.U. Klimyk, *On spectral properties of q-oscillator operators*, Lett. Math. Phys. **29** (1993), 13–18.

3. D.I. Fivel, *Interpolation between Fermi and Bose statistics using generalized commutators*, Phys. Rev. Lett. **65** (1990), 3361–3364.

4. P.E.T. Jørgensen, D.P. Proskurin and Yu. S. Samoĭlenko, *The kernel of Fock representations of Wick algebras with braided operator of coefficients*, accepted for publication in Pacific J. Math., math-ph/0001011.

5. A.J. Macfarlane, *On q-analogues of the quantum harmonic oscillator and the quantum group* $\mathrm{SU}(2)_q$, J. Phys. A **22** (1989), 4581–4588.

6. W. Pusz and S.L. Woronowicz, *Twisted second quantization*, Rep. Math. Phys. **27** (1989), 251–263.

7. R.F. Werner, *The free quon gas suffers Gibbs' paradox*, Phys. Rev. D (3) **48** (1993), 2929–2934.

8. M. Bożejko and R. Speicher, *An example of a generalized Brownian motion*, Commun. Math. Phys. **137** (1991), 519–531.

9. K. Dykema and A. Nica, *On the Fock representation of the q-commutation relations*, J. Reine Angew. Math. **440** (1993), 201–212.

10. P.E.T. Jørgensen, L.M. Schmitt and R.F. Werner, *q-canonical commutation relations and stability of the Cuntz algebra*, Pacific J. Math. **165** (1994), 131–151.

11. ———, *Positive representations of general commutation relations allowing Wick ordering*, J. Funct. Anal. **134** (1995), 33–99.

12. M. Bożejko and R. Speicher, *Completely positive maps on Coxeter groups, deformed commutation relations, and operator spaces*, Math. Ann. **300** (1994), 97–120.

13. W. Marcinek and R. Ralowski, *On Wick algebras with braid relations*, J. Math. Phys. **36** (1995), 2803–2820.

14. O.W. Greenberg, *Particles with small violations of Fermi or Bose statistics*, Phys. Rev. D (3) **43** (1991), 4111–4129.
15. W. Marcinek, *On commutation relations for quons*, Rep. Math. Phys. **41** (1998), 155–172.
16. P.E.T. Jørgensen and R.F. Werner, *Coherent states of the q-canonical commutation relations*, Commun. Math. Phys. **164** (1994), 455–471.
17. J. Cuntz, *Simple C^*-algebras generated by isometries*, Commun. Math. Phys. **57** (1977), 173–185.
18. D. Proskurin, *Stability of a special class of q_{ij}-CCR and extensions of higher-dimensional noncommutative tori*, to appear in Lett. Math. Phys.
19. D. Proskurin and Yu. Samoilenko, *Stability of a C^* algebra associated with the TCCR*, submitted to Algebras and Representation Theory.
20. V. Ostrovskyĭ and Yu. Samoĭlenko, *Introduction to the Theory of Representations of Finitely Presented ∗-Algebras, I: Representations by bounded operators*, The Gordon and Breach Publishing Group, London, 1999.
21. D. P. Proskurin, *Homogeneous ideals in Wick ∗-algebras*, Proc. Amer. Math. Soc. **126** (1998), 3371–3376.

NONSTANDARD QUANTIZATION OF THE ENVELOPING ALGEBRA U(so(n)) AND ITS APPLICATIONS

A. U. KLIMYK *
Institute for Theoretical Physics, Kiev, Ukraine

1. Introduction

Quantum orthogonal groups, quantum Lorentz groups and their corresponding quantum algebras are of special interest for modern mathematical physics (see, for example, [1] and [2]). M. Jimbo [3] and V. Drinfeld [4] defined q-deformations (quantum algebras) $U_q(g)$ for all simple complex Lie algebras g by means of Cartan subalgebras and root subspaces (see also [5] and [6]). Reshetikhin, Takhtajan and Faddeev [7] defined quantum algebras $U_q(g)$ in terms of the quantum R-matrix satisfying the quantum Yang–Baxter equation. However, these approaches do not give a satisfactory presentation of the quantum algebra $U_q(so(n, \mathbb{C}))$ from a viewpoint of some problems in quantum physics and representation theory. When considering representations of the quantum groups $SO_q(n + 1)$ and $SO_q(n, 1)$ we are interested in reducing them onto the quantum subgroup $SO_q(n)$. This reduction would give an analogue of the Gel'fand–Tsetlin basis for these representations. However, definitions of quantum algebras mentioned above do not allow the inclusions $U_q(so(n + 1, \mathbb{C})) \supset U_q(so(n, \mathbb{C}))$ and $U_q(so_{n,1}) \supset U_q(so_n)$. To be able to exploit such reductions we have to consider q-deformations of the Lie algebra $so(n + 1, \mathbb{C})$ defined in terms of the generators $I_{k,k-1} = E_{k,k-1} - E_{k-1,k}$ (where E_{is} is the matrix with elements $(E_{is})_{rt} = \delta_{ir}\delta_{st}$) rather than by means of Cartan subalgebras and root elements. To construct such deformations we have to deform trilinear relations for elements $I_{k,k-1}$ instead of Serre's relations (used in the case of Jimbo's quantum algebras). As a result, we obtain the associative algebra which will be denoted as $U_q'(so(n, \mathbb{C}))$. This q-deformation was first constructed in [8]. It permits one to construct the reductions of $U_q'(so(n + 1, \mathbb{C})$ onto $U_q'(so(n, \mathbb{C}))$.

---

* aklimyk@bitp.kiev.ua

S. Duplij and J. Wess (eds.), Noncommutative Structures in Mathematics and Physics, 331–342.
© 2001 *Kluwer Academic Publishers. Printed in the Netherlands.*

In the classical case, the imbedding $SO(n) \subset SU(n)$ (and its infinitesimal analogue) is of great importance for nuclear physics and in the theory of Riemannian symmetric spaces. It is well known that in the framework of Drinfeld–Jimbo quantum groups and algebras one cannot construct the corresponding embedding. The algebra $U_q'(\mathrm{so}(n, \mathbb{C}))$ allows to define such an embedding [9], that is, it is possible to define the embedding $U_q'(\mathrm{so}(n, \mathbb{C})) \subset U_q(\mathrm{sl}_n)$, where $U_q(\mathrm{sl}_n)$ is a Drinfeld-Jimbo quantum algebra.

As a disadvantage of the algebra $U_q'(\mathrm{so}(n, \mathbb{C}))$ we have to mention the difficulties with Hopf algebra structure. Nevertheless, $U_q'(\mathrm{so}(n, \mathbb{C}))$ turns out to be a coideal in $U_q(\mathrm{sl}_n)$ (see [9]) and this fact allows us to consider tensor products of finite dimensional irreducible representations of $U_q'(\mathrm{so}(n, \mathbb{C}))$ for many interesting cases (see [10] for the case $U_q'(\mathrm{so}(3, \mathbb{C}))$).

For convenience, below we denote the Lie algebra $\mathrm{so}(n, \mathbb{C})$ by so_n and the q-deformed algebra $U_q'(\mathrm{so}(n, \mathbb{C}))$ by $U_q'(\mathrm{so}_n)$.

Finite dimensional irreducible representations of the algebra $U_q'(\mathrm{so}_n)$ were constructed in [8]. The formulas of action of the generators of $U_q'(\mathrm{so}_n)$ upon the basis (which is a q-analogue of the Gel'fand–Tsetlin basis) are given there. A proof of these formulas and some their corrections were given in [11]. However, finite dimensional irreducible representations described in [8] and [11] are representations of the classical type. They are q-deformations of the corresponding irreducible representations of the Lie algebra so_n, that is, at $q \to 1$ they turn into representations of so_n.

The algebra $U_q'(\mathrm{so}_n)$ has other classes of finite dimensional irreducible representations which have no classical analogue. These representations are singular at the limit $q \to 1$. They are described in [12]. Note that the description of these representations for the algebra $U_q'(\mathrm{so}_3)$ is given in [10].

As in the case of Drinfeld–Jimbo quantum algebras, when q is a root of unity, then the representation theory of $U_q'(\mathrm{so}_n)$ is much more rich. In this case all irreducible representations of $U_q'(\mathrm{so}_n)$ are finite dimensional. The corresponding theorem is proved by means of an analogue of the Poincaré–Birkhoff–Witt theorem for $U_q'(\mathrm{so}_n)$ (this analogue was announced in [13]) and use central elements of this algebra for q a root of unity (they are derived in [14]).

2. The q-deformed algebra $U_q'(\mathrm{so}_n)$

The universal enveloping algebra $U(\mathrm{so}_n)$ of the Lie algebra so_n has two different structures. The first one is determined by roots and root elements of the Lie algebra so_n. A deformation of $U(\mathrm{so}_n)$ equipped with this structure leads to the Drinfeld–Jimbo quantum algebra $U_q(\mathrm{so}_n)$. The second structure of $U(\mathrm{so}_n)$ is related to the basis of the Lie algebra so_n consisting of skew-symmetric matrices. A deformation of $U(\mathrm{so}_n)$ equipped with this structure leads to the algebra $U_q'(\mathrm{so}_n)$ considered in this paper.

In order to obtain $U'_q(\mathrm{so}_n)$ we have to take determining relations for the generating elements $I_{21}, I_{32}, \cdots, I_{n,n-1}$ of $U(\mathrm{so}_n)$ and to deform these relations. The elements $I_{21}, I_{32}, \cdots, I_{n,n-1}$ belong to the basis I_{ij}, $i > j$, of the Lie algebra so_n. The matrices I_{ij}, $i > j$, are defined as $I_{ij} = E_{ij} - E_{ji}$, where E_{ij} is the matrix with entries $(E_{ij})_{rs} = \delta_{ir}\delta_{js}$. The universal enveloping algebra $U(\mathrm{so}_n)$ is generated by a part of the basis elements I_{ij}, $i > j$, namely, by the elements I_{21}, $I_{32}, \cdots, I_{n,n-1}$. These elements satisfy the relations

$$I_{i,i-1}^2 I_{i+1,i} - 2I_{i,i-1}I_{i+1,i}I_{i,i-1} + I_{i+1,i}I_{i,i-1}^2 = -I_{i+1,i},$$

$$I_{i,i-1}I_{i+1,i}^2 - 2I_{i+1,i}I_{i,i-1}I_{i+1,i} + I_{i+1,i}^2 I_{i,i-1} = -I_{i,i-1},$$

$$I_{i,i-1}I_{j,j-1} - I_{j,j-1}I_{i,i-1} = 0 \quad \text{for} \quad |i - j| > 1.$$

The following theorem is true [15] for the enveloping algebra $U(\mathrm{so}_n)$.

Theorem 1. *The universal enveloping algebra $U(\mathrm{so}_n)$ is isomorphic to the complex associative algebra (with a unit element) generated by the elements I_{21}, $I_{32}, \cdots, I_{n,n-1}$ satisfying the above relations.*

We make the q-deformation of these relations by $2 \to [2] := (q^2 - q^{-2})/(q - q^{-1}) = q + q^{-1}$. As a result, we obtain the complex associative algebra generated by elements $I_{21}, I_{32}, \cdots, I_{n,n-1}$ satisfying the relations

$$I_{i,i-1}^2 I_{i+1,i} - (q + q^{-1})I_{i,i-1}I_{i+1,i}I_{i,i-1} + I_{i+1,i}I_{i,i-1}^2 = -I_{i+1,i}, \qquad (1)$$

$$I_{i,i-1}I_{i+1,i}^2 - (q + q^{-1})I_{i+1,i}I_{i,i-1}I_{i+1,i} + I_{i+1,i}^2 I_{i,i-1} = -I_{i,i-1}, \qquad (2)$$

$$I_{i,i-1}I_{j,j-1} - I_{j,j-1}I_{i,i-1} = 0 \quad \text{for} \quad |i - j| > 1. \qquad (3)$$

This algebra was introduced by us in [8] and is denoted by $U'_q(\mathrm{so}_n)$. Here q takes any complex value such that $q \neq 0, \pm 1$.

Let us formulate for the algebra $U'_q(\mathrm{so}_n)$ an analogue of the Poincaré–Birkhoff–Witt theorem. For this we determine (see [16] and [17]) in $U'_q(\mathrm{so}_n)$ elements analogous to the matrices I_{ij}, $i > j$, of the Lie algebra so_n. In order to give them we use the notation $I_{k,k-1} \equiv I_{k,k-1}^+ \equiv I_{k,k-1}^-$. Then for $k > l + 1$ we define recursively

$$I_{kl}^+ := [I_{l+1,l}, I_{k,l+1}]_q \equiv q^{1/2}I_{l+1,l}I_{k,l+1} - q^{-1/2}I_{k,l+1}I_{l+1,l}, \qquad (4)$$

$$I_{kl}^- := [I_{l+1,l}, I_{k,l+1}]_{q^{-1}} \equiv q^{-1/2}I_{l+1,l}I_{k,l+1} - q^{1/2}I_{k,l+1}I_{l+1,l}.$$

The elements I_{kl}^+, $k > l$, satisfy the commutation relations

$$[I_{ln}^+, I_{kl}^+]_q = I_{kn}^+, \quad [I_{kl}^+, I_{kn}^+]_q = I_{ln}^+, \quad [I_{kn}^+, I_{ln}^+]_q = I_{kl}^+ \quad \text{for} \quad k > l > n, \qquad (5)$$

$$[I_{kl}^+, I_{nr}^+] = 0 \quad \text{for} \quad k > l > n > r \text{ and } k > n > r > l, \qquad (6)$$

$$[I_{kl}^+, I_{nr}^+]_q = (q - q^{-1})(I_{lr}^+ I_{kn}^+ - I_{kr}^+ I_{nl}^+) \quad \text{for} \quad k > n > l > r. \tag{7}$$

For I_{kl}^-, $k > l$, the commutation relations are obtained from these relations by replacing I_{kl}^+ by I_{kl}^- and q by q^{-1}.

The algebra $U_q'(so_n)$ can be considered as an associative algebra (with unit element) generated by I_{kl}^+, $1 \leq l < k \leq n$, satisfying the relations (5)–(7). Similarly, $U_q'(so_n)$ is an associative algebra generated by I_{kl}^-, $1 \leq l < k \leq n$, satisfying the corresponding relations. Now the Poincaré–Birkhoff–Witt theorem for the algebra $U_q'(so_n)$ can be formulated as follows.

Theorem 2. *The elements*

$$I_{21}^{+ \ m_{21}} I_{31}^{+ \ m_{31}} \cdots I_{n1}^{+ \ m_{n1}} I_{32}^{+ \ m_{32}} I_{42}^{+ \ m_{42}} \cdots I_{n2}^{+ \ m_{n2}} \cdots I_{n,n-1}^{+ \ m_{n,n-1}}, \quad m_{ij} \in \mathbb{N},$$

form a basis of the algebra $U_q'(so_n)$. This assertion is true if I_{ij}^+, $i < j$, are replaced by the corresponding elements I_{ij}^-.

The proof of this theorem is given in [18].

3. The embedding $U_q'(so_n) \to U_q(sl_n)$

The algebra $U_q'(so_n)$ can be embedded into the Drinfeld–Jimbo quantum algebra $U_q(sl_n)$ (see [9]). This quantum algebra is generated by the elements E_i, F_i, $K_i^{\pm 1} = q^{\pm H_i}$, $i = 1, 2, \cdots, n - 1$, satisfying the relations

$$K_i K_j = K_j K_i, \quad K_i K_i^{-1} = K_i^{-1} K_i = 1,$$

$$K_i E_j K_i^{-1} = q^{a_{ij}} E_j, \quad K_i F_j K_i^{-1} = q^{-a_{ij}} F_j, \quad [E_i, F_j] = \delta_{ij} \frac{K_i - K_i^{-1}}{q - q^{-1}},$$

$$E_i^2 E_{i \pm 1} - (q + q^{-1}) E_i E_{i \pm 1} E_i + E_{i \pm 1} E_i^2 = 0,$$

$$F_i^2 F_{i \pm 1} - (q + q^{-1}) F_i F_{i \pm 1} F_i + F_{i \pm 1} F_i^2 = 0,$$

$$[E_i, E_j] = 0, \quad [F_i, F_j] = 0 \quad \text{for} \quad |i - j| > 1,$$

where a_{ij} are elements of the Cartan matrix of the Lie algebra sl_n.

Let us introduce the elements

$$\tilde{I}_{j,j-1} = F_{j-1} - q q^{-H_{j-1}} E_{j-1}, \quad j = 2, 3, \cdots, n,$$

of $U_q(sl_n)$. It is proved in [9] that there exists the algebra homomorphism $\varphi : U_q'(so_n) \to U_q(sl_n)$ uniquely determined by the relations $\varphi(I_{i+1,i}) = \tilde{I}_{i+1,i}$, $i = 1, 2, \cdots, n - 1$. The following theorem states that this homomorphism is an isomorphism.

Theorem 3. *The homomorphism $\varphi : U_q'(so_n) \to U_q(sl_n)$ determined by the relations $\varphi(I_{i+1,i}) = \tilde{I}_{i+1,i}$, $i = 1, 2, \cdots, n - 1$, is an isomorphism of $U_q'(so_n)$ to $U_q(sl_n)$.*

In [16] the authors of that paper state that this homomorphism is an isomorphism and say that it can be proved by means of the Diamond Lemma. However, we could not restore their proof and found another one in [18]. Theorem 3 has the following important corollary, proved in [18]:

Corollary. *Finite dimensional irreducible representations of $U'_q(\mathrm{so}_n)$ separate elements of this algebra, that is, for any $a \in U'_q(\mathrm{so}_n)$ there exists a finite dimensional irreducible representation T of $U'_q(\mathrm{so}_n)$ such that $T(a) \neq 0$.*

This corollary is true for q not a root of unity as well as for q a root of unity.

Problems: We think that the algebra $U'_q(\mathrm{so}_n)$ is connected with some extension of the Drinfeld–Jimbo quantum algebra $\hat{U}_q(\mathrm{so}_n)$. This conjecture is proved in [10] for the case $n = 3$. It is shown there that there is an isomorphism $\varphi : U'_q(\mathrm{so}_3) \to \hat{U}_q(\mathrm{sl}_2)$, where $\hat{U}_q(\mathrm{sl}_2)$ is an extension of the quantum algebra $U_q(\mathrm{sl}_2)$.

4. Central elements of $U'_q(\mathrm{so}_n)$

Let us form the elements

$$J^{\pm}_{k_1,k_2,\ldots,k_{2r}} = q^{\mp \frac{r(r-1)}{2}} \sum_{s \in S_{2r}}{}' \varepsilon_{q^{\pm 1}}(s) I^{\pm}_{k_{s(2)},k_{s(1)}} I^{\pm}_{k_{s(4)},k_{s(3)}} \cdots I^{\pm}_{k_{s(2r)},k_{s(2r-1)}}, \quad (8)$$

of the algebra $U'_q(\mathrm{so}_n)$ (see [13]), where $1 \le k_1 < k_2 < \cdots < k_{2r} \le n$ and summation runs over all permutations s of indices $k_1, k_2, \cdots, k_{2r}$ such that

$$k_{s(2)} > k_{s(1)}, \quad k_{s(4)} > k_{s(3)}, \quad \ldots, k_{s(2r)} > k_{s(2r-1)},$$
$$k_{s(2)} < k_{s(4)} < \cdots < k_{s(2r)}.$$

The symbol $\varepsilon_{q^{\pm 1}}(s) \equiv (-q^{\pm 1})^{\ell(s)}$ stands for the q-analogue of Levi–Chivita antisymmetric tensor, $\ell(s)$ means the length of permutation s. Note that in the limit $q \to 1$ both sets in (8) reduce to the set of components of rank $2r$ antisymmetric tensor operator of Lie algebra so_n.

Theorem 4. *The elements*

$$C_n^{(2r)} = \sum_{1 \le k_1 < k_2 < \cdots < k_{2r} \le n} q^{k_1 + k_2 + \cdots + k_{2r} - r(n+1)} J^+_{k_1,k_2,\ldots,k_{2r}} J^-_{k_1,k_2,\ldots,k_{2r}}, \quad (9)$$

where $r = 1, 2, \cdots, \{n/2\}$ ($\{a\}$ means the integral part of a), are Casimir elements of $U'_q(\mathrm{so}_n)$, that is, they belong to the center of this algebra. If n is even, then the elements $C_n^{(n)+} \equiv J^+_{1,2,\cdots,n}$ and $C_n^{(n)-} \equiv J^-_{1,2,\cdots,n}$ also belong to the center of $U'_q(\mathrm{so}_n)$.

Central elements of this theorem are found in [13]. It was conjectured in [13] that for q not a root of unity the set of central elements $C_n^{(2r)}, r = 1, 2, \cdots, \{(n-$

1)/2}, and the element $C_n^{(n)+}$ (if n is even) generates the center of the algebra $U_q'(\mathrm{so}_n)$.

Let us give explicitly some central elements. For $U_q'(\mathrm{so}_3)$ and $U_q'(\mathrm{so}_4)$ we have

$$C_3^{(2)} = q^{-1}I_{21}^2 + I_{31}^+I_{31}^- + qI_{32}^2 = qI_{21}^2 + I_{31}^-I_{31}^+ + q^{-1}I_{32}^2,$$

$$C_4^{(2)} = q^{-2}I_{21}^2 + I_{32}^2 + q^2I_{43}^2 + q^{-1}I_{31}^+I_{31}^- + qI_{42}^+I_{42}^- + I_{41}^+I_{41}^-,$$

$$C_4^{(4)+}=C_4^{(4)-}=q^{-1}I_{21}I_{43} - I_{31}^+I_{42}^+ + qI_{32}I_{41}^+=qI_{21}I_{43} - I_{31}^-I_{42}^- + q^{-1}I_{32}I_{41}^-.$$

The quadratic central element of $U_q'(\mathrm{so}_n)$ is of the form

$$C_n^{(2)} = \sum_{1 \le i < j \le n} q^{i+j-n-1}I_{ji}^+I_{ji}^-.$$

If q is a root of unity, then (as in the case of Drinfeld–Jimbo quantum algebras) there exist additional central elements of $U_q'(\mathrm{so}_n)$ which are given by the following theorem, proved in [14]:

Theorem 5. *Let $q^k = 1$ for $k \in \mathbb{N}$ and $q^j \ne 1$ for $0 < j < k$. Then the elements*

$$C^{(k)}(I_{rl}^+) = \sum_{j=0}^{\{(k-1)/2\}} \binom{k-j}{j} \frac{1}{k-j}\left(\frac{i}{q-q^{-1}}\right)^{2j} I_{rl}^{+k-2j}, \quad r > l, \qquad (10)$$

where $\{(k-1)/2\}$ is the integral part of the number $(k-1)/2$, belong to the center of $U_q'(\mathrm{so}_n)$.

It is well-known that a Drinfeld–Jimbo algebra $U_q(g)$ for q a root of unity ($q^k = 1$) is a finite dimensional vector space over the center of $U_q(g)$. The same assertion is true for the algebra $U_q'(\mathrm{so}_n)$. By Theorem 5, any element $(I_{ij}^+)^s$, $s \ge k$, can be reduced to a linear combination of $(I_{ij}^+)^r$, $r < k$, with coefficients from the center $\mathcal{C}$ of $U_q'(\mathrm{so}_n)$. Now our assertion follows from this sentence and from Poincaré–Birkhoff–Witt theorem for $U_q'(\mathrm{so}_n)$. Using this assertion, it is proved the following theorem [18]:

Theorem 6. *If q is a root of unity, then any irreducible representation of $U_q'(\mathrm{so}_n)$ is finite dimensional.*

It can be proved more strong assertion: there exists a fixed positive integer r such that dimension of any irreducible representation of $U_q'(\mathrm{so}_n)$ at q a root of unity does not exceed r. Of course, the number r depends on k (recall that k is defined by $q^k = 1$).

5. Irreducible representations of $U_q'(\mathrm{so}_n)$

We first assume that q is not a root of unity. Then the algebra $U_q'(\mathrm{so}_n)$ has two types of irreducible finite dimensional representations:

(a) representations of the classical type (at $q \to 1$ they give the corresponding finite dimensional irreducible representations of the Lie algebra so_n);

(b) representations of the nonclassical type (they do not admit the limit $q \to 1$ since in this point the representation operators are singular).

Let us describe the classical type representations of the algebras $U'_q(so_n)$, $n \geq 3$, which are q-deformations of the finite dimensional irreducible representations of the Lie algebra so_n. As in the case of irreducible representations of the Lie algebra so_n, they are given by sets $\mathbf{m}_n$ consisting of $\{n/2\}$ numbers $m_{1,n}, m_{2,n}, ..., m_{\{n/2\},n}$ (here $\{n/2\}$ denotes integral part of $n/2$) which are all integral or all half-integral and satisfy the dominance conditions

$$m_{1,2p+1} \geq m_{2,2p+1} \geq ... \geq m_{p,2p+1} \geq 0, \tag{11}$$

$$m_{1,2p} \geq m_{2,2p} \geq ... \geq m_{p-1,2p} \geq |m_{p,2p}| \tag{12}$$

for $n = 2p + 1$ and $n = 2p$, respectively. These representations are denoted by $T_{\mathbf{m}_n}$. For a basis in a representation space we can take the q-analogue of the Gel'fand–Tsetlin basis which is obtained by successive reduction of the representation $T_{\mathbf{m}_n}$ to the subalgebras $U'_q(so_{n-1})$, $U'_q(so_{n-2})$, $\cdots$, $U'_q(so_3)$, $U'_q(so_2) := U(so_2)$. As in the classical case, its elements are labeled by Gel'fand–Tsetlin tableaux

$$\{\xi_n\} \equiv \begin{bmatrix} \mathbf{m}_n \\ \mathbf{m}_{n-1} \\ \cdots \\ \mathbf{m}_2 \end{bmatrix}, \tag{13}$$

where the components of $\mathbf{m}_r$ and $\mathbf{m}_{r-1}$ satisfy the betweenness conditions

$$m_{1,2p+1} \geq m_{1,2p} \geq m_{2,2p+1} \geq m_{2,2p} \geq ... \geq m_{p,2p+1} \geq m_{p,2p} \geq -m_{p,2p+1}, \tag{14}$$

$$m_{1,2p} \geq m_{1,2p-1} \geq m_{2,2p} \geq m_{2,2p-1} \geq ... \geq m_{p-1,2p-1} \geq |m_{p,2p}|. \tag{15}$$

The explicit formulas for the operators $T_{\mathbf{m}_n}(I_{j,j-1})$, $j = 2,3,\cdots,n$, of the representation $T_{\mathbf{m}_n}$ of $U'_q(so_n)$ and their proofs are given in [11].

The representations, described above, are called representations of the classical type, since under the limit $q \to 1$ the operators $T_{\mathbf{m}_n}(I_{j,j-1})$ turn into the corresponding operators $T_{\mathbf{m}_n}(I_{j,j-1})$ for irreducible finite dimensional representations with highest weights $\mathbf{m}_n$ of the Lie algebra so_n.

Let us give the explicit expressions for Casimir operators (corresponding to the central elements, described in Theorem 4) in the classical type irreducible representations of $U'_q(so_n)$. For this we define the generalized factorial elementary symmetric polynomials. Fixing an arbitrary sequence of complex numbers $\mathbf{a} = (a_1, a_2, \cdots)$, for each $r = 0, 1, 2, \cdots, N$, we introduce these polynomials in N variables $z_1, z_2, \cdots, z_N$ by the formula

$$e_r(z_1, z_2, \ldots, z_N | \mathbf{a}) =$$

$$= \sum_{1 \le p_1 < p_2 < \cdots < p_r \le N} (z_{p_1} - a_{p_1})(z_{p_2} - a_{p_2-1}) \cdots (z_{p_r} - a_{p_r-r+1}).$$

By Schur Lemma, Casimir operators in the irreducible finite dimensional representations, characterized by the numbers $(m_{1,n}, m_{2,n}, \ldots, m_{N,n})$, $N = \{n/2\}$, are multiple to the identity operator: $T_{\mathbf{m}_n}(C_n^{(2r)}) = \chi_{\mathbf{m}_n}^{(2r)} \mathbf{1}$.

Theorem 7 [13]. *The eigenvalue of the operator $T_{\mathbf{m}_n}(C_n^{(2r)})$ is*

$$\chi_{\mathbf{m}_n}^{(2r)} = (-1)^r e_r([l_{1,n}]^2, [l_{2,n}]^2, \ldots, [l_{N,n}]^2 | \mathbf{a}),$$

where $\mathbf{a} = ([\epsilon]^2, [\epsilon+1]^2, [\epsilon+2]^2, \ldots)$, $l_{k,n} = m_{k,n} + N - k + \epsilon$. (Here $\epsilon = 0$ for $n = 2N$ and $\epsilon = \frac{1}{2}$ for $n = 2N + 1$.) If $n = 2N$ is even, then

$$T_{\mathbf{m}_n}(C_n^{(n)+}) = T_{\mathbf{m}_n}(C_n^{(n)-}) = (\sqrt{-1})^N [l_{1,n}][l_{2,n}] \cdots [l_{N,n}] \mathbf{1}.$$

The algebra $U_q'(\mathrm{so}_n)$ has also irreducible finite dimensional representations T of nonclassical type, that is, such that the operators $T(I_{j,j-1})$ have no classical limit $q \to 1$. They are given by sets $\epsilon := (\epsilon_2, \epsilon_3, \cdots, \epsilon_n)$, $\epsilon_i = \pm 1$, and by sets $\mathbf{m}_n$ consisting of $\{n/2\}$ **half-integral** (but not integral) numbers $m_{1,n}, m_{2,n}, \cdots, m_{\{n/2\},n}$ (here $\{n/2\}$ denotes the integral part of $n/2$) that satisfy the dominance conditions

$$m_{1,2p+1} \ge m_{2,2p+1} \ge \ldots \ge m_{p,2p+1} \ge 1/2,$$

$$m_{1,2p} \ge m_{2,2p} \ge \ldots \ge m_{p-1,2p} \ge m_{p,2p} \ge 1/2$$

for $n = 2p + 1$ and $n = 2p$, respectively. These representations are denoted by $T_{\epsilon,\mathbf{m}_n}$.

For a basis in the representation space we can use an analogue of the basis (13). Its elements are labeled by the tableaux

$$\{\xi_n\} \equiv \begin{bmatrix} \mathbf{m}_n \\ \mathbf{m}_{n-1} \\ \cdots \\ \mathbf{m}_2 \end{bmatrix},$$

where the components of $\mathbf{m}_{2p+1}$ and $\mathbf{m}_{2p}$ satisfy the betweenness conditions

$$m_{1,2p+1} \ge m_{1,2p} \ge m_{2,2p+1} \ge m_{2,2p} \ge \ldots \ge m_{p,2p+1} \ge m_{p,2p} \ge 1/2,$$

$$m_{1,2p} \ge m_{1,2p-1} \ge m_{2,2p} \ge m_{2,2p-1} \ge \ldots \ge m_{p-1,2p-1} \ge m_{p,2p}.$$

Explicit formulas for the operator $T_{\epsilon,\mathbf{m}_n}(I_{j,j-1})$, $j = 2, 3, \cdots, n$, of the representation $T_{\epsilon,\mathbf{m}_n}$ of $U_q(\mathrm{so}_n)$ are given in [12].

Theorem 8. *The representations $T_{\epsilon,\mathbf{m}_n}$ are irreducible. The representations $T_{\epsilon,\mathbf{m}_n}$ and $T_{\epsilon',\mathbf{m}_n'}$ are pairwise nonequivalent for $(\epsilon, \mathbf{m}_n) \ne (\epsilon', \mathbf{m}_n')$. For any* *admissible $(\epsilon, \mathbf{m}_n)$ and $\mathbf{m}'_n$ the representations $T_{\epsilon, \mathbf{m}_n}$ and $T_{\mathbf{m}'_n}$ are pairwise nonequivalent.*

The algebra $U'_q(so_n)$ has non-trivial one-dimensional representations. They are special cases of the representations of the nonclassical type. They are described as follows. Let $\epsilon := (\epsilon_2, \epsilon_3, \cdots, \epsilon_n)$, $\epsilon_i = \pm 1$, and let $\mathbf{m}_n = (m_{1,n}, m_{2,n}, \cdots, m_{\{n/2\},n}) = (\frac{1}{2}, \frac{1}{2}, \cdots, \frac{1}{2})$. Then the corresponding representations $T_{\epsilon, \mathbf{m}_n}$ are one-dimensional and are given by the formulas

$$T_{\epsilon, \mathbf{m}_n}(I_{k+1,k})|\xi_n\rangle = \frac{\epsilon_{k+1}}{q^{1/2} - q^{-1/2}}|\xi_n\rangle.$$

Thus, to every $\epsilon := (\epsilon_2, \epsilon_3, \cdots, \epsilon_n)$, $\epsilon_i = \pm 1$, there corresponds a one-dimensional representation of $U'_q(so_n)$.

Conjecture. *If q is not a root of unity, then every irreducible finite dimensional representation of $U'_q(so_n)$ is equivalent to one of the representations $T_{\mathbf{m}_n}$ of the classical type or to one of the representations $T_{\epsilon, \mathbf{m}_n}$ of the nonclassical type.*

This conjecture is proved for the algebra $U'_q(so_3)$ (see [19]).

Irreducible representations of the algebra $U'_q(so_n)$ for q a root of unity are described in [18]. For construction of these irreducible representations of $U'_q(so_n)$, it is used the method of D. Arnaudon and A. Chakrabarti [20] for construction of irreducible representations of the quantum algebra $U_q(sl_n)$ when q is a root of unity. If $q^p = 1$ and p is an odd integer, then there exists the series of irreducible representations of $U'_q(so_n)$ which act on p^N-dimensional vector space (where N is the number of positive roots of the Lie algebra so_n) and are given by $r = \dim so_n$ complex parameters. These representations are irreducible for generic values of these parameters. These representations constitute the main class of irreducible representations of $U'_q(so_n)$. For some special values of the representation parameters in $\mathbb{C}^r$ the representations are reducible. These reducible representations give many other classes of (degenerate) irreducible representations which are given by less number of parameters or by parameters, values of which cover subsets of $\mathbb{C}^r$ of Lebesgue measure 0. As in the case of irreducible representations of the quantum algebra $U_q(sl_n)$, it is difficult to enumerate all irreducible representations of these classes. However, the most important classes of these degenerate representations can be constructed. In particular, in [18] we give 2^{n-1} classes of these representations, which are an analogue of the nonclassical type irreducible representations of $U'_q(so_n)$ for q not a root of unity.

6. Restriction of representations of $U_q(\mathrm{sl}_n)$ to $U'_q(\mathrm{so}_n)$

In this section we assume that q is not a root of unity. The algebra $U'_q(\mathrm{so}_n)$ is a subalgebra of the quantum algebra $U_q(\mathrm{sl}_n)$. Therefore, we may restrict irreducible finite dimensional representations of the algebra $U_q(\mathrm{sl}_n)$ to the subalgebra $U'_q(\mathrm{so}_n)$. Generally speaking, such a restriction leads to reducible representations of the subalgebra. It was proved in [16] that each irreducible finite dimensional representation of $U_q(\mathrm{sl}_n)$ under restriction to $U'_q(\mathrm{so}_n)$ decomposes into a direct sum of irreducible representations of this subalgebra. N. Iorgov has proved (will be published) that such a decomposition contains only irreducible representations of the classical type. However, explicit formula for the decomposition is known only for the restriction $U_q(\mathrm{sl}_3) \to U'_q(\mathrm{so}_3)$.

Irreducible finite dimensional representations of $U_q(\mathrm{sl}_3)$ are given by three integers $\ell = (l_1, l_2, l_3)$ such that $l_1 \geq l_2 \geq l_3$. We denote such the representation by R_ℓ. Irreducible finite dimensional classical type representations of $U'_q(\mathrm{so}_3)$ are denoted by T_k, where k is a nonnegative integral or half-integral number.

In order to find which irreducible representations of $U'_q(\mathrm{so}_3)$ are contained in the decomposition of $R_\ell \downarrow_{U'_q(\mathrm{so}_3)}$ we split in [21] the spectrum $\mathrm{Spec}\, R_\ell(I_{21})$ of the representation operator $R_\ell(I_{21})$ into spectra of operators $T_k(I_{21})$ of irreducible representations T_k of $U'_q(\mathrm{so}_3)$. (It is proved in [21] that such splitting is unique.) As a result, we have that

$$R_\ell \downarrow_{U'_q(\mathrm{so}_3)} = \sum_s{}' \sum_{k=s}^{s+l_2-l_3} T_k$$

if $l_1 - l_2$ is odd and

$$R_\ell \downarrow_{U'_q(\mathrm{so}_3)} = \sum_s{}' \sum_{k=s}^{s+l_2-l_3} T_k \oplus \sum_r{}' T_r$$

if $l_1 - l_2$ is even, where $\sum_s'$ means the summation over the values $l_1 - l_2, l_1 - l_2 - 2, l_1 - l_2 - 4, \cdots, 1$ (or 2) and the last sum $\sum_r'$ is over the values $l_2 - l_3, l_2 - l_3 - 2, l_2 - l_3 - 4, \cdots, 0$ (or 1). Note that these decompositions coincide with the corresponding decompositions for the reduction $SU(3) \to SO(3)$.

7. Applications

There are the following main applications of the algebra $U'_q(\mathrm{so}_n)$ and its irreducible representations:

1. The theory of orthogonal polynomials and special functions (especially, the theory of q-orthogonal polynomials and basic hypergeometric functions). This

direction is not good worked out. Some ideas of such applications can be found in [22].

2. The algebra $U'_q(so_n)$ (espesially its particular case $U'_q(so_3)$) is related to the algebra of observables in 2+1 quantum gravity on the Riemmanian surfaces (see the papers [23]–[25]).

3. A q-analogue of the Riemannian symmetric space $SU(n)/SO(n)$ is constructed by means of the algebra $U'_q(so_n)$. This construction is fulfilled in the paper [9].

4. A q-analogue of the theory of harmonic polynomials (q-harmonic polynomials on quantum vector space $\mathbb{R}^n_q$) is constructed by using the algebra $U'_q(so_n)$. In particular, a q-analogue of different separations of variables for the q-Laplace operator is given by means of this algebra and its subalgebras. This theory is contained in the papers [16] and [26].

5. The algebra $U'_q(so_n)$ also appear in the theory of links in the algebraic topology (see [27]).

Acknowledgment

The research contained in this paper was supported in part by Award No. UP1-2115 of the Civilian Research and Development Foundation for the Independent States of the Former Soviet Union (CRDF).

References

1. R. Gielerak, J. Lukierski, and Z. Popowicz (ed.), *Quantum Groups and Related Topics*, Kluwer, Dordrecht, 1992.

2. W. B. Schmidke, J. Wess, and B. Zumino, *A q-deformed Lorentz algebra*, Z. Phys. **C52** (1991), 471–476.

3. M. Jimbo, *A q-analogue of U(g) and the Yang–Baxter equation*, Lett. Math. Phys. **10** (1985), 63–69.

4. V. G. Drinfeld, *Hopf algebras and the quantum Yang–Baxter equation*, Sov. Math. Dokl. **32** (1985), 354–258.

5. J. C. Jantzen, *Lectures on Quantum Groups*, Aner. Math. Soc., Providence, RI, 1996.

6. A. Klimyk and K. Schmüdgen, *Quantum Groups and Their Representations*, Springer, Berlin, 1997.

7. N. Ya. Reshetikhin, L. A. Takhtajan, and L. D. Faddeev, *Quantization of Lie groups and Lie algebras*, Leningrad Math. J. **1** (1990), 193–225.

8. A. M. Gavrilik and A. U. Klimyk, *q-Deformed orthogonal and pseudo-orthogonal algebras and their representations*, Lett. Math. Phys. **21** (1991), 215–220.

9. M. Noumi, *Macdonald's symmetric polynomials as zonal spherical functions on quantum homogeneous spaces*, Adv. Math. **123** (1996), 16–77.

10. M. Havlíček, A. U. Klimyk, and S. Pošta, *Representations of the cyclically symmetric q-deformed algebra* $so_q(3)$, J. Math. Phys. **40** (1999), 2135–2161.

11. A. M. Gavrilik and N. Z. Iorgov, *q-Deformed algebras* $U_q(so_n)$ *and their representations*, Methods of Funct. Anal. Topology 3, No. 4 (1997), 51–63.

12. N. Z. Iorgov and A. U. Klimyk, *Nonclassical type representations of the q-deformed algebra* $U_q'(so_n)$, Czech. J. Phys. **50** (2000), 85–90.

13. A. M. Gavrilik and N. Z. Iorgov, *Higher Casimir operators of the nonstandard q-deformed algebras* $U_q'(so_n)$ *and their eigenvalues in representations*, Heavy Ion Physics **11**, No. 1–2 (2000), 33–38.

14. M. Havlíček, A. U. Klimyk, and S. Pošta, *Central elements of the algebra* $U_q'(so_n)$ *and* $U_q(iso_n)$, Czech. J. Phys. **50** (2000), 79–84.

15. A. U. Klimyk, *Nonstandard q-deformation of the universal enveloping algebra* $U(so_n)$, in Proc. Int. Conf. "Quantum Theory and Symmetry" (H.-D. Doebner *et al*, eds.), World Scientific, Singapore, 459–463.

16. M. Noumi, T. Umeda, and M. Wakayama, *Dual pairs, spherical harmonics and a Capelli identity in quantum group theory*, Compos. Math. **104** (1996), 227–277.

17. A. M. Gavrilik and N. Z. Iorgov, *Representations of the nonstandard algebras* $U_q(so_n)$ *and* $U_q(so_{n,1})$ *in Gel'fand–Tsetlin basis*, Ukr. J. Phys. **43** (1998), 791–797.

18. N. Z. Iorgov and A. U. Klimyk, *The nonstandard deformation* $U_q'(so_n)$ *for q a root of unity*, Methods Funct. Anal. Topol. **6**, No. 3 (2000), 15–29.

19. M. Havlíček and S. Pošta, *On the classification of irreducible finite-dimensional representations of* $U_q'(so_3)$ *algebra*, J. Math. Phys., submitted for publication.

20. D. Arnaudon and A. Chakrabarti, *Periodic and partially periodic representations of* $SU(n)_q$, Commun. Math. Phys. **139** (1991), 461–478.

21. A. U. Klimyk and I. I. Kachurik, *The embedding* $U_q'(so_3) \subset U_q(sl_3)$, J. Phys. A: Math. Gen., submitted for publication.

22. A. U. Klimyk and I. I. Kachurik, *Spectra, eigenvectors and overlap functions for representation operators of q-deformed algebras*, Commun. Math. Phys. **175** (1996), 89–111.

23. J. Nelson and T. Regge, *2+1 gravity for genus* $s > 1$, Commun. Math. Phys. **141** (1991), 211–223.

24. A. M. Gavrilik, *The use of quantum algebras in quantum gravity*, Proc. Inst. Math. NAS Ukraine **30** (2000), 304–309.

25. L. Chekhov and V. Fock, *2+1 Geometry and quantum Teichmüller spaces*, Czech. J. Phys., to be published.

26. N. Z Iorgov and A. U. Klimyk, *The q-Laplace operator and q-harmonic polynomials on the quantum vector space*, J. Math. Phys., submitted for publication.

27. D. Bullock and J. H. Przytycki, *Multiplicative structure of Kauffman bracket skein module quantization*, e-print: math.QA/9902117.

CAN THE CABIBBO MIXING ORIGINATE FROM

NONCOMMUTATIVE EXTRA DIMENSIONS?

ALEXANDRE GAVRILIK *
Bogolyubov Institute for Theoretical Physics, Kiev, Ukraine

Abstract. Treating hadronic flavor symmetries with quantum algebras $U_q(su_n)$ leads to interesting consequences such as: new mass sum rules for hadrons 1^-, $\frac{1}{2}^+$, $\frac{3}{2}^+$ of improved accuracy; possibility to label different flavors topologically - by torus winding number; properly fixed deformation parameter q in case of baryons is linked in a simplest way to the Cabibbo angle θ_C, that suggests for θ_C the exact value $\frac{\pi}{14}$. In this connection, we discuss the possibility that this angle and the Cabibbo mixing as a whole take its origin in noncommutativity of some additional, with regard to 3+1, space-time dimensions.

1. Introduction

The problem of fermion flavors, mixings and masses (see e.g., [1]) belongs to most puzzling ones in particle physics. The Cabibbo mixing first introduced for three lightest flavors in the context of weak decays [2] involves the angle θ_C. Importance of this concept was further confirmed after its generalization to mixing of 3 families [3]. Due to Wolfenstein parametrization [4] of CKM matrix, the Cabibbo angle now plays a prominent role: not only CKM matrix elements V_{ij}, but also the quark (and even lepton) mass ratios are often expressed as powers of small parameter $\lambda = \sin\theta_C \approx 0.22$. No doubt, it is necessary to know the value of λ as precise as possible. In this respect, the main bonus of our approach to flavor symmetries, based on quantum algebras, is that it suggests *theoretically motivated* exact value for θ_C, namely, $\theta_C = \frac{\pi}{14}$. As further implication, it leads us to a conjecture of possible noncommutative-geometric origin of the Cabibbo mixing, and our aim here is to argue this may indeed be the case. Below, when treating baryon masses, we restrict ourselves with 4 flavors including u-, d-, s-, and c- quarks. Basic tool of the approach used is the representation theory of quantum algebras [5] $U_q(su_n)$ adopted, instead of conventional $SU(n)$, to describe flavor symmetries classifying hadrons into multiplets.

---

* omgavr@bitp.kiev.ua

S. Duplij and J. Wess (eds.), Noncommutative Structures in Mathematics and Physics, 343–355.

2. Vector meson masses: q-deformation replaces (singlet) mixing

We use[1] Gelfand-Tsetlin basis vectors for meson states from $(n^2 - 1)$-plet of
'n-flavor' $U_q(u_n)$ embedded into $\{(n+1)^2 - 1\}$-plet of 'dynamical' $U_q(u_{n+1})$;
construct mass operator $\hat{M}_n$ invariant under the 'isospin+hypercharge' q-algebra
$U_q(u_2)$ from generators of 'dynamical' algebra $U_q(u_{n+1})$ (e.g., $\hat{M}_3 = M_0 1 +$
$\gamma_3 A_{34} A_{43} + \delta_3 A_{43} A_{34}$); calculate the expressions for masses $m_{V_i} \equiv \langle V_i | \hat{M}_3 | V_i \rangle$
- these involve symmetry breaking parameters γ_3, δ_3 and the q-parameter. In
particular, for $n = 3$ we obtain

$$m_\rho = M_o, \qquad m_{K^*} = M_o - \gamma_3, \qquad m_{\omega_8} = M_o - 2\frac{[2]_q}{[3]_q}\gamma_3, \qquad (1)$$

where $[x]_q \equiv \frac{q^x - q^{-x}}{q - q^{-1}}$ is the q-number that 'deforms' a number x and, to have equal
masses for particles and their anti's, $\delta_3 = \gamma_3$ was set. q-Dependence appears only
in the mass of ω_8 (isosinglet in $U_q(su_3)$-octet). Excluding M_0, γ_3, the q-analog of
Gell–Mann - Okubo (GMO) relation is [8] :

$$m_{\omega_8} + \left(2\frac{[2]_q}{[3]_q} - 1\right)m_\rho = 2\frac{[2]_q}{[3]_q}m_{K^*} . \qquad (2)$$

In the limit $q = 1$ (i.e., at $\frac{[2]}{[3]} = \frac{2}{3}$), this reduces to usual GMO formula $3m_{\omega_8} +$
$m_\rho = 4m_{K^*}$ which needs singlet mixing [9]. However, it also yields

$$m_{\omega_8} + m_\rho = 2m_{K^*} \qquad \text{if} \quad q = e^{i\pi/5} \quad \text{(then, } [2]_q = [3]_q). \qquad (3)$$

With $m_{\omega_8} \equiv m_\phi$, and no mixing, eq.(3) coincides with nonet mass formula of
Okubo [10] agreeing *ideally* with data [11].

For $3 \leq n \leq 6$ mass operator is constructed analogously. Again, calcula-
tions show: only isosinglets ω_{15}, ω_{24}, ω_{35} of $(n^2 - 1)$-plets of $U_q(u_n)$ contain
q-dependence. As result, we get the q-deformed mass relations [8, 6, 7]:

$$[n]_{(q)} m_{\omega_{n^2-1}} + (b_{n;q} + 2n - 4) m_\rho = 2 m_{D_n^*} + (c_{n;q} + 2) \sum_{r=3}^{n-1} m_{D_r^*}, \qquad (4)$$

$$b_{n;q} \equiv n c_{n;q} - 6 [n]_{(q)}^2 + \left(\frac{24}{[2]_q} - 1\right)[n]_{(q)}, \qquad c_{n;q} \equiv 2 [n]_{(q)}^2 - \frac{8}{[2]_q}[n]_{(q)},$$

where $[n]_{(q)} \equiv [n]_q/[n-1]_q$. Then, natural fixation by setting $[n]_q = [n-1]_q$,
$n = 4, 5, 6$, leads to the higher analogs of Okubo's sum rule:

$$m_{\omega_{15}} + (5 - 8/[2]_{q_4})m_\rho = 2 m_{D^*} + (4 - 8/[2]_{q_4})m_{K^*}$$

$$(5)$$

---

[1] For more details concerning this approach see refs. [6, 7, 13].

$$m_{\omega_{24}} + (9 - 16/[2]_{q5})m_\rho = 2\,m_{D_b*} + (4 - 8/[2]_{q5})(m_{D*} + m_{K*})$$

$$\tag{6}$$

$$m_{\omega_{35}} + (13 - 24/[2]_{q6})m_\rho = 2\,m_{D_t*} + (4 - 8/[2]_{q6})(m_{D_b*} + m_{D*} + m_{K*}).$$

$$\tag{7}$$

Here $q_n = e^{i\pi/(2n-1)}$ are the values that solve eqns. $[n]_q - [n-1]_q = 0$. Like in the case with $m_{\omega_8} \equiv m_\phi$, it is meant in (5)-(7) that J/ψ is put in place of ω_{15}, Υ in place of ω_{24}, toponium in place of ω_{35} (i.e., no mixing!).

The q-polynomials $[n]_q - [n-1]_q$ have a topological meaning.

3. Torus knots and topological labelling of flavors

Polynomials $[n]_q - [n-1]_q \equiv P_n(q)$, by their roots, reduce q-analogs (2), (4) to realistic mass sum rules (MSR) (3), (5)-(7). And, due to property (i) $P_n(q) = P_n(q^{-1})$, (ii) $P_n(1) = 1$, they coincide [8, 7] with such knot invariants as Alexander polynomials $\Delta(q)\{(2n-1)_1\}$ of $(2n-1)_1$-torus knots. E.g.,

$$[3]_q - [2]_q = q^2 + q^{-2} - q - q^{-1} + 1 \equiv \Delta(q)\{5_1\},$$

$$[4]_q - [3]_q = q^3 + q^{-3} - q^2 - q^{-2} + q + q^{-1} - 1 \equiv \Delta(q)\{7_1\}$$

correspond to the 5_1- and 7_1-knots. Since the q-deuce in (4) can be linked to the trefoil (or 3_1-) knot: $[2]_q - 1 = q + q^{-1} - 1 \equiv \Delta(q)\{3_1\}$, *all the q-dependence in masses of* ω_{n^2-1} *and in coefficients in* (2),(4) *is expressible through Alexander polynomials*. Namely, $\frac{[3]_q}{[2]_q} = 1 + \frac{\Delta\{5_1\}}{[2]_q} = 1 + \frac{\Delta\{5_1\}}{\Delta\{3_1\}+1}$,

$$\frac{[n]_q}{[n-1]_q} = 1 + \frac{\Delta\{(2n-1)_1\}}{[n-1]_q} = 1 + \frac{\Delta\{(2n-1)_1\}}{1 + \sum_{r=2}^{n-1}\Delta\{(2r-1)_1\}}, \quad n = 4, 5, 6.$$

$$\tag{8}$$

The values q_n are thus roots of respective Alexander polynomials. For each n, the 'senior' (numerator) polynomial in $\frac{[3]_q}{[2]_q}$ and (8) is specified: by its root, it 'singles out' the corresponding MSR from q-deformed analog.

Thus, the q-parameter for each n is fixed in a rigid way as a root q_n of $\Delta\{(2n-1)_1\}$, contrary to the choice of q by fitting in other phenomenological applications [12]. Moreover, using flavor q-algebras along with 'dynamical' q-algebras according to $U_q(u_n) \subset U_q(u_{n+1})$, we gain: the torus knots 5_1, 7_1, 9_1, 11_1 are put into correspondence [6, 7] with vector quarkonia $s\bar{s}$, $c\bar{c}$, $b\bar{b}$, and $t\bar{t}$ respectively. In a sense, the polynomial $P_n(q) \equiv [n]_q - [n-1]_q$ by its root $q(n)$ determines the value of q (deformation strength) for each n and thus serves as *defining polynomial* for the MSR/quarkonium/flavor corresponding to n. Hence, the applying of q-algebras suggests a possibility of *topological labeling of flavors*: fixed number n corresponds to $2n-1$ overcrossings of 2-strand braids whose closure gives these $(2n-1)_1$-torus knots. With the form $(2n-1, 2)$ of same torus knots this means the correspondence $n \leftrightarrow w \equiv 2n-1$, w being the winding number around tube of torus (winding number around hole is 2).

4. Defining q-polynomials for octet baryon mass sum rules

Analogous scheme was applied to baryons $\frac{1}{2}^+$ too. Excluding undetermined constants M_0, α, β from final obtained expressions for M_N, M_Ξ, M_Λ, M_Σ leads to the q-deformed mass relations (MRs) of the form [6, 7, 13]

$$[2]M_N + \frac{[2]}{[2]-1}M_\Xi = [3]M_\Lambda + \left(\frac{[2]^2}{[2]-1} - [3]\right)M_\Sigma$$

$$+ \frac{A_q}{B_q}\left(M_\Xi + [2]M_N - [2]M_\Sigma - M_\Lambda\right) \quad (9)$$

where A_q and B_q are certain polynomials of $[2]_q$ with non-overlapping sets of zeros. It is important that different dynamical representations produce differing pairs A_q, B_q. Any A_q possesses the factor $([2]_q - 2)$ and thus the 'classical' zero $q = 1$. In the limit $q = 1$ each q-deformed mass relation reduces to the standard GMO sum rule $M_N + M_\Xi = \frac{1}{2}M_\Sigma + \frac{3}{2}M_\Lambda$ for octet baryons (its accuracy is 0.58%). At some values of q which are zeros of particular A_q other than $q = 1$, we obtain MSRs which hold with better accuracy than the GMO one. The two new MSRs

$$q = e^{i\pi/6} \quad \Rightarrow \quad M_N + \frac{1+\sqrt{3}}{2}M_\Xi = \frac{2}{\sqrt{3}}M_\Lambda + \frac{9-\sqrt{3}}{6}M_\Sigma \quad (0.22\%) \quad (10)$$

$$q = e^{i\pi/7} \quad \Rightarrow \quad M_N + \frac{1}{[2]_{q7}-1}M_\Xi = \frac{1}{[2]_{q7}-1}M_\Lambda + M_\Sigma \quad (0.07\%) \quad (11)$$

result [6,7,13] from two different dynamical representations $D^{(1)}$ and $D^{(2)}$ whose respective polynomials $A_q^{(1)}$ and $A_q^{(2)}$ possess zeros $q = e^{i\pi/6}$ and $q = e^{i\pi/7}$. The choice with $q = e^{i\pi/7}$ turns out to be the best possible one.[2]

The sum rule (10) was first derived [6] from a specific dynamical representation (irrep) $D^{(1)}$ of $U_q(u_{4,1})$. However, the 'compact' dynamical $U_q(u_5)$ is equally well suited. Among the admissible dynamical irreps there exist an entire series of

---

[2] In sec. 8 we argue that this value of q is linked to the Cabibbo angle: $\theta_8 = \frac{\pi}{7} = 2\theta_C$.

irreps (numbered by integer m, $6 \leq m < \infty$) which produce the corresponding infinite set of MSRs:

$$M_N + \frac{1}{[2]_{q_m} - 1} M_\Xi = \frac{[3]_{q_m}}{[2]_{q_m}} M_\Lambda + \left(\frac{[2]_{q_m}}{[2]_{q_m} - 1} - \frac{[3]_{q_m}}{[2]_{q_m}} \right) M_\Sigma \qquad (12)$$

with $q_m = e^{i\pi/m}$. *Each of these shows better agreement with data* than the classical GMO one. Few of them, including the MSRs (10), (11) and the 'classical' GMO which corresponds to $q_\infty = 1$, are shown in the table.

$\theta = \frac{\pi}{m}$	(RHS−LHS), MeV	$\frac{|RHS-LHS|}{RHS}$, %
π/∞	26.2	0.58
$\pi/30$	25.42	0.56
$\pi/12$	20.2	0.44
$\pi/8$	10.39	0.23
$\pi/7$	3.26	0.07
$\pi/6$	-10.47	0.22

Comparing (12) with (9) shows that the vanishing of $\frac{A_q}{B_q}$ is crucial for obtaining this discrete set of MSRs and for providing a kind of 'discrete fitting'. Hence, A_q serves as *defining* polynomial for the corresponding MSR.

Since $[2]_{q_7} = q_7 + \frac{1}{q_7} = 2\cos\frac{\pi}{7}$, the MSR (11) takes the equivalent form

$$M_\Xi - M_N + M_\Sigma - M_\Lambda = (2\cos\frac{\pi}{7})(M_\Sigma - M_N) \qquad (13)$$

which exhibits some similarity with decuplet mass formula given below.

5. Decuplet baryons: universal q-deformed mass relation

In the case of $SU(3)$-decuplet baryons $\frac{3}{2}^+$, the convensional 1st order symmetry breaking yields [9] equal spacing rule (ESR) for isoplet members in **10**-plet. Empirical data show for $M_{\Sigma^*} - M_\Delta$, $M_{\Xi^*} - M_{\Sigma^*}$ and $M_\Omega - M_{\Xi^*}$ noticeable deviation from ESR: 152.6 MeV ↔ 148.8 MeV ↔ 139.0 MeV. Use of the q-algebras $U_q(su_n)$ instead of $SU(n)$ provides natural improvement. From evaluations of decuplet masses in two particular irreps of the dynamical algebra $U_q(u_{4,1})$, the q-deformed mass relation

$$(1/[2]_q)(M_{\Sigma^*} - M_\Delta + M_\Omega - M_{\Xi^*}) = M_{\Xi^*} - M_{\Sigma^*}, \qquad [2]_q \equiv q + q^{-1}, \qquad (14)$$

was derived [14]. As proven there, this mass relation is <u>universal</u> - it results from each admissible irrep (which contains $U_q(su_3)$-decuplet embedded in **20**-plet of

$U_q(su_4)$) of the dynamical $U_q(u_{4,1})$. With empirical masses [11], the formula (14) is successful if $[2]_q \simeq 1.96$. *Pure phase* $q = e^{i\theta}$ (or $[2]_q = 2\cos\theta$) with $\theta = \theta_{10} \simeq \frac{\pi}{14}$ provides excellent agreement with data (below, we argue that $\theta_{10} = \theta_C$). Notice a similarity of eq.(14) with the MR

$$(1/2)(M_{\Sigma^*} - M_\Delta + M_\Omega - M_{\Xi^*}) = M_{\Xi^*} - M_{\Sigma^*} \qquad (15)$$

obtained earlier in diverse contexts [15]: by tensor method, in additive quark model with general pair interaction, in a diquark–quark model, in modern chiral perturbation theory. Such model-independence of (15) stems because each of these approaches accounts 1st and 2nd order of $SU(3)$-breaking.

The q-deformed MSR (14) is universal even in a wider sense: it results from admissible irreps (containing $U_q(su_4)$ 20-plet) of both $U_q(su_{4,1})$ and the 'compact' dynamical $U_q(su_5)$. Say, within a dynamical irrep $\{4000\}$ of $U_q(su_5)$ calculation yields: $M_\Delta = M_{10} + \beta$, $M_{\Sigma^*} = M_{10} + [2]\beta + \alpha$, $M_{\Xi^*} = M_{10} + [3]\beta + [2]\alpha$, $M_\Omega = M_{10} + [4]\beta + [3]\alpha$, from which (14) stems. On the other hand, these four masses can be comprised by single formula

$$M_{D_i} = M(Y(D_i)) = M_{10} + \alpha[1-Y(D_i)] + \beta[2-Y(D_i)] \qquad (16)$$

with explicit dependence on Y (hypercharge). If $q = 1$, this reduces to $M_{D_i} = \tilde{M}_{10} + a\,Y(D_i)$, i.e., *linear dependence on hypercharge* Y (or strangeness) where $a = -\alpha - \beta$, $\tilde{M}_{10} = M_{10} + \alpha + 2\beta$.

6. Nonpolynomial $SU(3)$-breaking effects in baryon masses

Formula (16) *involves highly nonlinear dependence* of mass on hypercharge (it is Y that causes $SU(3)$-breaking for decuplet). Since for q-number $[N]$ we have $[N] = q^{N-1} + q^{N-3} + \ldots + q^{-N+3} + q^{-N+1}$ (N terms) this shows exponential Y-dependence of masses. Such high nonlinearity makes (14) and (16) radically different from the abovementioned result (15) of traditional treatment that accounts for effects linear and quadratic in Y.

For octet baryon masses, high nonlinearity (*nonpolynomiality*) in $SU(3)$-breaking effectively accounted by the model was demonstrated in [13]. For this, the expressions for (isoplet members of) octet masses with explicit dependence on hypercharge Y and isospin I, through $I(I + 1)$, are used. The typical matrix element (μ_1, μ_2 are functions of irrep labels m_{15}, m_{55}):

$$\langle B_i | A_{34}A_{45}A_{54}A_{43} | B_i \rangle = [2]^{-1}[3]^{-1}\left([Y/2][Y/2+1] - [I][I+1]\right)\mu_1(m_{15}, m_{55})$$

$$-[2]^{-1}[5]^{-1}\left([Y/2 - 1][Y/2 - 2] - [I][I + 1]\right)\mu_2(m_{15}, m_{55}),$$

contributing to octet baryon masses, illustrates the dependence. From definition of q-bracket $[n] = \frac{\sin(nh)}{\sin(h)}$, $q = \exp(ih)$, it is clearly seen that baryon masses depend on hypercharge Y and isospin I (hence, on $SU(3)$-breaking effects) in highly nonlinear - *nonpolynomial* - fashion.

The ability to take into account highly nontrivial symmetry breaking effects by applying q-analogs $U_q(su_n)$ of flavor symmetries is much alike the fact demonstrated in [16] that, by exploiting appropriate *free* q-deformed structure one is able to efficiently study the properties of (undeformed) quantum-mechanical systems with complicated interactions.

7. To use or not to use the Hopf-algebra structure

An alternative, as regards (9), version of q-deformed analog can be derived [13] using for the symmetry breaking part of mass operator a component of *q-tensor operator* - this clearly implies [17] the Hopf algebra structure (comultiplication, antipode) of the $U_q(su_n)$ quantum algebras. Let us briefly discuss such version. We use q-tensor operators (V_1, V_2, V_3) resp. $(V_{\bar{1}}, V_{\bar{2}}, V_{\bar{3}})$ formed from elements of $U_q(su_4)$ and transforming as **3** resp. **3*** under the adjoint action of $U_q(su_3)$. With H_1, H_2 as Cartan elements and with notation $[X, Y]_q \equiv XY - qYX$, the components (V_1, V_2, V_3) read

$$V_1 = [E_1^+, [E_2^+, E_3^+]_q]_q q^{-H_1/3 - H_2/6}, \quad V_2 = [E_2^+, E_3^+]_q q^{H_1/6 - H_2/6},$$

$$V_3 = E_3^+ q^{H_1/6 + H_2/3}, \tag{17}$$

and similarly for $(V_{\bar{1}}, V_{\bar{2}}, V_{\bar{3}})$ (see [13]), of which we here only give

$$V_{\bar{3}} = q^{H_1/6 + H_2/3} E_3^-. \tag{18}$$

Clearly, $U_q(su_3)$ is broken to $U_q(su_2)$. Like in the nondeformed case of $su(3)$ broken to its isospin subalgebra $su(2)$, the form of mass operator is

$$\hat{M} = \hat{M}_0 + \hat{M}_8 \tag{19}$$

where $\hat{M}_0$ is $U_q(su_3)$-invariant and $\hat{M}_8$ transforms as $I = 0, Y = 0$ component of tensor operator of 8-irrep of $U_q(su_3)$. If $|B_i\rangle$ is a basis vector of carrier space of **8** which corresponds to some baryon B_i, the mass of B_i is given by $M_{B_i} = \langle B_i|\hat{M}|B_i\rangle$. The irrep **8** occurs twice in the decomposition of $\mathbf{8} \otimes \mathbf{8}$. This, and the Wigner-Eckart theorem for $U_q(su_n)$ [18] applied to q-tensor operators under irrep **8** of $U_q(su_3)$, lead to the mass operator of the form $\hat{M} = M_0 \mathbf{1} + \alpha V_8^{(1)} + \beta V_8^{(2)}$ and thus to

$$M_{B_i} = \langle B_i|(M_0\mathbf{1} + \alpha V_8^{(1)} + \beta V_8^{(2)})|B_i\rangle \tag{20}$$

where $V_8^{(1)}$ and $V_8^{(2)}$ are two dictinct tensor operators which both transform as $I = 0, Y = 0$ component of irrep **8** of $U_q(su_3)$; M_0, α, β - undetermined constants depending on details of dynamics. From $\mathbf{3} \otimes \mathbf{3}^* = \mathbf{1} \oplus \mathbf{8}$, $\mathbf{3}^* \otimes \mathbf{3} = \mathbf{1} \oplus \mathbf{8}$ it is seen that the operators $V_3 V_{\bar{3}}$ and $V_{\bar{3}} V_3$ from (17),(18) are just the isosinglets needed in eq.(20). As result, mass operator in (20) with redefined M_0, α, β is $\hat{M} = M_0 \mathbf{1} + \alpha V_3 V_{\bar{3}} + \beta V_{\bar{3}} V_3$, or

$$\hat{M} = M_0 \mathbf{1} + \alpha E_3^+ E_3^- q^Y + \beta E_3^- E_3^+ q^Y \tag{21}$$

where $Y = (H_1 + 2H_2)/3$ is hypercharge. Matrix elements (20) with $\hat{M}$ from (21) are evaluated by embedding **8** in a particular representation of $U_q(su_4)$. Say, if one takes the adjoint **15** of $U_q(su_4)$, the evaluation of baryon masses yields:
$M_N = M_0 + \beta q$, $M_\Sigma = M_0$, $M_\Lambda = M_0 + \frac{[2]}{[3]}(\alpha + \beta)$, $M_\Xi = M_0 + \alpha q^{-1}$.
Excluding M_0, α and β, we finally obtain

$$[3]M_\Lambda + M_\Sigma = [2](q^{-1} M_N + q M_\Xi). \tag{22}$$

This alternative q-analog of octet mass relation looks much simpler than the former q-analog (9). This same q-relation (22) results from embedding **8** in any other admissible dynamical representation. What concerns empirical validity [11] of (22), there is no other way to fix the q-parameter as by usual fitting (for each of the values $q_{1,2} = \pm 1.035$, $q_{3,4} = \pm 0.903\sqrt{-1}$, the q-MR (22) indeed holds within experimental uncertainty). This is in sharp contrast with the q-analogs (9) for which there exists an appealing possibility to fix q in a rigid way by zeros of relevant polynomial A_q.

Summarizing we should stress that, although the use of Hopf-algebra structure leads to simple and mathematically appealing result eq.(22), from the physical (phenomenological) viewpoint the version (9) of q-analog obtained by applying only the tools of representation theory of quantum algebras and not strictly q-covariant symmetry breaking part in mass operator, provides much more interesting results. Among these is the degeneracy lifting and the possibility to choose among a variety of dynamical representations, defining polynomials and, thus, within discrete set of viable mass sum rules. That led us to the best MSR (11) (or (13)) for octet baryons.

8. On the connection: deformation parameter ↔ Cabibbo angle

In 3-flavor case of vector mesons, the deformation angle $\frac{\pi}{5}$ that determines ϕ-meson in (3) coincides remarkably with ω-ϕ mixing angle (known [11] to be $\theta_{\omega\phi} = 36°$) of traditional $SU(3)$-based scheme. In other words, the concept of q-deformed flavor symmetries *is closely related* with the issue of singlet mixing.

For pseudoscalar (PS) mesons, the generalization [19] of GMO-formula

$$f_\pi^2 m_\pi^2 + 3f_\eta^2 m_\eta^2 = 4f_K^2 m_K^2 \qquad \text{with} \qquad 1/f_\pi^2 + 3/f_\eta^2 = 4/f_K^2, \tag{23}$$

involves decay constants as coefficients. Presented in the equivalent form[3]

$$m_\pi^2 + \frac{9f_K^2/f_\pi^2}{4 - f_K^2/f_\pi^2}m_\eta^2 = 4\frac{f_K^2}{f_\pi^2}m_K^2, \tag{24}$$

it is to be compared with our *q-analog* (2) of GMO rewritten for PS mesons (with masses squared), namely

$$m_\pi^2 + \frac{[3]}{2[2] - [3]}m_{\eta_8}^2 = \frac{2[2]}{2[2] - [3]}m_K^2. \tag{25}$$

Without singlet mixing, it is satisfied for (the mass of) *physical* η-meson put instead of η_8 *at properly fixed* $q = q_{PS}$, and just this is meant below.

The two generalizations (24) resp. (25) yield the standard GMO mass formula in the corresponding limit of single parameter, $\frac{f_K}{f_\pi} \to 1$ resp. $q \to 1$. Moreover, the following identification is valid:

$$\frac{f_K^2}{f_\pi^2} \longleftrightarrow \frac{\frac{1}{2}[2]}{2[2] - [3]}, \qquad \frac{3f_K^2/f_\pi^2}{4 - f_K^2/f_\pi^2} \longleftrightarrow \frac{\frac{1}{3}[3]}{2[2] - [3]}, \tag{26}$$

from which, using $[3]_q = [2]_q^2 - 1$, we get

$$[2]_\pm = 1 - \xi_{\pi,K} \pm \sqrt{(1 - \xi_{\pi,K})^2 + 1}, \qquad \xi_{\pi,K} \equiv (4f_K^2/f_\pi^2)^{-1}. \tag{27}$$

The ratio f_K/f_π is related to the Cabibbo angle. This is evident either from the formula (see [20]): $\tan^2 \theta_C = \frac{m_\pi^2}{m_K^2}\left[\frac{f_K}{f_\pi} - \frac{m_\pi^2}{m_K^2}\right]^{-1}$, or from the formula

$$\frac{\Gamma_{K \to \mu\nu}}{\Gamma_{\pi \to \mu\nu}} = (\tan \theta_C)^2 \frac{f_K^2}{f_\pi^2} \frac{M_K}{M_\pi}\left(\frac{1 - (M_\mu/M_K)^2}{1 - (M_\mu/M_\pi)^2}\right)^2$$

for the ratio of weak decay rates usually applied to determine [21, 11] f_K/f_π in terms of the Cabibbo angle, with known empirical data on decay rates and masses. Thus, the value of f_K/f_π is expressible through θ_C. Together with (26), (27) this implies: within our scheme, the (realistic value q_{PS} of) *deformation parameter is directly connected with the Cabibbo angle.*

Similar conclusion can be arrived at in another, more general context. In [22], the q-deformed lagrangian for gauge fields of the Weinberg - Salam (WS) model invariant under the quantum-group valued gauge transformations was constructed. The obtained formula [22]

$$F_{\mu\nu}^0 = \mathrm{Tr}_q(F_{\mu\nu})[2(q^2 + q^{-2})]^{-1/2} = B_{\mu\nu}\cos\theta + F_{\mu\nu}^3\sin\theta, \tag{28}$$

---

[3] Note that having used the additional constraint in (23) we are led to the single dimensionless quantity $\frac{f_K}{f_\pi}$ involved in the multipliers of masses.

$$F^3_{\mu\nu} = \partial_\mu A^3_\nu - \partial_\nu A^3_\mu + ie^{ab3}(A^a_\mu A^b_\nu - A^a_\nu A^b_\mu) + [A^3_\mu, B_\nu] - [A^3_\nu, B_\mu],$$

$$B_{\mu\nu} = \partial_\mu B_\nu - \partial_\nu B_\mu + [B_\mu, B_\nu] + [A^a_\mu, A^a_\nu]$$

where

$$\tan\theta = (1 - q^2)/(1 + q^2), \tag{29}$$

exhibits a mixing of the $U(1)$-component B_μ with nonabelian components A^a_μ (the third one). Introducing the new potentials $\tilde{A}_\mu = B_\mu\cos\theta + A^3_\mu\sin\theta$, $Z_\mu = -B_\mu\sin\theta + A^3_\mu\cos\theta$ yields nothing but definition of physical photon $\tilde{A}_\mu$ and Z-boson of WS model, where θ coincides with the Weinberg angle, $\theta = \theta_W$. Since at $\theta = 0$ the potentials B_μ and A^3_μ get completely unmixed whereas nonzero θ (i.e., nontrivial q-deformation) provides proper mixing as a characteristic feature of the WS model, it is thus seen that the *weak mixing is adequately modelled by the q-deformation*. Moreover, formula analogous to (29), i.e., $\tan\theta_W = q\sqrt{[4]/([2][3])}\,[1/2]\,[3/2]$, was obtained [23] within somewhat different approach to q-deforming the standard model.

Hence, the q-deformation realizes proper mixing in the sector of gauge fields, thus providing explicit connection between the weak angle and the deformation parameter q.

On the other hand, the relation found in [24], namely

$$\theta_W = 2(\theta_{12} + \theta_{23} + \theta_{13}), \tag{30}$$

connects θ_W with the Cabibbo angle $\theta_{12} \equiv \theta_C$ (and two other Kobayashi-Maskawa angles θ_{13}, θ_{23}; as we deal with two lightest families, we have to discard θ_{13}, θ_{23}). The importance of (30) consists in that it links two apparently different mixings: one involved in *bosonic* (interaction) sector, the other in *fermionic* (matter) sector of the electroweak standard model.

Combining (29) and (30) (θ_{23}, θ_{13} omitted) we conclude: the Cabibbo angle *should be connected* with the q-parameter of a quantum-group (or quantum-algebra) based structure applied in the fermion sector.

It remains to recall that all our treatment in secs.4-7 using the q-algebras $U_q(su_n)$ concerned just the fermion sector although at the level of baryons as 3-quark bound states of fundamental fermions. Hence, it is natural to assert that there exists direct connection of the q-parameter involved in (13), (14) with fermion mixing angle. Setting $\theta_{10} = g(\theta_C)$ and $\theta_8 = h(\theta_C)$ we find for the functions $g(\theta_C)$ and $h(\theta_C)$ remarkably simple explicit form:

$$\theta_{10} = \theta_C, \qquad\qquad \theta_8 = 2\,\theta_C. \tag{31}$$

With $\theta_8 = \frac{\pi}{7}$ (see (11)) this suggests for Cabibbo angle the exact value $\frac{\pi}{14}$.

9. Discussion

Quantum groups and their Hopf dual counterpart - quantum universal enveloping algebras (QUEA) incorporate transformation/covariance properties of related quantum vector spaces [25]. In the context of quantum homogeneous spaces (see e.g., [26]) the corresponding quantum groups act (say, on their noncommuting 'coordinates') in a nonlinear way, as it was exemplified [27] with quantum CP_q^n. Both quantum groups and their dual QUEA provide necessary tools in constructing [28, 17] covariant differential calculi and particular noncommutative geometry on quantum spaces.

In the case at hand the *internal* symmetries, underlying our treatment of baryon mass sum rules in secs. 4-7 and based on the broken $U_q(su_n)$ $(n \geq 3)$ as well as unbroken isospin $U_q(su_2)$ q-algebras, are closely related to certain internal or extra (as regards the Minkowski space $M^{3,1}$) spacetime dimensions. From this we infer the following. The above justified direct link (31) between the Cabibbo angle $\theta_C = \frac{\pi}{14}$ and the q-parameter, which measures strength of q-deformation for the q-algebras $U_q(su_n)$ of flavor symmetry, can be viewed as an indication of noncommutative-geometric origin of fermion mixing. In this context, the value $\theta_C = \frac{\pi}{14}$ of the Cabibbo angle would serve as the noncommutativity measure of relevant quantum space (responsible for the mixing and explicitly as yet unknown) in extra dimensions. Concerning the latter, one can assert that their number is not less than 2.

Acknowledgements. I would like to thank the organizers for creating stimulating and warm atmosphere at this NATO workshop. The research contained in this paper was supported in part by Award No. UP1-2115 of the U.S. Civilian Research and Development Foundation for the Independent States of the Former Soviet Union (CRDF).

References

1. R. Peccei, *The mystery of flavor*, preprint UCLA/97/TEP/31, hep-ph/9712422. H. Fritzsch and Z.-Z. Xing, *Mass and flavor mixing schemes of quarks and leptons*, Prog. Part. Nucl. Phys. **45** (2000), 1–81, hep-ph/9912358.

2. N. Cabibbo, *Unitary symmetry and leptonic decays*, Phys. Rev. Lett. **10** (1963), 531–533.

3. M. Kobayashi and T. Maskawa, *CP violation in the renormalizable theory of weak interaction*, Prog. Theor. Phys. **49** (1973), 652–657.

4. L. Wolfenstein, *Parametrization of the Kobayashi - Maskawa matrix*, Phys. Rev. Lett. **51** (1983), 1945–1947.

5. V. G. Drinfeld, *Hopf algebra and Yang-Baxter equation*, Sov. Math. Dokl. **32** (1985), 254–259. M. Jimbo, *A q-difference analogue of U(g) and the Yang-Baxter equation*, Lett. Math. Phys. **10** (1985), 63-69.

6. A. M. Gavrilik, in "*Symmetries in Science VIII*" (Proc. Int. Conf., ed. by B.Gruber), Plenum, N.Y., 1995, pp. 109–123. A.M. Gavrilik, I.I. Kachurik, and A. Tertychnyj, *Representations*

of the quantum algebra $U_q(u_{4,1})$ *and a q-polynomial that determines baryon mass sum rules*, Kiev preprint ITP-94-34E, 1994, hep-ph/9504233.

7. A. M. Gavrilik, *Quantum Groups in Hadron Phenomenology*, "*Non-Euclidean Geometry in Modern Physics*" (Proc. Int. Conf.), Kiev, 1997, 183–192, hep-ph/9712411.

8. A. M. Gavrilik, *q-Serre relations in* $U_q(u_N)$, *q-deformed meson mass sum rules and Alexander Polynomials*, J. Phys. **A 27** (1994), L91–L95.

9. M. Gell-Mann and Y. Ne'eman, *The Eightfold Way*, Benjamin, New York, 1964.

10. S. Okubo, φ-*meson and unitary symmetry model*, Phys. Lett. **5** (1963), 165–169.

11. Particle Data Group: Caso C. *et al.*, The Europ. Phys. J. **C 3** (1998) 1.

12. S. Iwao, *Knot and conformal field theory approach in molecular and nuclear physics*, Progr. Theor. Phys. **83** (1990), 363–368; D. Bonatsos et al., $SU_q(2)$ *description of rotational spectra and its relation to the variable moment of inertia model*, Phys. Lett. **251B** (1990), 477–482.

13. A. M. Gavrilik and N. Z. Iorgov, *Quantum groups as flavor symmetries: account of non-polynomial SU(3)-breaking effects in baryon masses*, Ukr. J. Phys. **43** (1998), 1526–1533, hep-ph/9807559.

14. A. M. Gavrilik, I. I. Kachurik, and A. V. Tertychnyj, *Baryon decuplet masses from the viewpoint of q-equidistance*, Ukr. J. Phys. **40** (1995), 645–649.

15. S. Okubo, *Some consequences of unitary symmetry model*, Phys. Lett. **4** (1963), 14–16. I. Kokkedee, *The Quark Model*, Benjamin, New York, 1968. D. B. Lichtenberg, L. J. Tassie, and P. J. Keleman, *Quark-diquark model of baryons and SU(6)*, Phys. Rev. **167** (1968), 1535–1542. R. Dashen, E. Jenkins, and A. Manohar, *The 1/N expansion for baryons*, Phys. Rev. D **49** (1994), 4713–4738.

16. A. Lorek and J. Wess, *Dynamical symmetries in q-deformed quantum mechanics*, Z. Phys. C **67** (1995), 671–680, q-alg/9502007.

17. A. Klimyk and K. Schmüdgen, *Quantum Groups and Their Representations*, Springer, Berlin, 1997.

18. A. Klimyk, *Wigner-Eckart theorem for tensor operators of the quantum group* $U_q(n)$, J. Phys. **A 25** (1992) 2919–2927.

19. S. Okubo, in "*Symmetries and quark models*" (Proc. Int. Conf., ed. by R.Chand), Gordon and Breach, N.Y., 1970.

20. R. J. Oakes, $SU(2) \times SU(2)$ *breaking and the Cabibbo angle*, Phys. Lett. **29** (1969), 683–685.

21. H. Leutwyler and M. Roos, *Determination of the elements* V_{us} *and* V_{ud} *of the Kobayashi - Maskawa matrix*, Z. Phys. **C 25** (1984), 91–101.

22. A. P. Isaev and Z. Popowicz, *q-trace for quantum groups and q-deformed Yang - Mills theory*, Phys. Lett. **B 281** (1992), 271–278.

23. P. Watts, *Toward a q-deformed standard model*, Univ. of Miami preprint UMTG-189, hep-th/9603143.

24. D. Palle, *On the broken gauge, conformal and discrete symmetries in particle physics*, Nuovo Cim. **109 A** (1996), 1535–1554, hep-ph/9706266.

25. Yu. I. Manin, *Quantum groups and noncommutative geometry*, Publ. CRM, Universitè de Montrèal (1988).
    L. D. Faddeev, N. Reshetikhin, and L. Takhtajan, *Quantization of Lie groups and Lie algebras*, Leningrad Math. J. **1** (1990), 193–225.

26. M. Noumi, H. Yamada, and K. Mimachi, *Zonal spherical functions on the quantum homogeneous space* $SU_q(n+1)/SU_q(n)$, Proc. Japan Acad. Ser. A Math. Sci. **65** (1989), 169–171.
    L. L. Vaksman and Ya. I. Soibelman, *Algebra of functions on the quantum group* $SU_q(n+1)$ *and odd-dimensional quantum spheres*, Leningrad Math. J. **2** (1991), 1023–1042.

27. M. Arik, *Unitary quantum groups, quantum projective spaces and q-oscillators*, Z. Phys. C **59** (1993), 99–103.

28. S. L. Woronowicz, *Differential calculus on compact matrix pseudogroups (quantum groups)*,

Commun. Math. Phys. **122** (1989), 125-152.
J. Wess and B. Zumino, *Covariant differential calculus on the quantum hyperplane*, Nucl. Phys. B Proc. Suppl. **18 B** (1991), 302–312.

NONCLASSICAL TYPE REPRESENTATIONS OF NONSTANDARD QUANTIZATION OF ENVELOPING ALGEBRAS U(so(n)), U(so(n,1)) AND U(iso(n))

NIKOLAI IORGOV *
Bogoliubov Institute for Theoretical Physics, Kiev, Ukraine

1. Introduction

Quantum orthogonal groups, quantum Lorentz group and their corresponding quantum algebras are of special interest for modern physics [1]. M. Jimbo [2] and V. Drinfeld [3] defined q-deformations (quantum algebras) $U_q(g)$ for all simple complex Lie algebras g by means of Cartan subalgebras and root subspaces (see also [4]). However, this approach does not give a satisfactory presentation of the quantum algebra $U_q(so(n, \mathbf{C}))$ from a viewpoint of some problems in quantum physics and representation theory. When considering representations of the quantum algebras $U_q(so_{n+1})$ and $U_q(so_{n,1})$ we are interested in reducing them onto the quantum subalgebra $U_q(so_n)$. This reduction would give the analogue of the Gel'fand-Tsetlin basis for these representations. However, definitions of quantum algebras mentioned above do not allow the inclusions $U_q(so(n + 1, \mathbf{C})) \supset U_q(so(n, \mathbf{C}))$ and $U_q(so_{n,1}) \supset U_q(so_n)$. To be able to exploit such reductions we have to consider q-deformations of the Lie algebra $so(n + 1, \mathbf{C})$ defined in terms of the generators $I_{k,k-1} = E_{k,k-1} - E_{k-1,k}$ (where E_{is} is the matrix with elements $(E_{is})_{rt} = \delta_{ir}\delta_{st}$) rather than by means of Cartan subalgebras and root elements. To construct such deformations we have to deform trilinear relations for elements $I_{k,k-1}$ instead of Serre's relations (as in the case of Jimbo's quantum algebras). As a result, we obtain the associative algebra which will be denoted as $U_q'(so(n, \mathbf{C}))$.

These q-deformations were first constructed in [5]. They permit one to construct the reductions of $U_q'(so_{n+1})$ and $U_q'(so_{n,1})$ onto $U_q'(so_n)$. The q-deformed

* mmtpitp@bitp.kiev.ua

S. Duplij and J. Wess (eds.), Noncommutative Structures in Mathematics and Physics, 357–368.
© 2001 *Kluwer Academic Publishers. Printed in the Netherlands.*

algebra $U_q'(so(n, \mathbf{C}))$ leads for $n = 3$ to the q-deformed algebra $U_q'(so(3, \mathbf{C}))$ defined by A. Odesskii [6] and D. Fairlie [7].

In the classical case, the embedding $SO(n) \subset SU(n)$ (and its infinitesimal analogue) is of great importance for nuclear physics and in the theory of Riemannian symmetric spaces. It is well known that in the framework of Drinfeld–Jimbo quantum groups and algebras one cannot construct the corresponding embedding. The algebra $U_q'(so(n, \mathbf{C}))$ allows to define such an embedding [8,9], that is, it is possible to define the embedding $U_q'(so(n, \mathbf{C})) \subset U_q(sl_n)$, where $U_q(sl_n)$ is the Drinfeld–Jimbo quantum algebra.

As a disadvantage of the algebra $U_q'(so(n, \mathbf{C}))$ we have to mention the difficulties with Hopf algebra structure. Nevertheless, $U_q'(so(n, \mathbf{C}))$ turns out [8,9] to be a coideal in $U_q(sl_n)$.

Finite dimensional irreducible representations of algebra $U_q'(so(n,\mathbf{C}))$ were constructed in [5]. The formulas of action of the generators of the algebra upon the q-analogue of the Gel'fand–Tsetlin basis are given there. A proof of these formulas and some their corrections were given in [10]. However, finite dimensional irreducible representations described in [5] and [10] are representations of the classical type. They are q-deformations of the corresponding irreducible representations of the Lie algebra $so(n, \mathbf{C})$, that is, at $q \to 1$ they turn into representations of $so(n, \mathbf{C})$.

The algebra $U_q'(so(n, \mathbf{C}))$ has other classes of finite dimensional irreducible representations which have no classical analogue. These representations are singular at the limit $q \to 1$. They were described in [11]. Note that the description of these representations for the algebra $U_q'(so(3, \mathbf{C}))$ is given in [12]. A classification of irreducible $*$-representations of real forms of the algebra $U_q'(so(3, \mathbf{C}))$ is given in [13].

There exists an algebra, closely related to the algebra $U_q'(so(n, \mathbf{C}))$, which is a q-deformation of the universal enveloping algebra $U(iso_n)$ of the Lie algebra iso_n of the Euclidean group $ISO(n)$ (see [14]). It is denoted as $U_q(iso_n)$. Irreducible representations of the classical type of the algebra $U_q(iso_n)$ were described in [14]. A proof of the corresponding formulas was given in [15]. However, the algebra $U_q(iso_n)$, $q \in \mathbf{R}$, has irreducible representations of the nonclassical type. A description of these representations is the aim of this paper. Note that the description of these representations for $U_q(iso_2)$ is given in [16]. The second aim of this paper is to describe irreducible representations of nonclassical type of the algebra $U_q'(so_{n,1})$ which is a real form of the algebra $U_q'(so(n + 1, \mathbf{C}))$. Representations of the classical type of this algebra are described in [5] and [17].

We assume throughout the paper that q is a fixed positive number. Thus, we give formulas for representations for these values of q. However, these representations can be considered for any values of q not coinciding with a root of unity. For this we have to treat appropriately square roots in formulas for representations or to rescale basis vector in such a way that formulas for representations would not contain square roots.

For convenience, we denote the Lie algebra so$(n, \mathbf{C})$ by so$_n$ and the algebra $U_q'(\mathrm{so}(n, \mathbf{C}))$ by $U_q'(\mathrm{so}_n)$.

2. The q-deformed algebras $U_q'(\mathrm{so}_n)$ and $U_q(\mathrm{iso}_n)$

In our approach [5] to the q-deformation of the algebras $U(\mathrm{so}_n)$ we define the q-deformed algebra $U_q'(\mathrm{so}(n, \mathbf{C}))$ as the associate algebra (with a unit) generated by the elements $I_{i,i-1}$, $i = 2, 3, ..., n$ satisfying the defining relations

$$I_{i,i-1}I_{i-1,i-2}^2 - (q + q^{-1})I_{i-1,i-2}I_{i,i-1}I_{i-1,i-2} + I_{i-1,i-2}^2 I_{i,i-1} = -I_{i,i-1}, \quad (1)$$

$$I_{i,i-1}^2 I_{i-1,i-2} - (q + q^{-1})I_{i,i-1}I_{i-1,i-2}I_{i,i-1} + I_{i-1,i-2}I_{i,i-1}^2 = -I_{i-1,i-2}, \quad (2)$$

$$I_{i,i-1}I_{j,j-1} = I_{j,j-1}I_{i,i-1}, \quad |i - j| > 1. \quad (3)$$

In the limit $q \to 1$ formulas (1)–(3) give the relations defining the universal enveloping algebra $U(\mathrm{so}_n)$. Note also that relations (1) and (2) principally differ from the q-deformed Serre relations in the approach of Jimbo [2] and Drinfeld [3] to quantum orthogonal algebras by a presence of nonzero right hand side and by possibility of the reduction

$$U_q'(\mathrm{so}_n) \supset U_q'(\mathrm{so}_{n-1}) \supset \cdots \supset U_q'(\mathrm{so}_3).$$

Recall that in the standard Jimbo–Drinfeld approach to the definition of quantum algebras, the algebras $U_q(\mathrm{so}_{2m})$ and the algebras $U_q(\mathrm{so}_{2m+1})$ are distinct series of quantum algebras which are constructed independently of each other.

Various real forms of the algebras $U_q'(\mathrm{so}_n)$ are obtained by imposing corresponding $*$-structures. The compact real form $U_q'(\mathrm{so}(n))$ is defined by the $*$-structure

$$I_{i,i-1}^* = -I_{i,i-1}, \quad i = 2, 3, ..., n.$$

The noncompact q-deformed algebras $U_q'(\mathrm{so}_{p,r})$ where $r = n - p$ are singled out respectively by means of the $*$-structures

$$I_{i,i-1}^* = -I_{i,i-1}, \quad i \neq p + 1, \quad i \leq n, \quad I_{p+1,p}^* = I_{p+1,p}.$$

Among the noncompact real q-algebras $U_q'(\mathrm{so}_{p,r})$, the algebras $U_q'(\mathrm{so}_{n-1,1})$ (a q-analogue of the Lorentz algebras) are of special interest.

We also define the algebra $U_q(\mathrm{iso}_n)$ which is a nonstandard deformation of the universal enveloping algebra of the Lie algebra iso$_n$ of the Euclidean Lie group $ISO(n)$. It is the associative algebra (with a unit) generated by the elements $I_{21}, I_{32}, \cdots, I_{n,n-1}, T_n$ such that the elements $I_{21}, I_{32}, \cdots, I_{n,n-1}$ satisfy the defining relations of the subalgebra $U_q'(\mathrm{so}_n)$ and the additional defining relations are

$$I_{n,n-1}^2 T_n - (q + q^{-1})I_{n,n-1}T_n I_{n,n-1} + T_n I_{n,n-1}^2 = -T_n,$$

$$T_n^2 I_{n,n-1} - (q + q^{-1}) T_n I_{n,n-1} T_n + I_{n,n-1} T_n^2 = 0,$$

$$I_{k,k-1} T_n = T_n I_{k,k-1} \quad \text{if} \quad k < n$$

(see [14]). If $q = 1$, then these relations define the classical algebra $U(\text{iso}_n)$. Let us note that the defining relations for $U_q(\text{iso}_n)$ can be expressed by bilinear relations [15].

3. Finite dimensional classical type representations of $U_q'(\text{so}_n)$

In this section we describe (in the framework of a q-analogue of Gel'fand–Tsetlin formalism) irreducible finite dimensional representations of the algebra $U_q'(\text{so}_n)$, $n \geq 3$, which are q-deformations of the finite dimensional irreducible representations of the Lie algebra so_n. They are given by the sets $\mathbf{m}_n$ consisting of $\lfloor n/2 \rfloor$ numbers $m_{1,n}, m_{2,n}, \ldots, m_{\lfloor n/2 \rfloor, n}$ (here $\lfloor n/2 \rfloor$ denotes integral part of $n/2$) which are all integral or all half-integral and satisfy the dominance conditions

$$m_{1,2p+1} \geq m_{2,2p+1} \geq \ldots \geq m_{p,2p+1} \geq 0,$$

$$m_{1,2p} \geq m_{2,2p} \geq \ldots \geq m_{p-1,2p} \geq |m_{p,2p}|$$

for $n = 2p + 1$ and $n = 2p$, respectively. These representations are denoted by $T_{\mathbf{m}_n}$. For a basis in a representation space we take the q-analogue of Gel'fand–Tsetlin basis which is obtained by successive reduction of the representation $T_{\mathbf{m}_n}$ to the subalgebras $U_q'(\text{so}_{n-1}), U_q'(\text{so}_{n-2}), \cdots, U_q'(\text{so}_3), U_q'(\text{so}_2) := U(\text{so}_2)$. As in the classical case, its elements are labelled by Gel'fand–Tsetlin tableaux

$$\{\xi_n\} \equiv \left\{ \begin{array}{c} \mathbf{m}_n \\ \mathbf{m}_{n-1} \\ \cdots \\ \mathbf{m}_2 \end{array} \right\} \equiv \{\mathbf{m}_n, \xi_{n-1}\} \equiv \{\mathbf{m}_n, \mathbf{m}_{n-1}, \xi_{n-2}\}, \qquad (4)$$

where the components of $\mathbf{m}_k$ and $\mathbf{m}_{k-1}$ satisfy the "betweenness" conditions

$$m_{1,2p+1} \geq m_{1,2p} \geq m_{2,2p+1} \geq m_{2,2p} \geq \ldots \geq m_{p,2p+1} \geq m_{p,2p} \geq -m_{p,2p+1},$$

$$m_{1,2p} \geq m_{1,2p-1} \geq m_{2,2p} \geq m_{2,2p-1} \geq \ldots \geq m_{p-1,2p-1} \geq |m_{p,2p}|.$$

The basis element defined by tableau $\{\xi_n\}$ is denoted as $|\xi_n\rangle$.

It is convenient to introduce the so-called l-coordinates

$$l_{j,2p+1} = m_{j,2p+1} + p - j + 1, \qquad l_{j,2p} = m_{j,2p} + p - j, \qquad (5)$$

for the numbers $m_{i,k}$. In particular, $l_{1,3} = m_{1,3} + 1$ and $l_{1,2} = m_{1,2}$. The operator $T_{\mathbf{m}_n}(I_{2p+1,2p})$ of the representation $T_{\mathbf{m}_n}$ of $U_q'(\text{so}_n)$ acts upon Gel'fand–Tsetlin basis elements, labeled by (4), by the formula

$$T_{mn}(I_{2p+1,2p})|\xi_n\rangle = \sum_{j=1}^{p} \frac{A_{2p}^j(\xi_n)}{q^{l_{j,2p}} + q^{-l_{j,2p}}}|(\xi_n)_{2p}^{+j}\rangle - \sum_{j=1}^{p} \frac{A_{2p}^j((\xi_n)_{2p}^{-j})}{q^{l_{j,2p}} + q^{-l_{j,2p}}}|(\xi_n)_{2p}^{-j}\rangle \tag{6}$$

and the operator $T_{mn}(I_{2p,2p-1})$ of the representation T_{mn} acts as

$$T_{mn}(I_{2p,2p-1})|\xi_n\rangle = \sum_{j=1}^{p-1} \frac{B_{2p-1}^j(\xi_n)}{[2l_{j,2p-1} - 1][l_{j,2p-1}]}|(\xi_n)_{2p-1}^{+j}\rangle$$

$$-\sum_{j=1}^{p-1} \frac{B_{2p-1}^j((\xi_n)_{2p-1}^{-j})}{[2l_{j,2p-1} - 1][l_{j,2p-1} - 1]}|(\xi_n)_{2p-1}^{-j}\rangle + i\,C_{2p-1}(\xi_n)|\xi_n\rangle. \tag{7}$$

In these formulas, $(\xi_n)_k^{\pm j}$ means the tableau (4) in which j-th component $m_{j,k}$ in m_k is replaced by $m_{j,k}\pm 1$. The coefficients A_{2p}^j, B_{2p-1}^j, C_{2p-1} in (6) and (7) are given by the expressions

$$A_{2p}^j(\xi_n) = \left(\frac{\prod_{i=1}^{p}[l_{i,2p+1} + l_{j,2p}][l_{i,2p+1} - l_{j,2p} - 1]}{\prod_{i\neq j}^{p}[l_{i,2p} + l_{j,2p}][l_{i,2p} - l_{j,2p}]} \right.$$

$$\left. \times \frac{\prod_{i=1}^{p-1}[l_{i,2p-1} + l_{j,2p}][l_{i,2p-1} - l_{j,2p} - 1]}{\prod_{i\neq j}^{p}[l_{i,2p} + l_{j,2p} + 1][l_{i,2p} - l_{j,2p} - 1]} \right)^{1/2}, \tag{8}$$

and

$$B_{2p-1}^j(\xi_n) = \left(\frac{\prod_{i=1}^{p}[l_{i,2p} + l_{j,2p-1}][l_{i,2p} - l_{j,2p-1}]}{\prod_{i\neq j}^{p-1}[l_{i,2p-1} + l_{j,2p-1}][l_{i,2p-1} - l_{j,2p-1}]} \right.$$

$$\left. \times \frac{\prod_{i=1}^{p-1}[l_{i,2p-2} + l_{j,2p-1}][l_{i,2p-2} - l_{j,2p-1}]}{\prod_{i\neq j}^{p-1}[l_{i,2p-1} + l_{j,2p-1} - 1][l_{i,2p-1} - l_{j,2p-1} - 1]} \right)^{1/2}, \tag{9}$$

$$C_{2p-1}(\xi_n) = \frac{\prod_{i=1}^{p}[l_{i,2p}] \prod_{i=1}^{p-1}[l_{i,2p-2}]}{\prod_{i=1}^{p-1}[l_{i,2p-1}][l_{i,2p-1} - 1]}, \tag{10}$$

where numbers in square brackets mean q-numbers defined by

$$[a] := \frac{q^a - q^{-a}}{q - q^{-1}}.$$

It is seen from (5) that C_{2p-1} in (10) identically vanishes if $m_{p,2p} \equiv l_{p,2p} = 0$. A proof of the fact that formulas (6)-(10) indeed determine a representation of $U_q'(so_n)$ is given in [10].

4. Finite dimensional nonclassical type representations of $U'_q(\mathrm{so}_n)$

The representations of the previous section are called representations of the classical type, because at $q \to 1$ the operators $T_{\mathbf{m}_n}(I_{j,j-1})$ turn into the corresponding operators $T_{\mathbf{m}_n}(I_{j,j-1})$ for irreducible finite dimensional representations with highest weights $\mathbf{m}_n$ of the Lie algebra so_n.

The algebra $U'_q(\mathrm{so}_n)$ also has irreducible finite dimensional representations T of nonclassical type, that is, such that the operators $T(I_{j,j-1})$ have no classical limit $q \to 1$. They are given by sets $\epsilon := (\epsilon_2, \epsilon_3, \cdots, \epsilon_n)$, $\epsilon_i = \pm 1$, and by sets $\mathbf{m}_n$ consisting of $\lfloor n/2 \rfloor$ **half-integral** numbers $m_{1,n}, m_{2,n}, \ldots, m_{\lfloor n/2 \rfloor, n}$ (here $\lfloor n/2 \rfloor$ denotes integral part of $n/2$) that satisfy the dominance conditions

$$m_{1,2p+1} \geq m_{2,2p+1} \geq \ldots \geq m_{p,2p+1} \geq 1/2,$$

$$m_{1,2p} \geq m_{2,2p} \geq \ldots \geq m_{p-1,2p} \geq m_{p,2p} \geq 1/2$$

for $n = 2p+1$ and $n = 2p$, respectively. These representations are denoted by $T_{\epsilon,\mathbf{m}_n}$.

For a basis in the representation space we use the analogue of the basis of the previous section. Its elements are labeled by tableaux

$$\{\xi_n\} \equiv \left\{ \begin{array}{c} \mathbf{m}_n \\ \mathbf{m}_{n-1} \\ \cdots \\ \mathbf{m}_2 \end{array} \right\} \equiv \{\mathbf{m}_n, \xi_{n-1}\} \equiv \{\mathbf{m}_n, \mathbf{m}_{n-1}, \xi_{n-2}\}, \qquad (11)$$

where the components of $\mathbf{m}_k$ and $\mathbf{m}_{k-1}$ satisfy the "betweenness" conditions

$$m_{1,2p+1} \geq m_{1,2p} \geq m_{2,2p+1} \geq m_{2,2p} \geq \ldots \geq m_{p,2p+1} \geq m_{p,2p} \geq 1/2,$$

$$m_{1,2p} \geq m_{1,2p-1} \geq m_{2,2p} \geq m_{2,2p-1} \geq \ldots \geq m_{p-1,2p-1} \geq m_{p,2p}.$$

The basis element defined by tableau $\{\xi_n\}$ is denoted as $|\xi_n\rangle$.

It is convenient to introduce the l-coordinates as in (5) The operator $T_{\epsilon,\mathbf{m}_n}(I_{2p+1,2p})$ of the representation $T_{\epsilon,\mathbf{m}_n}$ of $U_q(\mathrm{so}_n)$ acts upon our basis elements, labeled by (11), by the formulas

$$T_{\epsilon,\mathbf{m}_n}(I_{2p+1,2p})|\xi_n\rangle = \delta_{m_{p,2p},1/2} \frac{\epsilon_{2p+1}}{q^{1/2} - q^{-1/2}} D_{2p}(\xi_n)|\xi_n\rangle$$

$$+ \sum_{j=1}^{p} \frac{A_{2p}^j(\xi_n)}{q^{l_{j,2p}} - q^{-l_{j,2p}}} |(\xi_n)_{2p}^{+j}\rangle - \sum_{j=1}^{p} \frac{A_{2p}^j((\xi_n)_{2p}^{-j})}{q^{l_{j,2p}} - q^{-l_{j,2p}}} |(\xi_n)_{2p}^{-j}\rangle,$$

where the summation in the last sum must be from 1 to $p-1$ if $m_{p,2p} = 1/2$, and the operator $T_{\mathbf{m}_n}(I_{2p,2p-1})$ of the representation $T_{\mathbf{m}_n}$ acts as

$$T_{\epsilon,\mathbf{m}_n}(I_{2p,2p-1})|\xi_n\rangle = \sum_{j=1}^{p-1} \frac{B_{2p-1}^j(\xi_n)}{[2l_{j,2p-1} - 1][l_{j,2p-1}]_+} |(\xi_n)_{2p-1}^{+j}\rangle$$

$$-\sum_{j=1}^{p-1} \frac{B_{2p-1}^j((\xi_n)_{2p-1}^{-j})}{[2l_{j,2p-1}-1][l_{j,2p-1}-1]_+}|(\xi_n)_{2p-1}^{-j}\rangle + \epsilon_{2p}\hat{C}_{2p-1}(\xi_n)|\xi_n\rangle,$$

where

$$[a]_+ = \frac{q^a + q^{-a}}{q - q^{-1}}.$$

In these formulas, $(\xi_n)_k^{\pm j}$ means the tableau (11) in which j-th component $m_{j,k}$ in $\mathbf{m}_k$ is replaced by $m_{j,k} \pm 1$. Matrix elements A_{2p}^j and B_{2p-1}^j are given by the same formulas as in (6) and (7) (that is, by the formulas (8) and (9)) and

$$\hat{C}_{2p-1}(\xi_n) = \frac{\prod_{s=1}^p [l_{s,2p}]_+ \prod_{s=1}^{p-1}[l_{s,2p-2}]_+}{\prod_{s=1}^{p-1}[l_{s,2p-1}]_+[l_{s,2p-1}-1]_+}.$$

$$D_{2p}(\xi_n) = \frac{\prod_{i=1}^p [l_{i,2p+1} - \frac{1}{2}]\prod_{i=1}^{p-1}[l_{i,2p-1} - \frac{1}{2}]}{\prod_{i=1}^{p-1}[l_{i,2p} + \frac{1}{2}][l_{i,2p} - \frac{1}{2}]}.$$

Theorem 1. *The representations $T_{\epsilon,\mathbf{m}_n}$ are irreducible. The representations $T_{\epsilon,\mathbf{m}_n}$ and $T_{\epsilon',\mathbf{m}_n'}$ are pairwise nonequivalent for $(\epsilon,\mathbf{m}_n) \neq (\epsilon',\mathbf{m}_n')$. For any admissible $(\epsilon,\mathbf{m}_n)$ and $\mathbf{m}_n'$ the representations $T_{\epsilon,\mathbf{m}_n}$ and $T_{\mathbf{m}_n'}$ are pairwise nonequivalent.*

The algebra $U_q'(\mathrm{so}_n)$ has non-trivial one-dimensional representations. They are special cases of the representations of the nonclassical type. They are described as follows.

Let $\epsilon = (\epsilon_2, \epsilon_3, \cdots, \epsilon_n)$, $\epsilon_i = \pm 1$, and let $\mathbf{m}_n = (m_{1,n}, m_{2,n}, \cdots, m_{\lfloor n/2\rfloor,n}) = (\frac{1}{2}, \frac{1}{2}, \cdots, \frac{1}{2})$. Then the corresponding representations $T_{\epsilon,\mathbf{m}_n}$ are one-dimensional and are given by the formulas

$$T_{\epsilon,\mathbf{m}_n}(I_{k+1,k})|\xi_n\rangle = \frac{\epsilon_{k+1}}{q^{1/2} - q^{-1/2}}|\xi_n\rangle.$$

Thus, to every $\epsilon := (\epsilon_2, \epsilon_3, \cdots, \epsilon_n)$, $\epsilon_i = \pm 1$, there corresponds a one-dimensional representation of $U_q'(\mathrm{so}_n)$.

5. Definition of representations of $U_q'(\mathrm{so}_{n,1})$ and $U_q(\mathrm{iso}_n)$

Let us recall that we assume that q is a positive number. We give the following definition of infinite dimensional representations of the algebras $U_q'(\mathrm{so}_{n,1})$ and $U_q(\mathrm{iso}_n)$ (we denote these algebras by $\mathcal{A}$). It is a homomorphism $R : \mathcal{A} \to \mathcal{L}(\mathcal{H})$ of $\mathcal{A}$ to the space $\mathcal{L}(\mathcal{H})$ of linear operators (bounded or unbounded) on a Hilbert space $\mathcal{H}$ such that

(a) operators $R(a)$, $a \in \mathcal{A}$, are defined on an invariant everywhere dense subspace $\mathcal{D} \subset \mathcal{H}$;

(b) $R \downarrow U_q'(so_n)$ decomposes into a direct sum of irreducible finite dimensional representations of $U_q'(so_n)$ (with finite multiplicities if R is irreducible);

(c) subspaces of irreducible representations of $U_q'(so_n)$ belong to $\mathcal{D}$.

Two infinite dimensional irreducible representations R and R' of $\mathcal{A}$ on spaces $\mathcal{H}$ and $\mathcal{H}'$, respectively, are called (algebraically) equivalent if there exists an everywhere dense invariant subspaces $V \subset \mathcal{D}$ and $V' \subset \mathcal{D}'$ and a one-to-one linear operator $A : V \to V'$ such that $AR(a)v = R'(a)Av$ for all $a \in \mathcal{A}$ and $v \in V$.

Remark that our definition of infinite dimensional representations of $U_q'(so_{n,1})$ and $U_q(iso_n)$ corresponds to the definition of Harish-Chandra modules for the pairs $(so_{n,1}, so_n)$ and (iso_n, so_n), respectively. Thus, modules determined by representations of the above definition can be called q-Harish-Chandra modules of the pairs $(U_q'(so_{n,1}), U_q'(so_n))$ and $(U_q(iso_n), U_q'(so_n))$, respectively.

6. Representations of $U_q(iso_n)$

There are the following classes of irreducible representations of $U_q(iso_n)$:

(a) Finite dimensional irreducible representations R of $U_q'(so_n)$. They are irreducible representations of $U_q(iso_n)$ with $R(T_n) = 0$.

(b) Infinite dimensional irreducible representations of the classical type.

(c) Infinite dimensional irreducible representations of the nonclassical type.

Representations $R_{\lambda,m}$ of class (b) are given in [14,15]. Let us describe representations of class (c), that is, representations R for which there exists no limit $q \to 1$ for the operators $R(T_n)$ and $R(I_{i,i-1})$. These representations are given by $\epsilon := (\epsilon_2, \epsilon_3, \cdots, \epsilon_{n+1})$, non-zero complex parameter λ and by numbers $\mathbf{m} = (m_{2,n+1}, m_{3,n+2}, \cdots, m_{\lfloor (n+1)/2 \rfloor, n+1})$, $m_{2,n+1} \geq m_{3,n+2} \geq \cdots \geq m_{\lfloor (n+1)/2 \rfloor, n+1} \geq 1/2$, describing irreducible representations of the nonclassical type of the subalgebra $U_q'(so_{n-1})$ (see section 4). We denote the corresponding representations of $U_q(iso_n)$ by $R_{\epsilon,\lambda,m}$.

In order to describe the space of the representation $R_{\epsilon,\lambda,m}$ we note that

$$R_{\epsilon,\lambda,m} \downarrow U_q'(so_n) = \bigoplus_{m_n} T_{\epsilon',m_n}, \quad m_n = (m_{1,n}, \cdots, m_{\lfloor n/2 \rfloor, n}), \qquad (12)$$

where $\epsilon' = (\epsilon_2, \cdots, \epsilon_n)$ is the part of the set ϵ, the summation is over all irreducible nonclassical type representations T_{ϵ',m_n} of $U_q'(so_n)$ for which the components of m_n satisfy the "betweenness" conditions

$$m_{1,2k} \geq m_{2,2k+1} \geq m_{2,2k} \geq \ldots \geq m_{k,2k+1} \geq m_{k,2k} \geq 1/2 \text{ if } n = 2k,$$

$$m_{1,2k-1} \geq m_{2,2k} \geq m_{2,2k-1} \geq \ldots \geq m_{k-1,2k-1} \geq m_{k,2k} \text{ if } n = 2k - 1.$$

The carrier space $\hat{\mathcal{H}}_{\epsilon,\mathbf{m}}$ of the representation $R_{\epsilon,\lambda,\mathbf{m}}$ decomposes as $\hat{\mathcal{H}}_{\epsilon,\mathbf{m}} = \bigoplus_{\mathbf{m}_n} \mathcal{H}_{\epsilon',\mathbf{m}_n}$, where the summation is such as in (12) and $\mathcal{H}_{\epsilon',\mathbf{m}_n}$ are the subspaces, where the representations $T_{\epsilon',\mathbf{m}_n}$ of $U_q'(so_n)$ are realized. We choose a basis in every subspace $\mathcal{H}_{\epsilon',\mathbf{m}_n}$ as in section 4. The set of all these bases gives a basis of the space $\hat{\mathcal{H}}_{\epsilon,\mathbf{m}}$. We denote the basis elements by $|\mathbf{m}_n, M\rangle$, where M are the corresponding tableaux. The numbers m_{ij} from $|\mathbf{m}_n, M\rangle$ determine the numbers l_{ij} as in section 3. The numbers $m_{i,n+1}$ determine the numbers

$$l_{i,2k+1} = m_{i,2k+1} + k - i + 1, \quad n = 2k, \quad l_{i,2k} = m_{i,2k} + k - i, \quad n = 2k - 1.$$

The operators $R_{\epsilon,\lambda,\mathbf{m}}(I_{i,i-1})$ are given by formulas of the nonclassical type representations of the algebra $U_q'(so_n)$ from section 4. For the operators $R_{\epsilon,\lambda,\mathbf{m}}(T_{2k})$ and $R_{\epsilon,\lambda,\mathbf{m}}(T_{2k-1})$ we have the expressions

$$R_{\epsilon,\lambda,\mathbf{m}}(T_{2k-1})|\mathbf{m}_{2k-1}, M\rangle = \lambda \sum_{j=1}^{k-1} \frac{\tilde{B}_{2k-1}^j(\mathbf{m}_{2k-1}, M)}{[2l_{j,2k-1} - 1][l_{j,2k-1}]_+} |\mathbf{m}_{2k-1}^{+j}, M\rangle$$

$$+ \lambda \sum_{j=1}^{k-1} \frac{\tilde{B}_{2k-1}^j(\mathbf{m}_{2k-1}^{-j}, M)}{[2l_{j,2k-1} - 1][l_{j,2k-1} - 1]_+} |\mathbf{m}_{2k-1}^{-j}, M\rangle$$

$$+ i\epsilon_{2k}\lambda \hat{C}_{2k-1}(\mathbf{m}_{2k-1}, M)|\mathbf{m}_{2k-1}, M\rangle,$$

$$R_{\epsilon,\lambda,\mathbf{m}}(T_{2k})|\mathbf{m}_{2k}, M\rangle = i\lambda \delta_{m_p,2p,1/2} \frac{\epsilon_{2k+1}}{q^{1/2} - q^{-1/2}} D_{2k}|\mathbf{m}_{2k}, M\rangle$$

$$+ \lambda \sum_{j=1}^{k} \frac{\tilde{A}_{2k}^j(\mathbf{m}_{2k}, M)}{q^{l_{j,2k}} - q^{-l_{j,2k}}} |\mathbf{m}_{2k}^{+j}, M\rangle + \lambda \sum_{j=1}^{k} \frac{\tilde{A}_{2k}^j(\mathbf{m}_{2k}^{-j}, M)}{q^{l_{j,2k}} - q^{-l_{j,2k}}} |\mathbf{m}_{2k}^{-j}, M\rangle,$$

where the summation in the last sum must be from 1 to $k - 1$ if $m_{k,2k} = 1/2$, and

$$\tilde{A}_{2k}^j(\mathbf{m}_{2k}, M) = \left(\frac{\prod_{i=2}^{k}[l_{i,2k+1} + l_{j,2k}][l_{i,2k+1} - l_{j,2k} - 1]}{\prod_{i \neq j}[l_{i,2k} + l_{j,2k}][l_{i,2k} - l_{j,2k}]} \right.$$

$$\left. \times \frac{\prod_{i=1}^{k-1}[l_{i,2k-1} + l_{j,2k}][l_{i,2k-1} - l_{j,2k} - 1]}{\prod_{i \neq j}[l_{i,2k} + l_{j,2k} + 1][l_{i,2k} - l_{j,2k} - 1]} \right)^{1/2}, \quad (13)$$

$$\tilde{B}_{2k-1}^j(\mathbf{m}_{2k-1}, M) = \left(\frac{\prod_{i=2}^{k}[l_{i,2k} + l_{j,2k-1}][l_{i,2k} - l_{j,2k-1}]}{\prod_{i \neq j}[l_{i,2k-1} + l_{j,2k-1}][l_{i,2k-1} - l_{j,2k-1}]} \right.$$

$$\left. \times \frac{\prod_{i=1}^{k-1}[l_{i,2k-2} + l_{j,2k-1}][l_{i,2k-2} - l_{j,2k-1}]}{\prod_{i \neq j}[l_{i,2k-1} + l_{j,2k-1} - 1][l_{i,2k-1} - l_{j,2k-1} - 1]} \right)^{1/2}, \quad (14)$$

$$\hat{C}_{2k-1}(M) = \frac{\prod_{s=2}^{k}[l_{s,2k}]_+ \prod_{s=1}^{k-1}[l_{s,2k-2}]_+}{\prod_{s=1}^{k-1}[l_{s,2k-1}]_+[l_{s,2k-1} - 1]_+}, \quad (15)$$

$$D_{2k} = \frac{\prod_{i=2}^{k} [l_{i,2k+1} - \frac{1}{2}] \prod_{i=1}^{k-1} [l_{i,2k-1} - \frac{1}{2}]}{\prod_{i=1}^{k-1} [l_{i,2k} + \frac{1}{2}][l_{i,2k} - \frac{1}{2}]}. \tag{16}$$

Theorem 2. *The representations $R_{\epsilon,\lambda,\mathbf{m}}$ are irreducible. The representations $R_{\epsilon,\lambda,\mathbf{m}}$ and $R_{\epsilon',\lambda',\mathbf{m}'}$ are equivalent if and only if $\epsilon = \epsilon'$, $\mathbf{m} = \mathbf{m}'$ and $\lambda = \pm\lambda'$. The operators $R_{\epsilon,\lambda,\mathbf{m}}(T_n)$ are bounded. The representation $R_{\epsilon,\lambda,\mathbf{m}}$ is equivalent to no of the representations $R_{\lambda',\mathbf{m}'}$ of classical type.*

7. Representations of $U_q'(\mathrm{so}_{n,1})$

Irreducible representations of classical type of algebra $U_q'(\mathrm{so}_{n,1})$ are given in [5,17]. Here we describe irreducible representations of nonclassical type (that is, representations R for which there exists no limit $q \to 1$ for the operators $R(I_{i,i-1})$). These representations are given by the set $\epsilon := (\epsilon_2, \epsilon_3, \cdots, \epsilon_{n+1})$, by a complex parameter c and by the set $\mathbf{m} = (m_{2,n+1}, m_{3,n+1}, \cdots, m_{\lfloor (n+1)/2 \rfloor, n+1})$, $m_{2,n+1} \geq m_{3,n+2} \geq \cdots \geq m_{\lfloor (n+1)/2 \rfloor, n+1} \geq 1/2$, describing irreducible representations of the nonclassical type of the subalgebra $U_q'(\mathrm{so}_{n-1})$ (see section 4). We denote the corresponding representations of $U_q(\mathrm{so}_{n,1})$ by $R_{\epsilon,c,\mathbf{m}}$.

In order to describe the space of the representation $R_{\epsilon,c,\mathbf{m}}$ we note that

$$R_{\epsilon,\lambda,\mathbf{m}} \downarrow U_q'(\mathrm{so}_n) = \bigoplus_{\mathbf{m}_n} T_{\epsilon',\mathbf{m}_n}, \quad \mathbf{m}_n = (m_{1,n}, \cdots, m_{\lfloor n/2 \rfloor, n}), \tag{17}$$

where $\epsilon' = (\epsilon_2, \cdots, \epsilon_n)$, the summation is over all irreducible nonclassical type representations $T_{\epsilon',\mathbf{m}_n}$ of the subalgebra $U_q'(\mathrm{so}_n)$ for which the components of $\mathbf{m}_n$ satisfy the "betweenness" conditions

$$m_{1,2k} \geq m_{2,2k+1} \geq m_{2,2k} \geq \dots \geq m_{k,2k+1} \geq m_{k,2k} \geq 1/2 \text{ if } n = 2k$$

$$m_{1,2k-1} \geq m_{2,2k} \geq m_{2,2k-1} \geq \dots \geq m_{k-1,2k-1} \geq m_{k,2k} \text{ if } n = 2k - 1.$$

The carrier space $\hat{\mathcal{H}}_{\epsilon,\mathbf{m}}$ of the representation $R_{\epsilon,c,\mathbf{m}}$ decomposes as $\hat{\mathcal{H}}_{\epsilon,\mathbf{m}} = \bigoplus_{\mathbf{m}_n} \mathcal{H}_{\epsilon,\mathbf{m}_n}$, where the summation is such as in (17) and $\mathcal{H}_{\epsilon',\mathbf{m}_n}$ are the subspaces, where the representations $T_{\epsilon',\mathbf{m}_n}$ of $U_q'(\mathrm{so}_n)$ are realized. We choose the basis in every subspace $\mathcal{H}_{\epsilon,\mathbf{m}_n}$ as in section 4. The set of all these bases gives a basis of the space $\hat{\mathcal{H}}_{\epsilon,\mathbf{m}}$. We denote the basis elements by $|\mathbf{m}_n, M\rangle$, where M are the corresponding tableaux. The numbers m_{ij} from $|\mathbf{m}_n, M\rangle$ determine the numbers l_{ij} as in section 3. The numbers $m_{i,n+1}$ determine the numbers $l_{i,n+1}$ as in section 6. The operators $R_{\epsilon,c,\mathbf{m}}(I_{i,i-1})$, $i \leq n$, are given by formulas of the nonclassical type representations of the algebra $U_q'(\mathrm{so}_n)$ as in section 4. For the operators $R_{\epsilon,c,\mathbf{m}}(I_{2k+1,2k})$ if $n = 2k$ and $R_{\epsilon,c,\mathbf{m}}(T_{2k,2k-1})$ if $n = 2k - 1$ we have the expressions

$$R_{\epsilon,c,\mathbf{m}}(I_{2k,2k-1})|\mathbf{m}_{2k-1}, M\rangle$$

$$= \sum_{j=1}^{k-1} ([c + l_{j,2k-1}][c - l_{j,2k}])^{1/2} \frac{\tilde{B}_{2k-1}^j(\mathbf{m}_{2k-1}, M)}{[2l_{j,2k-1} - 1][l_{j,2k-1}]_+} |\mathbf{m}_{2k-1}^{+j}, M\rangle$$

$$- \sum_{j=1}^{k-1} ([c + l_{j,2k-1} + 1][c - l_{j,2k} + 1])^{1/2} \frac{\tilde{B}_{2k-1}^j(\mathbf{m}_{2k-1}^{-j}, M)}{[2l_{j,2k-1} - 1][l_{j,2k-1} - 1]_+} |\mathbf{m}_{2k-1}^{-j}, M\rangle$$

$$+ \epsilon_{2k} [c]_+ \hat{C}_{2k-1}(\mathbf{m}_{2k-1}, M) |\mathbf{m}_{2k-1}, M\rangle,$$

$$R_{\epsilon,c,\mathbf{m}}(I_{2k+1,2k}) |\mathbf{m}_{2k}, M\rangle$$

$$= \delta_{m_{k,2k}, 1/2} [c - 1/2] \frac{\epsilon_{2k+1}}{q^{1/2} - q^{-1/2}} D_{2k}(\mathbf{m}_{2k}, M) |\mathbf{m}_{2k}, M\rangle$$

$$+ \sum_{j=1}^{k} ([c + l_{j,2k}][c - l_{j,2k} - 1])^{1/2} \frac{\tilde{A}_{2k}^j(\mathbf{m}_{2k}, M)}{q^{l_{j,2k}} - q^{-l_{j,2k}}} |\mathbf{m}_{2k}^{+j}, M\rangle -$$

$$- \sum_{j=1}^{k} ([c + l_{j,2k} - 1][c - l_{j,2k}])^{1/2} \frac{\tilde{A}_{2k}^j(\mathbf{m}_{2k}^{-j}, M)}{q^{l_{j,2k}} - q^{-l_{j,2k}}} |\mathbf{m}_{2k}^{-j}, M\rangle,$$

where the summation in the last sum must be from 1 to $k - 1$ if $m_{k,2k} = 1/2$, and $\tilde{A}_{2k}^j$, $\tilde{B}_{2k-1}^j$, $\hat{C}_{2k-1}$, D_{2k} are such as in (13)–(16).

Theorem 3. *The representation $R_{\epsilon,c,\mathbf{m}}$ of $U'_q(so_{2k,1})$ is irreducible if and only if c is not half-integer or one of the numbers c, $1 - c$ coincides with one of the numbers $l_{j,2k+1}$, $j = 2, 3, \cdots, k$. The representation $R_{\epsilon,c,\mathbf{m}}$ of $U'_q(so_{2k-1,1})$ is irreducible if and only if c is not half-integer or $|c|$ coincides with one of the numbers $l_{j,2k}$, $j = 2, 3, \cdots, k$, or $|c| < l_{k,2k}$.*

Acknowledgements

The author is thankful to Prof. A. U. Klimyk for the fruitful discussions. The research contained in this paper was supported in part by Award No. UP1-2115 of the Civilian Research and Development Foundation for the Independent States of the Former Soviet Union (CRDF).

References

1. W. B. Schmidke, J. Wess, and B. Zumino, *A q-deformed Lorentz algebra*, Z. Phys. C **52** (1991), 471–476.

2. M. Jimbo, *A q-difference analogue of U(g) and the Yang–Baxter Equation*, Lett. Math. Phys. **10** (1985), 63–69.

3. V. G. Drinfeld, *Hopf algebra and Yang–Baxter equation*, Sov. Math. Dokl. **32** (1985), 254–259.

4. A. U. Klimyk and K. Schmüdgen, *Quantum Groups and Their Representations*, Springer, Berlin, 1997.

5. A. M. Gavrilik and A. U. Klimyk, *q-Deformed orthogonal and pseudo-orthogonal algebras and their representations*, Lett. Math. Phys. **21** (1991), 215–220.

6. A. Odesskii, *An analogue of the Sklyanin algebra*, Funct. Anal. Appl. **20** (1986), 152–154.

7. D. B. Fairlie, Quantum deformation of $SU_q(2)$, J. Phys. A **23** (1990), L183–L187.

8. M. Noumi, *Macdonald's symmetric polynomials as zonal spherical functions on some quantum homogeneous spaces*, Adv. Math. **123** (1996), 16–77.

9. M. Noumi, T. Umeda, and M. Wakayama, *Dual pairs, spherical harmonics and a Capelli identity in quantum group theory*, Compos. Math. **104** (1996), 227–277.

10. A. M. Gavrilik and N. Z. Iorgov, *q-Deformed algebras $U_q(\mathrm{so}_n)$ and their representations*, Methods of Funct. Anal. Topology **3**, No. 4 (1997), 51–63.

11. N. Z. Iorgov and A. U. Klimyk, *Nonclassical type representations of the q-deformed algebra $U'_q(\mathrm{so}_n)$*, Czech. J. Phys. **50**, No. 1 (2000), 85–90.

12. M. Havlíček, A. U. Klimyk, and S. Pošta, *Representations of the cyclically symmetric q-deformed algebra $\mathrm{so}_q(3)$*, J. Math. Phys. **40** (1999), 2135–2161.

13. Yu. Samoilenko and L. Turovska, *Semilinear relations and *-representations of deformations of $\mathrm{so}(3)$*, in *Quantum Groups and Quantum Spaces*, Banach Center Publications, Warsaw, vol. **40**, 1997, pp. 21–40.

14. A. U. Klimyk, *Quantum inhomogeneous unitary and orthogonal algebras and their representations*, Inst. for Theor. Phys. preprint, ITP-90-27E, Kiev, 1990.

15. A. M. Gavrilik and N. Z. Iorgov, *q-Deformed inhomogeneous algebras $U_q(\mathrm{iso}_n)$ and their representations*, in Proc. of II Int. Conf. "*Symmetry in Nonlinear Mathematical Physics*", Kiev, vol.**2**, 1997, pp. 384–392.

16. M. Havlíček, A. U. Klimyk, and S. Pošta, *Representations of the q-deformed algebra $U_q(\mathrm{iso}_2)$*, J. Phys. A **32** (1999), 4681–4690.

17. A. M. Gavrilik and N. Z. Iorgov, *Representations of the nonstandard algebras $U_q(\mathrm{so}(n))$ and $U_q(\mathrm{so}(n-1,1))$ in Gel'fand–Tsetlin basis*, Ukr. J. Phys. **43** (1998), 791–797.

QUASIPARTICLES IN NON-COMMUTATIVE FIELD THEORY

KARL LANDSTEINER *
Theory Division CERN, 1211 Geneva 23, Switzerland

Abstract. After a short introduction to the UV/IR mixing in non-commutative field theories we review the properties of scalar quasiparticles in non-commutative supersymmetric gauge theories at finite temperature. In particular we discuss the appearance of superluminous wave propagation.

1. Introduction

Given the experience of quantum mechanics it seems a rather natural idea that spacetime at very small distance-scales might be described by non-commuting coordinates [1, 2]. Keeping the example of quantum mechanics in mind one is lead to write down a commutation relation for the coordinates such as

$$[x^m, x^n] = i\theta^{mn} . \tag{1}$$

In order to study quantum field theory on such non-commuting spaces it is useful to make some further simplifying assumptions, in particular we will take θ^{mn} to be an element of the center of the algebra defined by (1).

A convenient way of thinking about non-commutativity is by deformation of the product on the space of ordinary function. Using θ^{mn} as deformation parameter we define the so-called Moyal product (or star-product) by

$$f(x) * g(x) := \lim_{y \to x} e^{\frac{i}{2}\theta^{mn}\partial_m^x \partial_n^y} f(x)g(y) . \tag{2}$$

In momentum space it takes the form

$$f(x) * g(x) = \int \frac{d^n k}{(2\pi)^n} \int \frac{d^n q}{(2\pi)^n} \tilde{f}(k)\tilde{g}(q)e^{-i(k+q)x}e^{-\frac{i}{2}k_m\theta^{mn}q_n} . \tag{3}$$

An immediate consequence is that we can always delete one star under the integral because the additional terms by which the Moyal product differs from the usual

---

* karl.landsteiner@cern.ch

S. Duplij and J. Wess (eds.), *Noncommutative Structures in Mathematics and Physics*, 369–378.

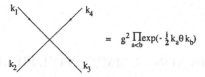

Figure 11.   Feynman rule for non-commutative Φ^4 vertex.

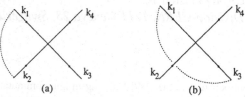

Figure 12.   Corrections to the two-point function can be either planar as in (a) or non-planar as in (b)

product are total derivatives thanks to the antisymmetry of θ^{mn}

$$\int f(x) * g(x) d^n x = \int \left(f(x).g(x) + \frac{i}{2}\theta^{mn}\partial_m f(x)\partial_n g(x) + \cdots \right) d^n x. \quad (4)$$

This furthermore implies cyclic symmetry under integral

$$\int f * g * h = \int f.g * h = \int g * h.f = \int g * h * f. \quad (5)$$

We have now all the ingredients do start discussing field theory. Before doing so we will introduce one further simplification, namely we will assume from that time is an ordinary commuting coordinate, i.e. $\theta^{m0} = 0$. This has the advantage that we are still dealing with a system with a finite number of time derivatives. Although a canonical formalism for theories with an infinite number of time derivatives can be developed [3] it turns out that quantum field theory on spaces with time-space non-commutativity are not unitary at the one-loop level [4][1].

Non-commutative field theories can be viewed as non-local deformations of local field theories. For fields of spin zero or one-half we can take a Lagrangian of an ordinary field theory and deform the product of fields according to the Moyal product (4), i.e. we replace the ordinary product by the star product. For spin one-fields we also have to consider that the gauge symmetry is deformed, $\delta A_m = \partial_m \lambda + i\{A_m, \lambda\}_*$, where $\{.,.\}_*$ denotes the Moyal bracket $\{f, g\}_* = f*g - g*f$. The non-commutative field strength of a gauge field is defined accordingly as $F_{mn} = \partial_m A_n - \partial_n A_m + i\{A_m, A_n\}_*$ and the covariant derivative as $D_m. = \partial_m. + i\{A_m, .\}_*$ [5].

---

[1] This applies to the time-like case, i.e. in all coordinate systems with $\theta^{mn} = const$ the commutator (1) involves the time coordinate.

Let us consider now a scalar Φ^4 theory on in four dimensions. Without further loss of generality we assume $\theta^{23} = -\theta^{32} = \theta$. Because one can drop one star-product in the Lagrangian the free theory is unchanged with respect to the one on ordinary $\mathbb{R}^n$. The tree level propagator is then the usual one

$$\langle \Phi(p)\Phi(-p) \rangle = \frac{i}{p^2 - m^2}. \tag{6}$$

The one-loop corrections to the two point function that arise from the Φ^4 vertex are shown in figure 2(a) and 2(b). Because of the cyclic symmetry of the vertex we have two distinct classes of graphs [6]. If we connect neighbouring lines of the vertex in figure (1) the dependence of the exponential on the internal momentum $k = k_1 = -k_2$ cancels. Thus the diagram 2(a) gives rise to a quadratic divergence in the same way as it happens in ordinary Φ^4 theory.

If we contract however non-neighbouring lines the dependence on the internal momentum of the exponent does not cancel. The distinct classes of Feynman diagrams in non-commutative field theories are called planar if they are of type 2(a) and non-planar if they are of type 2(b).

The divergence is regulated by the rapid oscillation of the exponential function at large internal momentum and we find

$$4g^2 \int \frac{d^4 k}{(2\pi)^4} \frac{e^{i\tilde{p}k}}{k^2 - m^2} = \frac{ig^2}{4\pi^2 \tilde{p}^2} + \cdots, \tag{7}$$

Where we introduced the notation $\tilde{p}^n = p_m \theta^{mn}$ and the dots indicate terms that are less singular for $\tilde{p} \to 0$. Resummation gives rise to a corrected two-point function on the one loop level of the form

$$\Gamma^2(p) = p^2 - m_R^2 + \frac{g^2}{\pi^2 \tilde{p}^2}. \tag{8}$$

The quadratic divergence in the planar graph gives rise to a renormalization of the mass. The non-planar graph results in a dramatic change of the infrared behaviour of the theory. On a technical level the origin of this infrared divergence is easily understood. The non-planar diagram is regulated by the phase factor stemming from the star product. This phase is absent if the external momentum flowing into the diagram vanishes. Thus the ultraviolet divergence has been converted into an infrared divergence. This phenomenon UV/IR mixing has first been discussed in [7] and has been further investigated in [9]- [33]. Notice also that the IR-singularity is present even in the massive theory. Since it is induced by modes in the far UV circling in the loop it is insensitive to the presence of a massterm.

It should be emphasized that there are usually also subleading logarithmic infrared divergencies. In the infrared these become important at momenta of the order of $p = \mathcal{O}(e^{-\frac{1}{g^2}})$. Down to these non-perturbatively small momenta the infrared behaviour is dominated by the effects stemming from the quadratic divergencies. In the following we will always concentrate on the leading order IR-behaviour and thus neglect the contributions from the logarithms.

In supersymmetric theories quadratic divergencies in four dimensions are absent. However at finite temperature supersymmetry is broken and the one-loop dispersion relation will again show effects from UV/IR mixing in non-planar graphs. Because temperature acts as a cutoff no IR-singularities are to be expected. The next section reviews these effects in the example of $\mathcal{N} = 4$ supersymmetric Yang-Mills theory.

2. Quasiparticles in non-commutative $\mathcal{N} = 4$ SYM

We limit ourselves to the study of a non-commutative $U(1)$ $\mathcal{N} = 4$ gauge theory. The spectrum of the theory consists of six scalars, four Majorana Fermions and a vector field. The Lagrangian takes the form

$$\mathcal{L} = \tfrac{1}{g^2} \int \left(-\tfrac{1}{4} F_{mn} F^{mn} + \tfrac{1}{2} D_m \Phi^{ab} D^m \Phi_{ab} + \tfrac{1}{4} \{\Phi^{ab}, \Phi^{cd}\}_* \{\Phi_{ab}, \Phi_{cd}\}_* + \right.$$
$$\left. + i\lambda_a \sigma^m D_m \bar{\lambda}^a + i\{\lambda_a, \lambda_b\}_* \Phi^{ab} + i\{\bar{\lambda}^a, \bar{\lambda}^b\}_* \Phi_{ab} \right) .$$

$$(9)$$

The theory has a global $SU(4)$ symmetry under which the fermions transform under the 4, $\bar{4}$. The 6 scalars transform in the antisymmetric. This symmetry is indicated by indices a, b.

We will study the dispersion relation of the $\mathcal{N} = 4$ scalars at finite temperature and one loop level. Finite temperature is implemented in the Matsubara formalism by considering the theory on $S^1 \times \mathbb{R} \times \mathbb{R}^2_{nc}$. The last factor indicates the two-dimensional non-commutative plane. The fermions are taken to have antiperiodic boundary conditions on the S^1 factor. Non-commutative field theories at finite temperature have been investigated in [34]-[37]

The scalar self-energy is given by

$$\Sigma_T = 32g^2 \int \frac{d^3 k}{(2\pi)^3} \frac{\sin^2 \frac{\tilde{p} \cdot k}{2}}{k} \left(n_B(k) + n_F(k) \right) + 4g^2 P^2 \bar{\Sigma}, \qquad (10)$$

$n_B(k)$ and $n_F(k)$ denote Bose-Einstein and Fermi-Dirac distributions. Four momentum is denoted by $P^2 = p_0^2 - p^2$, lowercase denotes three-momentum. Momenta along the non-commutative directions as will be called transverse.

The first term in (10) vanishes at $T = 0$ because of supersymmetry. The second term contributes to the finite temperature wave-function renormalization of the scalar field. It affects the position of the pole only to $\mathcal{O}(g^4)$ and we will drop it in the sequel.

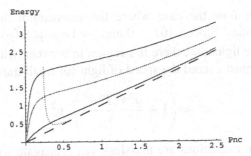

Figure 13.   Dispersion relation for scalars in $\mathcal{N} = 4$ Yang-Mills for different temperatures. The momentum p is taken to lie entirely in the non-commutative directions. The dashed line shows the light cone $\omega = p$. The dotted line shows the momentum p_c below which the group velocity $\frac{\partial \omega}{\partial p}$ is bigger than one.

Using the relation $\sin^2 \frac{\tilde{p}k}{2} = \frac{1}{2}(1 - \cos \tilde{p}k)$ we can separate the planar and non-planar contributions to the self-energy. The dispersion relation becomes

$$\omega^2 = p^2 + 2g^2 T^2 - \frac{4g^2 T}{\pi |\tilde{p}|} \tanh \frac{\pi |\tilde{p}| T}{2}. \tag{11}$$

A plot of the dispersion relation is shown in figure (3). The hyperbolic tangent arises solely from the non-planar contribution to the dispersion relation.

For large transverse external momenta the non-planar contribution is subleading with respect to the planar one,

$$\omega^2 \approx p^2 + 2g^2 T^2 - 4g^2 \frac{T}{\pi |\tilde{p}|}, \quad T\tilde{p} \gg 1. \tag{12}$$

The second term comes from the planar diagrams and gives a mass to the scalar excitations. The subdominant term linear in T arises solely from soft bosons in non-planar diagrams. These are modes with characteristic momentum $k \ll T$ and large occupation number $n_B \approx T/k \gg 1$,

$$\Sigma_{np} \sim \int d^3 k \frac{1}{k} \cos \tilde{p}k \frac{T}{k} \sim \frac{T}{\tilde{p}}. \tag{13}$$

In usual space-time the approximation $n_B \approx T/k \gg 1$ results in the well known ultraviolet catastrophe of classical field theory. In the non-planar sector of non-commutative space this does not happen as long as $\tilde{p}$ is different from zero. This is yet another manifestation of the UV/IR mixing of non-commutative field theories: to leading order at high temperature, the non-planar contribution is effectively purely classical [36]. At low transverse external momenta, the non-planar contribution tends to cancel the planar one. For zero external transverse momentum the interaction switches off. The theory becomes a free, gapless $U(1)$ gauge theory with $\omega^2 \approx p_3^2$.

Let us consider now the case where the momentum lies along the non-commutative directions. Since $\omega(0) = 0$ and for large p, $\omega(p) \approx \sqrt{p^2 + 2g^2T^2}$, which lies above the lightcone, there is a region in between with $\frac{\partial \omega(p)}{\partial p} > 1$. Thus the group velocity must exceed the speed of light for small transverse momenta!

$$\omega^2 \approx \left(1 + \frac{g^2\pi^2T^4\theta^2}{6} \right) p^2 . \tag{14}$$

The low momentum excitations are massless, but propagate with an index of refraction $n = p/\omega$ smaller than one. Because the interactions switch off at low momenta, we expect these modes to be long-lived. In figure (3) the momentum p_c below which the group velocity exceeds one is depicted by a dotted line. The dashed line shows the light cone $\omega = p$.

Let us emphasize that these qualitative features should be quite general and not an artifact of our one loop approximation, as they simply arise from the fact that the theory is non-interacting at zero transverse momentum and develops a mass gap otherwise[2].

We now investigate the consequences of the dispersion relation (14) for wave propagation. Imagine that some disturbance of the scalar field is created in the thermal bath at time $t = 0$. To simplify matters we will consider only a one dimensional problem with momentum pointing in a non-commutative direction. The fastest moving modes are the ones with longest wavelength. These are also the modes which are long lived in the thermal bath. For these it is possible to obtain the exact asymptotic behaviour by noting that the dispersion relation around $k = 0$ is

$$\omega(k) = c_0 k - \gamma k^3 + O(k^5) , \tag{15}$$

with $c_0 = \sqrt{1 + \frac{g^2\pi^2T^4\theta^2}{6}}$ and $\gamma = \frac{g^2\pi^4\theta^4T^6}{120c_0}$. This is the dispersion relation of the linearised Korteweg-deVries equation whose solution is expressed in terms of the Airy function $Ai(z)$. We can express the solution for the head of a wavetrain by [38]

$$\Phi = \frac{A}{2(3\gamma t)^{\frac{1}{3}}} Ai \left(\frac{x - c_0 t}{(3\gamma t)^{\frac{1}{3}}} \right) . \tag{16}$$

The Airy function has oscillatory behaviour for negative argument and decays exponentially for positive argument. Thus the wavetrain decays exponentially ahead of $x = c_0 t$. Behind the wave becomes oscillatory. In this region one can

---

[2] One might also be worried if these effects are gauge dependent. A model without gauge symmetry can be obtained if one sets the gauge field and one fermion (the field content of an $\mathcal{N} = 1$ vector multiplet) to zero. This would result in a $\mathcal{N} = 1$ Wess-Zumino model with Moyal bracket interactions. It would only change the overall factor in (10).

match the Airy function with the asymptotics obtained from a stationary wave approximation. In between the oscillatory region and the exponential decay there is a transition region of width proportional to $(\gamma t)^{\frac{1}{3}}$ around $x = c_0 t$. In this region the wavetrain has its first crest which therefore is moving with a velocity approximately given by c_0.

Group velocities faster than the speed of light do also appear in conventional physics, e.g. it is well-known that this happens for light propagation in media close to an absorption line. Since the dispersive effects are however large, the group velocity loses its meaning as the velocity of signal transportation. In our case, it is interesting to notice that as the temperature increases, not only c_0 but also γ grows. This implies that at high temperatures the soft transverse momenta become very dispersive. In such situations it is useful to introduce the concept of a front velocity which is the velocity of the head of the wavetrain. For the propagation of light in a medium it can be shown that this front velocity never exceeds the speed of light even if the group velocity can be faster than the speed of light [39]. In our case the front velocity can be defined as the velocity of the first crest of the wavetrain. According to (14) and (15) this is always bigger that the speed of light. The advance of the first crest with respect to an imagined light front is $(c_0 - 1)t$. Since its spread grows as $(\gamma t)^{\frac{1}{3}}$, the first crest is well defined outside the lightcone for large enough time, $t > t_0$ where $t_0 = \sqrt{\frac{\gamma}{(c_0-1)^3}}$.

The question arises if this superluminosity implies a violation of causality. This is not necessarily the case. Violation of causality needs both ingredients: superluminosity and the relativity principle. Imagine an observer A emitting some signal with superluminous velocity c_0. If the relativity principle is valid another observer B in a boosted frame relative to A could then catch the signal. B could send an answer also with superluminous speed c_0. The answer would reach observer A before he sent out the original signal. The crucial point is of course that in the non-commutative space-time we are considering boosts are not anymore symmetries. In particular only in the frame of observer A time is ordinary, commuting time. Any other frame involving a boost in a non-commutative direction implies that also time is non-commutative. To obtain an answer if causality is violated one would have to calculate the dispersion relation also in such a frame and study wave propagation then. Finite temperature field theory with non-commutative time is however difficult to formulate due to the infinite number of time derivatives appearing in the star product. This is an open question though progress could possibly be achieved along the lines in [3].

3. Discussion and Outlook

We have concentrated on reviewing the properties of scalar quasiparticles at finite temperature in non-commutative $\mathcal{N} = 4$ gauge theory. Another system that has been studied in [37] is the non-commutative Wess-Zumino model with star-product interactions instead of Moyal-brackets. The one-loop self-energy is given by a similar expression as (10) except that $\sin\frac{\tilde{p}k}{2}$ is substituted by $\cos\frac{\tilde{p}k}{2}$. It turns out that this has the effect that for temperatures $T > T_0 \approx \frac{1}{\sqrt{g\theta}}$ the minimum of the dispersion relation is displaced from $p = 0$! It has been argued that this makes Bose-condensation of scalar modes impossible for temperatures higher than T_0 [37] [3].

Another system that has been studied in [37] was $\mathcal{N} = 2$ gauge theory at finite density. The results are qualitatively analogous to the case with temperature. The role of the temperature is then played by the chemical potential.

Non-commutative field theories in the setup discussed here appear also in string theory. In [41] it was shown that the physics of D-branes in a B-field background in a particular scaling limit with $\alpha' \to 0$ is described by non-commutative supersymmetric gauge theories. It has been suggested that the effects of UV/IR mixing could be understood from a string perspective [7]. The UV/IR mixing in this stringy context has been considered in [43]-[49]. Since the model considered here arises as the scaling limit of a D-3-brane in a B-field background it would be very interesting to reconsider the one-loop dispersion relations from a string theory perspective.

References

1. H. S. Snyder, *Quantized Space-Time* , Phys. Rev. **71** (1947), 38.
2. A. Connes, *Non-commutative Geometry*, Academic Press, 1994.
   J. Madore, *An Introduction to Noncommutative Geometry and its Physical Applications*, Cambridge University Press 1995.
3. J. Gomis, K. Kamimura and J. Llosa, *Hamiltonian Formalism for Space-time Non-commutative Theories* , *e-print* hep-th/0006235.
4. J. Gomis and T. Mehen, *Space-Time Noncommutative Field Theories And Unitarity* , *e-print* hep-th/0005129.
5. A. Connes and M. Rieffel, *Yang-Mils for Noncommutative Two-Tori*, in Operator Algebras and Mathematical Physics, Contemp. Math. Oper. Alg. Math. Phys. 62, AMS, 1998, pp. 237.
6. A. Gonzalez-Arroyo and C.P. Korthals Altes, *Reduced Model for Large N Continuum Field Theories* , Phys. Lett. **B31** (1983), 396.
   T. Filk, *Divergencies In a Field Theory on Quantum Space*, Phys. Lett. **B376** (1996), 53.
7. S. Minwalla, M. Van Raamsdonk and N. Seiberg, *Noncommutative Perturbative Dynamics* , *e-print* hep-th/9912072.
8. I. Chepelev and R. Roiban, *Renormalization of Quantum Field Theories on Noncommutative $\mathbb{R}^d$, I. Scalars* , J. High Energy Phys. **0005** (2000), 037, *e-print* hep-th/9911098.
9. M. Hayakawa, *Perturbative analysis on infrared and ultraviolet aspects of noncommutative QED on $\mathbb{R}^4$* , *e-print* hep-th/9912167.
10. S. Iso, H. Kawai and Y. Kitazawa, *Bi-local Fields in Noncommutative Field Theory*, Nucl. Phys. **B576** (2000), 375 ,*e-print* hep-th/0001027.

---

[3] The phase structure of a non-commutative scalar field model in four dimensions has been investigated in [40] where it has been argued that condensation to stripe phases occurs.

11. H. Grosse, T. Krajewski and R. Wulkenhaar, *Renormalization of noncommutative Yang-Mills theories: A simple example* , *e-print* hep-th/0001182.

12. I. Ya. Aref'eva, D. M. Belov and A. S.Koshelev *A Note on UV/IR for Noncommutative Complex Scalar Field* , *e-print* hep-th/0001215.

13. A. Matusis, L. Susskind and N. Toumbas, *The IR/UV Connection in the Non-Commutative Gauge Theories* , *e-print* hep-th/0002075.

14. M. Van Raamsdonk and Nathan Seiberg, *Comments on Noncommutative Perturbative Dynamics* J. High Energy Phys. **0003** (2000), 035, *e-print* hep-th/0002186.

15. Chong-Sun Chu, *Induced Chern-Simons and WZW action in Noncommutative Spacetime* , Nucl. Phys. **B580** (2000), 352 ,*e-print* hep-th/0003007.

16. B. A. Campbell and K. Kaminsky, *Noncommutative Field Theory and Spontaneous Symmetry Breaking* , Nucl. Phys. **B581** (2000), 240 ,*e-print* hep-th/0003137.

17. I. Ya. Aref'eva, D. M. Belov and A. S.Koshelev *UV/IR Mixing for Noncommutative Complex Scalar Field Theory, II* , *e-print* hep-th/0003176.

18. A. A. Bichl, J. M. Grimstrup, V. Putz and M. Schweda, *Perturbative Chern-Simons Theory on Noncommutative $\mathbb{R}^3$* , J. High Energy Phys. **0007** (2000), 046 ,*e-print* hep-th/0004071.

19. F. Zamora *On the Operator Product Expansion in Noncommutative Quantum Field Theory* , J. High Energy Phys. **0005** (2000), 002 ,*e-print* hep-th/0004085.

20. J. Gomis, K. Landsteiner and Esperanza Lopez *Non-Relativistic Non-Commutative Field Theory and UV/IR Mixing* , Phys. Rev. **D62** (2000), 105006 ,*e-print* hep-th/0004115.

21. J. Ambjorn, Y.M. Makeenko, J. Nishimur and R.J. Szabo, *Lattice Gauge Fields and Discrete Noncommutative Yang-Mills Theory* J. High Energy Phys. **0005** (2000), 023 ,*e-print* hep-th/0004147.

22. Adi Armoni, *Comments on Perturbative Dynamics of Non-Commutative Yang-Mills Theory* ,*e-print* hep-th/00053028.

23. D. Bak, S. Ku Kim, K.-S. Soh and J. Hyung Yee, *Exact Wavefunctions in a Noncommutative Field Theory* Phys.Rev.Lett. **85** (2000), 3087 ,*e-print* hep-th/0005253.

24. H. O. Girotti, M. Gomes, V. O. Rivelles and A. J. da Silva , *A Consistent Noncommutative Field Theory: the Wess-Zumino Model*, *e-print* hep-th/0005272.

25. D. Bak, S. Ku Kim, K.-S. Soh and J. Hyung Yee, *Noncommutative Field Theories and Smooth Commutative Limits*, *e-print* hep-th/0006087.

26. G.-H. Chen, Y.-S. Wu , *One-loop Shift in Noncommutative Chern-Simons Coupling*, *e-print* hep-th/0006114.

27. A. A. Bichl, J. M. Grimstrup, H. Grosse, L. Popp, M. Schweda and R. Wulkenhaar , *The Superfield Formalism Applied to the Noncommutative Wess-Zumino Model*, *e-print* hep-th/0007050.

28. A. Micu and M. M. Sheikh-Jabbari , *Noncommutative Φ^4 Theory at Two Loops* , *e-print* hep-th/0008057.

29. I.F. Riad and M.M. Sheikh-Jabbari, *Noncommutative QED and Anomalous Dipole Moments* J. High Energy Phys. **0008** (2000), 045 ,*e-print* hep-th/0008132.

30. H. Liu and J. Michelson , **-TREK: The One-Loop N=4 Noncommutative SYM Action*, *e-print* hep-th/0008205.

31. W.-H. Huang, *Two-loop effective potential in noncommutative scalar field theory* , *e-print* hep-th/0009067.

32. T. Pengpan and X. Xion, *A Note on the Non-Commutative Wess-Zumino Model*, *e-print* hep-th/0009070.

33. E. T. Akhmedov, P. DeBoer and G. W. Semenoff, *Running couplings and triviality of field theories on non-commutative spaces*, *e-print* hep-th/0010003.

34. G. Arcioni and M.A. Vazquez-Mozo, *Thermal effects in perturbative noncommutative gauge theories* J. High Energy Phys. **0001** (2000), 028 ,*e-print* hep-th/9912140.

35. W. Fischler, Joaquim Gomis, E. Gorbatov, A. Kashani-Poor, S. Paban and P. Pouliot *Evidence for Winding States in Noncommutative Quantum Field Theory* J. High Energy Phys. **0005** (2000), 024 ,*e-print* hep-th/002067.

36. W. Fischler, E. Gorbatov, A. Kashani-Poor, R. McNees, S. Paban, P. Pouliot *The Interplay Between θ and T* J. High Energy Phys. **0006** (2000), 032 ,*e-print* hep-th/003216.

37. K. Landsteiner, E. Lopez and Michel H.G. Tytgat *Excitations in Hot Non-Commutative Theories* J. High Energy Phys. **0009** (2000), 027 ,*e-print* hep-th/006210.

38. G. B. Whitham, *Linear and nonlinear Waves*, John Wiley & Sons, 1998, pp. 438-444.

39. L. Brillouin, *Wave propagation and Group Velocity*, New York Academic Press, 1960.

40. S. S. Gubser and S. L. Sondhi, *Phase structure of non-commutative scalar field theories*, *e-print* hep-th/0006119.

41. N. Seiberg and E. Witten, *String Theory and Noncommutative Geometry* J. High Energy Phys. **9909** (1999), 032 ,*e-print* hep-th/9908142.

42. O. Andreev and H. Dorn *Diagrams of Noncommutative Phi-Three Theory from String Theory* Nucl. Phys. **B583** (2000), 145 ,*e-print* hep-th/0003113.

43. Y. Kiem and S. Lee, *UV/IR Mixing in Noncommutative Field Theory via Open String Loops* Nucl. Phys. **B586** (2000), 303 ,*e-print* hep-th/0003145.

44. A. Bilal, C.-S. Chu and R. Russo, *String Theory and Noncommutative Field Theories at One Loop* Nucl. Phys. **B582** (2000), 65 ,*e-print* hep-th/0003180.

45. J. Gomis, M. Kleban, T. Mehen, M. Rangamani and S. Shenker, *Noncommutative Gauge Dynamics From The String Worldsheet* J. High Energy Phys. **0008** (2000), 011 ,*e-print* hep-th/0003215.

46. A. Rajaraman and M. Rozali, *Noncommutative Gauge Theory, Divergences and Closed Strings* J. High Energy Phys. **0004** (2000), 033 ,*e-print* hep-th/0003227.

47. H.Liu and J. Michelson, *Stretched Strings in Noncommutative Field Theory* Phys. Rev. **D62** (2000), 066003 ,*e-print* hep-th/0004013.

48. G. Arcioni, J.L.F. Barbon, J. Gomis and M.A. Vazquez-Mozo, *On the stringy nature of winding modes in noncommutative thermal field theories* J. High Energy Phys. **0006** (2000), 038 ,*e-print* hep-th/0004080.

49. C.-S. Chu, R. Russo and Stefano Sciuto, *Multiloop String Amplitudes with B-Field and Noncommutative QFT* Nucl. Phys. **B585** (2000), 193 ,*e-print* hep-th/0004183.

TIME DEPENDENCE AND (NON)COMMUTATIVITY
OF SYMMETRIES OF EVOLUTION EQUATIONS

ARTUR SERGYEYEV *
Mathematical Institute, Silesian University at Opava,
Bezručovo nám. 13, 746 01 Opava, Czech Republic

1. Introduction

Nearly all known today integrable systems are homogeneous with respect to some scaling. For such systems no generality is lost in assuming the homogeneity of symmetries, master symmetries, recursion operators, etc., and this considerably simplifies their finding and study, see e.g. [1]–[10].

In the present paper we combine this well-known idea with our new results on the structure of *time-dependent* (cf. e.g. [6, 11–14] for the time-independent case) formal symmetries for a natural generalization of the systems, considered in [11, 12, 15], namely, for (1+1)-dimensional nondegenerate weakly diagonalizable (NWD) evolution systems with constraints. This enables us to find simple sufficient conditions for the commutativity and time-independence of higher order symmetries and for the existence of infinite number of such symmetries for *homogeneous* NWD systems with constraints. Note that the majority of known [8, 10, 12] and recently found, see e.g. [7, 16, 17], integrable evolution systems in (1+1) dimensions fit into this class. Moreover, our results, unlike the majority of already known ones, are valid for the systems with time-dependent coefficients as well, cf. e.g. [18], and are not restricted to scalar equations.

Let us stress that the proofs and the application of our results involve just an easy verification of some weight-related conditions and do *not* rely on the existence of a master symmetry or e.g. (hereditary) recursion operator. Hence, the results of present paper (except for those on existence of infinitely many symmetries) can be applied to non-integrable systems as well. On the other hand, the simplicity of use makes our results particularly helpful in the study of new

* artur.sergyeyev@math.slu.cz, arthurser@imath.kiev.ua

S. Duplij and J. Wess (eds.), Noncommutative Structures in Mathematics and Physics, 379–390.
© 2001 *Kluwer Academic Publishers. Printed in the Netherlands.*

integrable systems for which only a few higher order symmetries and (sometimes) a 'candidate' for the master symmetry are known, but no recursion operator is yet found. Indeed, we show that the check of a small number of conditions for the low order symmetries can replace tedious checks, cf. [19], that time-independent symmetries of sufficiently high order commute, that a 'candidate' for master symmetry is a nontrivial master symmetry and that its action yields the symmetries being well-defined (cf. [10] for recursion operators and *local* symmetries) functions of local variables $x, t, \mathbf{u}, \mathbf{u}_1, \ldots$ and of nonlocal variables ω_γ defined below.

Note that, unlike [4, 5, 19], in order to prove the existence of infinitely many symmetries we do not make *a priori* extra assumptions, say, about the existence of "negative" master symmetries τ_j, $j < 0$ [19]: all we need is a suitable 'candidate' τ for the master symmetry and a higher order time-independent symmetry. We also show that in order to verify the commutativity of *all* higher order time-independent homogeneous symmetries at once, it suffices to check only a small number of conditions for the time-independent symmetries of order lower than two. Moreover, checks of this kind are almost entirely algorithmic, so computer algebra software can be readily applied to perform them.

The paper is organized as follows. In Section 2 we give some definitions and facts, being the straightforward extension of those from [11, 12, 15] to the case of explicitly time-dependent evolution systems with constraints. In Section 3 we present the sufficient conditions of well-definiteness of the symmetries generated by means of master symmetry for the general evolution systems with constraints. In Section 4 we define nondegenerate weakly diagonalizable (NWD) systems with constraints and present some results on structure of their formal symmetries. In Section 5 we find the sufficient conditions for commutativity and time-independence of higher order symmetries and for the existence of infinite hierarchies of time-independent higher order symmetries for homogeneous NWD systems with constraints.

2. Basic definitions and structures

Let us consider a system of evolution equations with constraints (cf. [15])

$$\partial \mathbf{u}/\partial t = \mathbf{F}(x, t, \mathbf{u}, \ldots, \mathbf{u}_{n'}, \vec{\omega}) \tag{1}$$

for the vector function $\mathbf{u} = (u^1, \ldots, u^s)^T$. Here $\mathbf{u}_j = \partial^j \mathbf{u}/\partial x^j$, $\mathbf{u}_0 \equiv \mathbf{u}$ and $\mathbf{F} = (F^1, \ldots, F^s)^T$; $\vec{\omega} = (\omega_1, \ldots, \omega_c)^T$; T denotes the matrix transposition. The nonlocal variables ω_α are defined here by means of the relations [15, 20]

$$\partial \omega_\alpha/\partial x = X_\alpha(x, t, \mathbf{u}, \mathbf{u}_1, \ldots, \mathbf{u}_h, \vec{\omega}), \tag{2}$$

$$\partial \omega_\alpha/\partial t = T_\alpha(x, t, \mathbf{u}, \mathbf{u}_1, \ldots, \mathbf{u}_h, \vec{\omega}). \tag{3}$$

We shall denote by Ω the set of nonlocal variables ω_γ, $\gamma = 1, \ldots, c$.

Let $\mathcal{A}_{j,k}(\Omega)$ be the algebra of all locally analytic scalar functions of $x, t, \mathbf{u}, \mathbf{u}_1, \ldots, \mathbf{u}_j, \omega_1, \ldots, \omega_k$ with respect to the standard multiplication, $\mathcal{A} \equiv \mathcal{A}(\Omega) = \bigcup_{k=1}^{c} \bigcup_{j=0}^{\infty} \mathcal{A}_{j,k}(\Omega)$, and let $\mathcal{A}_{\mathrm{loc}} = \{ f \in \mathcal{A} \mid \partial f / \partial \vec{\omega} = 0 \}$ be the subalgebra of *local* functions in $\mathcal{A}$. Note that we do not exclude the case $c = \infty$. The operators of total x- and t-derivatives on $\mathcal{A}$ have the form

$$
D \equiv D_x = \frac{\partial}{\partial x} + \sum_{i=0}^{\infty} u_{i+1} \frac{\partial}{\partial u_i} + \sum_{\alpha=1}^{c} X_\alpha \frac{\partial}{\partial \omega_\alpha},
$$
$$
D_t = \frac{\partial}{\partial t} + \sum_{i=0}^{\infty} D^i(\mathbf{F}) \frac{\partial}{\partial u_i} + \sum_{\alpha=1}^{c} T_\alpha \frac{\partial}{\partial \omega_\alpha}.
$$

Following [15, 20], we require that $[D_x, D_t] = 0$ or, equivalently, $D_t(X_\alpha) = D_x(T_\alpha)$ for $\alpha = 1, \ldots, c$.

We shall denote by $\operatorname{Im} D$ the image of $\mathcal{A}$ under D. Throughout this paper except for Section 3 we make a *blanket assumption* that the kernel of D in $\mathcal{A}$ consists solely of functions of t.

Consider the set $\operatorname{Mat}_p(\mathcal{A})[\![D^{-1}]\!]$ of *formal series* in powers of D of the form $\mathfrak{H} = \sum_{j=-\infty}^{q} h_j D^j$, where h_j are $p \times p$ matrices with entries from $\mathcal{A}$, cf. e.g. [11, 12]. We shall write for short $\mathcal{A}[\![D^{-1}]\!]$ instead of $\operatorname{Mat}_1(\mathcal{A})[\![D^{-1}]\!]$.

The greatest $m \in \mathbb{Z}$ such that $h_m \neq 0$ is called the *degree* of $\mathfrak{H} \in \operatorname{Mat}_p(\mathcal{A})[\![D^{-1}]\!]$ and is denoted as $m = \deg \mathfrak{H}$. We assume that $\deg 0 = -\infty$, cf. e.g. [1]. The formal series $\mathfrak{H}$ of degree m is called *nondegenerate* [12], if $\det h_m \neq 0$. For $\mathfrak{H} = \sum_{j=-\infty}^{m} h_j D^j \in \mathcal{A}[\![D^{-1}]\!]$, $h_m \neq 0$, its *residue* and *logarithmic residue* are defined as $\operatorname{res} \mathfrak{H} = h_{-1}$ and $\operatorname{res} \ln \mathfrak{H} = h_{m-1}/h_m$ [11, 12].

The set $\operatorname{Mat}_p(\mathcal{A})[\![D^{-1}]\!]$ is an algebra under the multiplication law, given by the "generalized Leibniz rule", cf. [1],

$$
aD^i \circ bD^j = a \sum_{q=0}^{\infty} \frac{i(i-1)\cdots(i-q+1)}{q!} D^q(b) D^{i+j-q}
$$

for monomials $aD^i, bD^j, a, b \in \operatorname{Mat}_p(\mathcal{A})$, and extended by linearity to the whole $\operatorname{Mat}_p(\mathcal{A})[\![D^{-1}]\!]$. The commutator $[\mathfrak{A}, \mathfrak{B}] = \mathfrak{A} \circ \mathfrak{B} - \mathfrak{B} \circ \mathfrak{A}$ makes $\operatorname{Mat}_p(\mathcal{A})[\![D^{-1}]\!]$ into a Lie algebra. Below we omit $\circ$ if this is not confusing.

$\mathbf{G} \in \mathcal{A}^s$ is called, see e.g. [1–3], a *symmetry* for (1)–(3), if

$$
\partial \mathbf{G}/\partial t + [\mathbf{F}, \mathbf{G}] = 0, \tag{4}
$$

where $[\cdot, \cdot]$ is the Lie bracket $[\mathbf{K}, \mathbf{H}] = \mathbf{H}'[\mathbf{K}] - \mathbf{K}'[\mathbf{H}]$. The directional derivative of any (smooth) function $f \in \mathcal{A}^q$ along $\mathbf{H} \in \mathcal{A}^s$ is defined here as $f'[\mathbf{H}] = (df(x, t, \mathbf{u} + \epsilon \mathbf{H}, \mathbf{u}_1 + \epsilon D_x(\mathbf{H}), \ldots)/d\epsilon)_{\epsilon=0}$. Extending the technique of [15] to the case of time-dependent systems (1)–(3), we can easily show that for any $f \in \mathcal{A}$ we have $f' \in \mathcal{A}[\![D^{-1}]\!]$.

For any $f \in \mathcal{A}^q$ we shall define its *formal order* as $\operatorname{ford} f = \deg f'$. This naturally generalizes the notion of order for local functions, cf. e.g. [1, 12].

Let $S_F(\mathcal{A})$ be the set of all symmetries $\mathbf{G} \in \mathcal{A}^s$ for (1)–(3), $S_F^{(k)}(\mathcal{A}) = \{\mathbf{G} \in S_F(\mathcal{A}) \mid \text{ford }\mathbf{G} \leq k\}$, $\text{Ann}_F(\mathcal{A}) = \{\mathbf{G} \in S_F(\mathcal{A}) \mid \partial\mathbf{G}/\partial t = 0\}$. In general, for $\mathcal{A} \neq \mathcal{A}_{\text{loc}}$ neither $\mathcal{A}^s$ nor $S_F(\mathcal{A})$ are closed under the Lie bracket, but if $[\mathbf{P}, \mathbf{Q}] \in \mathcal{A}^s$ for some $\mathbf{P}, \mathbf{Q} \in S_F(\mathcal{A})$, then we have $[\mathbf{P}, \mathbf{Q}] \in S_F(\mathcal{A})$.

A formal series $\mathfrak{R} = \sum_{j=-\infty}^r \eta_j D^j \in \text{Mat}_s(\mathcal{A})[[D^{-1}]]$ is called [1, 11, 15] the *formal symmetry* of rank m for (1) (or, rather, for (1)–(3)), provided

$$\deg(D_t(\mathfrak{R}) - [\mathbf{F}', \mathfrak{R}]) \leq \deg \mathbf{F}' + \deg \mathfrak{R} - m. \qquad (5)$$

The derivative $D_t(\mathfrak{R})$ is defined here as $D_t(\mathfrak{R}) = \sum_{j=-\infty}^r D_t(\eta_j) D^j$.

The set $FS_F^{(q)}(\mathcal{A})$ of all formal symmetries of rank not lower than q of system (1)–(3) is a Lie algebra, because for the formal symmetries $\mathfrak{P}$ and $\mathfrak{Q}$ of ranks p and q we have $[\mathfrak{P}, \mathfrak{Q}] \in FS_F^{(r)}(\mathcal{A})$ for $r = \min(p, q)$, cf. [12].

Eq.(4) is well known to be nothing but the compatibility condition for (1) and $\partial \mathbf{u}/\partial \tau = \mathbf{G}$. Provided $\mathbf{G} \in \mathcal{A}^s$, we have $\partial(\partial \mathbf{u}/\partial \tau)\partial t = D_t(\mathbf{G})$ and $\partial(\partial \mathbf{u}/\partial t)\partial \tau = \mathbf{F}'[\mathbf{G}]$. Hence Eq.(4) may be rewritten as $D_t(\mathbf{G}) = \mathbf{F}'[\mathbf{G}]$.

Let $\mathbf{F}' \equiv \sum_{i=-\infty}^n \phi_i D^i$ and $n_0 = \begin{cases} 1 - j, \text{ if } \phi_i = \phi_i(x, t), i = n - j, \ldots, n, \\ 2 \text{ otherwise.} \end{cases}$

Since $D_t(\mathbf{G}) = \mathbf{F}'[\mathbf{G}]$ implies $D_t(\mathbf{G}') - [\mathbf{F}', \mathbf{G}'] - \mathbf{F}''[\mathbf{G}] = 0$, and $\deg \mathbf{F}''[\mathbf{G}] \leq \deg \mathbf{F}' + n_0 - 2$, we readily see that $\mathbf{G}' \in FS_F^{(\text{ford }\mathbf{G} - n_0 + 2)}(\mathcal{A})$.

3. Action of master symmetries on time-independent symmetries

As we have already mentioned above, for $\mathbf{P}, \mathbf{Q} \in \mathcal{A}^s$ in general $[\mathbf{P}, \mathbf{Q}] \notin \mathcal{A}^s$. In particular, when we repeatedly commute a master symmetry $\tau \in \mathcal{A}^s$ with some time-independent symmetry $\mathbf{Q} \in \text{Ann}_F(\mathcal{A})$, it is by no means obvious that $\mathbf{Q}_i = \text{ad}_\tau^i(\mathbf{Q}) = [\tau, \mathbf{Q}_{i-1}]$ belong to $\mathcal{A}^s$, except for the case $\mathcal{A} = \mathcal{A}_{\text{loc}}$. In some cases we can make the conditions $[\tau, \mathbf{Q}_i] \in \mathcal{A}^s$ or $[\mathbf{P}, \mathbf{Q}] \in \mathcal{A}^s$ hold by introducing new nonlocal variables $\tilde{\omega}_\kappa$ and thus replacing $\mathcal{A}$ by a larger algebra $\tilde{\mathcal{A}}$. But in order that $[\mathbf{P}, \mathbf{Q}] \in \mathcal{A}^s$ for $\mathbf{P}, \mathbf{Q} \in \mathcal{A}^s$ it obviously suffices to require that $\omega_\mu'[\mathbf{P}] \in \mathcal{A}$ for those ω_μ on which $\mathbf{Q}$ actually depends and $\omega_\nu'[\mathbf{Q}] \in \mathcal{A}$ for those ω_ν on which $\mathbf{P}$ actually depends, cf. Ch. 6 in [20].

Moreover, we have

Proposition 1. *Let* $\tau, \mathbf{Q} \in \mathcal{A}^s$, $\omega_\gamma'[\mathbf{Q}] \in \mathcal{A}$ *and* $\omega_\gamma'[\tau] \in \mathcal{A}$ *for* $\gamma = 1, \ldots, c.$ *Then* $\mathbf{Q}_l = \text{ad}_\tau^l(\mathbf{Q}) \in \mathcal{A}^s$ *for all* $l = 1, 2, \ldots$.

Proof. Let us use induction. To prove that $[\tau, \mathbf{Q}_l] \in \mathcal{A}^s$, if $\mathbf{Q}_l = [\tau, \mathbf{Q}_{l-1}] \in \mathcal{A}^s$, it suffices to prove that $\omega_\nu'([\tau, \mathbf{Q}_{l-1}]) \in \mathcal{A}$ for all ω_ν which τ depends on and that $\omega_\delta'[\tau] \in \mathcal{A}$ for all ω_δ which $[\tau, \mathbf{Q}_{l-1}]$ depends on. As $\omega_\nu'([\tau, \mathbf{Q}_{l-1}]) = (\omega_\nu'[\mathbf{Q}_{l-1}])'[\tau] - (\omega_\nu'[\tau])'[\mathbf{Q}_{l-1}]$, it suffices that $\omega_\gamma'[\tau] \in \mathcal{A}$ for all ω_γ, which $[\tau, \mathbf{Q}_{l-1}]$ and $\omega_\nu'[\mathbf{Q}_{l-1}]$ depend on, and $\omega_\kappa'[\mathbf{Q}_{l-1}] \in \mathcal{A}$ for all ω_κ which τ and $\omega_\nu'[\tau]$ depend on, in order that $[\tau, \mathbf{Q}_l] \in \mathcal{A}^s$. $\square$

It appears that nearly all known master symmetries of integrable systems (1)–(3) satisfy the conditions of Proposition 1 for a suitably chosen set Ω of nonlocal variables ω_γ, so their action indeed yields the symmetries from $\mathcal{A}^s$. For instance, if $\partial \mathbf{F}/\partial \vec{\omega} = 0$ and $\mathcal{A} = \mathcal{A}(\Omega_{\mathrm{UAC},F})$, then by virtue of the results of [20] Proposition 1 holds true for any $\tau, \mathbf{Q} \in S_F(\mathcal{A})$. Here $\Omega_{\mathrm{UAC},F}$ is the set of all nonlocal variables ω_γ associated with the universal abelian covering (see [20] for its definition) over (1). Let us stress that Proposition 1 is valid for any τ and $\mathbf{Q}$ meeting the relevant requirements, no matter whether τ is a master symmetry and $\mathbf{Q}$ is a symmetry for (1)–(3).

Note that Proposition 1 is obviously valid for more general systems of PDEs with constraints than (1)–(3), if we suitably redefine for them the Lie bracket, the directional derivative and the algebra $\mathcal{A}$.

4. The structure of formal symmetries for NWD systems

Consider a particular class of evolution systems with constraints (1)–(3) such that $n \equiv \mathrm{ford}\,\mathbf{F} \geq 2$ and the leading coefficient Φ of the formal series $\mathbf{F}'$ (i.e., $\mathbf{F}' \equiv \Phi D^n + \ldots$) has s distinct eigenvalues λ_i and can be diagonalized by means of a matrix $\Gamma = \Gamma(x, t, \mathbf{u}, \ldots, \mathbf{u}_{n'}, \vec{\omega})$, i.e., the matrix $\Lambda = \Gamma\Phi\Gamma^{-1}$ is diagonal, cf. [11, 12]. For these systems there exists a unique formal series $\mathfrak{T} = \Gamma + \Gamma\sum_{j=1}^{\infty}\Gamma_j D^{-j} \in \mathrm{Mat}_s(\mathcal{A})[[D^{-1}]]$ such that all coefficients of the formal series $\mathfrak{V} = \mathfrak{T}\mathbf{F}'\mathfrak{T}^{-1} + (D_t(\mathfrak{T}))\mathfrak{T}^{-1}$ are diagonal matrices and the diagonal entries of matrices $\Gamma_j, j = 1, 2, \ldots$, are equal to zero. The proof is essentially the same as for Proposition 3.1 from [11]. We shall call the systems with constraints (1)–(3) having the above properties and such that $\det\Phi \neq 0$ *nondegenerate weakly diagonalizable (NWD)*. Note that when $\mathbf{u}$ is scalar, i.e., $s = 1$, any system (1)–(3) with $n \equiv \mathrm{ford}\,\mathbf{F} \geq 2$ obviously is an NWD system with constraints, having $\mathfrak{T} = 1$ and $\mathfrak{V} = \mathbf{F}'$.

Below in this section (1)–(3) will be an NWD system with constraints.

Eq.(5) yields [11, 12] $\deg(D_t(\tilde{\mathfrak{R}}) - [\mathfrak{V}, \tilde{\mathfrak{R}}]) \leq \deg\mathfrak{V} + \deg\tilde{\mathfrak{R}} - m$, where $\tilde{\mathfrak{R}} = \mathfrak{T}\mathfrak{R}\mathfrak{T}^{-1}$, whence we find (cf. [6, 12, 13]) that any $\mathfrak{R} \in FS_F^{(n+1)}(\mathcal{A})$ can be represented in the form

$$\mathfrak{R} = \mathfrak{T}^{-1}\left(\sum_{j=r-n+1}^{r} c_j(t)\mathfrak{V}^{j/n}\right)\mathfrak{T} + \frac{1}{n}\mathfrak{T}^{-1}\left(D^{-1}\left(\dot{c}_r(t)\Lambda^{-1/n}\right.\right.$$

$$\left.\left. -rc_r(t)D_t(\Lambda^{-1/n})\right)\right)\mathfrak{V}^{\frac{r-n+1}{n}}\mathfrak{T} + \mathfrak{N}, \ \deg\mathfrak{N} < r - n + 1. \tag{6}$$

Likewise, for $\mathfrak{R} \in FS_F^{(m)}(\mathcal{A})$ with $m = 2, \ldots, n$ we have

$$\mathfrak{R} = \mathfrak{T}^{-1}\left(\sum_{j=r-m+2}^{r} c_j(t)\mathfrak{V}^{j/n}\right)\mathfrak{T} + \mathfrak{N}, \ \deg\mathfrak{N} < r - m + 2. \tag{7}$$

Here $r = \deg \mathfrak{R}$, $\mathfrak{N} = bD^\nu + \cdots \in \text{Mat}_s(\mathcal{A})[[D^{-1}]]$ is some formal series, $\nu = r - n$ in (6) and $\nu = r - m + 1$ in (7); $c_j(t)$ and $\Gamma b \Gamma^{-1}$ are diagonal $s \times s$ matrices; for $\mathfrak{V} \equiv \text{diag}(\mathfrak{V}_1, \ldots, \mathfrak{V}_s)$, $\mathfrak{V}_i \in \mathcal{A}[[D^{-1}]]$, we set $\mathfrak{V}^{j/n} = \text{diag}(\mathfrak{V}_1^{j/n}, \ldots, \mathfrak{V}_s^{j/n})$ [11]; dot stands for the *partial* derivative w.r.t. t.

In this section we assume that any function $\tilde{h} + a(t)$, where $a(t)$ is an arbitrary function of t, can be taken for $D^{-1}(h)$, if $h = D(\tilde{h})$ and $h, \tilde{h} \in \mathcal{A}$.

For $m = 2, \ldots, n + 1$ Eqs. (6), (7) represent a general solution of (5) for any NWD system with constraints (1)–(3). Hence, if at least one entry of the matrix $(\dot{c}_r(t)\Lambda^{-1/n} - rc_r(t)D_t(\Lambda^{-1/n}))$ does not belong to $\text{Im}\,D$, then (1)–(3) has no formal symmetries from $FS_F^{(n+1)}(\mathcal{A})$ with a given $c_r(t)$.

For any $\mathfrak{P} \equiv \mathfrak{T}^{-1}c_p(t)\mathfrak{V}^{p/n}\mathfrak{T} + \cdots$ and $\mathfrak{Q} \equiv \mathfrak{T}^{-1}d_q(t)\mathfrak{V}^{q/n}\mathfrak{T} + \cdots$ we have

$$[\mathfrak{P}, \mathfrak{Q}] = \mathfrak{T}^{-1}(1/n)(pc_p(t)\dot{d}_q(t) - qd_q(t)\dot{c}_p(t))\mathfrak{V}^{\frac{p+q-n}{n}}\mathfrak{T} + \mathfrak{K} \qquad (8)$$

by virtue of (6), provided $\mathfrak{P}, \mathfrak{Q} \in FS_F^{(n+1)}(\mathcal{A})$. Here $\mathfrak{K} \in \text{Mat}_s(\mathcal{A})[[D^{-1}]]$ is some formal series, $\deg \mathfrak{K} < p + q - n$.

Let $\mathbf{P}, \mathbf{Q} \in \mathcal{A}^s$, $\mathbf{R} \equiv [\mathbf{P}, \mathbf{Q}]$. Then $\mathbf{R}' = \mathbf{Q}''[\mathbf{P}] - \mathbf{P}''[\mathbf{Q}] - [\mathbf{P}', \mathbf{Q}']$. If $\mathbf{P}, \mathbf{Q} \in S_F(\mathcal{A})$, then (5) and (6) for $\mathfrak{R} = \mathbf{P}'$ and $\mathfrak{R} = \mathbf{Q}'$ imply $\deg \mathbf{P}''[\mathbf{Q}] \leq p + n_0 - 2 < p + q - n$ and $\deg \mathbf{Q}''[\mathbf{P}] \leq q + n_0 - 2 < p + q - n$ for $p, q > n + n_0 - 2$, $p \equiv \text{ford}\,\mathbf{P}$, $q \equiv \text{ford}\,\mathbf{Q}$. This result and (8) for $\mathfrak{P} = \mathbf{P}'$, $\mathfrak{Q} = \mathbf{Q}'$ yield

$$[\mathbf{P}, \mathbf{Q}]' = -\mathfrak{T}^{-1}(1/n)(pc_p(t)\dot{d}_q(t) - qd_q(t)\dot{c}_p(t))\mathfrak{V}^{\frac{p+q-n}{n}}\mathfrak{T} + \tilde{\mathfrak{K}}, \qquad (9)$$

where $\tilde{\mathfrak{K}} \in \text{Mat}_s(\mathcal{A})[[D^{-1}]]$ is some formal series, $\deg \tilde{\mathfrak{K}} < p + q - n$.

So, if $\mathbf{P}, \mathbf{Q} \in S_F(\mathcal{A})$, $p, q > n + n_0 - 2$, then ford $\mathbf{R} \leq p + q - n$. If $\mathbf{R} \in \mathcal{A}^s$, then $\mathbf{R} \in S_F^{(p+q-n)}(\mathcal{A})$, and $\mathbf{R} \in S_F^{(p+q-n-1)}(\mathcal{A})$, if $pc_p(t)\dot{d}_q(t) = qd_q(t)\dot{c}_p(t)$.

Let (1)–(3) have a nondegenerate formal symmetry $\mathfrak{R} \in \text{Mat}_s(\mathcal{A})[[D^{-1}]]$, $r \equiv \deg \mathfrak{R} \neq 0$, of rank $q > n$. Then $D_t(\rho_j^a) \in \text{Im}\,D$, i.e., ρ_j^a are conserved densities, for $a = 1, \ldots, s$ and $j = -1, 0, \ldots, q - n - 2$, where $\rho_0^a = \text{res}\ln((\mathfrak{T}\mathfrak{R}\mathfrak{T}^{-1})^{1/r})_{aa}$ and $\rho_j^a = \text{res}((\mathfrak{T}\mathfrak{R}\mathfrak{T}^{-1})^{j/r})_{aa}$ for $j \neq 0$, cf. [11]. For $n_0 < 2$ we have $\rho_j^a \in \text{Im}\,D$ for all $a = 1, \ldots, s$ and $j = -1, 0, \ldots, -n_0$.

Proposition 2. *Let an NWD system with constraints (1)–(3) have a nondegenerate formal symmetry* $\mathfrak{R} \in \text{Mat}_s(\mathcal{A})[[D^{-1}]]$, $r \equiv \deg \mathfrak{R} \neq 0$, $q \equiv \text{rank}\,\mathfrak{R} > n$; *let for* $a = 1, \ldots, s$ *there exist* $m_a \in \{-1, 1, 2, \ldots, \min(n - 2, q - n - 2)\}$ *such that* $m_a \neq 0$, $\rho_{m_a}^a \notin \text{Im}\,D$ *and* $\rho_j^a \in \text{Im}\,D$ *for* $j = -1, 1 \ldots, m_a - 1$, $j \neq 0$. *Then for each* $\mathfrak{P} \in FS_F^{(m+n+2)}(\mathcal{A})$, $m = \max_a m_a$, *there exists a constant* $s \times s$ *diagonal matrix* c *such that* $\mathfrak{P} = \mathfrak{T}^{-1}c\mathfrak{R}^{p/r}\mathfrak{T} + \cdots$, $p \equiv \deg \mathfrak{P}$.

Proof. Since $\mathfrak{R} \in FS_F^{(n+1)}(\mathcal{A})$, by (6) we have $\mathfrak{R} = \mathfrak{T}^{-1}h(t)\mathfrak{V}^{r/n}\mathfrak{T} + \cdots$.
For any $\mathfrak{P} \in FS_F^{(n+1)}(\mathcal{A})$ we can (cf. [6, 14] and (6)) represent $\tilde{\mathfrak{P}} \equiv \mathfrak{T}\mathfrak{P}\mathfrak{T}^{-1}$ as

$$\tilde{\mathfrak{P}} = \sum_{j=p-n+1}^{p} c_j(t)\tilde{\mathfrak{R}}^{j/r} + \frac{1}{n}\left(D^{-1}\left(\dot{c}_p(t)(h(t))^{n/r}\rho_{-1}\right)\right)\tilde{\mathfrak{R}}^{\frac{p-n+1}{r}} + \tilde{\mathfrak{N}}. \tag{10}$$

Here $\tilde{\mathfrak{N}} \equiv \sum_{j=-\infty}^{p-n} \tilde{b}_j D^j \in \mathrm{Mat}_s(\mathcal{A})[[D^{-1}]]$, $c_j(t)$, $h(t)$, $\tilde{b}_{p-n}$ are diagonal $s \times s$ matrices, $\rho_{-1} \equiv \mathrm{diag}(\rho_{-1}^1, \ldots, \rho_{-1}^s)$, $\tilde{\mathfrak{R}} = \mathfrak{T}\mathfrak{R}\mathfrak{T}^{-1}$; the fractional powers $\tilde{\mathfrak{R}}^{j/r}$ are defined so that their first r coefficients are diagonal, cf. [11, 12].

For $\mathfrak{P} \in FS_F^{(d)}(\mathcal{A})$ we have $\deg(D_t(\tilde{\mathfrak{P}}) - [\mathfrak{V}, \tilde{\mathfrak{P}}]) \leq n + p - d$, and thus $\deg(D_t(\tilde{\mathfrak{P}}_i) - [\mathfrak{V}, \tilde{\mathfrak{P}}_i]) \leq n + p + i - \min(q, d)$ for $\tilde{\mathfrak{P}}_i \equiv \tilde{\mathfrak{P}}\tilde{\mathfrak{R}}^{i/r}$. Hence, for $-p - 2 < i < \min(q, d) - n - p - 1$ we have $\mathrm{res}(D_t(\tilde{\mathfrak{P}}_i) - [\mathfrak{V}, \tilde{\mathfrak{P}}_i]) = 0$.

Let us plug (10) into this equality for $-p - 2 < i < \min(q, d, 2n) - n - p - 1$ and break it into s scalar equations. Since $\mathrm{res}([\mathfrak{V}, \tilde{\mathfrak{P}}_i])_{aa} \in \mathrm{Im}\, D$ by Adler's formula, see e.g. [12], and $D_t(\rho_{j+i}^a) \in \mathrm{Im}\, D$ by assumption, we easily find that for any $\mathfrak{P} \in FS_F^{(m+n+2)}(\mathcal{A})$ we have $(\dot{c}_p(t))_{aa}\rho_{m_a}^a = 0$ modulo the terms from $\mathrm{Im}\, D$ for all $a = 1, \ldots, s$. So, $\dot{c}_p(t) = 0$, and the result follows. $\square$

Corollary 1. *Under the assumptions of Proposition 2, for any $\mathbf{G} \in S_F(\mathcal{A})$, $k \equiv$ ford $\mathbf{G} \geq m + n + n_0$, we have $\mathbf{G}' = \mathfrak{T}^{-1}c\mathfrak{R}^{k/r}\mathfrak{T} + \cdots$, where c is a constant $s \times s$ diagonal matrix.*

5. Symmetries of homogeneous NWD systems

Let (1)–(3) possess a scaling symmetry $\mathbf{D} = \alpha t\mathbf{F} + x\mathbf{u}_1 + \beta\mathbf{u}$, where $\beta = \mathrm{diag}(\beta_1, \ldots, \beta_s)$ is a diagonal matrix, $\alpha, \beta_j = \mathrm{const}$, and let the determining equations (2), (3) for ω_γ, $\gamma = 1, \ldots, c$, be homogeneous with respect to $\mathbf{D}$. Then we shall call the evolution system with constraints (1)–(3) *homogeneous* w.r.t. $\mathbf{D}$, cf. e.g. [7, 8, 10, 20]. If a formal vector field $\mathbf{G}\partial/\partial\mathbf{u}$ is homogeneous of weight κ w.r.t. $\mathbf{D}$, then we shall say for short that $\mathbf{G} \in \mathcal{A}^s$ itself is homogeneous of weight κ and write $\mathrm{wt}(\mathbf{G}) = \kappa$.

For homogeneous systems (1)–(3) there usually exists a basis in $S_F(\mathcal{A})$ made of homogeneous symmetries, and hence the requirement of homogeneity of $\mathbf{P}, \mathbf{Q}$ and τ below is by no means restrictive. So, the phrase like "for all (homogeneous) $\mathbf{H} \in \mathcal{M}$ the condition P is true" below means that there exists a basis in $\mathcal{M}$ such that all its elements are homogeneous w.r.t. $\mathbf{D}$, and for all of them the condition P holds true. We have an obvious

Lemma 1. *Let (1)–(3) be a homogeneous system with constraints, and homogeneous $\mathbf{P}, \mathbf{Q} \in S_F(\mathcal{A})$ be such that $[\mathbf{P}, \mathbf{Q}] \in \mathcal{M}$, where $\mathcal{M}$ is a subspace of $\mathcal{A}^s$. Suppose that $\mathrm{wt}(\mathbf{G}) \neq \mathrm{wt}([\mathbf{P}, \mathbf{Q}]) = \mathrm{wt}(\mathbf{P}) + \mathrm{wt}(\mathbf{Q})$ for all (homogeneous) $\mathbf{G} \in S_F^{(p+q)}(\mathcal{A}) \cap \mathcal{M}$, $p \equiv$ ford $\mathbf{P}$, $q \equiv$ ford $\mathbf{Q}$. Then $[\mathbf{P}, \mathbf{Q}] = 0$.*

This result, as well as other results below, allows to prove the commutativity for large *families* of symmetries at once. Examples below show that we can usually choose the subspaces like $\mathcal{M}$ large enough so that the condition $[\mathbf{P}, \mathbf{Q}] \in \mathcal{M}$ can be verified for all symmetries in the family without actually computing $[\mathbf{P}, \mathbf{Q}]$. On the other hand, by proper choice of these subspaces we can considerably reduce the number of weight-related conditions to be verified, and thus make the application of our results truly efficient.

Below in this section we assume that (1)–(3) is a homogeneous NWD system with constraints and $\mathbf{P}, \mathbf{Q} \in S_F(\mathcal{A})$ are its *homogeneous* symmetries, $p \equiv \operatorname{ford} \mathbf{P}$, $q \equiv \operatorname{ford} \mathbf{Q}$. Note that if $p, q > n + n_0 - 2$, then by (9) we should verify the conditions of Lemma 1 only for $\mathbf{G} \in S_F^{(p+q-n)}(\mathcal{A}) \cap \mathcal{M}$ (for $\mathbf{G} \in S_F^{(p+q-n-1)}(\mathcal{A}) \cap \mathcal{M}$, if in addition $p c_p(t) \dot{d}_q(t) - q d_q(t) \dot{c}_p(t) = 0$).

5.1. COMMUTATIVITY AND TIME DEPENDENCE OF SYMMETRIES

Corollary 2. *Let* $\alpha \neq 0$, $\partial \Phi / \partial t = 0$ *and* $\partial X_\gamma / \partial t = \partial T_\gamma / \partial t = 0$, $\gamma = 1, \ldots, c$. *Let homogeneous* $\mathbf{P}, \mathbf{Q} \in \operatorname{Ann}_F(\mathcal{A})$ *be such that* $[\mathbf{P}, \mathbf{Q}] \in \mathcal{L}$, *where* $\mathcal{L}$ *is a subspace of* $\mathcal{A}^s$. *Let* $p, q \geq b_F \equiv \min(\max(n_0, 0), n + n_0 - 1)$, *where* $p \equiv \operatorname{ford} \mathbf{P}$, $q \equiv \operatorname{ford} \mathbf{Q}$. *Suppose that* $\operatorname{wt}(\mathbf{G}) \neq (p + q)\alpha / n$ *for all* (*homogeneous*) $\mathbf{G} \in S_F^{(n_0-1)}(\mathcal{A}) \cap \operatorname{Ann}_F(\mathcal{A}) \cap \mathcal{L}$. *Then* $[\mathbf{P}, \mathbf{Q}] = 0$.

Proof. If $\mathbf{P}, \mathbf{Q} \in \operatorname{Ann}_F(\mathcal{A})$, $[\mathbf{P}, \mathbf{Q}] \in \mathcal{A}^s$, $p, q \geq b_F$, then, using (6), (7) and (9), we find that $[\mathbf{P}, \mathbf{Q}] \in \mathcal{N} \equiv S_F^{(p+q-1)}(\mathcal{A}) \cap \operatorname{Ann}_F(\mathcal{A})$. Eqs. (6) or (7) for $\mathfrak{R} = \mathbf{G}'$ imply $\operatorname{wt}(\mathbf{G}) = k\alpha / n \neq \operatorname{wt}([\mathbf{P}, \mathbf{Q}]) = (p+q)\alpha / n$ for all homogeneous $\mathbf{G} \in \mathcal{N}$ with $k \equiv \operatorname{ford} \mathbf{G} \geq n_0$. Hence, under our assumptions $\operatorname{wt}(\mathbf{G}) \neq (p+q)\alpha / n$ for all homogeneous $\mathbf{G} \in \mathcal{N} \cap \mathcal{L} \equiv \mathcal{M}$, and thus by Lemma 1 $[\mathbf{P}, \mathbf{Q}] = 0$. $\square$

For instance, for the integrable [21] equation $u_t = D^2(u_1^{-1/2}) + u_1^{3/2} \equiv K$ with $n_0 = 2$ and $\alpha = 3/2$ the space $S_K^{(1)}(\mathcal{A}_{\text{loc}}) \cap \operatorname{Ann}_K(\mathcal{A}_{\text{loc}})$ is spanned by 1 and u_1, and $\operatorname{wt}(1), \operatorname{wt}(u_1) \leq 1 < \alpha(p+q)/n = (p+q)/2$ for $p, q \geq b_K = 2$. Hence, by Corollary 2 all (homogeneous) time-independent local generalized symmetries of formal order $p > 1$ for this equation commute.

Likewise, using Corollary 2, we can easily show that for any λ-homogeneous integrable evolution equation with $\lambda \geq 0$ from [8] all its x, t-independent homogeneous local generalized symmetries of formal order $k > 0$ commute.

If $n_0 \leq 0$ and, in addition to the conditions of Corollary 2 for $\mathbf{P}$ and $\mathbf{Q}$, the commutator $[\mathbf{P}, \mathbf{Q}] \in S_F(\mathcal{A}_{\text{loc}})$, $[\mathbf{P}, \mathbf{Q}]$ is x, t-independent and $\operatorname{wt}([\mathbf{P}, \mathbf{Q}]) \neq 0$, then $[\mathbf{P}, \mathbf{Q}] = 0$. The weight-related conditions are automatically satisfied, as the only x, t-independent symmetries in $S_F^{(n_0-1)}(\mathcal{A}_{\text{loc}})$ are constant ones, and their weight is zero. In particular, for *any* homogeneous (with $\alpha \neq 0$) NWD system of the form $\mathbf{u}_t = \Phi(x)\mathbf{u}_n + \Psi(x, t)\mathbf{u}_{n-1} + \mathbf{f}(x, t, \mathbf{u}, \ldots, \mathbf{u}_{n-2})$, where Φ, Ψ are $s \times s$ matrices, *all* homogeneous x, t-independent local generalized symmetries of formal order $k > 0$ commute.

Let $\mathfrak{R} \in FS_F^{(2)}(\mathcal{A})$ be a nondegenerate formal symmetry for (1)–(3), $r \equiv \deg \mathfrak{R} \neq 0$. Then by (7) $\mathfrak{R} = \Gamma^{-1}h(t)\Lambda^{r/n}\Gamma D^r + \cdots$, where $h(t) \equiv \mathrm{diag}(h_1(t), \ldots, h_s(t))$ is a $s \times s$ diagonal matrix. Assume that $h(t)$ is homogeneous w.r.t. $\mathbf{D}$ and $\zeta_{\mathfrak{R}} \equiv (\alpha/n + \mathrm{wt}(h(t))/r) \neq 0$. Let $Z_{F,\mathfrak{R}}(\mathcal{A}) = \{\mathbf{G} \in S_F(\mathcal{A}) \mid k \equiv \mathrm{ford}\,\mathbf{G} \geq n_0;$ there exists a diagonal matrix $c(t)$, $\mathrm{wt}(c(t)) = 0$, such that $\mathbf{G}' = \Gamma^{-1}c(t)(h(t))^{k/r}\Lambda^{k/n}\Gamma D^k + \cdots \}$. We set here $(h(t))^{k/r} \equiv \mathrm{diag}((h_1(t))^{k/r}, \ldots, (h_s(t))^{k/r})$. Let also $\mathrm{St}_{F,\mathfrak{R}}(\mathcal{A}) = \{\mathbf{G} \in Z_{F,\mathfrak{R}}(\mathcal{A}) \mid c(t) \text{ is a constant matrix}\}$, and $\mathrm{N}_{F,\mathfrak{R}}^{(j)}(\mathcal{A})$ be the set of symmetries $\mathbf{G} \in S_F(\mathcal{A})$ such that $k \equiv \mathrm{ford}\,\mathbf{G} \geq n_0$, $k \leq j$, and $\mathbf{G}' = \Gamma^{-1}c(t)(h(t))^{k/r}\Lambda^{k/n}\Gamma D^k + \cdots$, where $c(t)$ is an $s \times s$ diagonal matrix, different for different $\mathbf{G}$ and k, and the entries of $c(t)$ are linear combinations of functions of t, say, $\psi_b(t)$, such that for all b we have $\mathrm{wt}(\psi_b(t)) < \zeta_{\mathfrak{R}}(j - k)$ for $\zeta_{\mathfrak{R}} > 0$ and $\mathrm{wt}(\psi_b(t)) > \zeta_{\mathfrak{R}}(j - k)$ for $\zeta_{\mathfrak{R}} < 0$. For any homogeneous $\mathbf{G} \in \mathrm{N}_{F,\mathfrak{R}}^{(j)}(\mathcal{A})$ we have $\mathrm{wt}(\mathbf{G}) < j\zeta_{\mathfrak{R}}$ for $\zeta_{\mathfrak{R}} > 0$ and $\mathrm{wt}(\mathbf{G}) > j\zeta_{\mathfrak{R}}$ for $\zeta_{\mathfrak{R}} < 0$, so $\mathrm{wt}(\mathbf{H}) \neq \mathrm{wt}(\mathbf{P})$ for any homogeneous $\mathbf{P} \in Z_{F,\mathfrak{R}}(\mathcal{A})$ and $\mathbf{H} \in \mathrm{N}_{F,\mathfrak{R}}^{(\mathrm{ford}\,\mathbf{P})}(\mathcal{A})$.

Let $\mathbf{P}, \mathbf{Q} \in S_F(\mathcal{A})$ be homogeneous, and $[\mathbf{P}, \mathbf{Q}] \in \mathcal{L}_1 \cup \mathcal{L}_2$, where $\mathcal{L}_1$ is a subspace of $\mathrm{N}_{F,\mathfrak{R}}^{(j)}(\mathcal{A})$ for some j and $\mathfrak{R}$, and $\mathcal{L}_2$ is a subspace of $S_F^{(d)}(\mathcal{A})$ for some d. Assume that $\mathfrak{R}$ satisfies the above conditions, $\mathrm{wt}([\mathbf{P}, \mathbf{Q}]) \geq j\zeta_{\mathfrak{R}}$ for $\zeta_{\mathfrak{R}} > 0$ and $\mathrm{wt}([\mathbf{P}, \mathbf{Q}]) \leq j\zeta_{\mathfrak{R}}$ for $\zeta_{\mathfrak{R}} < 0$, and $\mathrm{wt}(\mathbf{H}) \neq \mathrm{wt}([\mathbf{P}, \mathbf{Q}])$ for all (homogeneous) $\mathbf{H} \in \mathcal{L}_2/(\mathcal{L}_2 \cap \mathrm{N}_{F,\mathfrak{R}}^{(j)}(\mathcal{A}))$. Then by Lemma 1 $[\mathbf{P}, \mathbf{Q}] = 0$.

Suppose that, in addition to the above conditions for $[\mathbf{P}, \mathbf{Q}]$, we have $d < 0$, $\mathrm{wt}([\mathbf{P}, \mathbf{Q}]) > 0$ for $\zeta_{\mathfrak{R}} > 0$ and $\mathrm{wt}([\mathbf{P}, \mathbf{Q}]) < 0$ for $\zeta_{\mathfrak{R}} < 0$, and $[\mathbf{P}, \mathbf{Q}]$ belongs to $S_F(\mathcal{A}_{\mathrm{loc}})$ and can be represented (as function of t and x) as a polynomial in variables $\chi(t)$ and $\xi(x)$ such that $\mathrm{wt}(\chi(t)) < 0$ and $\mathrm{wt}(\xi(x)) < 0$ for $\zeta_{\mathfrak{R}} > 0$, and $\mathrm{wt}(\chi(t)) > 0$ and $\mathrm{wt}(\xi(x)) > 0$ for $\zeta_{\mathfrak{R}} < 0$. Then $[\mathbf{P}, \mathbf{Q}] = 0$, and there is no further weight-related conditions to verify. Indeed, $S_F^{(d)}(\mathcal{A}_{\mathrm{loc}})$ for any $d < 0$ is spanned by the symmetries of the form $\mathbf{G} = \mathbf{G}(x, t)$, and for any homogeneous symmetry $\mathbf{H} = \mathbf{H}(x, t)$ being a polynomial in $\chi(t)$ and $\xi(x)$ we obviously have $\mathrm{wt}(\mathbf{H}) \neq \mathrm{wt}([\mathbf{P}, \mathbf{Q}])$.

Note that under the assumptions of Proposition 2 all $\mathbf{G} \in S_F(\mathcal{A})$ with $\mathrm{ford}\,\mathbf{G} \geq m+n+n_0$ belong to $\mathrm{St}_{F,\mathfrak{R}}(\mathcal{A})$ by Corollary 1. Suppose that $\mathfrak{R}$ satisfies the conditions, given above. Let $d = \min(m + n + n_0 - 1, p + q)$. Then for any $\mathbf{P}, \mathbf{Q} \in S_F(\mathcal{A})$ such that $[\mathbf{P}, \mathbf{Q}] \in \mathcal{A}^s$ we have $[\mathbf{P}, \mathbf{Q}] \in \mathrm{N}_{F,\mathfrak{R}}^{(p+q)}(\mathcal{A}) \cup S_F^{(d)}(\mathcal{A})$. Then $[\mathbf{P}, \mathbf{Q}] = 0$ for homogeneous $\mathbf{P}, \mathbf{Q} \in Z_{F,\mathfrak{R}}(\mathcal{A})$, once $\mathrm{wt}(\mathbf{H}) \neq \mathrm{wt}([\mathbf{P}, \mathbf{Q}])$ for all (homogeneous) $\mathbf{H} \in S_F^{(d)}(\mathcal{A})/(S_F^{(d)}(\mathcal{A}) \cap \mathrm{N}_{F,\mathfrak{R}}^{(p+q)}(\mathcal{A}))$. If $p, q > n + n_0 - 2$, then by (9) we can take $d = \min(m + n + n_0 - 1, p + q - n)$ (or $d = \min(m + n + n_0 - 1, p + q - n - 1)$, if $pc_p(t)\dot{d}_q(t) - qd_q(t)\dot{c}_p(t) = 0$).

If $\partial \mathbf{F}/\partial t = \partial X_\gamma/\partial t = \partial T_\gamma/\partial t = 0$, $\gamma = 1, \ldots, c$, then $\mathbf{F} \in S_F(\mathcal{A})$, and $\partial \mathbf{P}/\partial t = [\mathbf{P}, \mathbf{F}] \in S_F^{(p)}(\mathcal{A})$ for $\mathbf{P} \in S_F(\mathcal{A})$. So, taking $\mathbf{Q} = \mathbf{F}$ and imposing the extra condition $d \leq p$ in three previous paragraphs yields valid results.

We also have the following

Proposition 3. *Let* $\alpha \neq 0$ *and* $\partial F/\partial t = 0$, $\partial X_\gamma/\partial t = \partial T_\gamma/\partial t = 0$, $\gamma = 1, \ldots, c$; *let homogeneous* $\mathbf{P} \in S_F(\mathcal{A})$ *be such that* $p \equiv \text{ford}\,\mathbf{P} \geq n_0$, $\text{ford}\,\partial \mathbf{P}/\partial t < p$ *and* $[\mathbf{P}, \mathbf{F}] \in \mathcal{L}$, *where* $\mathcal{L}$ *is a subspace of* $\mathcal{A}^s$. *Suppose that* $\text{wt}(\mathbf{G}) \neq (p + n)\alpha/n$ *for all (homogeneous)* $\mathbf{G} \in S_F^{(p-1)}(\mathcal{A}) \cap \mathcal{L}$ *such that* $\mathbf{G} \notin N_{F,\mathbf{F}}^{(p+n)}(\mathcal{A})$. *Then* $[\mathbf{P}, \mathbf{F}] = 0$, *and thus* $\partial \mathbf{P}/\partial t = 0$ *and* $\mathbf{P} \in \text{Ann}_F(\mathcal{A})$.

Proof. As $\text{ford}\,\partial \mathbf{P}/\partial t < p$, we have $\partial \mathbf{P}/\partial t = [\mathbf{P}, \mathbf{F}] \in S_F^{(p-1)}(\mathcal{A}) \cap \mathcal{L} \equiv \mathcal{M}$. The conditions $\text{ford}\,\partial \mathbf{P}/\partial t < p$ and $p \geq n_0$ by virtue of (6) or (7) for $\mathfrak{R} = \mathbf{P}'$ readily imply $\text{wt}(\mathbf{P}) = p\alpha/n$. Hence $\text{wt}([\mathbf{P}, \mathbf{F}]) = (p + n)\alpha/n$, and thus by Lemma 1 $[\mathbf{P}, \mathbf{F}] = 0$. $\square$

Let $\alpha > 0$, $\partial F/\partial t = 0$, $\partial X_\gamma/\partial t = \partial T_\gamma/\partial t = 0$, $\gamma = 1, \ldots, c$, and homogeneous $\mathbf{P}, \mathbf{Q} \in \text{St}_{F,\mathbf{F}'}(\mathcal{A})$, $p, q \geq n_0$, be polynomials in t. If we take the space of symmetries from $S_F(\mathcal{A})$ polynomial in time t, for $\tilde{\mathcal{L}}$, and set $\mathcal{L}_1 = N_{F,\mathbf{F}}^{(p+q)}(\mathcal{A}) \cap \tilde{\mathcal{L}}$, $\mathcal{L}_2 = S_F^{(n_0-1)}(\mathcal{A}) \cap \tilde{\mathcal{L}}$, $d = n_0 - 1$, then $[\mathbf{P}, \mathbf{Q}] \in \mathcal{L}_1 \cup \mathcal{L}_2 \equiv \mathcal{M}$, and thus the weight-related conditions of Lemma 1, Corollary 2, Proposition 3, etc., are to be checked only for (homogeneous) $\mathbf{G} \in \mathcal{L}_2$. Furthermore, if $n_0 \leq 0$, then $S_F^{(n_0-1)}(\mathcal{A}_{\text{loc}})$ contains only the symmetries $\mathbf{G} = \mathbf{G}(x, t)$, and so *any* homogeneous local generalized symmetry $\mathbf{K}$ of formal order $k > 0$ being polynomial in t and x and such that $\partial^2 \mathbf{K}/\partial \mathbf{u}_k \partial t = 0$ is in fact time-independent, and any two such symmetries commute. This result applies e.g. to *any* homogeneous NWD system with $\alpha > 0$ having the form $\mathbf{u}_t = \Phi(x)\mathbf{u}_n + \Psi(x)\mathbf{u}_{n-1} + \mathbf{f}(x, \mathbf{u}, \ldots, \mathbf{u}_{n-2})$, where Φ, Ψ are $s \times s$ matrices.

5.2. MASTER SYMMETRIES OF HOMOGENEOUS NWD SYSTEMS

Corollary 3. *Let* $\alpha \neq 0$, $\partial \Phi/\partial t = 0$ *and* $\partial X_\gamma/\partial t = \partial T_\gamma/\partial t = 0$, $\gamma = 1, \ldots, c$. *Suppose that there exist a homogeneous* $\mathbf{Q} \in \text{Ann}_F(\mathcal{A})$ *and a homogeneous* $\tau \in \mathcal{A}^s$ *such that* $\partial \tau/\partial t = 0$, $\partial[\tau, \mathbf{F}]/\partial t = 0$, $\mathbf{K} = \tau + t[\tau, \mathbf{F}] \in S_F(\mathcal{A})$, $q \equiv \text{ford}\,\mathbf{Q} > n + n_0 - 2$, $b \equiv \text{ford}[\tau, \mathbf{F}] > \max(\text{ford}\,\tau, n)$, *the formal series* $([\tau, \mathbf{F}])'$ *is nondegenerate,* $[[\tau, \mathbf{F}], \mathbf{Q}] \in \mathcal{L}$, *where* $\mathcal{L}$ *is a subspace of* $\mathcal{A}^s$, $[\tau, \mathbf{Q}] \in \mathcal{A}^s$. *Let* $\text{wt}(\mathbf{H}) \neq (b + q)\alpha/n$ *for all (homogeneous)* $\mathbf{H} \in \mathcal{L} \cap S_F^{(n_0-1)}(\mathcal{A}) \cap \text{Ann}_F(\mathcal{A})$. *Then* $\mathbf{Q}_1 = [\tau, \mathbf{Q}] \in \text{Ann}_F(\mathcal{A})$, *and* $\text{ford}\,\mathbf{Q}_1 > q$.

Proof. From (4) with $\mathbf{G} = \mathbf{K}$ it clear that $[\tau, \mathbf{F}] \in \text{Ann}_F(\mathcal{A})$, so by Corollary 2 $[[\tau, \mathbf{F}], \mathbf{Q}] = 0$, whence, using $[\mathbf{F}, \mathbf{Q}] = 0$ and the Jacobi identity, we find that $[\mathbf{F}, [\tau, \mathbf{Q}]] = 0$, so $[\tau, \mathbf{Q}] \in \text{Ann}_F(\mathcal{A})$. By (9) the nondegeneracy of $([\tau, \mathbf{F}])'$ readily implies $\text{ford}[\tau, \mathbf{Q}] = \text{ford}[\mathbf{K}, \mathbf{Q}] = b + q - n > q$. $\square$

Theorem 1. *Let the conditions of Corollary 3 be satisfied,* $\text{ad}_{[\tau, \mathbf{F}]}^j(\mathbf{Q}) \in \mathcal{L}_j$, *where* $\mathcal{L}_j$ *are some subspaces of* $\mathcal{A}^s$, $\mathbf{Q}_j \equiv \text{ad}_\tau^j(\mathbf{Q}) \in \mathcal{A}^s$, *and* $\text{wt}(\mathbf{H}) \neq$

$((b-n)j+q+n)\alpha/n$ for all (homogeneous) $\mathbf{H} \in \mathcal{L}_j \cap S_F^{(n_0-1)}(\mathcal{A}) \cap \mathrm{Ann}_F(\mathcal{A})$, $j = 2, \ldots, i$. Then $\mathbf{Q}_j \in \mathrm{Ann}_F(\mathcal{A})$ and ford $\mathbf{Q}_j > $ ford $\mathbf{Q}_{j-1}$, $j = 1, \ldots, i$.

The proof of the theorem consists in replacing in Corollary 3 the symmetry $\mathbf{Q}$ by $\mathbf{Q}_j = \mathrm{ad}_\tau^j(\mathbf{Q})$ and repeated use of this corollary for $j = 2, \ldots, i$. Note that we can easily verify that $\mathrm{ad}_\tau^j(\mathbf{Q}) \in \mathcal{A}^s$, using Proposition 1.

Thus, Proposition 1, Corollary 3 and Theorem 1 enable us to ensure that τ indeed is a nontrivial master symmetry, producing a sequence of symmetries of infinitely growing formal orders, without assuming *a priori* the existence of hereditary recursion operator [5] or e.g. of "negative" master symmetries τ_j, $j < 0$ [19]. So, our results provide a useful complement to the known general results on master symmetries, cf. e.g. [3–5, 19].

It is important to stress that in general the symmetries $\mathbf{Q}_i$ are not obliged to commute pairwise. The check of their commutativity and picking out the commutative subset in the sequence of $\mathbf{Q}_i$ can be performed using either the results of present paper or other methods, see e.g. [1, 4, 5, 19].

We often can take $[\tau, \mathbf{F}]$ or $\mathbf{F}$ for $\mathbf{Q}$, and then in order to use Theorem 1 it suffices to know only a suitable 'candidate' τ for the master symmetry.

For instance, integrable Harry Dym equation $u_t = u^3 u_3 \equiv H$, see e.g. [1, 14], satisfies the conditions of Proposition 1 and of Theorem 1 for all $i = 2, 3, \ldots$ with $\alpha = 3$, $b = 5$, $\mathcal{A} = \mathcal{A}(\Omega_{\mathrm{UAC},H})$, $\tau = u^3 D^3(u\omega_1) \equiv \tau_0 + u^3 u_3 \omega_1$, $\tau_0 \in \mathcal{A}_{\mathrm{loc}}$, $\mathbf{Q} = [\tau, u^3 u_3] = 3u^5 u_5 + \cdots \in \mathrm{Ann}_H(\mathcal{A})$. In particular, the nonlocal variable ω_1 in τ is defined by means of the relations $\partial \omega_1 / \partial t = -uu_2 - u_1^2/2$, $\partial \omega_1 / \partial x = u^{-1}$ (informally, $\omega_1 = D^{-1}(u^{-1})$). Thus, by Theorem 1 $\mathbf{Q}_j = \mathrm{ad}_\tau^j(\mathbf{Q}) \in \mathrm{Ann}_H(\mathcal{A})$, $j = 1, 2, \ldots$, together with $\mathbf{Q}_{-1} \equiv u^3 u_3 \in \mathrm{Ann}_H(\mathcal{A}_{\mathrm{loc}})$ and $\mathbf{Q}_0 \equiv \mathbf{Q}$ form the infinite hierarchy of time-independent symmetries for the Harry Dym equation. The commutativity of $\mathbf{Q}_j$, $j = -1, 0, 1, \ldots$, readily follows from Corollary 2. Note that it is possible to show that $\mathbf{Q}_j$, $j = 0, 1, \ldots$, are in fact *local* generalized symmetries of Harry Dym equation and coincide with the members of hierarchy generated by means of the recursion operator $\mathfrak{R} = u^3 D^3 \circ u \circ D^{-1} \circ u^{-2}$ from the seed symmetry $u^3 u_3$, up to the constant multiples.

Acknowledgements

I am sincerely grateful to the organizers of NATO ARW *Noncommutative Structures in Mathematics and Physics* for inviting me to participate and to give a talk and for their hospitality. It is also my great pleasure to thank Profs. M. Błaszak, B. Fuchssteiner, B. Kupershmidt, P. Olver and V. V. Sokolov for stimulating discussions and comments. Last but not least, I acknowledge with deep gratitude Dr. M. Marvan's reading the drafts of this paper and making a lot of precise remarks, which considerably improved it.

This research was supported by the Ministry of Education, Youth and Sports of Czech Republic, Grants CEZ:J10/98:192400002 and VS 96003.

References

1. P. Olver, *Applications of Lie Groups to Differential Equations*, Springer-Verlag, New York, 1993.
2. A.S. Fokas, *Symmetries and integrability*, Stud. Appl. Math. **77**: 3 (1987), 253–299.
3. M. Błaszak, *Multi-Hamiltonian Theory of Dynamical Systems*, Springer-Verlag, Heildelberg, 1998.
4. B. Fuchssteiner, *Mastersymmetries, higher order time-dependent symmetries and conserved densities of nonlinear evolution equations*, Progr. Theor. Phys. **70** (1983), 1508–1522.
5. W. Oevel, *A geometrical approach to integrable systems admitting time dependent invariants*, in Topics in Soliton Theory and Exactly Solvable Nonlinear Equations (Oberwolfach, 1986) (M. Ablowitz, B. Fuchssteiner, M. Kruskal, eds.), World Scientific Publishing, Singapore, 1987, pp. 108–124.
6. V.V. Sokolov, *On the symmetries of evolution equations*, Russ. Math. Surveys **43**: 5 (1988), 165–204.
7. P.J. Olver, V.V. Sokolov, *Integrable evolution equations on associative algebras*, Comm. Math. Phys. **193** (1998), 245–268.
8. J.A. Sanders and J.P. Wang, *On the integrability of homogeneous scalar evolution equations*, J. Differential Equations, **147** (1998), 410–434.
9. A.H. Bilge, *On the equivalence of linearization and formal symmetries as integrability tests for evolution equations*, J. Phys. A: Math. Gen. **26** (1993), 7511–7519.
10. J.P. Wang, *Symmetries and Conservation Laws of Evolution Equations*, Ph.D. Thesis, Vrije Universiteit van Amsterdam, 1998.
11. A.V. Mikhailov, A.B. Shabat, and R.I. Yamilov, *The symmetry approach to classification of nonlinear equations. Complete lists of integrable systems*, Russ. Math. Surveys **42**:4 (1987), 1–63.
12. A.V. Mikhailov, A.B. Shabat and V.V. Sokolov, *The symmetry approach to classification of integrable equations*, in What is Integrability? (V.E. Zakharov, ed.), Springer-Verlag, New York, 1991, pp. 115–184.
13. A.V. Mikhailov, R.I. Yamilov, *Towards classification of (2+1)-dimensional integrable equations. Integrability conditions. I*, J. Phys. A: Math. Gen. **31** (1998), 6707–6715.
14. N.H. Ibragimov, *Transformation Groups Applied to Mathematical Physics*, D. Reidel Publishing Co., Boston, 1985.
15. F.Kh. Mukminov, V.V. Sokolov, *Integrable evolution equations with constraints*, Mat. Sb.(N.S.) **133(175)**:3 (1987), 392–414.
16. V.V. Sokolov, T. Wolf, *A symmetry test for quasilinear coupled systems*, Inverse Problems **15** (1999), L5–L11.
17. M.V. Foursov, *Classification of certain type integrable coupled potential KdV and modified KdV-type equations*, J. Math. Phys. **41** (2000), 6173–6185.
18. B. Fuchssteiner, *Integrable nonlinear evolution equations with time-dependent coefficients*, J. Math. Phys. **34** (1993), 5140–5158.
19. I. Dorfman, *Dirac Structures and Integrability of Nonlinear Evolution Equations*, John Wiley & Sons, Chichester, 1993.
20. *Symmetries and Conservation Laws for Differential Equations of Mathematical Physics* (I.S. Krasil'shchik and A.M. Vinogradov, eds.), American Mathematical Society, Providence, 1999.
21. J.A. Cavalcante and K. Tenenblat, *Conservation laws for nonlinear evolution equations*, J. Math. Phys. **29** (1988), 1044–1049.

p-ADIC STRINGS AND NONCOMMUTATIVITY

BRANKO DRAGOVICH[1,2] * and IGOR V. VOLOVICH[1]
[1]*Steklov Mathematical Institute, Gubkin St. 8, 117966 Moscow, Russia*

[2]*Institute of Physics, P.O.Box 57, 11001 Belgrade, Yugoslavia*

Abstract. Some possible connections between p-adic string theory and noncommutativity are considered. Their relation to the uncertainty in space measurements at the Planck scale is discussed. Existence of new p-adic string amplitudes is pointed out. Some similarities between p-adic solitonic branes and noncommutative scalar solitons are emphasized. More explicit and deeper connections between string field theory and p-adic string theory could emerge in the near future.

1. Introduction

It is well-known (for a recent review, see [1]) that the interplay between quantum-mechanical and general relativity principles gives an uncertainty Δx on the measurements of distances x in the form

$$\Delta x \geq \ell_0 = \sqrt{\frac{\hbar G}{c^3}} \sim 10^{-33} cm, \tag{1}$$

where ℓ_0 is the Planck length. This fact requires reconsideration of many our basic concepts about the spacetime structure at the Planck scale. It leads to the investigation of some new and more fundamental mathematical notions. To this end, we will consider here two very natural approaches. From the one point of view, the uncertainty (1) means at least restriction on the dominance of real numbers and archimedean geometry in their applications at the Planck scale. Namely, this formula has been derived with the implicit use of the real numbers and any archimedean geometry. In this way we see that the usual physical theory predicts its breakdown at the Planck scale. A graceful exit from this situation should be in the use of adeles and adelic topology, which contain archimedean as well as nonarchimedean geometries. From the other point of view, the uncertainty (1) has

* dragovic@mi.ras.ru

S. Duplij and J. Wess (eds.), Noncommutative Structures in Mathematics and Physics, 391–399.

to be a consequence of some noncommutativity between space coordinates. This conclusion follows from the analogous situation in ordinary quantum mechanics: the uncertainty $\Delta x \Delta k \geq \frac{\hbar}{2}$ is a direct consequence of the noncommutativity in the form of the Heisenberg algebra $[\hat{x}, \hat{k}] = i\hbar$ between coordinates x and k of the phase space. Thus, we see that the uncertainty (1) leads to consider also noncommutative geometry at the Planck scale. M-theory is the best candidate to describe physics at this scale. It contains strings and branes. By now, it seems that an employment of nonarchimedean geometry based on p-adic numbers and noncommutative geometry given by the commutation relation

$$[\hat{x}^i, \hat{x}^j] = i\hbar\theta^{ij} \tag{2}$$

is unavoidable in a further progress of the "theory of everything". In the sequel we will mainly consider some aspects of p-adic strings and their possible connection with noncommutative geometry. A notion of p-adic string was introduced in [2], where the hypothesis on the existence of nonarchimedean geometry at the Planck scale was made, and string theory with p-adic numbers was initiated. In particular, generalization of the usual Veneziano and Virasoro-Shapiro amplitudes with complex valued multiplicative characters over various number fields was proposed and p-adic valued Veneziano amplitude was constructed by means of p-adic interpolation. Very successful p-adic analogues of the Veneziano and Virasoro-Shapiro amplitudes were proposed in [3] as the corresponding Gel'fand-Graev [4] beta functions. Using this approach, Freund and Witten obtained [5] an attractive adelic formula, which states that the product of the crossing symmetric Veneziano (or Virasoro-Shapiro) amplitude and its all p-adic counterparts equals unit (or a definite constant). This gives possibility to consider an ordinary four-point function, which is rather complicate, as an infinite product of its inverse p-adic analogues, which have simple forms. These first papers induced an interest in various aspects of p-adic string theory (for a review, see [6, 7]). A recent interest in p-adic string theory has been mainly related to the generalized adelic formulas for four-point string amplitudes [8], the tachyon condensation [9], and the new promising adelic approach [10]. In addition to the expression (1), one can motivate the application of p-adic numbers in physics by the fact that the field of rational numbers $\mathbb{Q}$ is dense not only in $\mathbb{R}$ but also in the field of p-adic numbers $\mathbb{Q}_p$ (p denotes any prime number). Another motivation may be a conjecture that fundamental physical laws should be invariant under change $\mathbb{R} \longleftrightarrow \mathbb{Q}_p$ [11]. One of the very interesting and fruitful recent developments in string theory (for a reviev, see [12, 13]) has been noncommutative geometry and the corresponding noncommutative field theory. This subject started to be very actual after Connes, Douglas and Schwarz shown [14] that gauge theory on noncommutative torus describes compactifications of M-theory to tori with constant background three-form field. Noncommutative field theory (see, e.g. [15]) may be regarded as a deformation of the ordinary one in which field multiplication is replaced by the

Moyal (star) product

$$(f \star g)(x) = \exp\left[\frac{i\hbar}{2}\theta^{ij}\frac{\partial}{\partial y^i}\frac{\partial}{\partial z^j}\right] f(y)g(z)|_{y=z=x}, \tag{3}$$

where $x^1, x^2, \cdots, x^d$ denote coordinates of noncommutative space, and $\theta^{ij} = -\theta^{ji}$ are noncommutativity parameters. There are many properties of D-brane dynamics which may be studied by noncommutative field theory. In particular, it enables to investigate a mixing of the UV and IR effects, and the tachyon condensation. Replacing the ordinary product between coordinates by the Moyal product (3) we have

$$x^i \star x^j - x^j \star x^i = i\hbar\theta^{ij}, \tag{4}$$

which resembles the usual Heisenberg algebra. In the next Section we provide reader with some very basic facts on *p*-adic analysis. Section 3 is devoted to the *p*-adic string amplitudes. After that we consider an effective field theory of bosonic *p*-adic strings and its connection with noncommutative scalar solitons. At the end we discuss the obtained results and possible prospects.

2. *p*-Adic numbers and their functions

When we wish to introduce *p*-adic numbers it is instructive to start from $\mathbb{Q}$, since $\mathbb{Q}$ is the simplest field of numbers of characteristic 0 and it contains results of all physical measurements. Any non-zero rational number can be presented as infinite expansions into the two quite different forms. The usual one is to the base 10, i.e.

$$\sum_{k=n}^{-\infty} a_k 10^k, \quad a_k = 0, \cdots, 9, \tag{5}$$

and the other one is to the base p (p is a prime number) and reads

$$\sum_{k=m}^{+\infty} b_k p^k, \quad b_k = 0, \cdots, p-1, \tag{6}$$

where n and m are some integers. These representations have the usual repetition of digits, but, in a sense, expansions are in the mutually opposite directions. The series (5) and (6) are convergent with respect to the usual absolute value $|\cdot|_\infty$ and *p*-adic absolute value $|\cdot|_p$, respectively. Allowing arbitrary combinations for digits, we obtain standard representation of real numbers (5) and *p*-adic numbers (6). $\mathbb{R}$ and $\mathbb{Q}_p$ exhaust all number fields which contain $\mathbb{Q}$ as a dense subfield. They have many distinct geometric and algebraic properties. Geometry of *p*-adic numbers is the nonarchimedean one. For much more on *p*-adic numbers and *p*-adic analysis one can see, e.g. [4, 7, 16]. There are mainly two kinds of analysis on $\mathbb{Q}_p$ based on two different mappings: $\mathbb{Q}_p \to \mathbb{Q}_p$ and $\mathbb{Q}_p \to \mathbb{C}$. We use both of them, in classical and quantum p-adic models, respectively. Elementary p-adic functions are given by the same series as in the real case, but their regions of convergence are usually different. For instance, $\exp x = \sum_{n=0}^{\infty} \frac{x^n}{n!}$ and $\ln x = \sum_{n=1}^{\infty} (-1)^{n+1} \frac{(x-1)^n}{n}$ converge if $|x|_p < |2|_p$ and $|x - 1|_p < 1$, respectively. Derivatives of p-adic valued functions are also defined as in the real case, but using p-adic norm instead of the absolute value. As a definite p-adic valued integral we take difference of the corresponding antiderivative in end points. Usual complex-valued p-adic functions are: *(i)* an additive character $\chi_p(x) = \exp 2\pi i \{x\}_p$, where $\{x\}_p$ is the fractional part of $x \in \mathbb{Q}_p$, *(ii)* a multiplicative character $\pi_s(x) = |x|_p^s$, where $s \in \mathbb{C}$, and *(iii)* locally constant functions with compact support, like, e.g. $\Omega(|x|_p) = 1$ if $|x|_p \leq 1$ and $\Omega(|x|_p) = 0$ otherwise. There is well defined Haar measure and integration. For example,

$$\int_{\mathbb{Q}_p} \chi_p(\alpha x^2 + \beta x)dx = \lambda_p(\alpha)|2\alpha|_p^{-\frac{1}{2}} \chi_p\left(-\frac{\beta^2}{4\alpha}\right), \quad \alpha \neq 0, \qquad (7)$$

where $\lambda_p(\alpha)$ is an arithmetic function [7]. An adele x [4] is an infinite sequence

$$x = (x_\infty, x_2, \cdots, x_p, \cdots),$$

where $x_\infty \in \mathbb{R}$ and $x_p \in \mathbb{Q}_p$ with the restriction that for all but a finite set S of primes p we have $x_p \in \mathbb{Z}_p$. Componentwise addition and multiplication can be applied to adeles. It is useful to present the ring of adeles $\mathcal{A}$ in the following form:

$$\mathcal{A} = \cup_S \mathcal{A}(S), \quad \mathcal{A}(S) = \mathbb{R} \times \prod_{p \in S} \mathbb{Q}_p \times \prod_{p \notin S} \mathbb{Z}_p,$$

where $\mathbb{Z}_p = \{x \in \mathbb{Q}_p : |x|_p \leq 1\}$ is the ring of p-adic integers. $\mathcal{A}$ is also locally compact topological space. There are two kinds of analysis over $\mathcal{A}$, which generalize the corresponding analysis over $\mathbb{R}$ and $\mathbb{Q}_p$.

3. p-Adic string amplitudes

Like in the ordinary string theory, the starting point in an investigation of p-adic strings is a construction of the corresponding scattering amplitudes. Recall that the ordinary crossing symmetric Veneziano amplitude can be presented in the following forms: $A_\infty(k_1, \cdots, k_4) \equiv$

$$A_\infty(a, b) = g^2 \int_{\mathbb{R}} |x|_\infty^{a-1} |1 - x|_\infty^{b-1} dx \qquad (8)$$

$$= g^2 \left[\frac{\Gamma(a)\Gamma(b)}{\Gamma(a+b)} + \frac{\Gamma(b)\Gamma(c)}{\Gamma(b+c)} + \frac{\Gamma(c)\Gamma(a)}{\Gamma(c+a)} \right] \qquad (9)$$

$$= g^2 \frac{\zeta(1-a)}{\zeta(a)} \frac{\zeta(1-b)}{\zeta(b)} \frac{\zeta(1-c)}{\zeta(c)} \tag{10}$$

$$= g^2 \int \mathcal{D}X \exp\left(-\frac{i}{2\pi} \int d^2\sigma \partial^\alpha X_\mu \partial_\alpha X^\mu\right) \prod_{j=1}^{4} \int d^2\sigma_j \exp\left(ik_\mu^{(j)} X^\mu\right), \tag{11}$$

where $\hbar = 1$, $T = 1/\pi$, and $a = -\alpha(s) = -1 - \frac{s}{2}$, $b = -\alpha(t)$, $c = -\alpha(u)$ with the condition $s + t + u = -8$, i.e. $a + b + c = 1$. To introduce the corresponding p-adic Veneziano amplitude there is a sense to consider p-adic analogs of all the above four expressions. p-Adic generalization of the first expression was proposed in [3] and it reads

$$A_p(a, b) = g_p^2 \int_{\mathbb{Q}_p} |x|_p^{a-1} |1 - x|_p^{b-1} dx, \tag{12}$$

where $|\cdot|_p$ denotes p-adic absolute value. In this case only string world-sheet parameter x is treated as p-adic variable, and all other quantities maintain their usual (real) valuation. An attractive adelic formula of the form

$$A_\infty(a, b) \prod_p A_p(a, b) = 1 \tag{13}$$

was found [5], where $A_\infty(a, b)$ denotes the usual Veneziano amplitude (8). A similar product formula holds also for the Virasoro-Shapiro amplitude. These infinite products are divergent, but they can be successfully regularized. Unfortunately, there is a problem to extend this formula to the higher-point functions. p-Adic analogs of (9) and (10) were also proposed in [2] and [17], respectively. In these cases, world-sheet, string momenta and amplitudes are manifestly p-adic. Since string amplitudes are p-adic valued functions, it is not so far enough clear their physical interpretation. Expression (11) is based on Feynman's functional integral method, which is generic for all quantum systems and has successful p-adic generalization [18]. Its p-adic counterpart, proposed in [10], has been elaborated [19] and deserves further study. Note that in this approach, p-adic string amplitude is complex valued, while not only the world-sheet parameters but also target space coordinates and string momenta are p-adic variables. Such p-adic generalization is a natural extension of the formalism of p-adic [20] and adelic [21] quantum mechanics to string theory. In the framework of this new approach we will present here some results concerning the p-adic Veneziano amplitude. Instead of the start with the very expression (11) we will take in the real case as a starting point the following formula

$$A_\infty(k_1, \cdots, k_4) = g_\infty^2 \prod_{j=1}^{4} \int dx_j \exp\left(\frac{2}{hT} \sum_{i<j} k_i k_j \ln |x_i - x_j|_\infty\right), \tag{14}$$

which can be derived from (11), and after some standard evaluation [22] one has

$$A_\infty(k_1, \cdots, k_4) = g_\infty^2 \int_{\mathbb{Q}_\infty} dx |x|_\infty^{\frac{2k_1 k_2}{hT}} |1 - x|_\infty^{\frac{2k_2 k_3}{hT}}. \tag{15}$$

In the construction of p-adic amplitude we take p-adic analogue of (14), which is

$$A_p(k_1, \cdots, k_4) = g_p^2 \int_{\mathbb{Q}_p} dx \, \chi_p \left(\frac{1}{hT} \sum_{i<j} k_i k_j \ln(x_i - x_j) \right). \tag{16}$$

Note that from (16) one cannot obtain (12) since logarithmic function ln is p-adic valued and additive character χ_p is complex valued function. Thus, we have here a new type of p-adic string amplitudes. When $\frac{k_i k_j}{hT} \in \mathbb{Q}_p \setminus \mathbb{Z}_p$ additive character will be different from 1 and we have non-trivial p-adic amplitude. The corresponding adelic string amplitude is

$$A(k^{(1)}, \cdots, k^{(4)})$$

$$= A_\infty(k_\infty^{(1)}, \cdots, k_\infty^{(4)}) \prod_{p \in S} A_p(k_p^{(1)}, \cdots, k_p^{(4)}) \prod_{p \notin S} A_p(k_p^{(1)}, \cdots, k_p^{(4)}), \tag{17}$$

where $k^{(i)}$ is an adele, i.e.

$$k^{(i)} = (k_\infty^{(i)}, k_2^{(i)}, \cdots, k_p^{(i)}, \cdots) \tag{18}$$

with the restriction that $k_p^{(i)} \in Z_p$ for all but a finite set S of primes p. The topological ring of adeles $\mathcal{A}$ provides a framework for simultaneous and unified consideration of real and p-adic numbers. Rational numbers are also embedded in the space of adeles. If $\frac{k_p^{(i)} k_p^{(j)}}{hT} \in \mathbb{Z}_p$ for all primes p then $A_p(k_p^{(1)}, \cdots, k_p^{(4)}) = g_p^2 \prod_{j=1}^4 \int dx_j$, since $\chi_p(a) = 1$ when $a \in Z_p$. In this case, p-adic effects contribute only to the effective coupling constant, and adelic amplitude is equal to the ordinary one. When $\frac{k_p^{(i)} k_p^{(j)}}{hT} \in \mathbb{Q}_p \setminus \mathbb{Z}_p$ then additive character may give non-trivial contributions to adelic amplitude, what also depends on adelic state of the world-sheet.

4. p-Adic solitonic branes and noncommutative scalar solitons

There is an effective tachyon field theory in terms of real numbers with an exact action which describes p-adic strings with amplitude (12). The corresponding Lagrangian [23, 24] in d-dimensional Minkowski space ($\hbar = 1$) is

$$\mathcal{L} = \frac{1}{g^2} \frac{p^2}{p-1} \left[-\frac{1}{2} \varphi p^{-\frac{1}{2} \Box} \varphi + \frac{1}{p+1} \varphi^{p+1} \right], \tag{19}$$

where $\square$ denotes the Laplacian, φ is the tachyon field and p is an arbitrary prime number. Note that this Lagrangian has been recently considered in the context of tachyon condensation and brane descent relations [9]. The above Lagrangian yields the equation of motion

$$p^{-\frac{1}{2}\square}\varphi = \varphi^p. \tag{20}$$

In addition to solutions $\varphi = 0$ and $\varphi = 1$ there is also solution of the form

$$\varphi(x) = p^{\frac{n}{2(p-1)}} \exp\left(-\frac{p-1}{2p \ln p} \sum_{i=1}^{n} x_i^2\right), \tag{21}$$

where $n \leq d - 1$. This configuration can be called the p-adic solitonic q-brane solution, where $q = d - n - 1$. In particular case, $n = 2$ and $p = 2$, one has

$$\varphi(x_1, x_2) = 2\exp\left(-\frac{1}{4\ln 2}(x_1^2 + x_2^2)\right). \tag{22}$$

On the other hand there is a noncommutative scalar soliton [25]

$$\phi(x_1, x_2) = 2\exp\left(-\frac{1}{\theta}(x_1^2 + x_2^2)\right) \tag{23}$$

which is the simplest nontrivial (trivial solutions are $\phi = 0$ and $\phi = 1$) solution of the equation

$$(\phi \star \phi)(x) = \phi(x), \tag{24}$$

where $\star$ denotes the Moyal product (3) with $\theta^{ij} = \theta\varepsilon^{ij}$. The solution (23) of the equation (24) extremises energy in noncommutative scalar field theory [25] with the potential

$$V(\phi) = \frac{1}{2}m^2\phi \star \phi - \frac{1}{3}\phi \star \phi \star \phi, \tag{25}$$

where $m = 1$ and the kinetic term is neglected in the limit $\theta \longrightarrow \infty$. It is evident that the above solitonic solutions (22) and (23) are equal if $\theta = 4\ln 2$. This noncommutative scalar field model can be extended to the more general case with

$$V(\phi) = \frac{1}{2}m^2\phi^2 - \frac{c_{k+1}}{k+1}\phi^{k+1}, \tag{26}$$

where fields are multiplied by the star product, and $\phi \equiv \phi(x^1, \cdots, x^n)$ with even n spatial directions. The corresponding equation

$$c_{k+1}\phi^k(x) = m^2\phi(x) \tag{27}$$

has the solution

$$\phi(x) = 2^{\frac{n}{2}} \left(\frac{m^2}{c_{k+1}} \right)^{\frac{1}{k}} \exp \left(-\frac{1}{\theta} \sum_{i=1}^{n} x_i^2 \right). \tag{28}$$

The solutions (21) and (28) may be identified taking the corresponding values for mass m and noncommutativity parameter θ. Thus, we see that there is an intriguing similarity between p-adic solitonic branes and noncommutative scalar solitons.

Discussion and concluding remarks

In the previous Section we considered two nonlocal scalar field theories. Their potentials involve infinitely many derivatives. The corresponding differential equations are of the infinite order, and they extremize the action and the energy, respectively. It seems that there is a sense to expect something noncommutative in the effective p-adic Lagrangian (19), as well as something p-adic (nonarchimedean) in noncommutative scalar field theory with potential (26). Moreover, some more explicit relations between string field theory and p-adic string theory could be found in the coming years (see also comments in [9]). We believe that there is an underlying principle, which connects the following three space properties: noncommutativity, nonarchimedean geometry and the uncertainty relation (1). Let us also mention that various aspects of possible connection between quantum groups, nonarchimedean geometry and p-adic strings are discussed in [26, 27]. On q-deformation of the Veneziano amplitude one can see [28] and references therein. It is worth noting that one can introduce [29] the Moyal product in p-adic quantum mechanics and it reads ($h = 1$)

$$(\hat{f} * \hat{g})(x) = \int_{\mathbb{Q}_p^d} \int_{\mathbb{Q}_p^d} dk dk' \, \chi_p(-(x^i k_i + x^j k'_j) + \frac{1}{2} k_i k'_j \theta^{ij}) \tilde{f}(k) \tilde{g}(k'), \tag{29}$$

where d denotes spatial dimensionality.

Acknowledgements. B.D. wishes to thank Profs. J. Wess and S. Duplij for their invitation to participate and give a talk at the ARW "Noncommutative Structures in Mathematics and Physics". The work on this paper was supported in part by RFFI grant 990100866.

References

1. L.J. Garay, Int. J. Mod. Phys. **A 10** (1995) 145.
2. I.V. Volovich, *p-Adic String*, Class. Quantum Grav. **4** (1987) L83.
3. P.G.O. Freund and M. Olson, Phys. Lett. **B199** (1987) 186.

4. I.M. Gel'fand, M.I. Graev and I.I. Pyatetski-Shapiro, *Representation Theory and Automorphic Functions*, Saunders, London, 1966.
5. P.G.O. Freund and E. Witten, *Adelic String Amplitudes*, Phys. Lett. **B 199** (1987) 191.
6. L. Brekke and P.G.O. Freund, *p-Adic numbers in physics*, Phys. Rep. **233** (1993) 1.
7. V.S. Vladimirov, I.V. Volovich and E.I. Zelenov, *p-Adic Analysis and Mathematical Physics*, World Scientific, Singapore, 1994.
8. V.S. Vladimirov, *Adelic Formulas for Gamma and Beta Functions of One-Class Quadratic Fields: Applications to 4-Particle Scattering string amplitudes*, in Proc. Steklov Math. Institute **228** (2000) 67, math-ph/0004017.
9. D. Ghoshal and A. Sen, *Tachyon Condensation and Brane Descent Relations in p-Adic String Theory*, hep-th/0003278.
10. B. Dragovich, *On Adelic Strings* , hep-th/0005200.
11. I.V. Volovich, *Number Theory as the Ultimate Physical Theory*, Preprint CERN-TH. 4781 (1987).
12. J.H. Schwarz, *Recent Progress in Superstring Theory*, hep-th/0007130.
13. A. Sen, *Recent Developments in Superstring Theory* , hep-lat/0011073.
14. A. Connes, M.R. Douglas and A. Schwarz, *Noncommutative Geometry and Matrix Theory: Compactification on Tori*, JHEP **9802** (1998) 003, hep-th/9711162.
15. I.Ya. Aref'eva and I.V. Volovich, *Noncommutative Gauge Fields on Poisson Manifolds*, hep-th/9907114.
16. W.H. Schikhof, *Ultrametric Calculus*, Cambridge U.P., Cambridge, 1984.
17. I.Ya. Aref'eva, B. Dragovich and I.V. Volovich, *On the adelic string amplitudes*, Phys. Lett. **209 B** (1988) 445.
18. G.S. Djordjević and B. Dragovich, *p-Adic Path Integral for Quadratic Actions*, Mod. Phys. Lett. **A 12** (1997) 1455.
19. B. Dragovich, P. Rodić and I.V. Volovich, *New Amplitudes for pAdic and Adelic Bosonic Strings*, in preparation.
20. V.S. Vladimirov and I.V. Volovich, *p-Adic Quantum Mechanics*, Commun. Math. Phys. **123** (1989) 659.
21. B. Dragovich, *Adelic Model of Harmonic Oscillator*, Theor. Math. Phys. **101** (1994) 1404; *Adelic Harmonic Oscillator*, Int. J. Mod. Phys. **A 10** (1995) 2349.
22. M.B. Green, J.H. Schwarz and E. Witten, *Superstring Theory, I*, Cambridge U.P., Cambridge, 1987.
23. L. Brekke, P.G.O. Freund, M. Olson and E. Witten, Nucl. Phys. **B 302** (1988) 365.
24. P.H. Frampton and Y. Okada, Phys. Rev. Lett. **60** (1988) 484.
25. R. Gopakumar, S. Minwalla and A. Strominger, *Noncommutative Solitons*, hep-th/0003160.
26. I.Ya. Aref'eva and I.V. Volovich, *Quantum group particles and non-archimedean geometry*, Phys. Lett. **B 268** (1991) 179.
27. P.G.O. Freund, *On the Quantum Group-p-Adics Connection*, Preprint EFI 90-90, December, 1990.
28. L.J. Romans, *Deforming the Veneziano Model ("q-strings")*, in the ICTP Series in Theoretical Physics - Volume 6: 1989 Summer School in High Energy Physics and Cosmology, pp. 278-287, World Scientific, Singapore.
29. G. Djordjević, B. Dragovich and Lj. Nešć, *Adelic Quantum Mechanics: Nonarchimedean and Noncommutative Aspects*, in the Proceedings of this Workshop.

ADELIC QUANTUM MECHANICS: NONARCHIMEDEAN AND NONCOMMUTATIVE ASPECTS

GORAN DJORDJEVIĆ[1,2] *,   BRANKO DRAGOVICH[3,4] and
LJUBIŠA NEŠIĆ[1]

[1] *Department of Physics, Faculty of Sciences, University of Niš, P.O. Box 91, 18001 Niš, Yugoslavia*

[2] *Sektion Physik, Universität München, Theresienstr. 37, D-80333 München, Germany*

[3] *Institute of Physics, P.O.Box 57, 11001 Belgrade, Yugoslavia*

[4] *Steklov Mathematical Institute, Gubkin St. 8, 117966, Moscow, Russia*

Abstract. We present a short review of adelic quantum mechanics pointing out its non-Archimedean and noncommutative aspects. In particular, p-adic path integral and adelic quantum cosmology are considered. Some similarities between p-adic analysis and q-analysis are noted. The p-adic Moyal product is introduced.

1. Introduction

There is now a common belief that the usual picture of spacetime as a smooth pseudo-Riemannian manifold should breakdown somehow at the Planck length $l_p \sim 10^{-33} cm$, due to the quantum gravity effects. We consider here two possibilities, which come from modern mathematics and mathematical physics: non-Archimedean geometry related to p-adic numbers, and noncommutative geometry with space coordinates given by noncommuting operators

$$[\hat{x}^i, \hat{x}^j] = i\hbar\theta^{ij} \qquad (1)$$

* gorandj@junis.ni.ac.yu

S. Duplij and J. Wess (eds.), Noncommutative Structures in Mathematics and Physics, 401–413.
© 2001 *Kluwer Academic Publishers. Printed in the Netherlands.*

or by q-deformation $x^i x^j = q x^j x^i$. Some noncommutativity of configuration space should not be a surprise in physics since quantum phase space with the canonical commutation relation (9) is the well-known example of noncommutative geometry. We will mostly review our recent results concerning adelic quantum mechanics. We illustrate some features of adelic quantum mechanics by its application in quantum cosmology. A few remarkable similarities between non-Archimedean and noncommutative structures are noted. The usual Moyal product is extended to p-adic and adelic quantum mechanics. Since 1987, there have been many interesting applications of p-adic numbers and non-Archimedean geometry in various parts of modern theoretical and mathematical physics (for a review, see [1–3]). However we restrict ourselves here to p-adic and adelic quantum mechanics as well as to some related topics. In particular, we review Feynman's p-adic path integral method. A fundamental role of integral approach to p-adic and adelic quantum mechanics (and adelic quantum cosmology) is emphasized. The obtained p-adic probability amplitude for one-dimensional systems with quadratic Lagrangians has the form as that one in ordinary quantum mechanics. It is well known that measurements give rational numbers $\mathbb{Q}$, whereas theoretical models traditionally use real $\mathbb{R}$ and complex $\mathbb{C}$ number fields. A completion of $\mathbb{Q}$ with respect to the p-adic norms gives the fields of p-adic numbers $\mathbb{Q}_p$ (p is a prime number) in the same way as completion with absolute value yields $\mathbb{R}$. The paper of Volovich [4] initiated a series of articles on p-adic string theory and many other branches of theoretical and mathematical physics. The metric introduced by p-adic norm is the non-Archimedean (ultrametric) one. Possible existence of such space around the Planck length is the main motivation to study p-adic quantum models. However, p-adic analysis also plays a role in some areas of "macroscopic physics" as, for example: spin glasses, quasicrystals and some other complex systems. In order to investigate possible p-adic quantum phenomena it is necessary to have the corresponding theoretical formalism. An important step in this direction is a formulation of p-adic quantum mechanics [5, 6]. Because of total disconnectedness of p-adic spaces and different valuations of variables and wave functions, the quantization is performed by the Weyl procedure. A unitary representation of the evolution operator $U_p(t)$ on the Hilbert space $\mathcal{L}_2(\mathbb{Q}_p)$ of complex-valued functions of a p-adic argument is an appropriate way to describe quantum dynamics of p-adic systems. Recently formulated adelic quantum mechanics [7] successfully unifies ordinary and all p-adic quantum mechanics. The appearance of space-time discreteness in adelic formalism (see, e.g. [8]) is an encouragement for the further investigations. This paper is organized as follows. We start with a short introduction to p-adic numbers, adeles and their functions. After that, p-adic and adelic quantum mechanics based on the Weyl quantization and Feynman's path integral are presented. In Section 4 we review our previuos results concerning one-dimensional p-adic propagator. In Section 5 we will see how adelic quantum mechanics can be useful in investigation of the very early universe, where in a natural way space-time discreteness emerges in minisuper-space models of adelic quantum cosmology. In the last Section we give some of interesting relations between non-Archimedean and noncommutative analysis. We also define and discuss the corresponding p-adic Moyal product.

2. p-Adic numbers and adeles

Any $x \in \mathbb{Q}_p$ can be presented in the form [9]

$$x = p^{\nu}(x_0 + x_1 p + x_2 p^2 + \cdots), \qquad \nu \in \mathbb{Z}, \tag{2}$$

where $x_i = 0, 1, \cdots, p - 1$ are digits. p-Adic norm of any term $x_i p^{\nu+i}$ in the canonical expansion (2) is $\mid x_i p^{\nu+i} \mid_p = p^{-(\nu+i)}$ and the strong triangle inequality holds, i.e. $\mid a + b \mid_p \leq \max\{\mid a \mid_p, \mid b \mid_p\}$. It follows that $\mid x \mid_p = p^{-\nu}$ if $x_0 \neq 0$. There is no natural ordering on $\mathbb{Q}_p$. However one can introduce a linear order on $\mathbb{Q}_p$ by the following definition: $x < y$ if $\mid x \mid_p < \mid y \mid_p$ or when $\mid x \mid_p = \mid y \mid_p$ there exists such index $m \geq 0$ that digits satisfy $x_0 = y_0, x_1 = y_1, \cdots, x_{m-1} = y_{m-1}, x_m < y_m$. Derivatives of p-adic valued functions $\varphi : \mathbb{Q}_p \to \mathbb{Q}_p$ are defined as in the real case, but with respect to the p-adic norm. There is no integral $\int \varphi(x)dx$ in a sense of the Lebesgue measure [2], but one can introduce $\int_a^b \varphi(x)dx = \Phi(b) - \Phi(a)$ as a functional of analytic functions $\varphi(x)$, where $\Phi(x)$ is an antiderivative of $\varphi(x)$. In the case of map $f : \mathbb{Q}_p \to \mathbb{C}$ there is well-defined Haar measure. We use here the Gauss integral

$$\int_{\mathbb{Q}_v} \chi_v(ax^2 + bx)dx = \lambda_v(a) \mid 2a \mid_v^{-\frac{1}{2}} \chi_v\left(-\frac{b^2}{4a}\right), \qquad a \neq 0, \tag{3}$$

where index v denotes real ($v = \infty$) and p-adic cases, i.e. $v = \infty, 2, 3, 5, \cdots$. χ_v is an additive character: $\chi_\infty(x) = \exp(-2\pi i x)$, $\chi_p(x) = \exp(2\pi i \{x\}_p)$, where $\{x\}_p$ is the fractional part of $x \in \mathbb{Q}_p$. $\lambda_v(a)$ is the complex-valued arithmetic function [2]. An adele [10] is an infinite sequence $a = (a_\infty, a_2, ..., a_p, ...)$, where $a_\infty \in \mathbb{R} \equiv \mathbb{Q}_\infty, a_p \in \mathbb{Q}_p$ with a restriction that $a_p \in \mathbb{Z}_p$ for all but a finite set S of primes p. The set of all adeles $\mathbb{A}$ may be regarded as a subset of direct topological product $\mathbb{Q}_\infty \times \prod_p \mathbb{Q}_p$ whose elements satisfy the above restriction, i.e.

$$\mathbb{A} = \cup_S \mathbb{A}(S), \qquad \mathbb{A}(S) = \mathbb{R} \times \prod_{p \in S} \mathbb{Q}_p \times \prod_{p \notin S} \mathbb{Z}_p. \tag{4}$$

$\mathbb{A}$ is a topological space, and can be considered as a ring with respect to the componentwise addition and multiplication. An elementary function on adelic ring $\mathbb{A}$ is

$$\varphi(x) = \varphi_\infty(x_\infty) \prod_p \varphi_p(x_p) = \prod_v \varphi_v(x_v) \tag{5}$$

with the main restriction that $\varphi(x)$ must satisfy $\varphi_p(x_p) = \Omega(|x_p|_p)$ for all but a finite number of p, where

$$\Omega(|\,x\,|_p) = \begin{cases} 1, & 0 \le |\,x\,|_p \le 1, \\ 0, & |\,x\,|_p > 1, \end{cases} \tag{6}$$

is a characteristic function on the set of p-adic integers $\mathbb{Z}_p = \{x \in \mathbb{Q}_p : |x|_p \le 1\}$. It should be noted that the Fourier transform of the characteristic function (vacuum state) $\Omega(|x_p|)$ is $\Omega(|k_p|)$. All finite linear combinations of elementary functions (5) make the set $\mathcal{D}(\mathbb{A})$ of the Schwartz-Bruhat functions. The Fourier transform of $\varphi(x) \in \mathcal{D}(\mathbb{A})$ (that maps $\mathcal{D}(\mathbb{A})$ onto $\mathcal{D}(\mathbb{A})$) is

$$\tilde{\varphi}(y) = \int_{\mathbb{A}} \varphi(x)\chi(xy)dx = \int_{\mathbb{R}} \varphi_\infty(x)\chi_\infty(xy)dx \prod_p \int_{\mathbb{Q}_p} \varphi_p(x)\chi_p(xy)dx, \tag{7}$$

where $dx = dx_\infty dx_2 \ldots dx_p \ldots$ is the Haar measure on $\mathbb{A}$. The Hilbert space $L_2(\mathbb{A})$ is a space of complex-valued functions $\psi_1(x), \psi_2(x), \ldots$, with the scalar product and norm

$$(\psi_1, \psi_2) = \int_{\mathbb{A}} \bar{\psi}_1(x)\psi_2(x)dx, \quad ||\psi|| = (\psi, \psi)^{1/2} < \infty. \tag{8}$$

A basis of the above space may be given by the orthonormal eigenfunctions of an evolution operator [7].

3. Adelic quantum mechanics

In foundations of standard quantum mechanics (over $\mathbb{R}$) one usually starts with a representation of the canonical commutation relation

$$[\hat{q}, \hat{k}] = i\hbar, \tag{9}$$

where q is a coordinate and k is the corresponding momentum. It is well known that the procedure of quantization is not unique. In formulation of p-adic quantum mechanics [5, 6] the multiplication $\hat{q}\psi \to x\psi$ has no meaning for $x \in \mathbb{Q}_p$ and $\psi(x) \in \mathbb{C}$. Also, there is no possibility to define p-adic "momentum" or "Hamiltonian" operator. In the real case they are infinitesimal generators of space and time translations, but, since $\mathbb{Q}_p$ is disconnected field, these infinitesimal transformations become meaningless. However, finite transformations remain meaningful and the corresponding Weyl and evolution operators are p-adically well defined. For the one dimensional systems which classical evolution can be described by

$$z_t = T_t z, \quad z_t = \begin{pmatrix} q(t) \\ k(t) \end{pmatrix}, \quad z = \begin{pmatrix} q(0) \\ k(0) \end{pmatrix}, \tag{10}$$

where $q(0)$ and $k(0)$, are initial position and momentum, respectively, and T_t is a matrix. Canonical commutation relation in p-adic case can be represented by the Weyl operators ($h = 1$)

$$\hat{Q}_p(\alpha)\psi_p(x) = \chi_p(\alpha x)\psi_p(x) \qquad (11)$$

$$\hat{K}_p(\beta)\psi(x) = \psi_p(x + \beta). \qquad (12)$$

Now, to the relation (9) in the real case, corresponds

$$\hat{Q}_p(\alpha)\hat{K}_p(\beta) = \chi_p(\alpha\beta)\hat{K}_p(\beta)\hat{Q}_p(\alpha) \qquad (13)$$

in the p-adic one. It is possible to introduce the family of unitary operators

$$\hat{W}_p(z) = \chi_p(-\frac{1}{2}qk)\hat{K}_p(\beta)\hat{Q}_p(\alpha), \quad z \in \mathbb{Q}_p \times \mathbb{Q}_p, \qquad (14)$$

that is a unitary representation of the Heisenberg-Weyl group. Recall that this group consists of the elements (z, α) with the group product

$$(z, \alpha) \cdot (z', \alpha') = (z + z', \alpha + \alpha' + \frac{1}{2}B(z, z')), \qquad (15)$$

where $B(z, z') = -kq' + qk'$ is a skew-symmetric bilinear form on the phase space. Dynamics of a p-adic quantum model is described by a unitary operator of evolution $U(t)$ without using the Hamiltonian. Instead of that, the evolution operator has been formulated in terms of its kernel $K_t(x, y)$

$$U_p(t)\psi(x) = \int_{\mathbb{Q}_p} K_t(x, y)\psi(y)dy. \qquad (16)$$

The next section will be devoted to the path integral formulation and calculation of the quantum propagator $K_t(x, y)$ on p-adic spaces. In this way [5] p-adic quantum mechanics is given by a triple

$$(L_2(\mathbb{Q}_p), W_p(z_p), U_p(t_p)). \qquad (17)$$

Keeping in mind that standard quantum mechanics can be also given as the corresponding triple, ordinary and p-adic quantum mechanics can be unified in the form of adelic quantum mechanics [7]

$$(L_2(A), W(z), U(t)). \qquad (18)$$

$L_2(A)$ is the Hilbert space on A, $W(z)$ is a unitary representation of the Heisenberg-Weyl group on $L_2(A)$ and $U(t)$ is a unitary representation of the evolution operator on $L_2(A)$. The evolution operator $U(t)$ is defined by

$$U(t)\psi(x) = \int_A K_t(x, y)\psi(y)dy = \prod_v \int_{\mathbb{Q}_v} K_t^{(v)}(x_v, y_v)\psi^{(v)}(y_v)dy_v. \qquad (19)$$

The eigenvalue problem for $U(t)$ reads

$$U(t)\psi_{\alpha\beta}(x) = \chi(E_\alpha t)\psi_{\alpha\beta}(x), \tag{20}$$

where $\psi_{\alpha\beta}$ are adelic eigenfunctions, $E_\alpha = (E_\infty, E_2, ..., E_p, ...)$ is corresponding energy, indices α and β denote energy levels and their degeneration. Note that any adelic eigenfunction has the form

$$\Psi(x) = \Psi_\infty(x_\infty) \prod_{p \in S} \Psi_p(x_p) \prod_{p \notin S} \Omega(|x_p|_p), \quad x \in \mathbb{A}, \tag{21}$$

where $\Psi_\infty \in L_2(\mathbb{R})$, $\Psi_p \in L_2(\mathbb{Q}_p)$. Adelic quantum mechanics takes into account also p-adic quantum effects and may be regarded as a starting point for construction of a more complete superstring and M-theory. In the low-energy limit adelic quantum mechanics becomes ordinary one.

4. p-Adic path integrals

A suitable way to calculate propagator in p-adic quantum mechanics is by p-adic generalization of Feynman's path integral. For the classical action $\bar{S}(x'', t''; x', t')$ which is a polynomial quadratic in x'' and x' it is well known that in ordinary quantum mechanics the Feynman path integral is

$$\mathcal{K}(x'', t''; x', t') = \left(\frac{i}{h}\frac{\partial^2 \bar{S}}{\partial x'' \partial x'}\right)^{1/2} \exp\left(\frac{2\pi i}{h}\bar{S}(x'', t''; x', t')\right). \tag{22}$$

p-Adic generalization of the Feynman path integral was suggested in [5] and can be written on a p-adic line as

$$K_p(x'', t''; x', t') = \int \chi_p\left(-\frac{S[q]}{h}\right)\mathcal{D}q = \int \chi_p\left(-\frac{1}{h}\int_{t'}^{t''} L(q, \dot{q}, t)dt\right)\prod_t dq(t). \tag{23}$$

In (23) we take $h \in \mathbb{Q}$ and $q, t \in \mathbb{Q}_p$. This path integral is elaborated, for the first time, for the harmonic oscillator [11]. It was shown that there exists the limit

$$K_p(x'', t''; x', t') = \lim_{n \to \infty} K_p^{(n)}(x'', t''; x', t') = \lim_{n \to \infty} N_p^{(n)}(t'', t')$$

$$\times \int_{\mathbb{Q}_p} \cdots \int_{\mathbb{Q}_p} \chi_p\left(-\frac{1}{h}\sum_{i=1}^n \bar{S}(q_i, t_i; q_{i-1}, t_{i-1})\right)dq_1 \cdots dq_{n-1}, \tag{24}$$

where $N_p^{(n)}(t'', t')$ is the corresponding normalization factor for the harmonic oscillator. The subdivision of p-adic time segment $t_0 < t_1 < \cdots < t_{n-1} < t_n$ is made according to linear order on $\mathbb{Q}_p$ and $|t_i - t_{i-1}|_v \to 0$ for every

$i = 1, 2, \cdots, n$, when $n \to \infty$. In the similar way we have calculated path integrals for: a particle in a constant external field [12], some minisuperspace cosmological models and a relativistic free particle [8], as well as for a harmonic oscillator with a time-dependent frequency [12]. p-Adic classical mechanics has the same analytic form as in the real case. If $q(t) = \bar{q}(t) + y(t)$ denotes a possible quantum path, with conditions $y(t') = y(t'') = 0$, where $\bar{q}(t)$ is a p-adic classical path with $\delta S[\bar{q}] = 0$, we have the following action for quadratic Lagrangians:

$$S[q] = S[\bar{q}] + \frac{1}{2!}\delta^2 S[\bar{q}] = S[\bar{q}] + \frac{1}{2}\int_{t'}^{t''}\left(y\frac{\partial}{\partial q} + \dot{y}\frac{\partial}{\partial \dot{q}}\right)^{(2)} L(q, \dot{q}, t)dt. \quad (25)$$

Putting (25) into (23), and using condition

$$\int_{\mathbb{Q}_p} K_p^*(x'', t''; x', t')K_p(z, t''; x', t')dx' = \delta_p(x'' - z), \quad (26)$$

with quadratic expansion of action as well as the general form of the normalization factor

$$N_p(t'', t') = \mid N_p(t'', t')\mid_\infty A_p(t'', t'),$$

we obtain general expression for the propagator (for some details, see [13])

$$K_p(x'', t''; x', t') = \lambda_p\left(-\frac{1}{2h}\frac{\partial^2 \bar{S}}{\partial x''\partial x'}\right)\left|\frac{1}{h}\frac{\partial^2 \bar{S}}{\partial x''\partial x'}\right|_p^{\frac{1}{2}}\chi_p\left(-\frac{1}{h}\bar{S}(x'', t''; x', t')\right). \quad (27)$$

This result exhibits some very important properties. For instance, replacing an index p with v in (27) we can write quantum-mechanical amplitude K in ordinary and all p-adic cases in the same compact form. It points out a generic behaviour of quantum propagation in Archimedean and non-Archimedean spaces and emphasizes the fundamental role of the Feynman path integral method in quantum theory. Also, considering the most general quadratic p-adic Lagrangian $L(x, \dot{x}, t) = a(t)\dot{x}^2 + 2b(t)\dot{x}x + c(t)x^2 + 2d(t)\dot{x} + 2e(t)x + f(t)$ with analytic coefficients, we found a connection [14] between these coefficients and the simplest p-adic quantum state $\Omega(|x|_p)$, that is necessary for existence of adelic quantum dynamics. For space-time discreteness in adelic models, see [8]. It is worth mentioning that this approach can be extended to systems with the two, three and more dimensions, and results will be presented elsewhere. The above results are also a starting point for a further elaboration of adelic quantum mechanics and for a semiclassical computation of the p-adic path integrals with non-quadratic Lagrangians.

5. Adelic quantum cosmology

Adelic quantum cosmology [15] is an application of adelic quantum mechanics to the universe as a whole. It unifies ordinary and p-adic quantum cosmology. Here, path integral formalism occurs to be quite appropriate tool to take integration over both Archimedean and non-Archimedean geometries on the equal footing. In this approach we introduce υ-adic complex-valued cosmological amplitudes by a functional integral

$$\langle h''_{ij}, \phi'', \Sigma'' | h'_{ij}, \phi', \Sigma' \rangle_{\upsilon} = \int \mathcal{D}(g_{\mu\nu})_{\upsilon} \mathcal{D}(\Phi)_{\upsilon} \chi_{\upsilon}(-S_{\upsilon}[g_{\mu\nu}, \Phi]). \qquad (28)$$

In practice, it is not possible to deal with full superspace (the space of all 3-metrics and matter field configurations). Instead, one exploits minisuperspace (a finite number of coordinates (h_{ij}, ϕ)). After this simplification, υ-factors of adelic minisuperspace propagator are given by the relation

$$\langle q^{\alpha''} | q^{\alpha'} \rangle_{\upsilon} = \int dN K_{\upsilon}(q^{\alpha''}, N | q^{\alpha'}, 0), \qquad (29)$$

where K_{υ} is an ordinary quantum-mechanical propagator with fixed minisuperspace coordinates q^{α} and the lapse function N. We illustrate adelic quantum cosmology by Bianchi I model ($k = 0$). Using Lorentz metric [16]

$$ds^2 = \sigma^2 \left[-\frac{N^2(t)}{a^2(t)} dt^2 + a^2(t) dx^2 + b^2(t) dy^2 + c^2(t) dz^2 \right] \qquad (30)$$

and replacements:

$$x = \frac{bc + a^2}{2}, \quad y = \frac{bc - a^2}{2}, \quad \dot{z}^2 = a^2 \dot{b} \dot{c}, \qquad (31)$$

we obtain the corresponding action

$$S_p[x, y, z] = \frac{1}{2} \int_0^1 dt \left[-\frac{1}{N} \left(\frac{\dot{x}^2 - \dot{y}^2}{2} + \dot{z}^2 \right) - \lambda N(x + y) \right], \qquad (32)$$

and equations of motion

$$\ddot{x} + \lambda N^2 = 0, \quad \ddot{y} - \lambda N^2 = 0, \quad \ddot{z} = 0. \qquad (33)$$

Taking into account conditions $x(0) = x'$, $y(0) = y'$, $z(0) = z'$, $x(1) = x''$, $y(1) = y''$, $z(1) = z''$, the quantum transition amplitude can be written as

$$K_p(x'', y'', z'', N | x', y', z', 0) = \frac{\lambda_p(-2N)}{\left| 4^{\frac{1}{3}} N \right|_p^{\frac{3}{2}}} \chi_p \left(-\bar{S}(x'', y'', z'', N | x', y', z', 0) \right).$$

$$(34)$$

Conditions for the existence of the vacuum state $\Omega(|x|_p)\Omega(|y|_p)\Omega(|z|_p)$ can be calculated from the equality

$$\int_{|x'|_p\leq 1}\int_{|y'|_p\leq 1}\int_{|z'|_p\leq 1}\mathcal{K}_p(x'',y'',z'',N|x',y',z',0)dx'\,dy'\,dz'$$

$$=\Omega(|x''|_p)\Omega(|y''|_p)\Omega(|z''|_p),$$

and the simplest vacuum state is

$$\Psi_p(x,y,z,N)=\begin{cases}\Omega(|x|_p)\Omega(|y|_p)\Omega(|z|_p), & |N|_p\leq 1,\ |\lambda|_p\leq 1,\ p\neq 2,\\ \Omega(|x|_2)\Omega(|y|_2)\Omega(|z|_2), & |N|_2\leq\frac{1}{2},\ |\lambda|_2\leq 2,\ p=2.\end{cases}$$
$$(35)$$

According to (21) adelic wave function $\Psi(x,t)$ offers more information on a physical system than only its standard part $\Psi_\infty(x,t)$. In quantum-mechanical experiments, as well as in all measurements, numerical results belong to the field of rational numbers $\mathbb{Q}$. For the Bianchi I model, as well as for any adelic quantum model, according to the usual interpretation of the wave function we have to consider $|\Psi(x,t)|_\infty^2$ at rational space-time points. In the above adelic case we get

$$|\Psi(x,y,z,N)|_\infty^2=|\Psi_\infty(x,y,z,N)|_\infty^2\prod_p\Omega(|x|_p)\Omega(|y|_p\Omega(|z|_p)$$

$$=\begin{cases}|\Psi_\infty(x,y,z,N)|_\infty^2\,, & x,y,z\in\mathbb{Z}\,,\\ 0\,, & x,y,z\in\mathbb{Q}\setminus\mathbb{Z}.\end{cases}\quad(36)$$

Here we used the following properties of the Ω-function: $\Omega^2(|x|_p)=\Omega(|x|_p)$, $\prod_p\Omega(|x|_p)=1$ if $x\in\mathbb{Z}$, and $\prod_p\Omega(|x|_p)=0$ if $x\in\mathbb{Q}\setminus\mathbb{Z}$. Thus, it means that positions x,y,z may have only discrete values: $x=0,\pm 1,\pm 2,\dots$. Since the Ω-function is invariant under the Fourier transformation, there is also discrete momentum space. When system is in some excited state, the sharpness of the discrete structure disappears and space demonstrates usual continuous properties. It is worth mentioning that a space-time discreteness is also noted in the framework of q-deformed quantum mechanics [17].

6. p-Adic analysis and q-analysis. The Moyal product

Some connections between p-adic analysis and quantum deformations has been noticed [18] in a variety of cases during the last ten years or so. It was shown [19] that the two parameter Sklyanin quantum algebra and its generalizations provide a promising connection between the p-adics and quantum deformation. A similar connection has been indicated by Macdonald's paper [20] on orthogonal polynomials associated with the root systems. In [19] it was also pointed out that elliptic quantum group and its generalizations unify the p-adic and real versions of a Lie group (e.g. $SL(2)$). This result is connected with adelic approach and the possibility of establishing q-deformed Euler products. In some other contexts it has been observed that the Haar measure on $SU_q(2)$ coincides with the Haar measure on the field of p-adic numbers $\mathbb{Q}_p$ if $q = \frac{1}{p}$ [21]. Namely, Tomea-Jackson integral in q-analysis

$$\int_0^1 f(x)d_q x = (1 - q) \sum_{n=0}^{\infty} f(q^n)q^n, \tag{37}$$

and the integral in p-adic analysis

$$\int_{|x|_p \leq 1} f(|x|_p)dx = (1 - \frac{1}{p}) \sum_{n=0}^{\infty} f(p^{-n})p^{-n}, \tag{38}$$

are equal if $q = \frac{1}{p}$, *i.e.*

$$\int_0^1 f(x)d_{1/p}x = \int_{|x|_p \leq 1} f(|x|_p)dx. \tag{39}$$

In q-analysis there is the following differential operator (related to the q-deformed momentum in the coordinate representation [21])

$$\partial_q f(x) = \frac{f(x) - f(qx)}{(1 - q)x}. \tag{40}$$

In p-adic analysis, when one considers a complex-valued function $f(x)$ depending on a p-adic variable x we are not able to use standard definition of differentiation. Instead of that it is possible to use Vladimirov's operator

$$D^\alpha \psi(x) = \frac{p - 1}{1 - p^{-1-\alpha}} \int \frac{f(x) - f(y)}{|x - y|_p^{\alpha+1}} dy \tag{41}$$

which in a sense resembles (40). Moreover, there is a potential such that the spectrum of the p-adic Schrödinger- like (diffusion) equation [22]

$$D\psi(x) + V(|x|_p)\psi(x) = E\psi(x) \tag{42}$$

is the same one as in the case of q-deformed oscillator found by Biedenharn [23] and Macfarlane [24] for $q = 1/p$. For more details, see [21]. Recently [25], it has been proposed a new pseudodifferential operator with rational part of p-adic numbers $\{x\}_p$. In such case, energy levels for p-adic free particle exhibit discrete dependence on the corresponding momentum: $\{E\}_p = \{k\}_p^2$. Note also a proposal for q-deformation of Vladimirov's operator [26]. We see that there are some interesting relations between p-adic and q-analysis, and in a sense between adelic quantum mechanics and noncommutative one. It would be fruitful to find some deeper reasons for these connections, between theories which pretend to give us more insights on the space-time structure at the Planck scale. By now it is not enough understood. It seems to be reasonable to formulate a noncommutative adelic quantum mechanics that may connect non-Archimedean and noncommutative effects and structures. As the first step in this direction one has to consider a p-adic and adelic generalization of the Moyal product. Let us consider D-dimensional classical space with coordinates $x^1, x^2, \cdots, x^D$. Let $f(x)$ be a classical function $f(x) = f(x^1, x^2, \cdots, x^D)$. Then, with the respect to the Fourier transformations, we have

$$\tilde{f}(k) = \int_{\mathbb{Q}_v^D} dx \, \chi_v(kx) f(x),$$ (43)

$$f(x) = \int_{\mathbb{Q}_v^D} dk \, \chi_v(-kx) \tilde{f}(k).$$ (44)

According to the usual Weyl quantization

$$\hat{f}(x) = \int_{\mathbb{Q}_\infty^D} dk \, \chi_\infty(-k\hat{x}) \tilde{f}(k) \equiv f(\hat{x}).$$ (45)

Let us now have two classical functions $f(x)$ and $g(x)$ with

$$\hat{f}(x) = \int_{\mathbb{Q}_\infty^D} dk \, \chi_\infty(-k\hat{x}) \tilde{f}(k),$$ (46)

$$\hat{g}(x) = \int_{\mathbb{Q}_\infty^D} dk \, \chi_\infty(-k\hat{x}) \tilde{g}(k).$$ (47)

In the coordinate representation we can write the same above expressions replacing $\hat{x}$ by x and extend it to all p-adic cases. Now we are interested in product $\hat{f}(x)\hat{g}(x)$. In the real case this operator product is of the form

$$(\hat{f} \cdot \hat{g})(x) = \int \int dk dk' \, \chi_\infty(-k\hat{x}) \chi_\infty(-k'\hat{x}) \tilde{f}(k) \tilde{g}(k').$$ (48)

Using the Baker-Campbell-Hausdorff formula, the relation (1) and then the coordinate representation one finds the Moyal product in the form

$$(f * g)(x) = \int \int dk dk' \, \chi_v \left(-(k + k')x + \frac{1}{2} k_i k_j' \theta^{ij} \right) \tilde{f}(k) \tilde{g}(k'),$$ (49)

where we already used our generalization from $\mathbb{Q}_\infty$ to $\mathbb{Q}_v$. Note that in the real case we use $k_i \to -(i/2\pi)(\partial/\partial x^i)$ and obtain the well known form

$$(f * g)(x) = \chi_\infty \left(-\frac{\theta^{ij}}{2(2\pi)^2} \frac{\partial}{\partial y^i} \frac{\partial}{\partial z^j} \right) f(y) g(z)|_{y=z=x}.$$ (50)

Thus, as the p-adic Moyal product we take

$$(\hat{f} * \hat{g})(x) = \int_{\mathbb{Q}_p^D} \int_{\mathbb{Q}_p^D} dk dk' \, \chi_p(-(x^i k_i + x^j k_j') + \frac{1}{2} k_i k_j' \theta^{ij}) \tilde{f}(k) \tilde{g}(k'). \quad (51)$$

As the first step in adelization one can consider the Moyal product on $\mathbb{R} \times \prod_{p \in S} \mathbb{Q}_p$ $\times \prod_{p \notin S} \mathbb{Z}_p$ space. Various adelic aspects of the Moyal product will be presented elsewhere.

Acknowledgments. Authors G.Dj. and B.D. wish to thank the co-Directors of ARW "Noncommutative Structures in Mathematics and Physics" Profs. J. Wess and S. Duplij for their invitation to participate and give a talk. G.Dj. is partially supported by DFG Project "Noncommutative space-time structure - Cooperation with Balkan Countries". The work of B.D. was supported in part by RFFI grant 990100866.

References

1. L. Brekke and P.G.O. Freund, *p-Adic numbers in physics*, Phys. Rep. **233**, 1 (1993).
2. V.S. Vladimirov, I.V. Volovich and E.I. Zelenov, *p-Adic Analysis and Mathematical Physics*, (World Scientific, Singapore, 1994).
3. A. Khrennikov, *p-Adic Valued Distributions in Mathematical Physics* (Kluwer Acad. Publ., Dordrecht, 1994).
4. I.V. Volovich, *p-Adic String*, Class. Quantum Grav. **4**, L83 (1987).
5. V.S. Vladimirov and I.V. Volovich, *p-Adic Quantum Mechanics*, Comm. Math. Phys. **123**, 659 (1989).
6. Ph. Ruelle, E. Thiran, D. Verstegen and J. Weyers, *Quantum Mechanics on p-Adic Fields*, J. Math. Phys. **30**, 2854 (1989).
7. B. Dragovich, *Adelic Harmonic Oscillator*, Int. J. Mod. Phys. **A10**, 2349 (1995).
8. B. Dragovich, *Adelic Wave Function of the Universe*, in: Proc. of the Third A. Friedmann Int. Seminar on Grav. and Cosmology, Friedmann Lab. Publishing, St. Petersburg, 1995, pp. 311–321; G.S. Djordjević, B. Dragovich, Lj. Nešić, *p-Adic and Adelic Free Relativistic Particle*, Mod. Phys. Lett. **A 14**, 317 (1999).
9. W.H. Schikhof, *Ultrametric Calculus*, (Cambridge U.P., Cambridge, 1984).
10. I.M. Gel'fand, M.I. Graev and I.I. Pyatetskii-Shapiro, *Representation Theory and Automorphic Functions*, (Saunders, London, 1966).
11. E.I. Zelenov, *p-Adic path integrals*, J. Math. Phys.**32**, 147 (1991).
12. G.S. Djordjević and B. Dragovich, *On p-Adic Functional Integration*, in Proc. of the II Math. Conf. in Priština, Priština (Yugoslavia), 1996.
13. G.S. Djordjević and B. Dragovich, *p-Adic Path Integral for Quadratic Actions*, Mod. Phys. Lett. **A12**, 1455 (1997).
14. G.S. Djordjević and B. Dragovich, *Adelic connection between classical and quantum dynamics*, in Proc. of the XIII Conf. on Appl. Mathematics, Igalo'98 (Yugoslavia) 23 (2000).
15. B. Dragovich and Lj. Nešić, *p-Adic and Adelic Generalization of Quantum Cosmology*, Grav. Cosm. **5**, 222 (1999).
16. A. Ishikawa and H. Ueda, *The Wave Function of the Universe by the New Euclidean Path-integral Approach in Quantum Cosmology*, Int. J. Mod. Phys. **D2**, 249 (1993).
17. J. Wess, *q-Deformed Heisenberg Algebras*, mat-ph/9910013.

18. P.G.O. Freund, *On the Quantum Group - p-Adics Connection*, in Quarks, Symmetries and Strings, (M. Kaku, A. Jevicki and K. Kikkawa, eds,), World Scientific, Singapore, 1991, pp. 267-275.

19. P.G.O. Freund and A.V. Zabrodin, *Macdonald Polynomials from Sklyanin Algebras: A Conceptual Basis for the p-Adics-Quantum Group Connection*, Commun. Math. Phys. **147**, 277 (1992).

20. I.G. Macdonald, in Orthogonal Polynomials: Theory and Practice, (P. Nevai ed.), Kluwer Acad. Publ., Dordrecht, 1990, p.311.

21. I.Ya. Aref'eva and I. V. Volovich, *Quantum Group Particles and Non-Archimedean Geometry*, Phys. Lett. **B268**, 179 (1991).

22. V. S. Vladimirov and I. V. Volovich, *p-Adic Schrödinger Type Equation*, Lett. Math. Phys. **18**, 43 (1989).

23. L.C. Biedenharn, *The quantum group $SU_q(2)$ and a q-analogue of the boson operators*, J. Phys. A: Math. Gen. **22**, L873 (1989).

24. A.J. Macfarlane, *On q-analogues of the quantum harmonic oscillator and the quantum group $SU(2)_q$*, J. Phys. A: Math. Gen. **22**, 4581 (1989).

25. D. Dimitrijević, G.S. Djordjević and B. Dragovich *On Schrödinger-type equation on p-adic spaces*, to be published in Bal. Phys. Lett.

26. S.V. Kozyrev, *The Elements of noncommutative analysis on $\mathbb{R}$ and $\mathbb{Q}_p$*, Ph.D. Thesis, Moscow, 1995 (in Russian).

GIBBS STATES OF A LATTICE SYSTEM OF QUANTUM ANHARMONIC OSCILLATORS

YURI KOZITSKY * †

Institute of Mathematics, Marie Curie-Sklodowska University, Lublin 20-031, Poland

1. Introduction

Gibbs states of interacting quantum lattice systems are constructed as positive functionals on von Neumann algebras whose elements (observables) represent physical quantities [8], [13]. For the systems, the algebra of observables of every subsystem in a finite subset of the lattice may be represented as the C^*-algebra of bounded operators on a Hilbert space, the theory of Gibbs states is quite well elaborated [8]. But if one needs to include into consideration also unbounded operators, the situation becomes much more complicated. In 1975 an approach to the construction of Gibbs states, which uses the integration theory in path spaces, has been initiated [1] (see also [5], [6], [7], [11], [13], [15]). Here the state at a temperature $T = \beta^{-1}$ is defined by means of a probability measure μ_β on a certain infinite-dimensional space, analogously to the Euclidean quantum field theory. That is the reason why μ_β is known as *the Euclidean Gibbs state*.

In this paper we consider the following model. To each point of the lattice $\mathbb{L} = \mathbb{Z}^d$, $d \in \mathbb{N}$ there is attached a quantum particle (oscillator) with the reduced mass $m = m_{ph}/\hbar^2$ (m_{ph} is the physical mass), which has an unstable equilibrium position at this point. Such particles perform D-dimensional oscillations around their equilibrium positions and interact via attractive potential. Similar objects have been studied for many years as quite realistic models of crystalline substance undergoing structural phase transitions (see e.g. [16]).

In Section 2, following [2], [3], [4], we summarize main aspects of the construction of the Euclidean Gibbs state for the model considered. In Section 3, we

* jkozi@golem.umcs.lublin.pl

† Supported in part by the Polish Scientific Research Committee under the Grant 2 P03A 02915

S. Duplij and J. Wess (eds.), Noncommutative Structures in Mathematics and Physics, 415–425.
© 2001 *Kluwer Academic Publishers. Printed in the Netherlands.*

provide a number of assertions describing such states. In particular, we show that strong zero-point oscillations suppress critical point anomalies. The latter result is a strengthening of similar ones given in [2], [14].

2. Euclidean Formalism for Quantum Gibbs States

The oscillations of the particle having its equilibrium position at $l \in \mathbb{L}$ are described by the momentum and displacement operators $\{p_l, q_l\}$, densely defined on the complex Hilbert space $\mathcal{H}_l = L^2(\mathbb{R}^D)$. The whole system is described by the formal Hamiltonian

$$H = \frac{1}{2} \sum_{l,l'} d_{ll'}(q_l, q_{l'}) + \sum_l H_l, \tag{1}$$

$$H_l = \frac{1}{2m}(p_l, p_l) + \frac{1}{2}(q_l, q_l) + V(q_l), \tag{2}$$

where $(\,.\,,\,.\,)$ stands for scalar product in $\mathbb{R}^D$ and $d_{ll'}$ form a dynamical matrix. The potential V is chosen as follows

$$V(x) = v((x, x)), \tag{3}$$

where v is a polynomial, convex on $\mathbb{R}_+ \stackrel{\text{def}}{=} [0, +\infty)$. Some of our results were obtained under assumption that

$$v(\xi) = \frac{1}{2}a\xi + \sum_{s=2}^r b_s \xi^s, \quad r \geq 2, \ a \in \mathbb{R}, \ b_s \geq 0, \ b_r > 0. \tag{4}$$

For $p \in \mathbb{Z}$, let

$$\mathcal{S}_p = \left\{ \{x_l, l \in \mathbb{L}\} \ \Big| \ \sum_l (1 + |l|)^{2p} x_l^2 < \infty \right\}, \tag{5}$$

where $|l|$ is the Euclidean norm on $\mathbb{L} = \mathbb{Z}^d \subset \mathbb{R}^d$. Let also

$$\mathcal{S} \stackrel{\text{def}}{=} \bigcap \mathcal{S}_p, \quad \mathcal{S}' \stackrel{\text{def}}{=} \bigcup \mathcal{S}_{-p}, \quad p \in \mathbb{N}_0 \stackrel{\text{def}}{=} \mathbb{N} \cup \{0\}. \tag{6}$$

The dynamical matrix is supposed to be invariant under translations on $\mathbb{L}$, and attractive ($d_{ll'} \leq 0$). We also suppose that for every $l \in \mathbb{L}$, the sequence $\{d_{ll'}, l' \in \mathbb{L}\}$ belongs to $\mathcal{S}$. Set

$$\Lambda = \{l = (l_1, \ldots, l_d) \mid l_j^0 \leq l_j \leq l_j^1, \ l_j^0 < l_j^1, \ l_j^0, l_j^1 \in \mathbb{Z}, \ j = 1, \ldots, d\}.$$

Given a box Λ, let $\mathcal{L}(\Lambda)$ denote the partition of $\mathbb{L}$ by the boxes which are obtained as translations of Λ. Let also $\mathfrak{G}$ be the group of all translations of $\mathbb{L}$, and $\mathfrak{G}(\Lambda) =$

$\{t \in \mathfrak{G} \mid t(\Lambda) \in \mathcal{L}(\Lambda)\}$, where $t(\Lambda) = \{t(l), \ l \in \Lambda\}$. Then the dynamical matrix $(d_{ll'}^{\Lambda})_{l,l' \in \Lambda}$ obeying periodic conditions on the boundaries of Λ and the periodic local Hamiltonian H_Λ are

$$d_{ll'}^{\Lambda} = \min\{d_{lt(l')} : t \in \mathfrak{G}(\Lambda)\}, \tag{7}$$

$$H_\Lambda = \frac{1}{2} \sum_{l,l' \in \Lambda} d_{ll'}^{\Lambda}(q_l, q_{l'}) + \sum_{l \in \Lambda} H_l. \tag{8}$$

The latter is an essentially self-adjoint lower bounded operator acting in $\mathcal{H}_\Lambda = L^2\left(\mathbb{R}^{D|\Lambda|}\right)$ ($|\cdot|$ stands for cardinality).

For a box Λ and an inverse temperature $\beta = T^{-1}$, a periodic Gibbs state $\gamma_{\beta,\Lambda}$ is the following functional

$$\gamma_{\beta,\Lambda}(A) = \frac{\operatorname{trace}(Ae^{-\beta H_\Lambda})}{\operatorname{trace}(e^{-\beta H_\Lambda})}, \tag{9}$$

defined on the C^*-algebra $\mathfrak{A}_\Lambda$ of linear bounded operators on $\mathcal{H}_\Lambda$. Given Λ and $t \in \mathbb{R}$, we define an automorphism of $\mathfrak{A}_\Lambda$

$$\mathfrak{a}_t^\Lambda(A) = \exp\left(itH_\Lambda\right) A \exp\left(-itH_\Lambda\right). \tag{10}$$

A significant role in the construction of the Gibbs states of our model is played by multiplication operators. Bounded multiplication operators form a commutative subalgebra of $\mathfrak{A}_\Lambda$. The components of the displacement operator $q_l^{(k)}$, $l \in \Lambda$ are multiplication operators, but they do not belong to $\mathfrak{A}_\Lambda$ since they are unbounded. In [12] there was proved the following assertion (see also [1], [11]).

Proposition 38. *Let $t_1, \ldots, t_n \in \mathbb{R}$ and $A_1, \ldots A_n$ be bounded continuous functions $A_j : \mathbb{R}^{D|\Lambda|} \to \mathbb{C}$. Then $\mathfrak{A}_\Lambda$ is the smallest strongly closed linear space containing all operators of the form*

$$\mathfrak{a}_{t_1}^\Lambda(A_1)\mathfrak{a}_{t_2}^\Lambda(A_2)\ldots \mathfrak{a}_{t_n}^\Lambda(A_n).$$

For $A_1, \ldots, A_n \in \mathfrak{A}_\Lambda$ and $t_1, \ldots t_n \in \mathbb{R}$, a temporal Green function corresponding to the periodic boundary conditions is

$$G_{A_1,\ldots,A_n}^{\beta,\Lambda}(t_1,\ldots,t_n) = \gamma_{\beta,\Lambda}\left(\mathfrak{a}_{t_1}^\Lambda(A_1)\ldots \mathfrak{a}_{t_n}^\Lambda(A_n)\right). \tag{11}$$

For an open subset $\mathcal{O} \subset \mathbb{C}^n$, let $Hol(\mathcal{O})$ stand for the set of all holomorphic in $\mathcal{O}$ complex valued functions. Let also

$$\mathcal{D}_n^\beta \stackrel{\text{def}}{=} \{(t_1,\ldots,t_n) \in \mathbb{C}^n \mid 0 < \Im(t_1) < \Im(t_2) \cdots < \Im(t_n) < \beta\}. \tag{12}$$

By means of the arguments which were used in a similar situation in [1], Sect. 3 and [12], Sect. 2, one can prove the following statement.

Lemma 39. *For every* $A_1, \ldots, A_n \in \mathfrak{A}_\Lambda$,

(a) $G^{\beta,\Lambda}_{A_1,\ldots,A_n}$ *may be extended to a holomorphic function on* $\mathcal{D}^\beta_n$;

(b)   *this extension (which will also be written as* $G^{\beta,\Lambda}_{A_1,\ldots,A_n}$)
is continuous on the closure $\overline{\mathcal{D}}^\beta_n$ *of* $\mathcal{D}^\beta_n$, *moreover,*
for all $(t_1, \ldots, t_n) \in \overline{\mathcal{D}}^\beta_n$,

$$\left| G^{\beta,\Lambda}_{A_1,\ldots,A_n}(t_1,\ldots,t_n) \right| \le \|A_1\| \cdot \cdots \cdot \|A_n\|, \tag{13}$$

where $\|\cdot\|$ *stands for operator norm;*

(c)   *for every* $\xi_1, \ldots, \xi_n \in \mathbb{R}$, *the set*

$$\mathcal{D}^\beta_n(\xi_1,\ldots,\xi_n) \stackrel{\text{def}}{=} \{(t_1,\ldots,t_n) \in \mathcal{D}^\beta_n \mid \Re(t_j) = \xi_j, \ j = 1,\ldots,n\},$$

is such that for arbitrary $F, G \in Hol(\mathcal{D}^\beta_n)$, *their equality on*
$\mathcal{D}^\beta_n(\xi_1,\ldots,\xi_n)$ *implies that* F *and* G *are equal on the* $\mathcal{D}^\beta_n$.

The restriction of the function (11) to $\mathcal{D}^\beta_n(0,\ldots,0)$, i.e.

$$\Gamma^{\beta,\Lambda}_{A_1,\ldots,A_n}(\tau_1,\ldots\tau_n) = G^{\beta,\Lambda}_{A_1,\ldots,A_n}(i\tau_1,\ldots i\tau_n), \tag{14}$$

is a temperature (Matsubara) Green function, which has such a property

$$\Gamma^{\beta,\Lambda}_{A_1,\ldots,A_n}(\tau_1 + \theta, \ldots \tau_n + \theta) = \Gamma^{\beta,\Lambda}_{A_1,\ldots,A_n}(\tau_1,\ldots\tau_n), \tag{15}$$

for every $\theta \in \mathcal{I}_\beta \stackrel{\text{def}}{=} [0,\beta]$, where addition is modulo β.

In view of Proposition 38, the Green functions, defined by (11) with bounded multiplication operators, fully determine the state $\gamma_{\beta,\Lambda}$. Claim (c) of the latter assertion yields in turn that this state is determined by the Matsubara functions (14). In the Euclidean approach these functions are obtained as moments of probability measures. We begin their construction with introducing corresponding measure spaces. Given $\beta > 0$ and Λ, we set

$$\Omega_{\beta,\Lambda} = \{\omega_\Lambda = (\omega_l)_{l\in\Lambda} \mid \omega_l \in C(\mathcal{I}_\beta \to \mathbb{R}^D), \ \omega_\Lambda(0) = \omega_\Lambda(\beta)\}. \tag{16}$$

In the sequel, C_β will stand for $\Omega_{\beta,\Lambda}$ with a one-point Λ. Let also $\mathcal{X}_\beta$ stand for the real Hilbert space $L^2(\mathcal{I}_\beta \to \mathbb{R}^D)$ equipped with scalar product and norm

$$\langle \omega, \omega' \rangle_\beta = \int_{\mathcal{I}_\beta} (\omega(\tau), \omega'(\tau)) d\tau, \quad \|\omega\|_\beta = \sqrt{\langle \omega, \omega \rangle_\beta}. \tag{17}$$

Further

$$\mathcal{X}_{\beta,\Lambda} = \{\omega_\Lambda = (\omega_l)_{l\in\Lambda} \mid \omega_l \in \mathcal{X}_\beta\}. \tag{18}$$

Since Λ is finite, $\Omega_{\beta,\Lambda}$ and $\mathcal{X}_{\beta,\Lambda}$ may be equipped with the usual Banach space and Hilbert space structures respectively. Let $\mathfrak{B}(\Omega_{\beta,\Lambda})$ stand for the Borel σ–algebra of the subsets of $\Omega_{\beta,\Lambda}$. Consider the following strictly positive trace class operator on $\mathcal{X}_{\beta}$

$$S_{\beta} = (-\mathfrak{m}\Delta_{\beta} + 1)^{-1}\mathbf{1}, \qquad (19)$$

where Δ_{β} is the Laplace operator in $L^2(\mathcal{I}_{\beta})$ and $\mathbf{1}$ is the identity operator in $\mathbb{R}^D$. It determines on $\mathcal{X}_{\beta}$ a $O(D)$–invariant Gaussian measure χ_{β}, for which

$$\int_{\mathcal{X}_{\beta}} \exp\{\langle \varphi, \omega\rangle_{\beta}\}\, \chi_{\beta}(d\omega) = \exp\left\{\frac{1}{2}\langle S_{\beta}\varphi, \varphi\rangle_{\beta}\right\}. \qquad (20)$$

This measure is concentrated on $\mathcal{C}_{\beta} \subset \mathcal{X}_{\beta}$ [1], [11]. It describes a D-dimensional quantum harmonic oscillator with the mass $\mathfrak{m}$. One can show (see e.g. [1]) that for any $\tau \in \mathcal{I}_{\beta}$,

$$\int_{\mathcal{X}_{\beta}} \exp\left[\alpha(\omega(\tau), \omega(\tau))\right] \chi_{\beta}(d\omega) < \infty, \quad \forall \alpha < \alpha_*, \qquad (21)$$

where

$$\alpha_* = 2\sqrt{\mathfrak{m}} \cdot \frac{\exp(\beta/\sqrt{\mathfrak{m}}) - 1}{\exp(\beta/\sqrt{\mathfrak{m}}) + 1}. \qquad (22)$$

Given a box Λ, we write

$$\chi_{\beta,\Lambda}(d\omega_{\Lambda}) = \bigotimes_{l\in\Lambda} \chi_{\beta}(d\omega_l), \qquad (23)$$

$$E_{\beta,\Lambda}^V(\omega_{\Lambda}) = \frac{1}{2}\sum_{l,l'\in\Lambda} d_{ll'}^{\Lambda}\langle \omega_l, \omega_{l'}\rangle_{\beta} + \sum_{l\in\Lambda}\int_{\mathcal{I}_{\beta}} V(\omega_l(\tau))d\tau. \qquad (24)$$

Under the assumptions regarding V and $d_{ll'}$, $E_{\beta,\Lambda}^V$ is a continuous function from $\Omega_{\beta,\Lambda}$ to $\mathbb{R}$. A periodic local Euclidean Gibbs measure is

$$\mu_{\beta,\Lambda}(d\omega_{\Lambda}) = \frac{1}{Z_{\beta,\Lambda}} \exp\left\{-E_{\beta,\Lambda}^V(\omega_{\Lambda})\right\} \gamma_{\beta,\Lambda}(d\omega_{\Lambda}). \qquad (25)$$

It is a probability measure on the Hilbert space $\mathcal{X}_{\beta,\Lambda}$, supported on $\Omega_{\beta,\Lambda}$. $Z_{\beta,\Lambda}$ is the normalizing constant. Therefore, the Green functions (14) constructed with multiplication operators $A_1, \ldots A_n \in \mathfrak{A}_{\Lambda}$ may be written follows [1], [11]

$$\Gamma_{A_1,\ldots,A_n}^{\beta,\Lambda}(\tau_1, \ldots, \tau_n) \qquad (26)$$

$$= \int_{\mathcal{X}_{\beta,\Lambda}} A_1(\omega_{\Lambda}(\tau_1))\ldots A_n(\omega_{\Lambda}(\tau_n))\mu_{\beta,\Lambda}(d\omega_{\Lambda}).$$

The Gibbs states of the whole system which correspond to the periodic bound-
ary conditions are constructed as limits of the above states $\gamma_{\beta,\Lambda}$ when $\Lambda \nearrow \mathbb{L}$.
More precisely, let $\mathcal{L}$ be a sequence of boxes ordered by inclusion and such that
$\cup_{\Lambda \in \mathcal{L}} \Lambda = \mathbb{L}$. For $\Lambda_1 \subset \Lambda_2$, one may introduce a natural norm-preserving embed-
ding $\mathfrak{A}_{\Lambda_1} \subset \mathfrak{A}_{\Lambda_2}$, which defines an increasing sequence of algebras $\{\mathfrak{A}_\Lambda, \Lambda \in \mathcal{L}\}$.
In a standard way [8], this sequence defines a quasi-local algebra of observables.
Two sequences $\mathcal{L}$, $\mathcal{L}'$ are called equivalent if the corresponding quasi-local algebras coincide. A standard sequence $\mathcal{L}$ is the sequence of boxes $\{\Lambda_L, L \in \mathbb{N}\}$,
$\Lambda_L = (-L, L]^d \cap \mathbb{Z}^d$. In the sequel, all (thermodynamic) limits $\Lambda \nearrow \mathbb{L}$ are
taken over a sequence $\mathcal{L}$, which is equivalent to the standard one. The existence
of periodic Gibbs states for similar models was shown in [7].

The great advantage of the Euclidean approach lies in the fact that due to the
above relationship between the Green functions and local Gibbs measures one
may apply to the quantum case the machinery of conditional distributions, which
form the base of modern classical equilibrium statistical physics (see e.g. [9], [10]
and the references therein). To this end we will employ the spaces $\Omega_{\beta,\Lambda}$, defined
by (16), (18), also for infinite subsets Λ. In particular, Ω_β will stand for $\Omega_{\beta,\Lambda}$
with $\Lambda = \mathbb{L}$. These spaces are equipped with the product topology and with the
σ-algebras $\mathfrak{B}(\Omega_{\beta,\Lambda})$ generated by cylinder subsets. For $\Delta \subset \Lambda \subset \mathbb{L}$, we write
$\omega_\Delta \times \zeta_{\Lambda \setminus \Delta}$ for the configuration $(\xi_l)_{l \in \Lambda}$ such that $\xi_l = \omega_l$ for $l \in \Delta$, and $\xi_l = \zeta_l$
for $l \in \Lambda \setminus \Delta$. Given a sequence of boxes $\mathcal{L}$, in order to have the collections
$\{\Omega_{\beta,\Lambda}, \Lambda \in \mathcal{L}\}$ ordered by inclusion, we introduce the following mappings. For
$\Delta \subset \Lambda$, we put $\omega_\Delta \mapsto \omega_\Delta \times 0_{\Lambda \setminus \Delta} \in \Omega_{\beta,\Lambda}$, where 0_Λ is the zero configuration in
$\Omega_{\beta,\Lambda}$. Hence we consider every configuration ω_Δ as an element of all $\Omega_{\beta,\Lambda}$ with
$\Delta \subset \Lambda$. Besides, we define

$$\Omega_{\beta,\Lambda} \ni \omega_\Lambda \mapsto (\omega_\Lambda)_{\Lambda'} \in \Omega_{\beta,\Lambda'},$$

as a configuration such that $\omega_l = 0$ for $l \in \Lambda' \setminus \Lambda$. Let

$$\Omega_\beta^t \overset{\text{def}}{=} \{\zeta \in \Omega_\beta \mid \{\|\zeta_l\|_\beta, l \in \mathbb{L}\} \in \mathcal{S}'\}. \tag{27}$$

For $\zeta \in \Omega_\beta$ and a box Λ, we define the local Gibbs measure, subject to ζ, as the
following conditional probability measure. We put

$$\mu_{\beta,\Lambda}(B|\zeta) = 0, \quad \zeta \in \Omega_\beta \setminus \Omega_\beta^t, \quad B \in \mathfrak{B}(\Omega_{\beta,\Lambda}), \tag{28}$$

and for every $\zeta \in \Omega_\beta^t$,

$$\mu_{\beta,\Lambda}(d\omega_\Lambda|\zeta) = \frac{1}{Z_{\beta,\Lambda}(\zeta)} \exp\left\{-E_{\beta,\Lambda}^V(\omega_\Lambda|\zeta)\right\} \chi_{\beta,\Lambda}(d\omega_\Lambda). \tag{29}$$

Here

$$Z_{\beta,\Lambda}(\zeta) \overset{\text{def}}{=} \int_{\Omega_{\beta,\Lambda}} \exp\left\{-E_{\beta,\Lambda}^V(\omega_\Lambda|\zeta)\right\} \chi_{\beta,\Lambda}(d\omega_\Lambda),$$

is the local partition function subject to the external boundary condition ζ_{Λ^c}, and

$$E_{\beta,\Lambda}(\omega_\Lambda|\zeta) = \frac{1}{2}\sum_{l,l'\in\Lambda} d_{ll'}\langle\omega_l,\omega_{l'}\rangle_\beta + \sum_{l\in\Lambda,l'\in\Lambda^c} d_{ll'}\langle\omega_l,\zeta_{l'}\rangle_\beta, \qquad (30)$$

$$E_{\beta,\Lambda}^V(\omega_\Lambda|\zeta) = E_{\beta,\Lambda}(\omega_\Lambda|\zeta) + \sum_{l\in\Lambda}\int_{\mathcal{I}_\beta} V(\omega_l(\tau))d\tau, \qquad (31)$$

where V is given by (3). Under the assumptions regarding V and $d_{ll'}$, both $E_{\beta,\Lambda}(\cdot|\zeta)$, $E_{\beta,\Lambda}^V(\cdot|\zeta)$ are continuous functions from $\Omega_{\beta,\Lambda}$ to $\mathbb{R}$ for all $\zeta \in \Omega_\beta^t$. The function $E_{\beta,\Lambda}(\cdot|\zeta)$ describes the interaction of the particles in Λ between themselves and with the fixed configuration ζ_{Λ^c}, $\Lambda^c = \mathbb{L}\setminus\Lambda$.

Thus, along with (26), one may introduce the temperature Green function which corresponds to the external boundary condition ζ_{Λ^c}

$$\Gamma_{A_1,\ldots,A_n}^{\zeta,\beta,\Lambda}(\tau_1,\ldots,\tau_n) \qquad (32)$$
$$= \int_{X_{\beta,\Lambda}} A_1(\omega_\Lambda(\tau_1))\ldots A_n(\omega_\Lambda(\tau_n))\mu_{\beta,\Lambda}(d\omega_\Lambda|\zeta).$$

Here $A_1,\ldots,A_n$ are multiplication operators such that for every $\tau_1,\ldots,\tau_n \in \mathcal{I}_\beta$, the function

$$\Omega_{\beta,\Lambda} \ni \omega_\Lambda \mapsto A_1(\omega_\Lambda(\tau_1))\ldots A_n(\omega_\Lambda(\tau_n)),$$

is $\mu_{\beta,\Lambda}(\cdot|\zeta)$ integrable for every $\zeta \in \Omega_\beta$, that holds for $A_1,\ldots,A_n \in \mathfrak{A}_\Lambda$. Note that the above temperature Green function is defined only for multiplication operators, there are no a priori information regarding its analytic and continuity properties (except for $\zeta = 0$), even in the case of bounded operators.

For $B \in \mathfrak{B}(\Omega_\beta)$ and $\omega \in \Omega_\beta$, let $\delta_B(\omega)$ take values 1, resp. 0, if ω belongs, resp. does not belong, to B. Then one can introduce a family of probability kernels $\{\pi_{\beta,\Lambda} \mid \Lambda \subset \mathbb{L}, \ |\Lambda| < \infty\}$, on $(\Omega_\beta, \mathfrak{B}(\Omega_\beta))$

$$\pi_{\beta,\Lambda}(B|\zeta) \stackrel{\text{def}}{=} \int_{\Omega_{\beta,\Lambda}} \delta_B(\omega_\Lambda \times \zeta_{\Lambda^c})\mu_{\beta,\Lambda}(d\omega_\Lambda|\zeta). \qquad (33)$$

They satisfy the consistency conditions (for more details see e.g. [10])

$$\pi_{\beta,\Lambda}\pi_{\beta,\Delta}(B|\zeta) \stackrel{\text{def}}{=} \int_{\Omega_\beta} \pi_{\beta,\Lambda}(d\omega|\zeta)\pi_{\beta,\Delta}(B|\omega) = \pi_{\beta,\Lambda}(B|\zeta), \qquad (34)$$

which holds for arbitrary pairs of finite subsets $\Delta \subset \Lambda \subset \mathbb{L}$ and any $B \in \mathcal{B}(\Omega_\beta)$, $\zeta \in \Omega_\beta^t$.

Definition 40.   A probability measure μ on the space $(\Omega_\beta, \mathcal{B}(\Omega_\beta))$ is said to be a Euclidean Gibbs state at the inverse temperature β if it satisfies the Dobrushin-Lanford-Ruelle (DLR) equilibrium equation

$$\int_{\Omega_\beta} \mu(d\omega)\pi_{\beta,\Lambda}(B|\omega) = \mu(B), \qquad (35)$$

for all finite $\Lambda \subset \mathbb{L}$ and $B \in \mathfrak{B}(\Omega_\beta)$.

3. The Results

By means of the representation (26) we extend the Green functions to unbounded multiplication operators.

Theorem 41. *Let the functions $A_1, \ldots, A_n : \mathbb{R}^{D|\Lambda|} \to \mathbb{C}$ be such that for every $\beta > 0$ and every $\tau \in \mathcal{I}_\beta$, the functions $\Omega_{\beta,\Lambda} \ni \omega_\Lambda \mapsto A_j(\omega_\Lambda(\tau))$, $j = 1, \ldots n$, are $\mu_{\beta,\Lambda}$–integrable. Then, for the corresponding multiplication operators $A_1, \ldots, A_n$, the Green function (26) may be analytically continued on the domain $\mathcal{D}_n^\beta$ defined by (12).*

In contrast to the case of bounded operators (c.f. claim (b) of Lemma 39), one cannot expect that such extended Green functions are uniformly bounded on $\overline{\mathcal{D}}_n^\beta$ and continuous on its boundaries.

Definition 42. A continuous function $A : \mathbb{R}^{D|\Lambda|} \to \mathbb{C}$ belongs to the family $\mathfrak{F}_\Lambda^{(D)}$ if for arbitrary $\alpha > 0$, the function

$$\mathbb{R}^{D|\Lambda|} \ni x_\Lambda \mapsto |A(x_\Lambda)| \exp\left\{ -\alpha \sum_{l \in \Lambda} |x_l|^2 \right\}, \qquad (36)$$

is bounded on $\mathbb{R}^{D|\Lambda|}$.

In the case of one-point boxes, i.e. for $|\Lambda| = 1$, we write $\mathfrak{F}^{(D)}$.

Corollary 43. For arbitrary $A_1, \ldots, A_n \in \mathfrak{F}_\Lambda^{(D)}$, the temperature Green function (26) may be continued analytically in accordance with Theorem 41.

Indeed, by (21), functions from $\mathfrak{F}_\Lambda^{(D)}$ are integrable. As it has been already mentioned, the above analyticity does not imply continuity of the temperature Green functions. To prove it we have used the tightness of the local Gibbs measures.

Theorem 44. *Given a box Λ, let $A_1, \ldots, A_n$ belong to $\mathfrak{F}_\Lambda^{(D)}$. Then for all $\zeta \in \Omega_\beta$, the Green functions (26), (32) are continuous on $\mathcal{I}_\beta^n \ni (\tau_1, \ldots, \tau_n)$.*

Theorem 45. [**FKG Inequality**] *Given* Λ *and* $\zeta \in \Omega_\beta$, *let* μ *stand for any of the local Gibbs measures (25), (29) with* $D = 1$. *Then for any functions* $F, G \in \mathfrak{F}_\Lambda^{(1)}$, *which grow when every chosen* $\omega_l(\tau)$ *increases, the following inequality holds*

$$< FG >_\mu \geq < F >_\mu < G >_\mu, \tag{37}$$

where $< \cdot >_\mu$ *stands for expectation with respect to the measure* μ.

Theorem 46. [**GKS Inequalities**] *Given* Λ, *let the local Gibbs measure be defined by (25) with* $D = 1$. *Let also the real valued functions* $A_1, \ldots, A_{n+m} \in \mathfrak{F}_\Lambda^{(1)}$, $n, m \in \mathbb{N}$ *have the following properties:*

(a)   *every* A_j *depends only on the values of* x_{l_j} *with certain* $l_j \in \Lambda$;

(b)   *every* A_j *is either an odd monotone growing function of* x_{l_j}
     or an even positive function, monotone growing on $[0, +\infty)$.

Then for the Green functions (26), (32), the following inequalities hold for arbitrary $\tau_1, \ldots, \tau_{n+m} \in \mathcal{I}_\beta$:

$$\Gamma_{A_1,\ldots,A_n}^{\beta,\Lambda}(\tau_1,\ldots,\tau_n) \geq 0, \quad \Gamma_{A_1,\ldots,A_n}^{0,\beta,\Lambda}(\tau_1,\ldots,\tau_n) \geq 0, \tag{38}$$

$$\Gamma_{A_1,\ldots,A_{n+m}}^{\beta,\Lambda}(\tau_1,\ldots,\tau_{n+m}) \geq$$
$$\Gamma_{A_1,\ldots,A_n}^{\beta,\Lambda}(\tau_1,\ldots,\tau_n) \times \tag{39}$$
$$\Gamma_{A_{n+1},\ldots,A_{n+m}}^{\beta,\Lambda}(\tau_{n+1},\ldots,\tau_{n+m})$$

$$\Gamma_{A_1,\ldots,A_{n+m}}^{0,\beta,\Lambda}(\tau_1,\ldots,\tau_{n+m}) \geq$$
$$\Gamma_{A_1,\ldots,A_n}^{0,\beta,\Lambda}(\tau_1,\ldots,\tau_n) \times$$
$$\Gamma_{A_{n+1},\ldots,A_{n+m}}^{0,\beta,\Lambda}(\tau_{n+1},\ldots,\tau_{n+m}).$$

Now the model (1) - (3) with $D \in \mathbb{N}$ will be compared with the scalar model described by the same local Hamiltonian with $D = 1$. In order to distinguish vector and scalar objects we will supply the latter ones with tilde, writing $\tilde{H}_\Lambda$, $\tilde{\gamma}_{\beta,\Lambda}$, $\tilde{\Gamma}^{\beta,\Lambda}$. In the sequel, the polynomial v is supposed to be of the form (4).

Theorem 47. [**Scalar Domination**]   *Given* $A_1, \ldots, A_n \in \mathfrak{F}_\Lambda^{(D)}$, *let there exist* $k = 1, \ldots, D$ *and the functions* $\tilde{A}_1, \ldots \tilde{A}_n \in \mathfrak{F}_\Lambda^{(1)}$, *satisfying the conditions of the above theorem, such that* $A_j(x_\Lambda) = \tilde{A}_j(x_\Lambda^{(k)})$, $j = 1, \ldots, n$. *Then for arbitrary* $\tau_1, \ldots, \tau_n \in \mathcal{I}_\beta$

$$0 \leq \Gamma_{A_1,\ldots,A_n}^{\beta,\Lambda}(\tau_1,\ldots,\tau_n) \leq \tilde{\Gamma}_{\tilde{A}_1,\ldots,\tilde{A}_n}^{\beta,\Lambda}(\tau_1,\ldots,\tau_n). \tag{40}$$

REMARK 3. Note that all A_j depend on $x_\Lambda^{(k)}$ with one and the same k. The first above inequality is a D-dimensional version of (38). The second inequality in (40) describes scalar domination.

In the model considered, the structural phase transition, breaking $O(D)$-symmetry, is associated with the appearance of large fluctuations of displacements of particles. To describe them we introduce *fluctuation* operators

$$Q_\Lambda = \frac{1}{\sqrt{|\Lambda|}} \sum_{l \in \Lambda} q_l, \tag{41}$$

corresponding to *normal* fluctuations. If the Green functions (14) constructed with $A = Q_\Lambda^{(k)}$, remain bounded when $\Lambda \nearrow \mathbb{L}$, the fluctuations are regarded as normal. At the critical point the fluctuations become so large that to preserve the boundedness of the Green functions one should use an *abnormal* normalization, i.e.

$$Q_{\lambda,\Lambda} = \lambda(\Lambda)Q_\Lambda = \frac{\lambda(\Lambda)}{\sqrt{|\Lambda|}} \sum_{l \in \Lambda} q_l,$$

where $\{\lambda(\Lambda) \in \mathbb{R}, \Lambda \in \mathcal{L}\}$ is a converging to zero sequence. Given $F_1, \ldots, F_n \in \mathfrak{F}^{(D)}$, let A_j^λ stand for $F_j(Q_{\lambda,\Lambda})$, $j = 1, \ldots, n$.

Definition 48. Given $\beta > 0$, let the convergence

$$\Gamma_{A_1^\lambda,\ldots,A_n^\lambda}^{\beta,\Lambda}(\tau_1, \ldots, \tau_n) \longrightarrow F_1(0) \ldots F_n(0), \quad \Lambda \nearrow \mathbb{L}, \tag{42}$$

hold for all $n \in \mathbb{N}$, all $\tau_1, \ldots, \tau_n \in \mathcal{I}_\beta$, all $F_1, \ldots, F_n \in \mathfrak{F}^{(D)}$, arbitrary $\mathcal{L}$, and any converging to zero sequence $\{\lambda(\Lambda), \Lambda \in \mathcal{L}\}$. Then the fluctuations of displacements of particles are said to be normal.

Set

$$J = -\sum_{l'} d_{ll'}, \quad T = \tilde{H}_l + J\left(q_l^{(1)}\right)^2, \tag{43}$$

where the sum is taken over the whole lattice $\mathbb{L}$. The operator T has a purely discrete non-degenerate spectrum. Denote

$$T\psi_n = \epsilon_n \psi_n, \quad \Delta = \min\{\epsilon_{n+1} - \epsilon_n, n \in \mathbb{N}\}.$$

Theorem 49. *Let the mass* m, *the spectral parameter* Δ, *and the interaction parameter* J *obey the condition*

$$\mathrm{m}\Delta^2 > 2J. \tag{44}$$

Then for any $D \in \mathbb{N}$, *the fluctuations of displacements of particles in the D-dimensional model remain normal at all temperatures.*

References

1. S. Albeverio, R. Høegh–Krohn, *Homogeneous Random Fields and Quantum Statistical Mechanics*, J. Funct. Anal., **19** (1975), 242–279.
2. S. Albeverio, Yu. Kondratiev, Yu. Kozitsky, *Suppression of Critical Fluctuations by Strong Quantum Effects in Quantum Lattice Systems*, Comm. Math. Phys., **194** (1998), 493–521.
3. S. Albeverio, Yu. Kondratiev, Yu. Kozitsky, M.Röckner, *Uniqueness for Gibbs Measures of Quantum Lattices in Small Mass Regime*, to appear in Ann. Inst. H. Poincaré Probab. Statist.
4. S. Albeverio, Yu. Kondratiev, M.Röckner, T.V. Tsikalenko, *Uniqueness of Gibbs States on Loop Lattices*, C.R. Acad. Sci. Paris, Probabilités/Probability Theory, **342**, Série 1 (1997), 1401–1406.
5. V.S. Barbulyak, Yu.G. Kondratiev, *Functional Integrals and Quantum Lattice Systems: I. Existence of Gibbs States*, Rep. Nat. Acad. Sci of Ukraine, No 9 (1991), 38-40.
6. V.S. Barbulyak, Yu.G. Kondratiev, *Functional Integrals and Quantum Lattice Systems: II. Periodic Gibbs States*, Rep. Nat. Acad. Sci of Ukraine, No 8 (1991), 31-34.
7. V.S. Barbulyak, Yu.G. Kondratiev, *A Criterion for the Existence of Periodic Gibbs States of Quantum Lattice Systems*, Selecta Math. formerly Sov., **12** (1993), 25–35.
8. O. Bratteli, D.W. Robinson, *Operator Algebras and Quantum Statistical Mechanics*, I, II, Springer, New York, 1981.
9. R.L. Dobrushin, *Prescribing a System of Random Variables by Conditional Distributions*, Theory Prob. Appl., **15** (1970), 458–486.
10. H.O. Georgii, *Gibbs Measures and Phase Transitions*. Vol 9, Walter de Gruyter, Springer, Berlin New York, 1988.
11. S.A. Globa, Yu.G. Kondratiev, *The Construction of Gibbs States of Quantum Lattice Systems*, Selecta Math. Sov., **9** (1990), 297–307 (1990).
12. R. Høegh–Krohn, *Relativistic Quanum Statistical Mechanics in Two-dimensional Space-time*, Comm. Math. Phys., **38**(1974), 195–224.
13. A. Klein, L. Landau, *Stochastic Processes Associated with KMS States*, J. Funct. Anal., **42** (1981), 368–428.
14. Yu. Kozitsky, *Quantum Effects in a Lattice Model of Anharmonic Vector Oscillators*, Letters Math. Phys., **51** (2000), 71–81.
15. B. Simon, *Functional Integration and Quantum Physics*, Academic Press, New York San Francisco London, 1979.
16. S. Stamenković, *Unified Model Description of Order-Disorder and Displacive Structural Phase Transitions*, Condensed Matter Physics, **1(14)** (1998), 257-309.

A METRIC-AFFINE FIELD MODEL FOR THE NEUTRINO

DMITRI VASSILIEV *
Department of Mathematical Sciences, University of Bath,
Bath BA2 7AY, UK

1. Main result

We define space-time as a real oriented 4-manifold M equipped with a non-degenerate metric g (not necessarily symmetric) and an affine connection Γ. We write space-time as a triple $\{M, g, \Gamma\}$. The 16 components of the metric tensor $g_{\mu\nu}$ and the 64 connection coefficients $\Gamma^\lambda{}_{\mu\nu}$ are the unknowns in our model, as is the manifold M itself.

This approach is known as the Einstein–Schrödinger metric-affine field theory; see, for example, Appendix II in [1], or [2]. During the period from the 1920s to the 1950s many mathematicians and physicists contributed to this subject, with the list of authors containing names such as M.Born, A.S.Eddington, L.Infeld, T.Levi-Civita and H.Weyl. In modern theoretical physics metric-affine field theories are not a mainstream subject; reviews of some of the more recent work in this area can be found in [3], [4], [5], [6].

The immediate motivation for our paper comes from [7] where it was shown that it is possible to give a sensible tensor interpretation of the Dirac equation in flat Minkowski 3-space by treating the electromagnetic field as an affine connection in the embedding Minkowski 4-space. The "electromagnetic" connection suggested in [7] is the metric compatible connection corresponding to torsion

$$T = e * A$$

where e is the electron charge, A is the (given) real-valued vector potential of the electromagnetic field, and $*$ is the Hodge star; here we use a system of units in which both the speed of light c and Planck's constant $\hbar$ have value 1. In particular, such an interpretation of electromagnetism resolves the problem of distinguishing

---

* d.vassiliev@bath.ac.uk

S. Duplij and J. Wess (eds.), Noncommutative Structures in Mathematics and Physics, 427–439.
© *2001 Kluwer Academic Publishers. Printed in the Netherlands.*

the electron from the positron without resorting to "negative frequencies". Regarding the affine connection itself as an unknown quantity is the next obvious step.

We construct our mathematical model for the neutrino as follows.

Firstly, we consider the Yang–Mills equation for the affine connection:

$$\delta_{YM} R = 0 \tag{1}$$

where R is the Riemann curvature tensor (10) and δ_{YM} is the divergence on curvatures (13).

Secondly, we consider the Einstein equation:

$$Ric = 0 \tag{2}$$

where Ric is the Ricci curvature tensor. Equation (2) describes the absence of sources of gravitation.

The objective of this paper is the study of the combined system (1), (2) which is a system of 80 real non-linear partial differential equations with 80 real unknowns $g_{\mu\nu}$, $\Gamma^\lambda{}_{\mu\nu}$. In other words, we are combining the basic equation of relativistic quantum mechanics (Yang–Mills equation) with the basic equation of general relativity (Einstein equation).

REMARK 4. If the metric is symmetric and the connection is that of Levi-Civita then (2) implies (1). In the general case (1) and (2) are independent.

We define Minkowski space $\mathbb{M}^4$ as a real 4-manifold which admits a global coordinate system (x^0, x^1, x^2, x^3) and is equipped with the metric

$$g_{\mu\nu} = \text{diag}(+1, -1, -1, -1). \tag{3}$$

Our definition of $\mathbb{M}^4$ specifies two elements of the triple $\{M, g, \Gamma\}$, namely, the manifold M and the metric g, but does not specify the connection Γ.

Our main result is

Theorem 50. *Let u be a complex-valued vector function which is a plane wave solution of the polarised Maxwell equation*

$$*du = \pm i du \tag{4}$$

in $\mathbb{M}^4$. Let Γ be the metric compatible connection corresponding to torsion

$$T^\lambda{}_{\mu\nu} = \text{Re}(u^\lambda (du)_{\mu\nu}). \tag{5}$$

Then the space-time $\{\mathbb{M}^4, \Gamma\}$ is a solution of (1), (2).

Note that the vector equation (4) forms the basis of the mathematical model in [7]. It is shown in [7] that under certain circumstances equation (4) produces effects normally attributed to spinors.

Let us rewrite (4) as

$$*du = i\alpha du, \tag{6}$$

$\alpha = \pm 1$. The non-trivial ($du \neq 0$) plane wave solutions of (6) can, of course, be written down explicitly: up to a proper Lorentz transformation they are

$$u(x) = w\, e^{-ik\cdot x} \tag{7}$$

where

$$w_\mu = C(0, 1, -i\alpha, 0), \qquad k_\mu = \beta(1, 0, 0, 1), \tag{8}$$

$\beta = \pm 1$, and $C \in \mathbb{R}_+$ is an arbitrary constant (amplitude).

Substitution of (7) into (5) produces

$$T^\lambda{}_{\mu\nu} = \mathrm{Re}(-iw^\lambda\, (k \wedge w)_{\mu\nu}\, e^{-2ik\cdot x}). \tag{9}$$

Thus, the space-time in Theorem 50 is a wave of torsion which, up to a proper Lorentz transformation, is given by the explicit formulae (9), (8).

The paper has the following structure.

In Section 2 we specify our notation.

Section 3 is a brief description of Yang–Mills theory in our particular setting (affine connection over vectors).

In Section 4 we prove Theorem 50. The crucial element of the proof is the linearisation ansatz (17), (16).

In Section 5 we establish general invariant properties of our solutions (3)–(5). It turns out that our Riemann curvature tensors possess *all* the symmetry properties of the "usual" curvature tensors generated by Levi-Civita connections. This means that in observing such connections we might be led to believe (mistakenly) that we live in a Levi-Civita universe.

In Section 6 we show that the Riemann curvature tensors corresponding to our solutions (3)–(5) have an algebraic structure which makes them equivalent to bispinors. It turns out that these bispinors satisfy the Weyl equation (Dirac equation for massless particle), which justifies our interpretation of space-times (3)–(5) as the neutrino and antineutrino. We show that our model explains the well known fact that neutrinos are always left-handed whereas antineutrinos are always right-handed.

In Section 7 we compare our results with those of Einstein who suggested [8] a double duality equation as a possible model for elementary particles. We show that our space-times (3)–(5) satisfy this equation. Here the crucial point is that we get the sign predicted by Einstein.

In Section 8 we vary the Yang–Mills Lagrangian (12) with respect to the *metric* and show that our solutions (3)–(5) provide stationary points. This fact is highly unusual and does not follow from abstract Yang–Mills theory which guarantees only conformal invariance.

2. Basic notation

We denote $\partial_\mu = \partial/\partial x^\mu$ and define the covariant derivative of a vector function as $\nabla_\mu v^\lambda := \partial_\mu v^\lambda + \Gamma^\lambda{}_{\mu\nu} v^\nu$. We define the torsion tensor as $T^\lambda{}_{\mu\nu} := \Gamma^\lambda{}_{\mu\nu} - \Gamma^\lambda{}_{\nu\mu}$, the Riemann curvature tensor as

$$R^\kappa{}_{\lambda\mu\nu} := \partial_\mu \Gamma^\kappa{}_{\nu\lambda} - \partial_\nu \Gamma^\kappa{}_{\mu\lambda} + \Gamma^\kappa{}_{\mu\eta}\Gamma^\eta{}_{\nu\lambda} - \Gamma^\kappa{}_{\nu\eta}\Gamma^\eta{}_{\mu\lambda}, \qquad (10)$$

and the Ricci curvature tensor as $Ric_{\lambda\nu} := R^\kappa{}_{\lambda\kappa\nu}$.

We define the contravariant metric tensor as the solution of the linear algebraic system $g^{\mu\nu} g_{\nu\kappa} = \delta^\mu{}_\kappa$. We have to take great care when raising or lowering tensor indices because in our statement of the problem the metric is not assumed to be symmetric and the connection is not assumed to be metric compatible. Only when it is clear that we are in a situation when the metric is symmetric and the connection is metric compatible we gain the full freedom of writing any tensor with either upper or lower indices (in any combinations), the raising or lowering being achieved via contraction with the contravariant or covariant symmetric metric tensor.

Given a scalar function f we write for brevity

$$\int f := \int_M f \sqrt{|\det g|}\, dx^0 dx^1 dx^2 dx^3, \quad \det g := \det(g_{\mu\nu}) \neq 0.$$

We define the Hodge star as $(*Q)_{\mu_{q+1}\ldots\mu_4} := (q!)^{-1} \sqrt{|\det g|}\, Q^{\mu_1\ldots\mu_q} \varepsilon_{\mu_1\ldots\mu_4}$ where ε is the totally antisymmetric quantity. We put $\varepsilon_{0123} := \pm 1$, where $+$ or $-$ is taken depending on whether the orientation of the coordinate system is positive or negative, respectively.

When dealing with a connection which is compatible with a given symmetric metric it is convenient to introduce the *contortion* tensor $K^\lambda{}_{\mu\nu} := \Gamma^\lambda{}_{\mu\nu} - \left\{ {\lambda \atop \mu\nu} \right\}$, where $\left\{ {\lambda \atop \mu\nu} \right\} := \frac{1}{2} g^{\lambda\kappa}(\partial_\mu g_{\nu\kappa} + \partial_\nu g_{\mu\kappa} - \partial_\kappa g_{\mu\nu})$ is the Christoffel symbol. Contortion has the antisymmetry property $K_{\lambda\mu\nu} = -K_{\nu\mu\lambda}$. A symmetric metric and contortion uniquely determine the metric compatible connection. Torsion and contortion are related as (see [9], formula (7.35))

$$T^\lambda{}_{\mu\nu} = K^\lambda{}_{\mu\nu} - K^\lambda{}_{\nu\mu}, \qquad K^\lambda{}_{\mu\nu} = (T^\lambda{}_{\mu\nu} + T_\mu{}^\lambda{}_\nu + T_\nu{}^\lambda{}_\mu)/2. \qquad (11)$$

A bispinor in $\mathbb{M}^4$ is a column of four complex numbers $(\,\xi^1\ \xi^2\ \eta_{\dot{1}}\ \eta_{\dot{2}}\,)^T$ which change under Lorentz transformations in a particular way, see Sections 18, 19 and 26 in [10] for details; a more compact exposition is given in the beginning of Section 3 in [11]. The Pauli and Dirac matrices are

$$I = \begin{pmatrix} 1 & 0 \\ 0 & 1 \end{pmatrix}, \qquad \sigma^1 = \begin{pmatrix} 0 & 1 \\ 1 & 0 \end{pmatrix}, \qquad \sigma^2 = \begin{pmatrix} 0 & -i \\ i & 0 \end{pmatrix}, \qquad \sigma^3 = \begin{pmatrix} 1 & 0 \\ 0 & -1 \end{pmatrix},$$

$$\gamma^0 = \begin{pmatrix} 0 & I \\ I & 0 \end{pmatrix}, \qquad \gamma^j = \begin{pmatrix} 0 & -\sigma^j \\ \sigma^j & 0 \end{pmatrix}, \qquad \gamma^5 = -i\gamma^0\gamma^1\gamma^2\gamma^3 = \begin{pmatrix} -I & 0 \\ 0 & I \end{pmatrix}.$$

3. The Yang–Mills equation

Put $R^\kappa{}_\lambda{}^\rho{}_\nu := g^{\rho\mu} R^\kappa{}_{\lambda\mu\nu}$ where $R^\kappa{}_{\lambda\mu\nu}$ is the Riemann curvature tensor (10). The Yang–Mills Lagrangian for the affine connection is

$$\mathcal{L}_{\mathrm{YM}} := -\frac{1}{2} \int R^\kappa{}_\lambda{}^\rho{}_\nu R^\lambda{}_\kappa{}^\nu{}_\rho. \tag{12}$$

The Yang–Mills equation (1) is the Euler–Lagrange equation obtained from (12) by varying the connection coefficients $\Gamma^\lambda{}_{\mu\nu}$ (but not the metric). The explicit formula for the differential operator δ_{YM} appearing in (1) is

$$(\delta_{\mathrm{YM}} R)^\rho = \frac{1}{2\sqrt{|\det g|}} (\partial_\sigma + [\Gamma_\sigma, \cdot]) \left(\sqrt{|\det g|}\, (g^{\rho\mu} g^{\nu\sigma} + g^{\mu\rho} g^{\sigma\nu}) R_{\mu\nu} \right). \tag{13}$$

In writing (13) we used matrix notation to hide two indices: $R_{\mu\nu} = R^\kappa{}_{\lambda\mu\nu}$, $\Gamma_\sigma = \Gamma^\kappa{}_{\sigma\lambda}$, with κ enumerating the rows and λ the columns. By $[\cdot, \cdot]$ we denote the commutator, i.e., $[L, N]^\tau{}_\lambda := L^\tau{}_\kappa N^\kappa{}_\lambda - N^\tau{}_\kappa L^\kappa{}_\lambda$.

Note that the operator (13) is invariant under the transposition of the metric, $g_{\mu\nu} \to \tilde{g}_{\mu\nu} := g_{\nu\mu}$. For more details concerning transposition invariance and its possible physical significance see [1] p. 142–143.

From now on, until Section 8, we work only in Minkowski space and only with metric compatible connections. This leads to a number of simplifications. Connection coefficients now coincide with contortion, for which we continue using matrix notation $K_\sigma = K^\kappa{}_{\sigma\lambda}$. Formula (10) becomes

$$R_{\mu\nu} = \partial_\mu K_\nu - \partial_\nu K_\mu + [K_\mu, K_\nu], \tag{14}$$

and the Yang–Mills equation (1), (13) becomes

$$(\partial_\nu + [K_\nu, \cdot]) R^{\mu\nu} = 0. \tag{15}$$

The Yang–Mills equation (15) appears to be overdetermined as it is a system of 64 equations with only 24 unknowns (24 is the number of independent components of the contortion tensor). However 40 of the 64 equations are automatically fulfilled. This is a consequence of the fact that the Lie algebra of real antisymmetric rank 2 tensors is a subalgebra of the general Lie algebra of real rank 2 tensors.

4. Proof of Theorem 50

The fundamental difficulty with the Yang–Mills equation (15) as well as with the Einstein equation (2) is that these equations are non-linear with respect to the unknown contortion K. The following lemma plays a crucial role in our construction by allowing us to get rid of the non-linearities.

Lemma 51. *Let L be a complex rank 2 antisymmetric tensor satisfying*

$$*L = \pm iL. \tag{16}$$

Then $[\mathrm{Re}L, \mathrm{Im}L] = 0$.

Proof. The result follows from the general formula $[*L, N] = *[L, N]$.

Lemma 51 can be rephrased in the following way: the Lie algebra of real antisymmetric rank 2 tensors has 2-dimensional abelian subalgebras which can be explicitly described in terms of the eigenvectors of the Hodge star.

Lemma 51 immediately implies the following linearisation ansatz.

Corollary 52. Suppose contortion is of the form

$$K^{\kappa}{}_{\nu\lambda}(x) = \mathrm{Re}(\, L^{\kappa}{}_{\lambda}\, v_{\nu}(x)\,) \tag{17}$$

where L is a constant complex antisymmetric tensor satisfying (16) and v is a complex-valued vector function. Then the non-linear terms in the formula for Riemann curvature (14) and in the Yang–Mills equation (15) vanish.

Substituting (17) into (14), (15) we reduce equations (1), (2) to

$$\delta dv = 0, \tag{18}$$

$$L^{\kappa}{}_{\lambda}\, (dv)_{\kappa\nu} = 0. \tag{19}$$

Here d is the exterior derivative and δ is its adjoint, so that (18) is the Maxwell equation.

Let us look for plane wave solutions, i.e.,

$$v(x) = -iw\, e^{-2ik\cdot x} \tag{20}$$

where $w \neq 0$ is a constant complex vector and $k \neq 0$ is a constant real vector. Here we put the extra factor $-i$ at w as well as the extra factor 2 in the exponent for the sake of convenience; the reason for doing this is to achieve agreement with (9). Substituting (20) into (18), (19) we get

$$k^{\nu}(k \wedge w)_{\mu\nu} = 0, \tag{21}$$

$$L^{\kappa}{}_{\lambda} (k \wedge w)_{\kappa\nu} = 0. \tag{22}$$

We have reduced our original system of partial differential equations (1), (2) to the purely algebraic problem (16), (21), (22). Straightforward analysis shows that the space-times described in Theorem 50 are solutions of (16), (21), (22), and, moreover, the only non-trivial ($R \not\equiv 0$) solutions.

5. Invariant properties of our solutions

It is known [4], [5], [6] that the 24-dimensional space of real torsions decomposes into the following 3 irreducible subspaces: tensor torsions, trace torsions, and axial torsions. The dimensions are 16, 4, and 4, respectively.

Lemma 53. *The torsions in Theorem 50 are purely tensor.*

Proof. The trace component of a torsion tensor $T_{\lambda\mu\nu}$ is zero iff $T^{\lambda}{}_{\lambda\nu} = 0$, and the axial component is zero iff $T_{\lambda\mu\nu} \, \varepsilon^{\lambda\mu\nu\kappa} = 0$. These identities are established by direct examination of the explicit formulae (9), (8).

Let us mention the following useful general result.

Lemma 54. *If the axial component of a torsion is zero then this torsion coincides, up to a natural reordering of indices, with the corresponding (see (11)) contortion:* $T_{\lambda\mu\nu} = K_{\mu\lambda\nu}$.

Lemma 54 explains why the torsion of our space-times has the simple structure (5), (4). Our linearisation ansatz (17), (16) required us to work with contortion rather than torsion, and in the end in order to calculate torsion we had to use the first formula (11). We did not get a cumbersome expression for torsion only because its axial component is zero.

Lemma 55. *The Riemann curvatures of space-times from Theorem 50 have all the symmetry properties of curvatures in the Levi-Civita setting, that is,*

$$R_{\kappa\lambda\mu\nu} = -R_{\lambda\kappa\mu\nu} = -R_{\kappa\lambda\nu\mu} = R_{\mu\nu\kappa\lambda}, \tag{23}$$

$$R_{\kappa\lambda\mu\nu} \, \varepsilon^{\kappa\lambda\mu\nu} = 0. \tag{24}$$

Proof. Let us define the complex Riemann curvature tensor

$$\mathbb{C}R_{\kappa\lambda\mu\nu} := F_{\kappa\lambda} \, F_{\mu\nu} \tag{25}$$

where

$$F := du \tag{26}$$

and u is from (6). Lemmas 53, 54 and Corollary 52 imply

$$R_{\kappa\lambda\mu\nu} = \mathrm{Re}(\mathbb{C}R_{\kappa\lambda\mu\nu}). \tag{27}$$

Direct examination of formulae (25)–(27), (7), (8) establishes (23), (24).

6. Weyl's equation

The torsions (and, therefore, space-times) from Theorem 50 are described, up to a proper Lorentz transformation and a scaling factor $C \in \mathbb{R}_+$, by a pair of indices $\alpha, \beta = \pm 1$; see (9), (8). It may seem that this gives us 4 essentially different space-times. However, formula (9) contains the operation of taking the real part and, as a result, the transformation $\{\alpha, \beta\} \to \{-\alpha, -\beta\}$ does not change our torsion. Thus, Theorem 50 provides us with only two essentially different space-times labeled by the index $\tau := \alpha\beta = \pm 1$. The purpose of this section is to show that it is natural to interpret these two space-times as the neutrino and antineutrino.

We base our interpretation on the analysis of the Riemann curvature tensor. We chose to analyse curvature rather than torsion because curvature is an accepted physical obervable.

In our analysis of the Riemann curvature tensor we will work with the complex curvature (25) rather than the real curvature (27) because the complex one has a simpler structure. Indeed, according to formula (25) the complex Riemann curvature tensor $\mathbb{C}R$ factorizes as the square of a rank 2 tensor F and is, therefore, completely determined by it.

Working with the rank 2 tensor F is much easier than with the original rank 4 tensor $\mathbb{C}R$, but one would like to simplify the analysis even further by factorizing F itself. It is impossible to factorize F as the square of a vector but it is possible to factorize F as the square of a bispinor.

Lemma 56. *A complex rank 2 antisymmetric tensor F satisfying*

$$F_{\mu\nu}F^{\mu\nu} = 0, \qquad (*F)_{\mu\nu}F^{\mu\nu} = 0 \tag{28}$$

is equivalent to a bispinor ψ, the relationship between the two being

$$F^{\mu\nu} = -\frac{i}{4}\psi^T\gamma^0\gamma^2\gamma^\mu\gamma^\nu\psi. \tag{29}$$

Proof. Formula (29) is a special case of the general equivalence relation between rank 2 antisymmetric tensors and rank 2 symmetric bispinors, see end of Section 19 in [10]. Conditions (28) are necessary and sufficient for the factorization of the symmetric rank 2 spinors as squares of rank 1 spinors.

REMARK 5. The corresponding text in the end of Section 19 in [10] contains mistakes. These can be corrected by replacing everywhere i by $-i$.

REMARK 6. For a given tensor F formula (29) defines the individual spinors $\xi = (\xi^1 \ \xi^2)^T$ and $\eta = (\eta_{\dot{1}} \ \eta_{\dot{2}})^T$ uniquely up to choice of sign. This is in agreement with the general fact that a spinor does not have a specific sign, see the beginning of Section 19 in [10].

REMARK 7. Conditions (28) are equivalent to $\det F = 0$, $\det * F = 0$.

REMARK 8. Formula (29) is invariant under proper Lorentz transformations and space inversion, but not under time inversion.

Our particular tensor F defined in accordance with formula (26) satisfies the conditions (28). Indeed, $F_{\mu\nu}F^{\mu\nu} = 0$ is the statement that the complex scalar curvature is zero (consequence of the complex Ricci curvature being zero), whereas $(*F)_{\mu\nu}F^{\mu\nu} = 0$ is the statement that the complex Riemann curvature tensor $\mathbb{C}R$ satisfies the cyclic sum identity, cf. (24).

Thus, the complex Riemann curvature tensor (25) has an algebraic structure which makes it equivalent to a bispinor. Direct calculations show that the corresponding bispinor function $\psi(x)$ satisfies the Weyl equation

$$\gamma^\mu \partial_\mu \psi = 0 \tag{30}$$

as well as the additional condition

$$\gamma^5 \psi = -\alpha\psi \tag{31}$$

where $\alpha = \pm 1$ is from (6). Conversely, any plane wave solution of (30), (31) generates a complex Riemann curvature tensor of the type (25).

A non-trivial ($\psi(x) \not\equiv$ const) plane wave solution of (30), (31) can, up to a proper Lorentz transformation, be written as $\psi(x) = \varphi e^{-\frac{i}{2}k \cdot x}$ where φ is a constant bispinor and k is given by (8). Recall that the formula for k contains the parameter $\beta = \pm 1$ which determines whether the wave vector k lies on the forward ($\beta = +1$) or backward ($\beta = -1$) light cone.

Non-trivial plane wave solutions of (30), (31) with $\beta = +1$ are called neutrinos whereas those with $\beta = -1$ are called antineutrinos. A neutrino is said to be left-handed if $\alpha = -1$ and right-handed if $\alpha = +1$. An antineutrino is said to be left-handed if $\alpha = +1$ and right-handed if $\alpha = -1$.

REMARK 9. The above definitions agree with the operation of charge conjugation (see formula (26.6) in [10]) in that the left-handed neutrino and left-handed antineutrino are charge conjugates of one another, as are the right-handed neutrino and right-handed antineutrino.

As explained in the beginning of this section, the transformation $\{\alpha, \beta\} \to \{-\alpha, -\beta\}$ does not change the resulting space-time. This means that in our model the left-handed neutrino is identical to the left-handed antineutrino, and the right-handed neutrino is identical to the right-handed antineutrino.

7. Einstein's double duality equation

The only *a priori* symmetry properties of the Riemann curvature tensor generated by a connection compatible with a symmetric metric are

$$R_{\kappa\lambda\mu\nu} = -R_{\lambda\kappa\mu\nu} = -R_{\kappa\lambda\nu\mu}. \tag{32}$$

Let $\mathcal{R}$ be the 36-dimensional linear space of real rank 4 tensors satisfying (32). We consider the following two endomorphisms in $\mathcal{R}$:

$$R \to R^T, \qquad (R^T)_{\kappa\lambda\mu\nu} := R_{\mu\nu\kappa\lambda}, \tag{33}$$

$$R \to {}^*R^*, \qquad ({}^*R^*)_{\kappa\lambda\mu\nu} := (|\det g|/4)\,\varepsilon_{\kappa'\lambda'\kappa\lambda}\,R^{\kappa'\lambda'\mu'\nu'}\,\varepsilon_{\mu'\nu'\mu\nu}. \tag{34}$$

REMARK 10. It is easy to see that the endomorphisms (33), (34) are well defined even if the manifold is not orientable. In the case of (34) this observation is a consequence of a much deeper fact established in [12]: the rank 8 tensor $(\det g)\varepsilon_{\kappa'\lambda'\kappa\lambda}\varepsilon_{\mu'\nu'\mu\nu}$ is a purely metrical quantity in that it is expressed via the metric tensor. This is a special feature of dimension 4.

The endomorphisms (33), (34) have the following properties: (i) they commute, (ii) their eigenvalues are ± 1, (iii) they have no associated eigenvectors. Therefore, $\mathcal{R}$ decomposes into a direct sum of 4 invariant subspaces

$$\mathcal{R} = \oplus_{a,b=\pm} \mathcal{R}_{ab}, \qquad \mathcal{R}_{ab} := \{R \in \mathcal{R} \mid R^T = aR,\ {}^*R^* = bR\}. \tag{35}$$

The decomposition (35) was suggested in [13] and developed in [8], [12]. Actually, the papers [13], [8], [12] deal only with the case of a Levi-Civita connection, but the generalization to the case of an arbitrary affine connection compatible with a symmetric metric is straightforward. Lanczos called tensors $R \in \mathcal{R}$ self-dual (respectively, antidual) if ${}^*R^* = -R$ (respectively, ${}^*R^* = R$). Such a choice of terminology is due to the fact that Einstein and Lanczos defined their double duality endomorphism as

$$R \to (\text{sgn} \det g)\,{}^*R^*. \tag{36}$$

The advantage of (36) is that this linear operator is expressed via the metric tensor as a rational function. The endomorphism (36) is, in a sense, even more invariant than (34) as it does not "feel" the signature of the metric.

Lemma 57. (Rainich [13]) *The subspaces $\mathcal{R}_{++}$ and $\mathcal{R}_{+-}$ have dimensions 9 and 12, respectively.*

REMARK 11. In Rainich's paper the dimensions are actually given as 9 and 11. The reason behind this is that Rainich imposed on curvatures the cyclic sum condition (24). This excludes from $\mathcal{R}_{+-}$ curvatures of the type $R_{\kappa\lambda\mu\nu} = \varepsilon_{\kappa\lambda\mu\nu}$ and, therefore, reduces the dimension by 1.

Lemma 58. (Einstein [8]) *Let $R \in \mathcal{R}_{++}$. Then the corresponding Ricci tensor is symmetric and trace free. Moreover, R is uniquely determined by its Ricci tensor and the metric tensor according to the formula*

$$R_{\kappa\lambda\mu\nu} = (g_{\kappa\mu}Ric_{\lambda\nu} + g_{\lambda\nu}Ric_{\kappa\mu} - g_{\kappa\nu}Ric_{\lambda\mu} - g_{\lambda\mu}Ric_{\kappa\nu})/2. \qquad (37)$$

Einstein's goal in [8] was to construct a mathematical model for the electron; note that this paper was published a year before Dirac discovered his equation. Einstein argued that the Riemann curvature tensor of the electron should lie in an eigenspace of the endomorphism (34). As in this particular paper Einstein restricted his analysis to the case of a Levi-Civita connection he had to make the choice between the invariant subspaces $\mathcal{R}_{++}$ and $\mathcal{R}_{+-}$. The difference between these two invariant subspaces is fundamental: it has nothing to do with the choice of forward and backward light cones or the choice of orientation of the coordinate system, and, as a consequence, it has nothing to do with the notions of "particle" and "antiparticle" or the notions of "left-handedness" and "right-handedness".

Lemmas 57 and 58 led Einstein to the conclusion that curvatures from $\mathcal{R}_{++}$ are too trivial and the dimension of the subspace too low (9 instead of the expected 10 which is the number of independent components of the energy–momentum tensor) to associate it with the electron. Einstein's conjecture was that the Riemann curvature tensor of the electron should lie in the invariant subspace $\mathcal{R}_{+-}$, that is, it should satisfy the equation

$$*R^* = -R. \qquad (38)$$

Formulae (25)–(27), (4) imply that our space-times (3)–(5) satisfy (38).

Our paper falls short of constructing an affine field model for the electron. Nevertheless, we find it encouraging that our affine field model for the neutrino agrees with Einstein's double duality equation (38).

8. Variation of the metric

Variation of the Yang–Mills Lagrangian (12) with respect to the metric produces the following Euler–Lagrange equation:

$$H - (\mathrm{tr}H/4)g = 0 \qquad (39)$$

where $H_{\mu\sigma} := R^{\kappa}{}_{\lambda\mu\nu}g^{\nu\rho}R^{\lambda}{}_{\kappa\rho\sigma}$, $\mathrm{tr}H := H_{\mu\sigma}g^{\sigma\mu}$. In deriving (39) we did not make any assumptions on the symmetry of the metric.

Note the fundamental difference between our original equations (1), (2) and equation (39): (1), (2) are linear in curvature, whereas (39) is quadratic.

Lemma 59. *Let the metric be symmetric and Lorentzian, and let R be of the form (27) where $\mathbb{C}R$ is a complex rank 4 tensor which factorises as the product of antisymmetric rank 2 tensors, $\mathbb{C}R_{\kappa\lambda\mu\nu} = F_{\kappa\lambda}\,G_{\mu\nu}$, such that $*F = i\alpha F$, $*G = i\alpha'G$, $\alpha, \alpha' = \pm 1$. Then R satisfies the equation (39).*

Proof. The Lemma is proved by a straightforward Maple^{TM} calculation.

Lemma 59 and formulae (25)–(27), (4) immediately imply

Corollary 60. Our space-times (3)–(5) provide stationary points of the Yang–Mills Lagrangian (12) with respect to the variation of the metric.

In order to illustrate how unusual Corollary 60 is let us examine what happens in the case of the Maxwell equation, which is the simplest example of a Yang–Mills equation. Straightforward calculations show that the Maxwell equation on a Lorentzian manifold does not have nontrivial solutions which provide stationary points of the Maxwell Lagrangian with respect to the variation of the metric.

We see that affine connections are very special in that they produce effects which are not manifest in the abstract Yang–Mills theory.

Acknowledgements

The author is indebted to D. V. Alekseevsky, F. E. Burstall and A. D. King for stimulating discussions. The author's research was supported by a Leverhulme Fellowship.

References

1. A. Einstein, *The meaning of relativity*, 6th edition, Methuen & Co, London, 1960.
2. E. Schrödinger, *Space-time structure*, Cambridge University Press, 1985.
3. E. W. Mielke, *Geometrodynamics of gauge fields*, Akademie-Verlag, Berlin, 1987.
4. F. W. Hehl, J. D.McCrea, E. W. Mielke, and Y. Ne'eman, *Metric-affine gauge theory of gravity: field equations, Noether identities, world spinors and breaking of dilation invariance*, Physics Reports **258** (1995), 1–171.
5. F. Gronwald, *Metric-affine gauge theory of gravity I. Fundamental structure and field equations*, International Journal of Modern Physics D **6** (1997), 263–303.
6. F. W. Hehl and A. Macias, *Metric-affine gauge theory of gravity II. Exact solutions*, International Journal of Modern Physics D **8** (1999), 399–416.
7. D. Vassiliev, *A tensor interpretation of the 2D Dirac equation*, preprint, http://xxx.lanl.gov/abs/math-ph/0006019, 2000.
8. A. Einstein, *Über die formale Beziehung des Riemannschen Krümmungstensors zu den Feldgleichungen der Gravitation*, Mathematische Annalen **97** (1927), 99–103.
9. M. Nakahara, *Geometry, Topology and Physics*, Institute of Physics, Bristol, 1998.
10. V. B. Berestetskii, E. M. Lifshitz and L. P. Pitaevskii, *Quantum Electrodynamics*, Course of Theoretical Physics vol. **4**, 2nd Edition, Pergamon Press, Oxford, 1982.

11. D. Elton and D. Vassiliev, *The Dirac equation without spinors*, in Rostock Conference on Functional Analysis, Partial Differential Equations and Applications (J.Rossmann, P.Takác, and G.Wildenhain eds.), series Operator Theory: Advances and Applications vol. **110**, Birkhäuser Verlag, Basel, 1999, 133–152.

12. C. Lanczos, *The splitting of the Riemann tensor*, Reviews of Modern Physics **34** (1962), 379–389.

13. G. Y. Rainich, *Electricity in curved space-time*, Nature **115** (1925), 498.

GENERALIZED TAUB-NUT METRICS AND KILLING-YANO TENSORS

MIHAI VISINESCU *
Department of Theoretical Physics,
National Institute for Physics and Nuclear Engineering,
P.O.Box M.G.-6, Magurele, Bucharest, Romania

Abstract. The relation between "hidden" symmetries encapsulated in the Stäckel-Killing tensors and the Killing-Yano tensors is investigated. A necessary condition that a Stäckel-Killing tensor of valence 2 be the contracted product of a Killing-Yano tensor of valence 2 with itself is re-derived for a Riemannian manifold. This condition is applied to the generalized Euclidean Taub-NUT metrics which admit a Kepler type symmetry. It is shown that in general the Stäckel-Killing tensors involved in the Runge-Lenz vector cannot be expressed as a product of Killing-Yano tensors. The only exception is the original Taub-NUT metric.

1. Introduction

It is known that spacetime isometries give rise to constants of motion along geodesics. However not all conserved quantities along geodesics arise from isometries of the manifold and associated Killing vector fields. Such integrals of motion are related to "hidden" symmetries of the manifold encapsulated in the Stäckel-Killing tensors.

A Stäckel-Killing tensor of valence r is a tensor $K_{\mu_1 \ldots \mu_r}$ which is completely symmetric and which satisfies a generalized Killing equation

$$K_{(\mu_1 \ldots \mu_r;\lambda)} = 0. \qquad (1)$$

On manifolds like the four-dimensional Kerr-Newman and Taub-NUT manifolds, the geodesic equations are integrable because of the existence of a Stäckel-Killing tensor $K_{\mu\nu}$ of valence 2 [1] allowing the construction of a constant of motion quadratic in particle's four-momentum p_μ:

$$k = \frac{1}{2} K_{\mu\nu}(x) p^\mu p^\nu = \frac{1}{2} K_{\mu\nu}(x) \dot{x}^\mu \dot{x}^\nu \qquad (2)$$

where the overdot denotes ordinary proper-time differentiation $\frac{d}{d\tau}$.

* mvisin@theor1.theory.nipne.ro

S. Duplij and J. Wess (eds.), Noncommutative Structures in Mathematics and Physics, 441–452.
© *2001 Kluwer Academic Publishers. Printed in the Netherlands.*

The Killing condition (1) is actually equivalent with the conservation of K, i.e. K commutes with the world-line Hamiltonian

$$H = \frac{1}{2} g_{\mu\nu} p^\mu p^\nu \tag{3}$$

in the sense of Poisson brackets.

Related to this, the Klein-Gordon, Schrödinger and Dirac equations are separable in Kerr-Newman [2, 3] and Taub-NUT spaces [4, 5].

Moreover Carter and McLenaghan [6] showed the existence of a Dirac-type linear differential operator which commutes with the standard Dirac operator in the Kerr-Newman space. The construction of this operator depends upon the remarkable fact that the Stäckel-Killing tensor of the Kerr-Newman geometry has a certain root

$$K_{\mu\nu} = f_{\mu\lambda} f^\lambda_\nu \tag{4}$$

where $f_{\mu\nu}$ is a Killing-Yano tensor. A tensor $f_{\mu_1 \ldots \mu_r}$ is called a Killing-Yano tensor of valence r [7] if it is totally antisymmetric and it satisfies the equation

$$f_{\mu_1 \ldots (\mu_r ; \lambda)} = 0. \tag{5}$$

The role of the Killing-Yano tensors can also be noticed for spinning manifolds [8, 9]. The configuration space of spinning particles (spinning space) is an extension of an ordinary Riemannian manifold, parametrized by local coordinates $\{x^\mu\}$, to a graded manifold parametrized by local coordinates $\{x^\mu, \psi^\mu\}$, with the first set of variables being Grassmann-even (commuting) and the second set Grassmann-odd (anticommuting). The equations of motion of the pseudo-classical Dirac particle can be derived from the action

$$S = \int_a^b d\tau \left(\frac{1}{2} g_{\mu\nu}(x) \, \dot{x}^\mu \, \dot{x}^\nu + \frac{i}{2} g_{\mu\nu}(x) \, \psi^\mu \, \frac{D\psi^\nu}{D\tau} \right). \tag{6}$$

where the covariant derivative of the Grassmann-valued spin variable ψ^μ is defined by

$$\frac{D\psi^\mu}{D\tau} = \dot{\psi}^\mu + \dot{x}^\lambda \, \Gamma^\mu_{\lambda\nu} \, \psi^\nu. \tag{7}$$

The action (6) is invariant under the supersymmetry

$$\delta x^\mu = -i\epsilon \psi^\mu \quad , \quad \delta \psi^\mu = \dot{x}^\mu \epsilon \tag{8}$$

where the infinitesimal parameter ϵ of the transformation is Grassmann-odd.

This supersymmetry transformation are obtained from the conserved supercharge

$$Q = \dot{x}_\mu \psi^\mu \quad , \quad \dot{Q} = 0 \tag{9}$$

by taking the bracket

$$\delta F = i\epsilon\{Q, F\}. \tag{10}$$

That Q is conserved and the above supertransformation represent a symmetry follows from the bracket relations

$$\{Q, Q\} - -2iH \quad , \quad \{Q, H\} = 0. \tag{11}$$

Additional conserved supercharges exist if the background geometry admits a Killing-Yano tensor $f_{\mu_1...\mu_r}$. In such a geometry there exist an additional superinvariant constant of motion Q_f defined by [10]

$$Q_f = f_{\mu_1...\mu_r}\Pi^{\mu_1}\psi^{\mu_2}\ldots\psi^{\mu_r} + \frac{i}{r+1}(-1)^{r+1}f_{[\mu_1...\mu_r;\mu_{r+1}]}\cdot\psi^{\mu_1}\ldots\psi^{\mu_{r+1}}. \tag{12}$$

which is superinvariant

$$\{Q_f, Q\} = 0. \tag{13}$$

The existence of a new supersymmetry of this kind implies automatically the existence of a new Grassmann-even constant of motion Z defined by the bracket of Q_f with itself

$$\{Q_f, Q_f\} = -2iZ. \tag{14}$$

The explicit form of Z is given in [9] for Killing-Yano tensors of valence 2.

This paper is devoted to the relations between the Stäckel-Killing and the Killing-Yano tensors for a 4-dimensional Riemannian manifold. The general results are applied to the case of the generalized Euclidean Taub-NUT metrics which admit a Kepler-type symmetry [11].

The Euclidean Taub-NUT metric is involved in many modern studies in physics. Hawking [12] has suggested that the Euclidean Taub-NUT metric might give rise to the gravitational analog of the Yang-Mills instanton. In this case Einstein's equations are satisfied with zero cosmological constant and the manifold is $\mathbb{R}^4$ with a boundary which is a twisted three-sphere S^3 possessing a distorted metric. The Kaluza-Klein monopole was obtained by embedding the Taub-NUT gravitational instanton into five-dimensional Kaluza-Klein theory. On the other hand, in the long-distance limit, neglecting radiation, the relative motion of two monopoles is described by the geodesics of this space [13].

From the symmetry viewpoint, the geodesic motion in Taub-NUT space admits a "hidden" symmetry of the Kepler type if a cyclic variable is gotten rid of [14]. Moreover in the Taub-NUT geometry there are four Killing-Yano tensors [7]. Three of these are complex structure realizing the quaternionic algebra and the Taub-NUT manifold is hyper-Kähler [14]. In addition to these three vector-like Killing-Yano tensors, there is a scalar one which has a non-vanishing field strength and it exists by virtue of the metric being type D.

For the geodesic motions in the Taub-NUT space, the conserved vector analogous to the Runge-Lenz vector of the Kepler type problem is quadratic in 4-velocities, its components are Stäckel-Killing tensors and they can be expressed as symmetrized products of Killing-Yano tensors [14–16, 10, 17].

In the last time, Iwai and Katayama [18–20] extended the Taub-NUT metric so that it still admits a Kepler-type symmetry. This class of metrics, of course, includes the original Taub-NUT metric.

In what follows we investigate if the Stäckel-Killing tensors involved in the conserved Runge-Lenz vector of the extended Taub-NUT metrics can also be expressed in terms of Killing-Yano tensors.

The relationship between Killing tensors and Killing-Yano tensors has been studied to the purpose of the Lorentzian geometry used in general relativity [21, 22]. In the next section we re-examine the conditions that a Killing tensor of valence 2 be the contracted product of a Killing-Yano tensor of valence 2 with itself. The procedure is quite simple and devoted to the Riemannian geometry appropriate to Euclidean Taub-NUT metrics.

In Section 3 we show that in general the Killing tensors involved in the Runge-Lenz vector cannot be expressed as a product of Killing-Yano tensors. The only exception is the original Taub-NUT metric.

Our comments and concluding remarks are presented in Section 4.

2. The relationship between Killing tensors and Killing-Yano tensors

We consider a $4-$dimensional Riemannian manifold M and a metric $g_{\mu\nu}(x)$ on M in local coordinates x^μ. We write the metric in terms of the local orthonormal vierbein frame $e^a{}_\mu$

$$ds^2 = g_{\mu\nu}(x)dx^\mu dx^\nu = \sum_{a=0,1,2,3} (e^a)^2 \tag{15}$$

where $e^a = e^a{}_\mu dx^\mu$. Greek indices $\mu, \nu, \dots$ are raised and lowered with $g_{\mu\nu}$ or its inverse $g^{\mu\nu}$, while Latin indices $a, b, \dots$ are raised and lowered by the flat metric $\delta_{ab}, a, b = 0, 1, 2, 3$. Vierbeins and inverse vierbeins inter-convert Latin and Greek indices when necessary.

Let Λ^2 be the space of two-forms $\Lambda^2 := \Lambda^2 T^*(\mathbb{R}^4 - \{0\})$. We define self-dual and anti-self dual bases for Λ^2 using the vierbein one-forms e^a:

$$basis\ of\ \Lambda^2_\pm = \begin{cases} \lambda^1_\pm = e^0 \wedge e^1 \pm e^2 \wedge e^3 \\ \lambda^2_\pm = e^0 \wedge e^2 \pm e^3 \wedge e^1 \\ \lambda^3_\pm = e^0 \wedge e^3 \pm e^1 \wedge e^2 \end{cases}, \quad *\lambda^i_\pm = \pm\lambda^i_\pm \tag{16}$$

Let f be a Killing-Yano tensor of valence 2 and $*f$ its dual. The symmetric combination of f and $*f$ is a self-dual two-form

$$f + *f = \sum_{i=1,2,3} y_i \lambda_+^i \tag{17}$$

while their difference is an anti-self-dual two-form

$$f - *f = \sum_{i=1,2,3} z_i \lambda_-^i. \tag{18}$$

An explicit evaluation shows that

$$(f + *f)^2 = - \sum_{i=1,2,3} (y_i)^2 \cdot \mathbf{1}, \tag{19}$$

$$(f - *f)^2 = - \sum_{i=1,2,3} (z_i)^2 \cdot \mathbf{1} \tag{20}$$

where $\mathbf{1}$ is 4×4 identity matrix.

Let us suppose that a Stäckel-Killing tensor $K_{\mu\nu}$ can be written as the contracted product of a Killing-Yano tensor $f_{\mu\nu}$ with itself:

$$K_{\mu\nu} = f_{\mu\lambda} \cdot f^\lambda{}_\nu = (f^2)_{\mu\nu}, \quad \mu,\nu = 0,1,2,3. \tag{21}$$

We infer from the last equations that:

$$K + \frac{1}{16} \left[\sum_i (y_i^2 - z_i^2) \right]^2 K^{-1} + \frac{1}{2} \sum_i (y_i^2 + z_i^2) \cdot \mathbf{1} = 0. \tag{22}$$

On the other hand the Killing tensor K is symmetric and it can be diagonalized with the aid of an orthogonal matrix. Its eigenvalues satisfy an equation of the second degree:

$$\lambda_\alpha^2 + \frac{1}{2} \sum_i (y_i^2 + z_i^2) \lambda_\alpha + \frac{1}{16} \left[\sum_i (y_i^2 - z_i^2) \right]^2 = 0 \tag{23}$$

with at most two distinct roots.

In conclusion a Stäckel-Killing tensor K which can be written as the square of a Killing-Yano tensor has at the most two distinct eigenvalues.

3. Generalized Taub-NUT metrics

For a special choice of coordinates the generalized Euclidean Taub-NUT metric considered by Iwai and Katayama [18–20] takes the form:

$$ds_G^2 = f(r)[dr^2 + r^2 d\theta^2 + r^2 \sin^2 \theta \, d\varphi^2] + g(r)[d\chi + \cos \theta \, d\varphi]^2 \qquad (24)$$

where $r > 0$ is the radial coordinate of $\mathbb{R}^4 - \{0\}$, the angle variables $(\theta, \varphi, \chi), (0 \leq \theta < \pi, 0 \leq \varphi < 2\pi, 0 \leq \chi < 4\pi)$ parameterize the unit sphere S^3, and $f(r)$ and $g(r)$ are arbitrary functions of r.

We decompose the metric (24) into the orthogonal vierbein basis:

$$\begin{aligned}
e^0 &= g(r)^{\frac{1}{2}}(d\chi + \cos \theta d\varphi), \\
e^1 &= rf(r)^{\frac{1}{2}}(\sin \chi d\theta - \sin \theta \cos \chi d\varphi), \\
e^2 &= rf(r)^{\frac{1}{2}}(-\cos \chi d\theta - \sin \theta \sin \chi d\varphi), \\
e^3 &= f(r)^{\frac{1}{2}} dr.
\end{aligned} \qquad (25)$$

Spaces with a metric of the form above have an isometry group $SU(2) \times U(1)$. There are four Killing vectors

$$D_A = R_A^\mu \, \partial_\mu, A = 0, 1, 2, 3, \qquad (26)$$

corresponding to the invariance of the metric (24) under spatial rotations ($A = 1, 2, 3$) and χ translations ($A = 0$).

Let us consider geodesic flows of the generalized Taub-NUT metric which has the Lagrangian L on the tangent bundle $T(\mathbb{R}^4 - \{0\})$

$$L = \frac{1}{2}f(r)[\dot{r}^2 + r^2(\dot{\theta}^2 + \sin^2 \theta \dot{\varphi}^2)] + \frac{1}{2}g(r)(\dot{\chi} + \cos \theta \dot{\varphi})^2 \qquad (27)$$

where $(\dot{r}, \dot{\theta}, \dot{\varphi}, \dot{\chi}, r, \theta, \varphi, \chi)$ stand for coordinates in the tangent bundle. Since χ is a cyclic variable

$$q = g(r)(\dot{\theta} + \cos \theta \dot{\varphi}) \qquad (28)$$

is a conserved quantity. This is known in the literature as the "relative electric charge".

Taking into account this cyclic variable, the dynamical system for the geodesic flow on $T(\mathbb{R}^4 - \{0\})$ can be reduced to a system on $T(\mathbb{R}^3 - \{0\})$. The reduced system admits manifest rotational invariance, and hence has a conserved angular momentum:

$$\vec{J} = \vec{r} \times \vec{p} + q\frac{\vec{r}}{r} \qquad (29)$$

where $\vec{r}$ denotes the three-vector $\vec{r} = (r, \theta, \varphi)$ and $\vec{p} = f(r)\dot{\vec{r}}$ is the mechanical momentum.

If $f(r)$ and $g(r)$ are taken to be

$$f(r) = \frac{4m + r}{r} \quad , \quad g(r) = \frac{16m^2 r}{4m + r} \tag{30}$$

the metric ds_G^2 becomes the original Euclidean Taub-NUT metric. The parameter m can be positive or negative, depending on the application; for $m > 0$ the four-dimensional Taub-NUT metric represents a non-singular solution of the self-dual Euclidean Einstein equation and as such is interpreted as a gravitational instanton.

As observed in [14], the Taub-NUT geometry also possesses four Killing-Yano tensors of valence 2. The first three are rather special: they are covariantly constant (with vanishing field strength)

$$f_i = 8m(d\chi + \cos\theta d\varphi) \wedge dx_i - \epsilon_{ijk}(1 + \frac{4m}{r})dx_j \wedge dx_k,$$
$$D_\mu f_{i\lambda}^\nu = 0, \quad i, j, k = 1, 2, 3. \tag{31}$$

They are mutually anticommuting and square the minus unity:

$$f_i f_j + f_j f_i = -2\delta_{ij}. \tag{32}$$

Thus they are complex structures realizing the quaternion algebra. Indeed, the Taub-NUT manifold defined by (24) and (30) is hyper-Kähler.

In addition to the above vector-like Killing-Yano tensors there also is a scalar one

$$f_Y = 8m(d\chi + \cos\theta d\varphi) \wedge dr + 4r(r + 2m)(1 + \frac{r}{4m}) \sin\theta d\theta \wedge d\varphi \tag{33}$$

which has a non-vanishing component of the field strength

$$f_{Y r\theta;\varphi} = 2(1 + \frac{r}{4m})r \sin\theta. \tag{34}$$

In the original Taub-NUT case there is a conserved vector analogous to the Runge-Lenz vector of the Kepler-type problem:

$$\vec{K} = \frac{1}{2}\vec{K}_{\mu\nu}\dot{x}^\mu\dot{x}^\nu = \vec{p} \times \vec{j} + \left(\frac{q^2}{4m} - 4mE\right)\frac{\vec{r}}{r} \tag{35}$$

where the conserved energy E, from eq. (3), is

$$E = \frac{\vec{p}^2}{2f(r)} + \frac{q^2}{2g(r)}. \tag{36}$$

The components $K_{i\mu\nu}$ involved with the Runge-Lenz type vector (35) are Killing tensors and they can be expressed as symmetrized products of the Killing-Yano tensors f_i (31) and f_Y (33) [16, 10]:

$$K_{i\mu\nu} - \frac{1}{8m}(R_{0\mu}R_{i\nu} + R_{0\nu}R_{i\mu}) = m\left(f_{Y\mu\lambda}f_i{}^\lambda{}_\nu + f_{Y\nu\lambda}f_i{}^\lambda{}_\mu\right). \tag{37}$$

Returning to the generalized Taub-NUT metric, on the analogy of eq. (35), Iwai and Katayama [18–20] assumed that in addition to the angular momentum vector there exist a conserved vector $\vec{S}$ of the following form:

$$\vec{S} = \vec{P} \times \vec{J} + \kappa\frac{\vec{r}}{r} \tag{38}$$

with an unknown constant κ.

It was found that the metric (24) still admits a Kepler type symmetry (38) if the functions $f(r)$ and $g(r)$ take, respectively, the form

$$f(r) = \frac{a + br}{r} \ , \quad g(r) = \frac{ar + br^2}{1 + cr + dr^2} \tag{39}$$

where a, b, c, d are constants. The constant κ involved in the Runge-Lenz vector (38) is

$$\kappa = -aE + \frac{1}{2}cq^2. \tag{40}$$

If $ab > 0$ and $c^2 - 4d < 0$ or $c > 0, d > 0$, no singularity of the metric appears in $\mathbb{R}^4 - \{0\}$. On the other hand, if $ab < 0$ a manifest singularity appears at $r = -a/b$ [19].

It is straightforward to verify that the components of the vector $\vec{S}$ are Stäckel-Killing tensors in the extended Taub-NUT space (24) with the function $f(r)$ and $g(r)$ given by (39). Moreover the Poisson brackets between the components of $\vec{J}$ and $\vec{S}$ are [18]:

$$\begin{aligned}
\{J_i, J_j\} &= \epsilon_{ijk}J_k, \\
\{J_i, S_j\} &= \epsilon_{ijk}S_k, \\
\{S_i, S_j\} &= (dq^2 - 2bE)\epsilon_{ijk}J_k
\end{aligned} \tag{41}$$

as it is expected from the same relations known for the original Taub-NUT metric.

Our task is to investigate if the components of the Runge-Lenz vector (38) can be the contracted product of Killing-Yano tensors of valence 2. On the model of eq.(37) from the original Taub-NUT case it is not required that a component S_i of the Runge-Lenz vector (38) to be directly expressed as a symmetrized product of Killing-Yano tensors. Taking into account that $\vec{S}$ transforms as a vector under rotations generated by $\vec{J}$, eq.(41), the components $S_{i\mu\nu}$ can be combined with trivial Stäckel-Killing tensors of the form $(R_{0\mu}R_{i\nu} + R_{0\nu}R_{i\mu})$ to get the appropriate tensor which has to be decomposed in a product of Killing-Yano tensors.

In order to use the results from the previous section, we shall write the symmetrized product of two different Killing-Yano tensors f' and f'' as a contracted product of $f' + f''$ with itself, extracting adequately the contribution of f'^2 and f''^2. Since the generalized Taub-NUT space (24) does not admit any other nontrivial Stäckel-Killing tensor besides the metric $g_{\mu\nu}$ and the components $S_{i\mu\nu}$ of (38), f'^2 and f''^2 should be connected with the scalar conserved quantities $E, \vec{J}^2, q^2$ through the tensors $g_{\mu\nu}, \sum_{A=1,2,3} R_{A\mu}R_{A\nu}$ and $R_{0\mu}R_{0\nu}$.

In conclusion we shall consider a general linear combination between a component S_i of the Runge-Lenz vector (38) and symmetrized pairs of Killing vectors of the form

$$S_{iab} + \alpha_1 \sum_{A=1}^{3} R_{Aa}R_{Ab} + \alpha_2 R_{0a}R_{0b} + \alpha_3(R_{0a}R_{ib} + R_{ia}R_{0b}) \qquad (42)$$

where α_i are constants. We are looking for the conditions the above tensor be the contracted product of a Killing-Yano tensor with itself. For this purpose we evaluate the eigenvalues of the matrix (42) and we get that it has at the most two distinct eigenvalues if and only if

$$\alpha_1 + \alpha_2 = 0,$$
$$\alpha_3 = -\frac{c}{4},$$
$$d = \frac{c^2}{4}. \qquad (43)$$

Hence the constants involved in the functions f, g are constrained, restricting accordingly their expressions. It is worth to mention that if relation (43) between the constants c and d is satisfied, the metric is conformally self-dual or anti-self-dual depending upon the sign of the quantity $2 + cr$ [19].

Finally the condition stated for a Stäckel-Killing tensor to be written as the square of a skew symmetric tensor in the form (21) must be supplemented with eq.(5) which defines a Killing-Yano tensor. To verify this last condition we shall use the Newman-Penrose formalism for Euclidean signature [23]. We introduce a tetrad which will be given as an isotropic complex dyad defined by the vectors l, m together with their complex conjugates subject to the normalization conditions

$$l_\mu \bar{l}^\mu = 1, \quad m_\mu \bar{m}^\mu = 1 \qquad (44)$$

with all others vanishing and the metric is expressed in the form

$$ds^2 = l \otimes \bar{l} + \bar{l} \otimes l + m \otimes \bar{m} + \bar{m} \otimes m. \qquad (45)$$

For a Stäckel-Killing tensor K with two distinct eigenvalues one can choose the tetrad in such that

$$K_{\mu\nu} = 2\lambda_1^2 l_{(\mu} \bar{l}_{\nu)} + 2\lambda_2^2 m_{(\mu} \bar{m}_{\nu)}. \tag{46}$$

The skew symmetric tensor $f_{\mu\nu}$ which enter decomposition (21) has the form

$$f_{\mu\nu} = 2\lambda_1 l_{[\mu} \bar{l}_{\nu]} + 2\lambda_2 m_{[\mu} \bar{m}_{\nu]}. \tag{47}$$

Again, a standard evaluation shows that the above quantity is a Killing-Yano tensor only if

$$c = \frac{2b}{a}. \tag{48}$$

With this constraint, together with (43), the extended metric (24) coincides, up to a constant factor, with the original Taub-NUT metric on setting $a/b = 4m$.

4. Concluding remarks

The aim of this paper is to show that the extensions of the Taub-NUT geometry do not admit a Killing-Yano tensor, even if they possess Stäckel-Killing tensors.

This result is not unexpected. The conserved quantities $K_{i\mu\nu}$ which enter eq.(37) are the components of the Runge-Lenz vector $\vec{K}$ given in (35). In the original Taub-NUT case these components $K_{i\mu\nu}$ are related to the symmetrized products between the Killing-Yano tensors f_i (31) and f_Y (33). Adequately the three Killing-Yano tensors f_i transform as vectors under rotations generated by $\vec{J}$ like the Runge-Lenz vector (41), while f_Y is a scalar.

The extended Taub-NUT metrics are not Ricci flat and, consequently, not hyper-Kähler. On the other hand the existence of the Killing-Yano tensors f_i is correlated with the hyper-Kähler, self-dual structure of the metric.

The non-existence of the Killing-Yano tensors makes the study of "hidden" symmetries more laborious in models of relativistic particles with spin involving anticommuting vectorial degrees of freedom. In general the conserved quantities from the scalar case receive a spin contribution involving an even number of Grassmann variables ψ^μ. For example, starting with a Killing vector K_μ, the conserved quantity in the spinning case is

$$J(x, \dot{x}, \psi) = K^\mu \dot{x}_\mu + \frac{i}{2} K_{[\mu;\nu]} \psi^\mu \psi^\nu. \tag{49}$$

The first term in the r.h.s. is the conserved quantity in the scalar case, while the last term represents the contribution of the spin.

The generalized Killing equations on spinning spaces in the presence of a Stäckel-Killing tensor are more involved. Unfortunately it is not possible to write closed, analytic expressions of the solutions of these equations using directly the components of the Stäckel-Killing tensors. However, assuming that the Stäckel-Killing tensors can be written as symmetrized products of pairs of Killing-Yano tensors, the evaluation of the spin corrections is feasible [9, 16, 10, 17].

If the Killing-Yano tensors are missing, to take up the question of the existence of extra supersymmetries and the relation with the constants of motion we are forced to enlarge the approach to Killing equations (5), (1). In fact, in ref.[9], supersymmetries are shown to depend on the existence of a tensor field $f_{\mu\nu}$ satisfying eq.(5) which will be referred to as the f-symbol. The general conditions for constants of motion were derived, and it was shown that one can have new supercharges which do not commute with the original supercharge Q (9) if one allows the f-symbols to have a symmetric part. It was shown that in this case the antisymmetric part does not satisfy the Killing-Yano condition (5). We should like to remark that the general conditions of ref.[9] allow more possibilities than Killing-Yano tensors for the construction of supercharges.

Summing up, we believe that the relation between the f-symbols and the Killing-Yano tensors could be fruitful and that it should deserve further studies. An analysis of the f-symbols in the generalized Taub-NUT geometry is under way.

Acknowledgements

I should like to acknowledge the generosity of NATO in its support for this workshop. It is a pleasure to thank the organizers of the NATO ARW, Kiev 2000, in particular to Julius Wess and Steven Duplij, for the extremely friendly and stimulating atmosphere.

References

1. B. Carter, *Killing tensor quantum numbers and conserved currents in curved space*, Phys.Rev. **16** (1977) 3395-3414.
2. B. Carter, *Global structure of the Kerr family of gravitational fields*, Phys.Rev. **174** (1968) 1559-1571.
3. S. Chandrasekhar, *The solution of Dirac's equation in Kerr geometry*, Proc. R. Soc. London A. **349** (1976) 571-575.
4. I. I. Cotaescu and M. Visinescu, *Schrödinger quantum modes on the Taub-NUT background*, Mod. Phys. Lett. A **15** (2000) 145-157.
5. I. I. Cotaescu and M. Visinescu, *The Dirac field in Taub-NUT background*, hep-th/0008181.
6. B. Carter and R. G. McLenaghan, *Generalized total angular momentum operator for the Dirac equation in curved space-time*, Phys.Rev. **D19** (1979) 1093-1097.
7. K.Yano, *Some remarks on tensor fields and curvature*, Ann.Math. **55** (1952) 328-347.
8. F. A. Berezin and M. S. Marinov, *Particle spin dynamics as the Grassmann variant of classical mechanics*, Ann.Phys.N.Y. **104** (1977) 336-362.

9. G. W. Gibbons, R. H. Rietdijk and J. W.van Holten, *SUSY in the sky*, Nucl.Phys. **B404** (1993) 42-64.

10. D. Vaman and M. Visinescu, *Supersymmetries and constants of motion in Taub-NUT spinning space*, Fortschr.Phys. **47** (1999) 493-514.

11. M. Visinescu, *Generalized Taub-NUT metrics and Killing-Yano tensors*, J. Phys. A: Math. Gen. **33** (2000) 4383-4391.

12. S. W. Hawking, *Gravitational instantons*, Phys.Lett. **60A** (1977) 81-83.

13. N. S. Manton, *A remark on the scattering of BPS monopoles* , Phys.Lett. **B110** (1985) 54-56; *Monopole interaction at long range*, Phys.Lett. **B154** (1985) 397-400.

14. G. W. Gibbons and P. J. Ruback, *The hidden symmetries of Taub-NUT and monopole scattering*, Phys.Lett. **B188** (1987) 226-230; Commun.Math.Phys. *The hidden symmetries of multi-centre metrics*, **115** (1988) 267-300.

15. J. W. van Holten, *Supersymmetry and the geometry of Taub-NUT*, Phys.Lett. **B342** (1995) 47-52.

16. D. Vaman and M. Visinescu, *Spinning particles in Taub-NUT space*, Phys.Rev. **D57** (1998) 3790-3793.

17. J. W. van Holten, *Killing-Yano tensors, non-standard supersymmetries and an index theorem*, gr-qc/9910035.

18. T. Iwai and N. Katayama, *On extended Taub-NUT metrics*, J.Geom.Phys. **12** (1993) 55-75.

19. T. Iwai and N. Katayama, *Two kinds of generalized Taub-NUT metrics and the symmetry of associated dynamical systems*, J.Phys. A: Math.Gen. **27** (1994) 3179-3190.

20. T. Iwai and N. Katayama, *Two classes of dynamical systems all of whose bounded trajectories are closed*, J.Math.Phys. **35** (1994) 2914-2933.

21. C. D. Collinson, *On the relationship between Killing tensors and Killing-Yano tensors*, Int.J.Theor.Phys. **15** (1976) 311-314.

22. W. Dietz and R. Rüdinger, *Space-times admitting Killing-Yano tensors. I*, Proc.R.Soc.Lond. **A375** (1981) 361-378.

23. A. N. Aliev and Y. Nutku, *Gravitational instantons admit hyper-Kähler structure*, Class.Quant.Grav. **16** (1999) 189-210.

AN EFFECTIVE MODEL OF THE SPACETIME FOAM

VLADIMIR DZHUNUSHALIEV *
Kyrgyz-Russian Slavic University, Bishkek, Kyrgyzstan

1. Introduction

The notion of a spacetime foam was introduced by Wheeler [1, 2] for the description of the possible complex structure of the spacetime on the Planck scale ($L_{Pl} \approx 10^{-33} cm$). This hypothesized spacetime foam is a set of quantum wormholes (WH) (handles) appearing in the spacetime on the Planck scale level (see Fig.1). For the macroscopic observer these quantum fluctuations are smoothed and we have an ordinary smooth manifold with the metric submitting to Einstein equations. The exact mathematical description of this phenomenon is very difficult and even though there is a doubt: does the Feynman path integral in the gravity contain a topology change of the spacetime ? This question spring up because

* dzhun@hotmail.kg

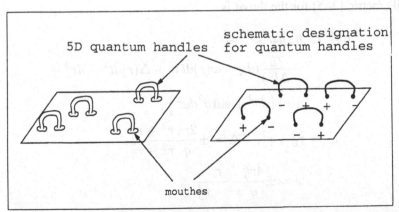

Figure 14. At the left side of the figure is presented a hypothesized spacetime foam. If we neglect of the cross section of handle then (at the right) hand we have a schematic designation for the spacetime foam.

S. Duplij and J. Wess (eds.), Noncommutative Structures in Mathematics and Physics, 453–463.
© 2001 *Kluwer Academic Publishers. Printed in the Netherlands.*

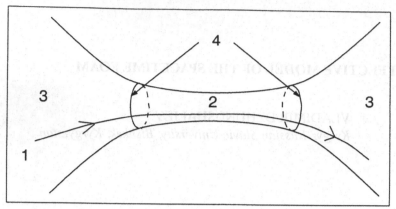

Figure 15. Here whole spacetime is 5D but in the external spacetime (3) G_{55} is nonvariable and we have Kaluza-Klein theory in its initial interpretation as 4D gravity + electromagnetism. In the throat (2) G_{55} component of the 5D metric is equivalent to 4D gravity+electromagnetism+scalar field. Near the event horizon (4) the metric is the Reissner-Nordstrom metric and the throat is a solution of the 5D Kaluza-Klein theory. We should join these metrics on the event horizons. (1) is the force line of the electric field.

(according to the Morse theory) the singular points must arise by the topology change. In such points the time arrow is undefined that leads in difficulties at definition of the Lorentzian metric, curvature tensor and so on. The main goal of this paper is to submit *an effective model of the spacetime foam.*

2. Model of a single quantum wormhole

At first we present a model of a single handle in the spacetime foam, see Fig(2). The 5D metric [3–5] for the throat is

$$ds^2 = \eta_{AB}\omega^A\omega^B =$$
$$-\frac{r_0^2}{\Delta(r)}(d\chi - \omega(r)dt)^2 + \Delta(r)dt^2 - dr^2 -$$
$$a(r)\left(d\theta^2 + \sin\theta^2 d\varphi^2\right), \tag{1}$$

$$a = r_0^2 + r^2, \quad \Delta = \pm\frac{2r_0}{q}\frac{r^2 + r_0^2}{r^2 - r_0^2},$$

$$\omega = \pm\frac{4r_0^2}{q}\frac{r}{r^2 - r_0^2}. \tag{2}$$

where χ is the 5th extra coordinate; $\eta_{AB} = (\pm, -, -, -, \mp)$, $A, B = 0, 1, 2, 3, 5$; r, θ, φ are the 3D polar coordinates; $r_0 > 0$ and q are some constants. We can see that there are two closed $ds_{(5)}^2(\pm r_0) = 0$ hypersurfaces at the $r = \pm r_0$. In some sense these hypersurfaces are like to the event horizon and in Ref.[6] such hypersurfaces are named as a D-holes. On these hypersurfaces we should join [7]:

-   the flux of the 4D electric field (defined by the Maxwell equations) with the flux of the 5D electric field defined by $R_{5t} = 0$ Kaluza-Klein equation.
-   the area of the Reissner-Nordström event horizon with the area of the $ds_{(5)}^2 (\pm r_0) = 0$ hypersurface.

It is necessary to note that both solutions (Reissner-Nordström black hole and 5D throat) have only two integration constants[1] and on the event horizon takes place an algebraic relation between these 4D and 5D integration constants. Another explanation of the fact that we use only two joining condition is the following (see Ref.[8] for the more detailed explanations): in some sense on the event horizon holds a "holography principle". This means that in the presence of the event horizon the 4D and 5D Einstein equations lead to a reduction of the amount of initial data. For example the Einstein - Maxwell equations for the Reissner-Nordström metric

$$ds^2 = \Delta dt^2 - \frac{dr^2}{\Delta} - r^2 \left(d\theta^2 + \sin^2 d\varphi^2 \right), \tag{3}$$

$$A_\mu = (\omega, 0, 0, 0)) \tag{4}$$

(where A_μ is the electromagnetic potential, κ is the gravitational constant) can be written as

$$-\frac{\Delta'}{r} + \frac{1 - \Delta}{r^2} = \frac{\kappa}{2}\omega'^2, \tag{5}$$

$$\omega' = \frac{q}{r^2}. \tag{6}$$

For the Reissner - Nordström black hole the event horizon is defined by the condition $\Delta(r_g) = 0$, where r_g is the radius of the event horizon. Hence in this case we see that on the event horizon

$$\Delta'_g = \frac{1}{r_g} - \frac{\kappa}{2}r_g\omega'^2_g, \tag{7}$$

here (g) means that the corresponding value is taken on the event horizon. Thus, Eq. (5), which is the Einstein equation, is a first-order differential equations in the whole spacetime $(r \geq r_g)$. The condition (7) tells us that the derivative of the metric on the event horizon is expressed through the metric value on the event horizon. This is the same what we said above: the reduction of the amount of initial data takes place by such a way that we have only two integration constants (mass m and charge e for the Reissner-Nordström solution and q and r_0 for the 5D throat).

---

[1] in fact, for the Reissner-Nordström black hole this leads to the "no hair" theorem.

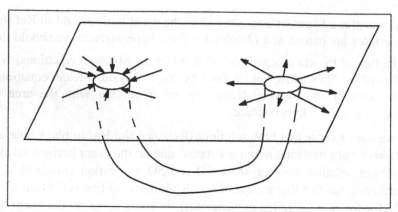

Figure 16. The left mouth of the quantum WH entraps the force lines of the electric field and looks as (-) electric charge. The force lines outcome from the right mouth of WH which one looks as (+) charge.

The 5D throat has an interesting property [9]. We see that the signs of the η_{55} and η_{00} are not defined. We remark that this 5D metric is located behind the event horizon therefore the 4D observer is not able to determine the signs of the η_{55} and η_{00}. Moreover this 5D metric can fluctuate between these two possibilities. Hence the external 4D observer is forced to describe such composite WH by means of something like spinor.

Another interesting characteristic property of this solution is that we have the flux of electric field through the throat, *i.e.* each mouth can entrap the electric force lines and this leads that this mouth is like to electric charge for the external 4D observer, see Fig.16. We can neglect the cross section of the throat and in this case each mouth is point-like and we can try to describe these mouths with help of some effective field. Taking into account the spinor-like properties of quantum handles, we assume that *spacetime foam can be described with help of an effective spinor field.*

3. Approximate model of the spacetime foam

The physical meaning of the spinor field depends on the method of attaching the quantum handles to the external space, see Fig.(17).

3.1. QUANTUM WORMHOLES WITH SEPARATED MOUTHS

In this case $|\psi|^2$ is a density of the mouths in the external space and $e|\psi|^2$ is a density of the electric charge [10].

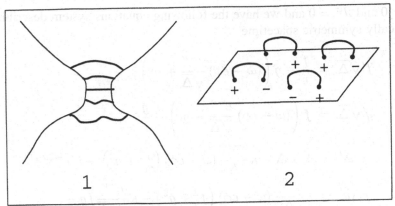

Figure 17. At the left hand of the figure quantum handles connect two spaces. At the right hand the mouths of quantum handles are separated in distance of the order l_{Pl}.

Following this way we write differential equations for the gravitational + electromagnetic fields in the presence of the spacetime foam (ψ) as follows

$$R_{\mu\nu} - \frac{1}{2}g_{\mu\nu}R = T_{\mu\nu}, \tag{8}$$

$$\left(i\gamma^\mu\partial_\mu + eA_\mu - \frac{i}{4}\omega_{\bar{a}\bar{b}\mu}\gamma^\mu\gamma^{[\bar{a}}\gamma^{\bar{b}]} - m\right)\psi = 0, \tag{9}$$

$$D_\nu F^{\mu\nu} = 4\pi e\left(\bar{\psi}\gamma^\mu\psi\right), \tag{10}$$

For our model we use the following ansatz: the spherically symmetric metric

$$ds^2 = e^{2\nu(r)}\Delta(r)dt^2 - \frac{dr^2}{\Delta(r)} - r^2\left(d\theta^2 + \sin^2 d\varphi^2\right), \tag{11}$$

the electromagnetic potential

$$A_\mu = (-\phi, 0, 0, 0), \tag{12}$$

and the spinor field

$$\tilde{\psi} = e^{-i\omega t}\frac{e^{-\nu/2}}{r\Delta^{1/4}}\left(f, 0, ig\cos\theta, ig\sin\theta e^{i\varphi}\right). \tag{13}$$

The following is *very important* for us: the ansatz (13) for the spinor field ψ has the $T_{t\varphi}$ component of the energy-momentum tensor and the $J^\varphi = 4\pi e(\bar{\psi}\gamma^\varphi\psi)$ component of the current. Let we remind that ψ determines the stochastical gas of the virtual WH's which can not have a preferred direction in the spacetime. This means that after substitution expression (11)-(13) into field equations they should be averaged by the spin direction of the ansatz (13). After this averaging we have

$T_{t\varphi} = 0$ and $J^\varphi = 0$ and we have the following equations system describing our spherically symmetric spacetime

$$f'\sqrt{\Delta} = \frac{f}{r} - g\left((\omega - e\phi)\frac{e^{-\nu}}{\sqrt{\Delta}} + m\right), \tag{14}$$

$$g'\sqrt{\Delta} = f\left((\omega - e\phi)\frac{e^{-\nu}}{\sqrt{\Delta}} - m\right) - \frac{g}{r}, \tag{15}$$

$$r\Delta' = 1 - \Delta - \kappa\frac{e^{-2\nu}}{\Delta}(\omega - e\phi)\left(f^2 + g^2\right) - r^2 e^{-2\nu}\phi'^2, \tag{16}$$

$$r\Delta\nu' = \kappa\frac{e^{-2\nu}}{\Delta}(\omega - e\phi)\left(f^2 + g^2\right) - \kappa\frac{e^{-\nu}}{r\sqrt{\Delta}}fg -$$
$$\frac{\kappa}{2}m\frac{e^{-\nu}}{\sqrt{\Delta}}\left(f^2 - g^2\right), \tag{17}$$

$$r^2\Delta\phi'' = -8\pi e\left(f^2 + g^2\right) - \left(2r\Delta - r^2\Delta\nu'\right)\phi', \tag{18}$$

where κ is some constant. This equations system was investigated in [11] and result is the following. A particle-like solution exists which has the following expansions near $r = 0$

$$f(r) = f_1 r + \mathcal{O}(r^2), \quad g(r) = \mathcal{O}(r^2), \tag{19}$$
$$\Delta(r) = 1 + \mathcal{O}(r^2), \quad \nu(r) = \mathcal{O}(r^2), \quad \phi(r) = \mathcal{O}(r^2) \tag{20}$$

and the following asymptotical behaviour

$$\Delta(r) \approx 1 - \frac{2m_\infty}{r} + \frac{(2e_\infty)^2}{r^2}, \quad \nu(r) \approx const, \tag{21}$$

$$\phi(r) \approx \frac{2e_\infty}{r}, \tag{22}$$

$$f \approx f_0 e^{-\alpha r}, \quad g \approx g_0 e^{-\alpha r},$$
$$\frac{f_0}{g_0} = \sqrt{\frac{m_\infty + \omega}{m_\infty - \omega}}, \quad \alpha^2 = m_\infty^2 - \omega^2, \tag{23}$$

where m_∞ is the mass for the observer at infinity and $2e_\infty$ is the charge of this solution.

The solution exists for both cases $(|e_\infty|/m_\infty) > 1$ and $(|e_\infty|/m_\infty) < 1$ but for us is essential the first case with $(|e_\infty|/m_\infty) > 1$. In this case the classical Einstein-Maxwell theory leads to the "naked" singularity. The presence of the spacetime foam drastically changes this result: *the appearance of the virtual wormholes can prevent the formation of the "naked" singularuty in the Reissner-Nordström solution with $|e|/m > 1$.*

Our interpretation of this solution is presented on the Fig.(18).

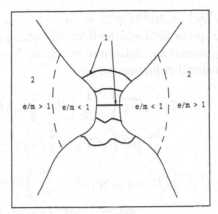

Figure 18. **1** are the quantum (virtual) WHs, **2** are two solutions with $|e_\infty|/m_\infty > 1$. Such object can be named as ***the wormhole with quantum throat.***

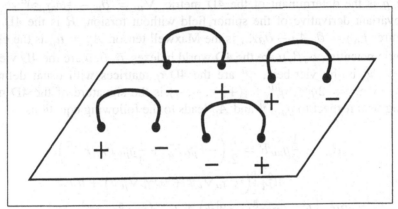

Figure 19. The distance between mouths of the quantum handle is of order l_{Pl}.

3.2. QUANTUM WORMHOLES WITH NON-SEPARATED MOUTHS

The second possibility [12] is presented on the Fig.(19).

We will consider the 5D Kaluza-Klein theory + torsion + spinor field. The Lagrangian in this case is

$$\mathcal{L} = \sqrt{-G} \left\{ -\frac{1}{2k} \left(R^{(5)} - S_{ABC}S^{ABC} \right) + \frac{\hbar c}{2} \left[i\bar{\psi} \left(\gamma^C \nabla_C - \frac{mc}{i\hbar} \right) \psi + h.c. \right] \right\} \tag{24}$$

where $\nabla_C = \partial_C - \frac{1}{4}(\omega_{\bar{A}\bar{B}C} + S_{\bar{A}\bar{B}C})\gamma^{[\bar{A}}\gamma^{\bar{B}]}$ is the covariant derivative, G is the determinant of the 5D metric, $R^{(5)}$ is the 5D scalar curvature, S_{ABC} is the antisymmetrical torsion tensor, A, B, C are the 5D world indexes, $\bar{A}, \bar{B}, \bar{C}$ are the 5-bein indexes, $\gamma^B = h_{\bar{A}}^B \gamma^{\bar{A}}$, $h_{\bar{A}}^B$ is the 5-bein, $\gamma^{\bar{A}}$ are the 5D γ matrices with usual definitions $\gamma^{\bar{A}}\gamma^{\bar{B}} + \gamma^{\bar{B}}\gamma^{\bar{A}} = 2\eta^{\bar{A}\bar{B}}$, $\eta^{\bar{A}\bar{B}} = (+,-,-,-,-)$ is the signature of the 5D metric, ψ is the spinor field which effectively and approximately describes the spacetime foam, $[]$ means the antisymmetrization, $\hbar$, c and m are the usual constants. After dimensional reduction we have

$$\mathcal{L} = \sqrt{-g} \left\{ -\frac{1}{2k} \left(R + \frac{1}{4} F_{\alpha\beta} F^{\alpha\beta} \right) + \right.$$

$$\frac{\hbar c}{2} \left[i\bar{\psi} \left(\gamma^\mu \tilde{\nabla}_\mu - \frac{1}{8} F_{\bar{\alpha}\bar{\beta}} \gamma^5 \gamma^{[\bar{\alpha}} \gamma^{\bar{\beta}]} - \right. \right.$$

$$\left. \frac{1}{4} l_{Pl}^2 \left(\gamma^{[\bar{A}} \gamma^{\bar{B}} \gamma^{\bar{C}]} \right) \left(i\bar{\psi} \gamma_{[\bar{A}} \gamma_{\bar{B}} \gamma_{\bar{C}]} \psi \right) - \frac{mc}{i\hbar} \right) \psi + h.c. \right] \right\} \tag{25}$$

$$S^{\bar{A}\bar{B}\bar{C}} = 2 l_{Pl}^2 \left(i\bar{\psi} \gamma^{[\bar{A}} \gamma^{\bar{B}} \gamma^{\bar{C}]} \psi \right) \tag{26}$$

where g is the determinant of the 4D metric, $\tilde{\nabla}_\mu = \partial_\mu - \frac{1}{4}\omega_{\bar{a}\bar{b}\mu}\gamma^{[\bar{a}}\gamma^{\bar{b}]}$ is the 4D covariant derivative of the spinor field without torsion, R is the 4D scalar curvature, $F_{\alpha\beta} = \partial_\alpha A_\beta - \partial_\beta A_\alpha$ is the Maxwell tensor, $A_\mu = h_\mu^{\bar{5}}$ is the electromagnetic potential, α, β, μ are the 4D world indexes, $\bar{\alpha}, \bar{\beta}, \bar{\mu}$ are the 4D vier-bein indexes, $h_\nu^{\bar{\mu}}$ is the vier-bein, $\gamma^{\bar{\mu}}$ are the 4D γ matrices with usual definitions $\gamma^{\bar{\mu}}\gamma^{\bar{\nu}} + \gamma^{\bar{\nu}}\gamma^{\bar{\mu}} = 2\eta^{\bar{\mu}\bar{\nu}}$, $\eta^{\bar{\mu}\bar{\nu}} = (+,-,-,-)$ is the signature of the 4D metric. Varying with respect to $g_{\mu\nu}$, $\bar{\psi}$ and A_μ leads to the following equations

$$R_{\mu\nu} - \frac{1}{2} g_{\mu\nu} R = \frac{1}{2} \left(-F_{\mu\alpha} F_\nu^\alpha + \frac{1}{4} g_{\mu\nu} F_{\alpha\beta} F^{\alpha\beta} \right) +$$

$$4 l_{Pl}^2 \left[\left(i\bar{\psi} \gamma_\mu \tilde{\nabla}_\nu \psi + i\bar{\psi} \gamma_\nu \tilde{\nabla}_\mu \psi \right) + h.c. \right] -$$

$$2 l_{Pl}^2 \left[F_{\mu\alpha} \left(i\bar{\psi} \gamma^5 \gamma_{[\nu} \gamma^{\alpha]} \psi \right) + F_{\nu\alpha} \left(i\bar{\psi} \gamma^5 \gamma_{[\mu} \gamma^{\alpha]} \psi \right) \right] -$$

$$2 g_{\mu\nu} l_{Pl}^4 \left(i\bar{\psi} \gamma^{[\bar{A}} \gamma^{\bar{B}} \gamma^{\bar{C}]} \psi \right) \left(i\bar{\psi} \gamma_{[\bar{A}} \gamma_{\bar{B}} \gamma_{\bar{C}]} \psi \right) , \tag{27}$$

$$D_\nu H^{\mu\nu} = 0, \quad H^{\mu\nu} = F^{\mu\nu} + \tilde{F}^{\mu\nu} ,$$

$$\tilde{F}^{\mu\nu} = 4 l_{Pl}^2 \left(i\bar{\psi} \gamma^5 \gamma^{[\mu} \gamma^{\nu]} \psi \right) = 4 l_{Pl}^2 E^{\mu\nu\alpha\beta} \left(i\bar{\psi} \gamma_{[\alpha} \gamma_{\beta]} \psi \right) , \tag{28}$$

$$i\gamma^\mu \tilde{\nabla}_\mu \psi - \frac{1}{8} F_{\bar{\alpha}\bar{\beta}} \left(i\gamma^5 \gamma^{[\bar{\alpha}} \gamma^{\bar{\beta}]} \psi \right) -$$

$$\frac{1}{2} l_{Pl}^2 \left(i\gamma^{[\bar{A}} \gamma^{\bar{B}} \gamma^{\bar{C}]} \psi \right) \left(i\bar{\psi} \gamma_{[\bar{A}} \gamma_{\bar{B}} \gamma_{\bar{C}]} \psi \right) = 0, \tag{29}$$

where $\omega_{\bar{a}\bar{b}\mu}$ is the 4D Ricci coefficients without torsion, $E^{\mu\nu\alpha\beta}$ is the 4D absolutely antisymmetric tensor. The most interesting for us is the Maxwell equation (28) which permits us to discuss the physical meaning of the spinor field. We would like to show that this equation in the given form is similar to the electrodynamic in the continuous media. Let we remind that for the electrodynamic in the continuous media two tensors $\bar{F}^{\mu\nu}$ and $\bar{H}^{\mu\nu}$ are introduced [13] for which we have the following equations system (in the Minkowski spacetime)

$$\bar{F}_{\alpha\beta,\gamma} + \bar{F}_{\gamma\alpha,\beta} + \bar{F}_{\beta\gamma,\alpha} = 0, \tag{30}$$

$$\bar{H}^{\alpha\beta}_{,\beta} = 0 \tag{31}$$

and the following relations between these tensors

$$\bar{H}_{\alpha\beta}u^{\beta} = \varepsilon \bar{F}_{\alpha\beta}u^{\beta}, \tag{32}$$

$$\bar{F}_{\alpha\beta}u_{\gamma} + \bar{F}_{\gamma\alpha}u_{\beta} + \bar{F}_{\beta\gamma}u_{\alpha} = \mu \left(\bar{H}_{\alpha\beta}u_{\gamma} + \bar{H}_{\gamma\alpha}u_{\beta} + \bar{H}_{\beta\gamma}u_{\alpha} \right) \tag{33}$$

where ε and μ are the dielectric and magnetic permeability respectively, u^{α} is the 4-vector of the matter. For the rest media and in the 3D designation we have

$$\varepsilon \bar{E}_i = \bar{E}_i + 4\pi \bar{P}_i = \bar{D}_i, \quad \text{where} \quad \bar{E}_i = \bar{F}_{0i}, \quad \bar{D}_{0i} = \bar{H}_{0i}, \tag{34}$$

$$\mu \bar{H}_i = \bar{H}_i + 4\pi \bar{M}_i = \bar{B}_i, \quad \text{where} \quad \bar{B}_i = \epsilon_{ijk}\bar{F}^{jk}, \quad \bar{H}_i = \epsilon_{ijk}\bar{H}^{jk}, \tag{35}$$

where P_i is the dielectric polarization and M_i is the magnetization vectors, ϵ_{ijk} is the 3D absolutely antisymmetric tensor. Comparing with the (28) Maxwell equation for the spacetime foam in the 3D form

$$E_i + \tilde{E}_i = D_i \quad \text{where} \quad E_i = F_{0i}, \quad \tilde{E}_i = \tilde{F}_{0i}, \quad D_i = H_{0i} \tag{36}$$

$$B_i + \tilde{B}_i = H_i \quad \text{where} \quad B_i = \epsilon_{ijk}F^{jk}, \quad \tilde{B}_i = \epsilon_{ijk}\tilde{F}^{jk}, \quad H_i = \epsilon_{ijk}H^{jk} \tag{37}$$

we see that the following notations can be introduced.

$$\tilde{E}_i = 4l_{Pl}^2 \epsilon_{ijk} \left(i\bar{\psi}\gamma^{[j}\gamma^{k]}\psi \right) \tag{38}$$

is the polarization vector of the spacetime foam and

$$\tilde{B}_i = -4l_{Pl}^2 \epsilon_{ijk} \left(i\bar{\psi}\gamma^5\gamma^{[j}\gamma^{k]}\psi \right) \tag{39}$$

is the magnetization vector of the spacetime foam.

The physical reason for this is evidently: each quantum WH is like to a moving dipole (see Fig.(20) which produces microscopical electric and magnetic fields.

4. Supergravity as a possible model of the spacetime foam

From the above-mentioned arguments we see that the most important for such kind models of the spacetime foam is the presence of the nonminimal interaction term (in Lagrangian) between spinor and electromagnetic fields. Let we note that the N=2 supergravity [14] which contains the vier-bein e_μ^a, Majorana Rarita-Schwinger field ψ_μ, photon A_μ and a second Majorana spin-$\frac{3}{2}$ field φ_μ has the following term in Lagrangian

$$\mathcal{L}_{se} = \frac{\kappa}{\sqrt{2}}\bar{\psi}_\mu \left(eF^{\mu\nu} + \frac{1}{2}\gamma_5\tilde{F}^{\mu\nu} \right) \varphi_\nu + \cdots, $$

$$\tilde{F}_{\mu\nu} = e_{\mu\nu\alpha\beta}F^{\alpha\beta} \tag{40}$$

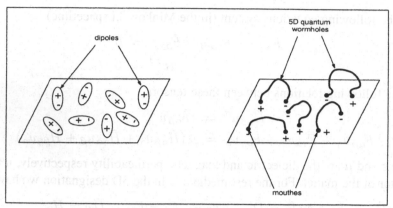

Figure 20.   For the 4D observer each mouth looks as a moving electric charge. This allows us in some approximation imagine the spacetime foam as a continuous media with a polarization.

The term like this usually occur in supergravities which have some gauge multiplet of supergravity and some matter multiplet. Taking into account the previous reasonings we can suppose that **supergravity theories can be considered as approximate models of the spacetime foam.**

5. Conclusions

Thus, here we have proposed the approximate model for the description of the spacetime foam. This model is based on the assumption that the whole spacetime is 5 dimensional but G_{55} is the dynamical variable only in the quantum topological handles (wormholes). In this case 5D gravity has the solution which we have used as a model of the single quantum wormhole. The properties of this solution is such that we can assume that the quantum topological handles (wormholes) can be approximately described by some effective spinor field.

The topological handles of the spacetime foam either can be attached to one space or connect two different spaces. In the first case we have something like to strings between two D-branes (or wormhole with the quantum throat) and such object can demonstrate a model of preventing the formation the naked singularity with relation $e > m$. In the second case the spacetime foam looks as a dielectric with quantum handles as dipoles.

Such model leads to the very interesting experimental consequences. We see that the spacetime foam has 5D structure and it connected with the electric field. This observation allows us to presuppose that the very strong electric field can open a door into 5 dimension! The question is: as is great should be this field ? The electric field E_i in the CGSE units and e_i in the "geometrized" units can be connected by formula

$$e_i = \frac{G^{1/2}}{c^2} E_i = \left(2.874 \times 10^{-25} \; cm^{-1}/gauss \right) E_i, \tag{41}$$

$$[e_i] = cm^{-1}, \quad [E_i] = V/cm \tag{42}$$

As we see the value of e_i is defined by some characteristic length l_0. It is possible that l_0 is a length of the 5th dimension. If $l_0 = l_{Pl}$ then $E_i \approx 10^{57} V/cm$ and this field strength is in the Planck region, and is will beyond experimental capabilities to create. But if l_0 has a different value it can lead to much more realistic scenario for the experimental capability to open door into 5th dimension.

Another interesting conclusion of this paper is that supergravity theories having nonminimal interaction between spinor and electromagnetic fileds can be considered as approximate and effective models of the spacetime foam.

6. Acknowledgment

I would like to acknowledge the generosity of NATO in its support for this workshop and ICTP (grant KR-154).

References

1. C. Misner and J. Wheeler, Ann. of Phys., **2**, 525 (1957); J. Wheeler, Ann. of Phys., **2**, 604(1957).
2. J. Wheeler, *Neutrinos, Gravitation and Geometry* (Princeton Univ. Press, 1960).
3. A. Chodos and S. Detweiler, *Gen. Rel. Grav.* **14** (1982) 879-890.
4. G. Clément, *Gen. Rel. Grav.* **16** (1984) 477-489; G. Clément, *Gen. Rel. Grav.* **16** (1984) 131-138.
5. V. Dzhunushaliev, Grav. Cosmol., **3**, 240(1997).
6. Bronnikov K., Int.J.Mod.Phys. **D4**, 491(1995), Grav. Cosmol., **1**, 67(1995).
7. V. Dzhunushaliev, Mod. Phys. Lett. A **13**, 2179 (1998).
8. V. Dzhunushaliev, "Matching condition on the event horizon and the holography principle", gr-qc/9907086, to be published in Int. J. Mod. Phys. D.
9. V. Dzhunushaliev, H.-J.Schmidt, Grav. Cosmol. **5**, 187 (1999).
10. V. Dzhunushaliev, "Wormhole with Quantum Throat", gr-qc/0005008, to be published in Grav. Cosmol.
11. F. Finster, J. Smoller, S.-T. Yau, Phys. Lett. **A259**, 431 (1999).
12. V. Dzhunushaliev, "An Approximate Model of the Spacetime Foam", gr-qc/0006016.
13. L.D. Landau and E.M. Lifshitz, "Electrodynamics of Continuous Media", (Pergamon Press, Oxford - London - New Jork - Paris, 1960).
14. S. Ferrara and P. V. Nieuwenhuizen, Phys. Rev. Lett. **37**, 1669 (1976).

POSSIBLE CONSTRAINTS ON STRING THEORY IN CLOSED SPACE WITH SYMMETRIES

ATSUSHI HIGUCHI *

Department of Mathematics, University of York, YORK,
YO10 5DD, United Kingdom

Abstract. It is well known that certain quadratic constraints have to be imposed on linearized gravity in closed space with symmetries. We review this phenomenon and discuss one of the constraints which arise in linearized gravity on static flat torus in detail. Then we point out that the mode with negative kinetic energy, which is necessary for satisfying this constraint, appears to be missing in the free bosonic string spectrum.

1. Introduction

(Super)string theory is the leading candidate for a unified theory including gravity. In particular, it contains and generalizes Einstein's general relativity [1–3]. Therefore, it is natural to expect that the theory incorporates diffeomorphism invariance. However, this invariance is not manifest in the perturbative definition of string theory starting from non-interacting string. Now, it is well known that a solution of linearized Einstein equations (with or without matter fields) in compact background space with Killing symmetries cannot be extended to an exact solution unless the linearized solution satisfies certain quadratic constraints [4, 5]. This phenomenon, called linearization instability, is a consequence of diffeomorphism invariance of the full theory. (This fact can be seen most clearly in the quantum context.) Therefore, one may obtain some insight into how diffeomorphism invariance is incorporated in string theory by investigating the way linearization instabilities manifest themselves.

In this article we review the phenomenon of linearization instability in general relativity with emphasis on the case with static flat torus space. In particular, we point out that in this space a mode with negative kinetic term is essential in satisfying one of the constraints and that this mode seems to be missing in the spectrum of free bosonic string theory. The rest of the article is organized

---

* ah28@york.ac.uk

S. Duplij and J. Wess (eds.), Noncommutative Structures in Mathematics and Physics, 465–473.

as follows. In Section 2 the phenomenon of linearization instability in classical and quantum general relativity is reviewed. In Section 3 one of the constraints occurring in flat torus space is discussed in detail and the importance of a mode with negative kinetic term is emphasized. In Section 4 it is pointed out that this mode is absent in a seemingly natural treatment of the zero-momentum sector of closed bosonic string in this space. In Section 5 a summary of this article is given. The metric signature is $(-+ + \cdots +)$ throughout this article.

2. Linearization instabilities in general relativity

Consider classical general relativity with any bosonic matter fields. Suppose we want to find a solution in this theory order by order in perturbation theory starting from a (globally-hyperbolic) background spacetime satisfying the vacuum Einstein equations $R_{ab} = 0$. To do do so we write the metric g_{ab} and the matter fields ϕ_i as

$$g_{ab} = g^{(0)}_{ab} + h^{(1)}_{ab} + h^{(2)}_{ab} + \cdots,$$

$$\phi_i = \phi^{(1)}_i + \phi^{(2)}_i + \cdots,$$

where $g^{(0)}_{ab}$ is the background metric and where $h^{(k)}_{ab}$ and $\phi^{(k)}_i$ are the fields obtained as the k-th order approximation. (The fields ϕ_i are assumed to vanish at zero-th order for simplicity.) The first-order approximation $(h^{(1)}_{ab}, \phi^{(1)}_i)$ corresponds to non-interacting waves in the background spacetime. The second-order perturbation of the metric, $h^{(2)}_{ab}$, can be regarded as the gravitational field generated by the free fields $h^{(1)}_{ab}$ and $\phi^{(1)}_i$.

Let the stress-energy tensor of the fields $h^{(1)}_{ab}$ and $\phi^{(1)}_i$ in the background spacetime with metric $g^{(0)}_{ab}$ be $T^{(1)}_{ab}$. We note first that the linear contribution to the Einstein tensor

$$E_{ab} = R_{ab} - \frac{1}{2} g_{ab} R$$

with $g_{ab} = g^{(0)}_{ab} + h_{ab}$ is

$$E^{(L)}_{ab}(h) = \frac{1}{2}(\nabla_c \nabla_b h^c{}_a + \nabla_c \nabla_a h^c{}_b - \nabla_c \nabla^c h_{ab} - \nabla_a \nabla_b h^c{}_c)$$

$$-\frac{1}{2} g^{(0)}_{ab}(\nabla_c \nabla_d h^{cd} - \nabla_c \nabla^c h^d{}_d).$$

Here the covariant derivatives are compatible with the metric $g^{(0)}_{ab}$ and indices are raised and lowered by this metric. The field $h^{(2)}_{ab}$ must satisfy

$$E^{(L)}_{ab}(h^{(2)}) = \kappa T^{(1)}_{ab}, \tag{1}$$

where κ is a constant. The stress-energy tensor $T_{ab}^{(1)}$ is divergence-free, i.e., $\nabla^a T_{ab}^{(1)} = 0$, if the linear equations of motion are satisfied. On the other hand the equation

$$\nabla^a E_{ab}^{(L)}(h) = 0 \qquad (2)$$

holds for *any* h_{ab}. This is a consequence of the Bianchi identity $\tilde{\nabla}^a E_{ab} = 0$, where $\tilde{\nabla}_a$ is the covariant derivative compatible with the full metric g_{ab}. For this reason Eq. (2) is called the background Bianchi identity.

Now, suppose that there is a Killing vector field X^a satisfying

$$\nabla_a X_b + \nabla_b X_a = 0 .$$

Then, it is easy to verify that the current $j_X^a \equiv T^{(1)ab} X_b$ is conserved. The corresponding conserved Noether charge is given by

$$Q_X \equiv \int_\Sigma d\Sigma \, n_a j_X^a ,$$

where the integration is over any Cauchy surface Σ and n_a is the unit normal to the Cauchy surface. (Since Q_X comes from a stress-energy tensor of the *free* fields $h_{ab}^{(1)}$ and $\phi_i^{(1)}$, it is quadratic in these fields.) If the vector X^a is a time-translation Killing vector, then the charge Q_X is nothing but the energy. If it is a space-translation Killing vector, then Q_X is a component of the momentum. We note that

$$E^{(L)ab}(h) X_b = \frac{1}{2} \nabla_b K^{ab}(h) ,$$

where $K_{ab}(h)$ is an anti-symmetric tensor given by

$$\begin{aligned} K_{ab}(h) = \; & X_a \nabla_b h^c{}_c - X_b \nabla_a h^c{}_c + X^c \nabla_a h_{bc} - X^c \nabla_b h_{ac} \\ & + X^c \nabla_a h_{bc} - X^c \nabla_b h_{ac} + h_{ca} \nabla_b X^c - h_{cb} \nabla_a X^c . \end{aligned}$$

Hence, the integral of $E^{(L)ab}(h) X_b$ over the Cauchy surface can be expressed as a surface integral as

$$\int_\Sigma d\Sigma \, n_a E^{(L)ab}(h) X_b = \frac{1}{2} \int_{\partial\Sigma} dS \, n_a r_b K^{ab}(h) ,$$

where $\partial\Sigma$ is the "boundary" of the Cauchy surface at infinity and r_a is the unit vector normal to the boundary along the Cauchy surface. By using this expression and Eq. (1) one can write the Noether charge Q_X as a surface integral:

$$Q_X = \frac{1}{2\kappa} \int_{\partial\Sigma} dS \, n_a r_b K^{ab}(h^{(2)}) . \qquad (3)$$

In asymptotically-flat spacetime this equation allows us to express energy and momentum of an isolated system as surface integrals at spacelike infinity [6].

Now, suppose that the Cauchy surface is compact, i.e., that the space is "closed". Then, the right-hand side of Eq. (3) must vanish for any h_{ab} because there is no surface term. Hence,

$$Q_X = 0. \tag{4}$$

Thus, the conserved charge Q_X is constrained to vanish. Note that this constraint cannot be derived from the linearized theory alone. It arises in the full theory when we try to find the correction to the linear theory. Solutions of the linearized field equations are not extendible to exact solutions unless they satisfy this constraint. (The background spacetime here is said to be linearization unstable because of the existence of spurious solutions to the linearized equations. The constraint (4) is sometimes called a linearization stability condition.)

Although we will concentrate on classical theory, it is interesting to note what the constraint (4) implies in quantum theory. In the Dirac quantization, constraints are imposed on the physical states. Thus, the quantum version of (4) reads

$$\mathcal{Q}_X |\text{phys}\rangle = 0, \tag{5}$$

where $|\text{phys}\rangle$ is any physical state and $\mathcal{Q}_X$ is the quantum operator corresponding to the conserved Noether charge Q_X. Since the operator $\mathcal{Q}_X$ generates the spacetime symmetry associated with the Killing vector field X^a, the constraint (5) implies that all physical states must be invariant under this spacetime symmetry [7]. This requirement might seem absurdly strong at first sight. For example, in linearized gravity in de Sitter spacetime *all* physical states are required to be de Sitter invariant.[1] However, in the (formal) Dirac quantization of full general relativity, the states are (roughly speaking) required to be diffeomorphism invariant. The constraint (5) can be interpreted to be inforcing the part of the diffeomorphism invariance of the physical states that has not been broken by the background metric.

3. The Hamiltonian constraint of linearized gravity on flat torus

In this section we discuss linearized gravity in static flat $(D - 1)$-dimensional torus space with all directions compactified. This spacetime has space- and time-translation invariance. Therefore, the energy and momentum of linearized gravity are conserved and are both constrained to vanish. Below we concentrate on the linearization stability condition which requires that energy be zero since it will

---

[1] The vacuum state is the only de Sitter invariant state if one insists on using the original Fock space of linearized gravity, but one can construct infinitely many invariant states by using a different Hilbert space [8].

be important later in the discussion of string theory. We find that there is a mode with negative kinetic term and that there would be no excitation as a result of the linearization stability condition if it were not for this mode. We consider only pure gravity for simplicity.

Let us impose the standard ("Lorenz" or Hilbert) gauge condition

$$\partial_a h^{ab} = \frac{1}{2}\partial^b h, \tag{6}$$

where $h = h^c{}_c$. Then the Hamiltonian density reads

$$\mathcal{H} = \frac{1}{4}\left[\partial_t \tilde{h}_{ab}\partial_t \tilde{h}^{ab} + \partial_i \tilde{h}_{ab}\partial^i \tilde{h}^{ab}\right] - \frac{D-2}{4D}\left[(\partial_t h)^2 + \partial_i h \partial^i h\right],$$

where $\tilde{h}_{ab} = h_{ab} - \frac{1}{D}g_{ab}h$ is the traceless part of h_{ab}. The index i runs from 1 to $D-1$, i.e., it is a spacelike index. The field equations are simply

$$\Box \tilde{h}_{ab} = 0, \quad \Box h = 0.$$

The modes with nonzero momentum $\mathbf{k}$ are proportional to $e^{-ik^0 t + i\mathbf{k}\cdot\mathbf{x}}$, where $(k^0)^2 - \mathbf{k}^2 = 0$. On the other hand, the modes with $\mathbf{k} = 0$ take the form

$$\tilde{h}_{ab}, \ h \propto At + B,$$

where A and B are constants.[2]

The Hamiltonian can be written as

$$H = \int d^{D-1}\mathbf{x}\,\mathcal{H} = H_0 + H',$$

where H_0 is the energy in the modes with $\mathbf{k} = 0$ and where H' is the energy in the modes with $\mathbf{k} \neq 0$. For the modes with $\mathbf{k} \neq 0$ the trace h can be gauged away and the physical modes have the form

$$\tilde{h}_{ab} \propto H_{ab}e^{-ik^0 t + i\mathbf{k}\cdot\mathbf{x}},$$

where H_{ab} is a constant symmetric tensor satisfying $H_{tb} = 0$, $H^i{}_i = 0$ and $k^i H_{ij} = 0$. Then we can easily see that $H' \geq 0$. The situation is rather different for the modes with $\mathbf{k} = 0$. Since these modes are constant in space, they satisfy $\partial_i \tilde{h}_{ab} = \partial_i h = 0$. Hence, the conditions coming from (6) are $\partial_t \tilde{h}_{ti} = 0$ and

$$\partial_t \tilde{h}_{tt} = -\frac{D-2}{2D}\partial_t h.$$

---

[2] Note that the energy corresponding to these modes would be infinite for $A \neq 0$ if the space were not compactified. This is why these modes would not be present in uncompactified space.

Let us write

$$\tilde{h}_{ab} = \tilde{h}_{ab}^{(0)} + \tilde{h}_{ab}',$$
$$h = h^{(0)} + h',$$

where $\tilde{h}_{ab}^{(0)}$ and $h^{(0)}$ are the zero-momentum parts of $\tilde{h}_{ab}$ and h. Then the zero-momentum Hamiltonian H_0 is given by

$$H_0 = \int d^{D-1}\mathbf{x} \left[\frac{1}{4} \partial_t \tilde{h}_{ij}^{(0)} \partial_t \tilde{h}^{(0)ij} - \frac{D^2 - 4}{8D} (\partial_t h^{(0)})^2 \right].$$

Notice that the trace mode $h^{(0)}$ has a negative kinetic term.

Since the Hamiltonian is the Noether charge corresponding to the time-translation symmetry of the background spacetime, the discussion in the previous section shows that

$$H = H_0 + H' = 0.$$

The solutions of the linearized equations which do not satisfy this condition cannot be extended to exact solutions. This equation can be re-expressed as

$$-\frac{D^2 - 4}{8D} \int d^{D-1}\mathbf{x} (\partial_t h^{(0)})^2 + H'' = 0, \tag{7}$$

where

$$H'' = H' + \frac{1}{4} \int d^{D-1}\mathbf{x} \, \partial_t \tilde{h}_{ij}^{(0)} \partial_t \tilde{h}^{(0)ij} \geq 0.$$

Now, the quantity $\frac{1}{2} h^{(0)} V$, where V is the volume of the background space, is the change in the volume of the space. Hence, Eq. (7) relates the expansion/contraction rate of space to the energy due to the excitation of the system. In fact this equation is the linearized version of a familiar equation in cosmology. Notice that the trace mode $h^{(0)}$ plays a vital role in satisfying Eq. (7). If this mode were absent, Eq. (7) would imply that there were no excitations on flat torus compactified in all directions.

4. Massless sector of bosonic string in the position representation

Massless excitations of closed string include gravitons, i.e., linearized gravity is present among the modes of free closed bosonic string in Minkowski spacetime.[3] This fact is one of the most important features of string theory as a unified theory. It is natural to expect that this feature persists in string theory in static flat torus compactified in all directions. Therefore, the total energy and momentum in string (field) theory are expected to vanish in this spacetime. We also expect that there

---

[3] This fact goes beyond the linearized level as is well known[1–3].

is a mode with negative kinetic term among the closed-string modes so that the linearization stability condition (7) can be satisfied by non-vacuum states (in string field theory). However, we will find in the "old covariant approach" that there is no massless string excitation which corresponds to the zero-momentum mode $h^{(0)}$ with negative kinetic energy if we treat the zero-momentum modes in a way which seems most natural.

Let us start with a discussion of open string in flat $(D-1)$-dimensional torus. The massless states in the old covariant approach are denoted by

$$\alpha^a_{-1}|0;p\rangle,$$

where the state $|0;p\rangle$ with momentum p^a has no string excitation (see, e.g., Ref. [9]). The creation operator α^a_{-1} creates the lowest harmonic-oscillator mode on the string in the a-direction and the annihilation operator α^a_1 annihilates it. As is well known, the physical state conditions lead to $p^2 = 0$ and $p \cdot \alpha_1|\text{phys}\rangle = 0$, where $[\alpha^a_1, \alpha^b_{-1}] = g^{ab}$ and $p \cdot \alpha_1 \equiv p_a \alpha^a_1$. [Here, $g_{ab} = \text{diag}(-1, 1, 1, \cdots, 1)$.] Let us consider a wave-packet state

$$|\psi\rangle = \int \frac{d^D p}{(2\pi)^D} \hat{A}_a(p)\alpha^a_{-1}|0;p\rangle,$$

where $\hat{A}_a(p)$ is a function of p^a. The physical state conditions then read $p^2 \hat{A}_a(p) = 0$ and $p^a \hat{A}_a(p) = 0$. Now, define the (spacetime) position representation of this wave packet as

$$A_a(x) = \int \frac{d^D p}{(2\pi)^D} \hat{A}_a(p)e^{-ip \cdot x}.$$

Then the physical state conditions become $\Box A_a = 0$ and $\partial^a A_a = 0$. Thus, we recover the equations satisfied by a non-interacting $U(1)$ gauge field in the Lorentz gauge. The zero-momentum modes in flat $(D-1)$-dimensional torus satisfy

$$\partial_t A_t = 0, \quad \partial_t^2 A_i = 0.$$

These imply that $A_t = \text{const}$ and $A_i = E_i t + A_i^{(0)}$. The constant A_t can be gauged away, but the constants E_i (the electric field) and $A_i^{(0)}$ represent physical degrees of freedom.

Next, we will apply the above procedure to a closed string on static flat torus and examine whether or not there is a mode with negative kinetic term. The massless excitations of a closed bosonic string are

$$\alpha^a_{-1}\tilde{\alpha}^b_{-1}|0;p\rangle.$$

The operator α^a_{-1} ($\tilde{\alpha}^a_{-1}$) creates the lowest left-moving (right-moving) mode on the string in the a-direction, and the operator α^a_1 and $\tilde{\alpha}^a_1$ annihilate them. The physical state conditions lead to $p^2 = 0$ and $p \cdot \alpha_1 |\text{phys}\rangle = p \cdot \tilde{\alpha}_1 |\text{phys}\rangle = 0$, where $[\alpha_1^a, \alpha_{-1}^b] = [\tilde{\alpha}_1^a, \tilde{\alpha}_{-1}^b] = g^{ab}$. We again consider a wave-packet state

$$|\Psi\rangle = \int \frac{d^D p}{(2\pi)^D} \hat{H}_{ab}(p)\alpha_{-1}^a \tilde{\alpha}_{-1}^b |0; p\rangle .$$

(Note here that the tensor $\hat{H}_{ab}(p)$ is not necessarily symmetric.) The physical state conditions read $p^2 \hat{H}_{ab}(p) = 0$ and $p^a \hat{H}_{ab} = p^b \hat{H}_{ab} = 0$. In the spacetime position representation,

$$H_{ab}(x) = \int \frac{d^D p}{(2\pi)^D} \hat{H}_{ab}(p)e^{-ip \cdot x} ,$$

the physical state conditions are $\Box H_{ab} = 0$ and

$$\partial^a H_{ab} = \partial^b H_{ab} = 0 . \tag{8}$$

The equation $\Box H_{ab} = 0$ naturally come from the following Lagrangian density:

$$\mathcal{L} = -\frac{1}{4}\partial_a H_{bc}\partial^a H^{bc} . \tag{9}$$

The constraints (8) can be imposed by hand. One finds the modes corresponding to gravitons, anti-symmetric tensor particles and dilatons in the nonzero momentum sector of this theory as in Minkowski spacetime. The constraints (8) for the zero-momentum sector read

$$\partial_t H_{ta} = \partial_t H_{at} = 0$$

for all a. The energy in the zero-momentum sector is

$$E_0 = \frac{1}{4} \int d^{D-1}\mathbf{x} \, \partial_t H_{ij}\partial_t H^{ij} ,$$

where $i, j = 1, 2, \cdots D - 1$. There is no mode with negative kinetic term in this expression, and E_0 is positive definite. Thus, the negative-energy mode, which is necessary for non-vacuum states to satisfy the constraint (7), does not appear in a seemingly natural position representation of the massless sector of closed bosonic string.

5. Summary

In this article, we reviewed the fact that quadratic constraints arise in linearized gravity if the background spacetime allows Killing symmetries and has compact Cauchy surfaces. This implies that the total energy and momentum in free string (field) theory should be constrained to vanish in flat torus space with all directions compactified. We examined one of these constraints in linearized gravity in this space, emphasizing that a mode with negative kinetic energy is essential in satisfying this constraint. Then we analyzed free closed bosonic string theory in this space and found that this mode does not appear in a seemingly natural treatment of the massless sector.

It is possible that the Lagrangian density (9) is wrong, and a more careful analysis may lead to a Lagrangian density describing the usual linearized gravity, anti-symmetric tensor gauge field and dilaton scalar field after all. It will be interesting to see how this can be achieved. The situation is rather puzzling, however, because string theory is formulated in terms of a physical object, i.e., a string, and does not seem to allow any negative-energy mode.

References

1. T. Yoneya, *Quantum gravity and the zero-slope limit of the generalized Virasoro model*, Nuovo Cim. Lett. **8** (1973), pp. 951–955.
2. T. Yoneya, *Connection of dual models to electrodynamics and gravidynamics*, Prog. Theor. Phys. **51** (1974), pp. 1907–1920.
3. J. Scherk and J. Schwarz, *Dual models for non-hadrons*, Nucl. Phys. **B81** (1974), pp. 118–144.
4. D. Brill and S. Deser, *Instability of closed spaces in general relativity*, Commun. Math. Phys. **32** (1973), pp. 291–304.
5. A. Fischer and J. Marsden, *Linearization stability of Einstein equations*, Bull. Am. Math. Soc. **79** (1973), pp. 997–1003.
6. R. Arnowitt, S. Deser and C. W. Misner, *The dynamics of general relativity*, in "Gravitation: an introduction to current research", ed. L. Witten, Wiley, New York, 1962, pp. 226–265.
7. V. Moncrief, *Invariant states and quantized gravitational perturbations*, Phys. Rev. D **18** (1978), pp. 983–989.
8. A. Higuchi, *Quantum linearization instabilities of de Sitter spacetime: II*, Class. Quantum Grav. **8** (1991), pp. 1983–2004.
9. M. B. Green, J. H. Schwarz and E. Witten, *Supersting theory: vol. 1. Introduction*, Cambridge University Press, Cambridge, 1987, pp. 113–116.

SEMICLASSICAL DYNAMICS OF $SU(2)$ MODELS

ADRIAN ALSCHER * and HERMANN GRABERT [†]
*Fakultät für Physik, Albert-Ludwigs-Universität Freiburg,
Hermann-Herder-Str. 3, D-79106 Freiburg, Germany*

Within the scope of simple quantum mechanics we present a semiclassical theory which is exact. While the semiclassical theory of canonical phase space path integrals is now well established [1, 2] we examine here the case where the classical phase space is the two-sphere. After summarizing some relevant features of a classical spin, we briefly discuss the localization of classical phase space integrals and then present an extension for a quantum spin. The semiclassical propagator is employed to solve the Jaynes-Cummings model.

1. Classical spin

A classical spin is described by a classical Bloch vector on the two-sphere

$$\vec{S} \in S^2(s) = \left\{ (S_x, S_y, S_z) \in \mathcal{R}^3 \,|\, S = s \right\}.$$

We make use of spherical coordinates

$$U = \{\Omega = (\vartheta, \varphi) \,|\, 0 < \vartheta < \pi,\, 0 < \varphi < 2\pi\}.$$

This coordinate system cannot be extended over the whole $S^2(s)$. However, as $S^2(s)$ is embedded in $\mathcal{R}^3$, an appropriate metric g and volume form ω are induced

$$g = s^2(d\vartheta \otimes d\vartheta + \sin^2(\vartheta)\, d\varphi \otimes d\varphi),$$
$$\omega = s\sin(\vartheta)\, d\vartheta \wedge d\varphi.$$

The symplectic volume form is closed and non degenerate. Hence, the pair $(S^2(s), \omega)$ generates a symplectic differential manifold. Now, Hamiltonian dynamics is determined by the Hamiltonian vector field X_H

$$\omega(X_H, \cdot) = dH,$$

* alscher@physik.uni-freiburg.de
† grabert@physik.uni-freiburg.de

S. Duplij and J. Wess (eds.), Noncommutative Structures in Mathematics and Physics, 475–480.

leading to the dynamical system

$$s \sin(\vartheta)\dot{\vartheta} = \frac{\partial H}{\partial \varphi},$$

$$s \sin(\vartheta)\dot{\varphi} = -\frac{\partial H}{\partial \vartheta}.$$

These classical equations of motion can also be derived by introducing the classical action

$$S[\Omega(t)] = \int_0^T dt \left[\theta_\vartheta \dot{\vartheta} + \theta_\varphi \dot{\varphi} - H \right],$$

with the symplectic potential

$$\theta = s[\cos(\vartheta) \, d\varphi + dG].$$

For classical spin dynamics the localization of oscillating phase space integrals was observed [3]. To see this we examine the symplectic form α of the external algebra of the cotangent bundle

$$\alpha = e^{-iT(H-\omega)},$$

which is equivariantly closed. The integral over the whole sphere can be written as

$$Z = \int_{S^2(s)} \alpha = \int_{S^2(s)} \alpha \, e^{-\nu D_H \beta}, \tag{1}$$

with the equivariant exact form $D_H\beta = dg(X_H, \cdot) + g(X_H, X_H)$. Now, the right hand side of Eq. (1) does not depend on ν, allowing for the localization of Z [4]

$$Z = \lim_{\nu \to \infty} \int_{S^2(s)} \alpha \, e^{-\nu D_H \beta}.$$

The stationary phase approximation results in the Berlinge-Vergue formula

$$Z = -2\pi \sum_{\Omega \in U_{\mathrm{fp}}} \frac{\alpha^{(0)}(\Omega)}{\sqrt{\det dX_H(\Omega)}},$$

and only the sum over the fix points $U_{\mathrm{fp}} = \{\Omega \in U \mid X_H(\Omega) = 0\}$ has to be considered. Therefore, the question arises whether there exists a similar saddle point approximation of path integrals for quantum mechanical spins.

2. Quantum spin

Niemi and Pasanen [5] have proposed a supersymmetric formulation of a path integral which leads to a semiclassical localization formula. However, it only describes correct quantum mechanics if the action is supersymmetrically exact, leading to the necessary condition $\theta(X_H) = H$. Another approach [6] is based on geometric quantization. Here we make use of a path integral in the spin coherent state representation of the quantum mechanical spin Hilbert space [7, 8]

$$|\psi_g\rangle = \mathcal{D}^s(g)|\uparrow\rangle,$$

where the $(2s + 1)$-dimensional irreducible representation of $g \in SU(2)$ acts on $|\uparrow\rangle = |s, m = s\rangle$. The spin coherent states $|\psi_g\rangle$ and $|\psi_{g'}\rangle$ describe the same physical state if

$$g \sim g' \Leftrightarrow g' \in gU(1),$$

which gives rise to the fiber bundle representation of $SU(2)$ over $S^2(s) \equiv SU(2)/U(1)$. Distinct spin coherent states are canonically isomorphic to the left cosets which becomes obvious if we parameterize any $g \in SU(2)$ with Euler angles $(\vartheta, \varphi, \chi)$:

$$|\Omega\rangle = |\psi_g\rangle = e^{-is\chi}e^{-i\varphi S_z}e^{-i\vartheta S_y}|\uparrow\rangle.$$

We make use of a section of the $SU(2)$ bundle and choose one special member in every left coset. In particular we fix $\chi = 0$ for every $|\Omega\rangle$. The scalar product

$$\langle\Omega''|\Omega'\rangle = \left[\cos(\vartheta''/2)\cos(\vartheta'/2)e^{\frac{i}{2}(\varphi''-\varphi')} + \sin(\vartheta''/2)\sin(\vartheta'/2)e^{-\frac{i}{2}(\varphi''-\varphi')}\right]^{2s}$$

gives rise to a gauge invariant metric and volume form which are identical to the geometric structures of $S^2(s)$ [9]. Hence, a representation of quantum states is found which is useful in order to understand quantum systems with discrete degrees of freedom in terms of classical mechanics.

We consider the most general $SU(2)$ model described by the Hamiltonian

$$H(t) = B_x(t)S_x + B_y(t)S_y + B_z(t)S_z. \tag{2}$$

Following the lines of [10] the propagator can be represented as the limit of a Wiener regularized phase space path integral

$$\langle\Omega''|U(T)|\Omega'\rangle = \lim_{\nu\to\infty}\int d\mu_w \, \exp\{iS[\Omega(t)]\} \tag{3}$$

with the spherical Wiener measure

$$d\mu_w = N \prod_{t=0}^{T} d\cos[\vartheta(t)]\, d\varphi(t)\, \exp\left\{-\frac{1}{2s\nu}\int_0^T dt\left[g_{\vartheta\vartheta}\dot{\vartheta}^2 + g_{\varphi\varphi}\dot{\varphi}^2\right]\right\}.$$

This enforces that only continuous Brownian motion paths contribute to the path integral. Now, the dominant path approximation of the right hand side of Eq. (3) can be shown to coincide with the exact quantum result [10]

$$\exp\{iS_{cl}[\Omega(t)]\} = \langle\Omega''|U(T)|\Omega'\rangle. \tag{4}$$

For $SU(2)$ models (2) no contributions of fluctuations around the dominant path have to be taken into account.

Apart from an extension of the localization of classical phase space integrals to the case of quantum propagators, the formula (4) is also useful to study spins coupled with other degrees of freedom. Here, we apply it to an exactly solvable model.

3. Jaynes-Cummings model

The Jaynes-Cummings model is characterized by the Hamiltonian [11, 12]

$$H = a^\dagger a + (1 + \Delta)S_z + \lambda(aS_+ + a^\dagger S_-),$$

where a is the canonical annihilation operator of a bosonic field mode and $S_\pm = S_x \pm iS_y$, S_z are operators of a spin-$\frac{1}{2}$. It is well known that the Jaynes-Cummings model allows apart from H for another time independent operator [14]

$$N = a^\dagger a + S_z.$$

Hence, the time evolution operator

$$U(T) = e^{-iHT} = e^{-iNT}e^{-iCT},$$

where $C = H - N$. Representing the spin operators in the eigenbasis of S_z formed by the eigenvectors $|\uparrow\rangle$ and $|\downarrow\rangle$

$$e^{-iNT} = e^{-ia^\dagger a T}\left(e^{-\frac{i}{2}T}|\uparrow\rangle\langle\uparrow| + e^{+\frac{i}{2}T}|\downarrow\rangle\langle\downarrow|\right).$$

Introducing further the eigenkets of $a^\dagger a$, invariant subspaces are distinguished. In particular the kets $|\uparrow n-1\rangle \equiv |\uparrow\rangle|n-1\rangle$ and $|\downarrow n\rangle \equiv |\downarrow\rangle|n\rangle$ span the subspace with $N = (n - \frac{1}{2})$. In this subspace the time independent operator C generates $SU(2)$ dynamics. In terms of the operators

$$J_x = \frac{1}{2}\left(|\uparrow n-1\rangle\langle\downarrow n| + |\downarrow n\rangle\langle\uparrow n-1|\right),$$

$$J_y = \frac{i}{2}\left(-|\uparrow n-1\rangle\langle\downarrow n| + |\downarrow n\rangle\langle\uparrow n-1|\right),$$

$$J_z = \frac{1}{2}\left(|\uparrow n-1\rangle\langle\uparrow n-1| - |\downarrow n\rangle\langle\downarrow n|\right),$$

we have

$$C = 2\lambda\sqrt{n}J_x + \Delta J_z.$$

Accordingly,

$$\langle\Omega''|e^{-iCT}|\Omega\rangle = \lim_{\nu\to\infty}\int d\mu_w \exp\{iS[\vartheta(t), \varphi(t)]\},$$

with the action

$$S[\vartheta(t), \varphi(t)] = \int_0^T dt \left[\frac{1}{2} \cos(\vartheta)\dot{\varphi} - C(\vartheta, \varphi) \right],$$

where

$$C(\vartheta, \varphi) = \langle \vartheta \, \varphi | C | \vartheta \, \varphi \rangle$$

$$= \lambda \sqrt{n} \sin(\vartheta) \cos(\varphi) + \frac{\Delta}{2} \cos(\vartheta).$$

Now the dominant path approximation (4) gives

$$\exp\{iS_{\text{cl}}[\Omega(t)]\} = \exp\left\{ -i \int_0^T dt \, C(\bar{\vartheta}''(t), \bar{\varphi}''(t)) \right\} \langle \Omega'' | \Omega' \rangle, \tag{5}$$

Introducing the complex variables

$$\zeta = \tan\left(\frac{\bar{\vartheta}}{2}\right) e^{i\bar{\varphi}},$$

$$\eta = \tan\left(\frac{\bar{\vartheta}}{2}\right) e^{-i\bar{\varphi}}, \tag{6}$$

the dominant path is determined by

$$\dot{\zeta} = -i\lambda\sqrt{n}(1 - \zeta^2) + i\Delta\zeta,$$
$$\dot{\eta} = i\lambda\sqrt{n}(1 - \eta^2) - i\Delta\eta,$$

with boundary conditions $\zeta(0) = \zeta'$ and $\eta(T) = \eta''$. Hence, the endpoint of the classical trajectory obeys

$$\zeta(T) = \frac{2\Omega_n \zeta' \cos(\Omega_n T) + i[\Delta\zeta' - \lambda\sqrt{n}]\sin(\Omega_n T)}{2\Omega_n \zeta' \cos(\Omega_n T) - i[\lambda\sqrt{n}\,\zeta' + \Delta]\sin(\Omega_n T)},$$

$$\eta(T) = \eta'',$$

with the Rabi frequency

$$\Omega_n = \sqrt{\lambda^2 n + \frac{\Delta^2}{4}}.$$

In terms of the complex variables (6) we get

$$C(\zeta(T), \eta'') = i\frac{d}{dT} \log\left\{ (1 + \zeta'\eta'') \cos(\Omega_n T) \right.$$

$$\left. - \frac{i}{\Omega_n} \left[\lambda\sqrt{n}(\zeta' + \eta'') + \frac{\Delta}{2}(1 - \zeta'\eta'') \right] \sin(\Omega_n T) \right\}.$$

Now, the integral in Eq.(5) is readily solved and the propagator takes the form

$$
\begin{aligned}
e^{iS_{cl}} = {} & a(T)\cos\left(\frac{\vartheta''}{2}\right)\cos\left(\frac{\vartheta'}{2}\right)e^{\frac{i}{2}(\varphi''-\varphi')} \\
& +a^*(T)\sin\left(\frac{\vartheta''}{2}\right)\sin\left(\frac{\vartheta'}{2}\right)e^{-\frac{i}{2}(\varphi''-\varphi')} \\
& +b(T)\cos\left(\frac{\vartheta''}{2}\right)\sin\left(\frac{\vartheta'}{2}\right)e^{\frac{i}{2}(\varphi''+\varphi')} \\
& -b^*(T)\sin\left(\frac{\vartheta''}{2}\right)\cos\left(\frac{\vartheta'}{2}\right)e^{-\frac{i}{2}(\varphi''+\varphi')},
\end{aligned}
$$

where

$$
a(T) = \cos(\Omega_n T) - i\frac{\Delta}{2\Omega_n}\sin(\Omega_n T),
$$

$$
b(T) = -i\frac{\lambda\sqrt{n}}{\Omega_n}\sin(\Omega_n T).
$$

This gives indeed the exact propagator [13] of the model.

This work was supported by a grant from the Deutsche Forschungsgemeinschaft (DFG).

References

1. I. Daubechies, J.R. Klauder, J.Math.Phys. **26** (1985), 2239.
2. J.R. Klauder, Phys.Rev.Lett. **56**, 897 (1986).
3. E. Keski-Vakkuri, A.J. Niemi, G. Semenoff and O. Tirkkonen, Phys.Rev. **D44** (1991), 3899.
4. N. Berline, E. Getzler and M. Vergne, *Heat Kernels and Dirac-Operators* Springer, Berlin, 1991.
5. A.J. Niemi and P. Pasanen, Phys.Lett. **B253** (1991), 349.
6. E.A. Kochetov, J.Phys. **A31** (1998), 4473.
7. J.M. Radcliffe, J.Phys. **A4** (1971), 313.
8. A.M. Perelomov, *Generalized Coherent States and Their Applications* Springer, Berlin, 1986.
9. J.P. Provost and G.Vallee, Comm.Math.Phys. **76** (1980), 289.
10. A. Alscher and H. Grabert, J.Phys. **A32** (1999), 4907.
11. E.T. Jaynes and F.W. Cummings, Proc.IEEE **51** (1963), 89.
12. S. Stenholm, Phys.Rep. **C6** (1973), 1.
13. B.W. Shore and P.L. Knight, J.Mod.Opt. **40** (1993), 1195.
14. J.R. Ackerhalt and K. Rzazewski, Phys.Rev. **A12** (1975), 2549.

LIST OF SPEAKERS AND THEIR E-PRINTS (ARW contributions are in bold)

1. **Adrian Alscher** alscher@pollux.physik.uni-freiburg.de
   quant-ph/0006072, quant-ph/0004046, quant-ph/9904102, nucl-th/9606011

2. **Andrzej Borowiec** borow@ift.uni.wroc.pl
   gr-qc/0011103, math.QA/0007151, math-ph/0007031, math.QA/9910018, gr-qc/9906043, math-ph/9906012, gr-qc/9806116, hep-th/9801126, q-alg/9710006, gr-qc/9705025, dg-ga/9612009, gr-qc/9611067, hep-th/9312023

3. **Friedemann Brandt** fbrandt@mis.mpg.de **hep-th/0010155**
   hep-th/0009133, hep-th/0006152, hep-th/0005086, hep-th/0002245, hep-th/9910177

4. **Alexander Burinskii** bur@ibrae.ac.ru **hep-th/0011188**
   hep-th/0008129, gr-qc/0008055, hep-th/9910045, hep-th/9908198, gr-qc/9904012, hep-th/9903032, hep-th/9802110, hep-th/9801177, hep-th/9704102, hep-th/9504139, hep-th/9503094, gr-qc/9501012, gr-qc/9303003

5. **Goran Djordjevic** gorandj@junis.ni.ac.yu
   hep-th/0005216, quant-ph/0005027, math-ph/0005026, math-ph/0005025

6. **Branko Dragovich** dragovic@mi.ras.ru
   math-ph/0010023, hep-th/0005216, hep-th/0005200, gr-qc/0005103, quant-ph/0005027, math-ph/0005026, math-ph/0005025, math-ph/0005020

7. **Steven Duplij** steven.a.duplij@univer.kharkov.ua **math-ph/0012039**
   physics/0008231, physics/0006062, math.FA/0006001, math-ph/0005033, math-ph/9910045, hep-th/9809089, q-alg/9609022, funct-an/9609002, alg-geom/9510013, alg-geom/9506004, hep-th/9505179

8. **Vladimir Dzhunushaliev** dzhun@hotmail.kg **gr-qc/0010029**
   hep-th/0010185, gr-qc/0006016, gr-qc/0005123, gr-qc/0005008, cond-mat/0001257, hep-th/9912194, gr-qc/9912018, gr-qc/9911120, gr-qc/9911080, gr-qc/9910092, gr-qc/9908076, gr-qc/9908074, gr-qc/9908049, gr-qc/9907086, gr-qc/9905104, gr-qc/9903075, hep-th/9902076, hep-th/9810094, gr-qc/9810050, hep-ph/9807239, gr-qc/9807086, gr-qc/9807080, hep-th/9806073, gr-qc/9806046, gr-qc/9805104, gr-qc/9712068, gr-qc/9711033, hep-th/9707039, cond-mat/9704062, gr-qc/9612047, hep-th/9611096, gr-qc/9607007, hep-th/9606125, hep-th/9606124, hep-th/9603120, gr-qc/9603007, gr-qc/9512014, hep-th/9510056, supr-con/9510001

9. **Andrzej Frydryszak** amfry@ift.uni.wroc.pl
   math-ph/9807036, hep-th/9601020

10. **Dmitri Galtsov** galtsov@grg.phys.msu.su **hep-th/0012059**
    gr-qc/0008076, hep-th/0007228, hep-th/0006242, gr-qc/0006087, hep-th/0005099, hep-th/9912127, hep-th/9910171, hep-th/9908133, hep-th/9908132, hep-th/9901130, hep-th/9810070, gr-qc/9808002, hep-th/9801160, gr-qc/9712024, gr-qc/9712003, hep-th/9709181, gr-qc/9706067, gr-qc/9706063, hep-th/9702039, gr-qc/9612067, gr-qc/9612007, gr-qc/9608023, gr-qc/9608021, hep-th/9607043, hep-th/9606042, hep-th/9606041, gr-qc/9606014, hep-th/9507164, hep-th/9507005, hep-th/9504155, hep-th/9503092, hep-th/9410217, hep-th/9409041, hep-th/9407155, hep-th/9308068, hep-th/9305112, hep-th/9212153, gr-qc/9209008

11. **Alexander Ganchev** ganchev@inrne.bas.bg
    hep-th/9906139, math.QA/9807106, physics/9803038, hep-th/9709103, hep-

th/9608018, hep-th/9407013, hep-th/9403075, hep-th/9402153, hep-th/9308038, hep-th/9308037, hep-th/9207032, hep-th/9201080, dg-ga/9606011

12. **Alexandre Gavrilik** omgavr@bitp.kiev.ua **hep-ph/0011057**
hep-ph/0010019, hep-ph/9912222, math.QA/9911201, hep-th/9911120, gr-qc/9911094, nucl-th/9906034, hep-ph/9807559, hep-ph/9712411, q-alg/9709036, hep-ph/9504233, q-alg/9511017

13. **Atsushi Higuchi** ah28@york.ac.uk
gr-qc/0011070, gr-qc/0011062, quant-ph/0006125, quant-ph/0005013, gr-qc/0004079, gr-qc/9901006, quant-ph/9812036, gr-qc/9806093, gr-qc/9804066, gr-qc/9609025, gr-qc/9605030, gr-qc/9603045, gr-qc/9508051, gr-qc/9505035, gr-qc/9505009, gr-qc/9412048, gr-qc/9407038, gr-qc/9406009

14. **Nikolay Iorgov** mmtpitp@bitp.kiev.ua
hep-ph/0010019, math.QA/0007105, hep-ph/9912222, math.QA/9911201, math.QA/9911129, nucl-th/9906034, math.QA/9905059, hep-ph/9807559, math.QA/9805032, q-alg/9709036, q-alg/9709036, q-alg/9511017

15. **Anatolij Klimyk** aklimyk@gluk.org
math.QA/0007105, math.QA/9911130, math.QA/9911129, math.QA/9911114, math.QA/9905059, math.QA/9901080, math.QA/9805048, math.QA/9805032, q-alg/9709035

16. **Yuri Kozitsky** jkozi@golem.umcs.lublin.pl
math.DS/9909182, math-ph/9812017

17. **Karl Landsteiner** landstei@mail.cern.ch **hep-th/0011003**
hep-th/0006210, hep-th/0004115, hep-th/9911124, hep-th/9909166, hep-th/9908010, hep-th/9901143, hep-th/9806137, hep-th/9805158, hep-th/9801002, hep-th/9708118, hep-th/9705199, hep-th/9609059, hep-th/9606146, hep-th/9507008, hep-th/9502147, hep-th/9412198, hep-th/9408033, hep-th/9309111

18. **Dimitry Leites** mleites@matematik.su.se
hep-th/9710045, hep-th/9702120, hep-th/9702073

19. **Jerzy Lukierski** lukier@ift.uni.wroc.pl **hep-th/0011053**
hep-th/0012056, hep-th/0011214, hep-th/0009120, hep-th/0007102, math.QA/0007065, math.QA/0005145, hep-th/0005112, hep-th/9912264, hep-th/9912051, hep-th/9907113, hep-th/9904109, gr-qc/9903066, hep-th/9902037, hep-th/9812074, hep-th/9812063, hep-th/9811022, math-ph/9807036, hep-th/9706031, hep-th/9612017, hep-th/9610230, hep-th/9606170, hep-th/9504110, hep-th/9412114, hep-th/9411115, hep-th/9405076, hep-th/9312153, hep-th/9312068, hep-th/9310117, hep-th/9204086, hep-th/9108018

20. **Volodymyr Lyubashenko** lub@imath.kiev.ua
q-alg/9510004, hep-th/9405168, hep-th/9405167, hep-th/9403189, hep-th/9311095

21. **John Madore** john.madore@th.u-psud.fr
hep-th/0009230, hep-th/0005273, math.QA/0004011, math.QA/0002215, math.QA/0002007, hep-th/0001203, math.QA/9907023, gr-qc/9906059, math.QA/9904027, hep-th/9903239, math.QA/9812141, math.QA/9809160, math.QA/9807123, math.QA/9806071, q-alg/9709007, gr-qc/9709002, gr-qc/9708053, gr-qc/9706047, gr-qc/9705083, q-alg/9702030, gr-qc/9611026, gr-qc/9607065, gr-qc/9607060, hep-th/9601169, hep-th/9601120, hep-th/9506183,

hep-th/9506041, hep-th/9502017, hep-th/9411127, hep-th/9410199, gr-qc/9307030, hep-ph/9209226

22. **Vladimir Mazorchuk** mazor@mail.univ.kiev.ua

23. **Jan Naudts** naudts@uia.ua.ac.be
hep-th/0012209, math-ph/0012051, cond-mat/0011225, math-ph/0009031, math-ph/9908025, math-ph/9907008, quant-ph/9904110, cond-mat/9904070, math-ph/9903002, quant-ph/9809061

24. **Irina Shchepochkina (Paramonova)** ira@paramonova.mccme.ru
physics/9703022, hep-th/9702122, hep-th/9702121, hep-th/9702120

25. **Christiane Quesne** cquesne@ulb.ac.be **math-ph/0012033**
math-ph/0008034, math-ph/0008020, math-ph/0007016, math-ph/0004027, quant-ph/0003085, math-ph/0003025, math-ph/9911004, math-ph/9908022, math-ph/9908021, math.QA/9903151, math-ph/9901016, math.QA/9811064, math.QA/9810161, solv-int/9808017, quant-ph/9802066, physics/9708004, hep-th/9706067, q-alg/9706002, quant-ph/9703037, q-alg/9701031, q-alg/9701030, q-alg/9701029, hep-th/9612173, hep-th/9607035, q-alg/9605041, hep-th/9604132, q-alg/9512032, hep-th/9510006, hep-th/9507078, hep-th/9505071, hep-th/9505011

26. **Yurii Samoilenko** yurii_sam@imath.kiev.ua
math.QA/0010308, math-ph/0001011, math-ph/9910018

27. **Alexander Sergeev** sergeev@bittu.org.ru
math.RT/9904079, math.RT/9810148, math.RT/9810113, math.RT/9810111, math.RT/9810110, math.RT/9810109

28. **Artur Sergyeyev** arthurser@imath.kiev.ua
solv-int/9902002

29. **Joan Simon** jsimon@ecm.ub.es
hep-th/0010242, hep-th/0007253, hep-th/0003211, hep-th/9910177, hep-th/9909005, hep-th/9907022, hep-th/9812095, hep-th/9807113, hep-th/9803196, hep-th/9803040, hep-th/9712125, hep-th/9707063

30. **Kellogg Stelle** k.stelle@ic.ac.uk
hep-th/0011167, hep-th/0007120, hep-th/9911156, hep-th/9907202, hep-th/9903057, hep-th/9812086, hep-th/9810159, hep-th/9807051, hep-th/9806051, hep-th/9803259, hep-th/9803235, hep-th/9803116, hep-th/9710244, hep-th/9708109, hep-th/9707207, hep-th/9706207, hep-th/9701088, hep-th/9608173, hep-th/9605082, hep-th/9602140, hep-th/9511203, hep-th/9508042, hep-th/9502108, hep-th/9412168, hep-th/9404170, hep-th/9401007, hep-th/9212037, hep-th/9212017, hep-th/9209111, hep-th/9206108, hep-th/9201020, hep-th/9110015

31. **Francesco Toppan** toppan@cbpf.br
hep-th/0010135, hep-th/0005035, hep-th/0005034, solv-int/9912003, hep-th/9907148, hep-th/9904134, hep-th/9810145, hep-th/9809003, hep-th/9805147, solv-int/9710001, hep-th/9705109, hep-th/9703224, hep-th/9612245, hep-th/9610038, hep-th/9608036, hep-th/9603187, hep-th/9506133, hep-th/9504138, hep-th/9503122, hep-th/9411046, hep-th/9409126, hep-th/9409125, hep-th/9405095, hep-th/9312045, hep-th/9310062, hep-th/9307106, hep-th/9303073, hep-th/9210020, hep-th/9208048

32. **Sergiu Vacaru** sergiu.vacaru@phys.asm.md **hep-th/0011221**
hep-th/0009163, gr-qc/0009039, gr-qc/0009038, gr-qc/0005025, gr-qc/0001060, gr-

qc/0001057, gr-qc/0001020, gr-qc/9905053, gr-qc/9811048, hep-th/9810229, hep-th/9807214, gr-qc/9806080, physics/9801016, physics/9706038, physics/9705030, physics/9704024, hep-th/9611091, hep-th/9611034, dg-ga/9609004, hep-th/9607196, hep-th/9607195, hep-th/9607194, gr-qc/9604017, gr-qc/9604016, gr-qc/9604015, gr-qc/9604014, gr-qc/9604013, gr-qc/9602010

33. **Leonid Vaksman** vaksman@ilt.kharkov.ua

math.QA/9904173, math.QA/9809018, math.QA/9808015, math.QA/9803074, math.QA/9909036, math.QA/9905035, math.QA/9904173, math.QA/9809038, math.QA/9809018, math.QA/9809002, math.QA/9808047, math.QA/9808037, math.QA/9808015, math.QA/9803110, math.QA/9803074, q-alg/9703005, q-alg/9603012, q-alg/9511007

34. **A. Van Proeyen** antoine.vanproeyen@fys.kuleuven.ac.be **hep-th/0012110**

hep-th/0010195, hep-th/0010194, hep-th/0007044, hep-th/0006179, hep-th/0003261, hep-th/0003023, math.DG/0002122, hep-th/9912049, hep-th/9910030, hep-th/9907124, hep-th/9904085, hep-th/9904066, hep-th/9902100, hep-th/9901060, hep-th/9812066, hep-th/9804177, hep-th/9804099, hep-th/9803228, hep-th/9801206, hep-th/9801140, hep-th/9801112, hep-th/9801102, hep-th/9712092, hep-th/9711161, hep-th/9710166, hep-th/9703082, hep-th/9703081, hep-th/9611112, hep-th/9606073, hep-th/9512139, hep-th/9510195, hep-th/9510186, hep-th/9509035, hep-th/9506075, hep-th/9505123, hep-th/9505097, hep-th/9503022, hep-th/9502072, hep-th/9412200, hep-th/9410162, hep-th/9407061, hep-th/9310067, hep-th/9307126, hep-th/9306147, hep-th/9210068, hep-th/9207091, hep-th/9206097, hep-th/9205027, hep-th/9112027

35. **Pierre VanHove** vanhove@spht.saclay.cea.fr

hep-th/0010182, hep-th/0010167, hep-th/9910056, hep-th/9910055, hep-th/9903050, hep-th/9809130, hep-th/9712079, hep-th/9707126, hep-th/9706175, hep-th/9704145

36. **Dmitri Vassiliev** masdv@bath.ac.uk

gr-qc/0012046, math-ph/0006019

37. **Mihai Visinescu** mvisin@theor1.theory.nipne.ro

hep-th/0008181, hep-th/9911126, hep-th/9911014, hep-th/9805116, hep-th/9707175, hep-th/9610097, hep-th/9602015, hep-th/9407130, hep-th/9401036, hep-th/9304022

38. **Julius Wess** julius.wess@physik.uni-muenchen.de

hep-th/0009230, hep-th/0006246, math.QA/0006179, hep-th/0005005, math.QA/0004011, hep-th/0001203, math-ph/9910013, math.QA/9809160, math.QA/9808024, math.QA/9807123, math.QA/9801104, hep-th/9605161, hep-th/9511106, hep-ph/9505291, q-alg/9502007

39. **Vladimir Zima** olefir@ftf.univer.kipt.kharkov.ua

hep-th/0009166, hep-th/9807192, hep-th/9802032, hep-th/9409117

40. **George Zoupanos** george.zoupanos@cern.ch

hep-ph/0010141, hep-ph/0010069, hep-ph/0006262, hep-ph/9910277, hep-ph/9812221, hep-th/9808178, hep-th/9804074, hep-ph/9803217, hep-th/9803095, hep-ph/9802280, hep-ph/9802267, hep-th/9711157, hep-ph/9708225, hep-ph/9707425, hep-ph/9704218, hep-ph/9703289, hep-ph 9702391, hep-ph/9609218, hep-ph/9606434, hep-ph/9604216, hep-ph/9512435, \ep-ph/9512400, hep-ph/9512258, hep-ph/9511304, hep-ph/9510279, hep-ph/9509434, hep-th/9506092, hep-th/9502017, hep-ph/9411222, hep-th/9409106, hep-th/9409032, hep-th/9409003, hep-ph/9210218